D1564785

Methods in Enzymology

Volume 365

DIFFERENTIATION OF EMBRYONIC STEM CELLS

METHODS IN ENZYMOLOGY

EDITORS-IN-CHIEF

John N. Abelson   Melvin I. Simon

DIVISION OF BIOLOGY
CALIFORNIA INSTITUTE OF TECHNOLOGY
PASADENA, CALIFORNIA

FOUNDING EDITORS

Sidney P. Colowick and Nathan O. Kaplan

Methods in Enzymology

Volume 365

Differentiation of Embryonic Stem Cells

EDITED BY

Paul M. Wassarman

MOUNT SINAI SCHOOL OF MEDICINE
DEPARTMENT OF BIOCHEMISTRY AND MOLECULAR BIOLOGY
1 GUSTAVE L. LEVY PLACE, BOX 1020
NEW YORK, NY 10029

Gordon M. Keller

MOUNT SINAI SCHOOL OF MEDICINE
CARL C. ICAHN CENTER FOR GENE THERAPY AND MOLECULAR MEDICINE
1 GUSTAVE L. LEVY PLACE, BOX 1496
NEW YORK, NY 10029

ELSEVIER
ACADEMIC
PRESS

Amsterdam Boston Heidelberg London New York Oxford
Paris San Diego San Francisco Singapore Sydney Tokyo

This book is printed on acid-free paper. 

Elsevier Academic Press.
525 B Street, Suite 1900, San Diego, California 92101-4495, USA
84 Theobald's Road, London WC1X 8RR, UK
http://www.academicpress.com

International Standard Book Number: 0-12-182268-0

PRINTED IN THE UNITED STATES OF AMERICA
03 04 05 06 07 08 9 8 7 6 5 4 3 2 1

Table of Contents

Section I. Differentiation of Mouse Embryonic Stem Cells

Early Commitment Steps and Generation of Chimeric Mice

Differentiation to Mesoderm Derivatives: Hematopoietic and Vascular

Other Derivatives of Mesoderm

Differentiation to Endoderm Derivatives

Contributors to Volume 365

Article numbers are in parentheses following the names of contributors.
Affiliations listed are current.

MICHAL AMIT (32), *Department of Physiology and Biophysics, Bruce Rappaport Faculty of Medicine, Technion-Israel Institute of Technology, 2 Efron Street, POB 9649, Haifa 31096, Israel*

HITOMI AOKI (24), *Department of Tissue and Organ Development, Regeneration and Advanced Medical Science, Gifu University Graduate School of Medicine, Gifu, 500-8705, Japan*

VICTORIA L. BAUTCH (6), *Department of Biology and Program in Genetics, University of North Carolina at Chapel Hill, CB#3280, Chapel Hill, North Carolina 27599*

NISSIM BENVENISTY (31), *Department of Genetics, The Silberman Institute of Life Sciences, The Hebrew University, Jerusalem 91904, Israel*

PRZEMYSLAW BLYSZCZUK (21), *In Vitro Differentiation Group, Institute of Plant Genetics and Crop Plant Research, Corrensstr. 3, Gatersleben, D-06466, Germany*

KENNETH R. BOHELER (16), *Molecular Cardiology Unit, Laboratory of Cardiovascular Science, National Institute on Aging, NIH 5600 Nathan Shock Drive, Baltimore, Maryland 21224*

SARAH K. BRONSON (17), *Department of Cellular & Molecular Physiology, Penn State College of Medicine, The Milton S. Hershey Medical Center, H166, 500, University Drive, Hershey, Pennsylvania 17033-0850*

LEIF CARLSSON (14), *Umeå Centre for Molecular Medicine, Umeå University, 901 87 Umeå, Sweden*

WEISHENG V. CHEN (26), *Program in Developmental Biology, Division of Basic Sciences, Fred Hutchinson Cancer Research Center, 1100 Fairview Avenue N, A2-025, P.O. Box 19024, Seattle, Washington 98109-1024*

YIJING CHEN (28), *Department of Genetics, University of North Carolina at Chapel Hill, Chapel Hill, North Carolina 27599*

SARAH K. CHO (11), *Division of Biological Sciences, University of California at San Diego, 9500 Gilman Drive, La Jolla, California 92093*

KYUNGHEE CHOI (13), *Department of Pathology & Immunology, Washington University School of Medicine, 660 S. Euclid Avenue, Box 8118, St. Louis, Missouri 63110*

YUN SHIN CHUNG (13), *Department of Pathology & Immunology, Washington University School of Medicine, 660 S. Euclid Avenue, Box 8118, St. Louis, Missouri 63110*

JAROSLAW CZYZ (21), *In Vitro Differentiation Group, Institute of Plant Genetics and Crop Plant Research Corrensstr. 3, Gatersleben, D-06466, Germany*

GEORGE Q. DALEY (8), *Whitehead Institute for Biomedical Research, Harvard Medical School, Cambridge, Massachusetts 02142-1479*

CHRISTIAN DANI (19), *Centre de Biochimie, Institut de Recherches Signalisation, Biologie du Developpement et Cancer, Université de Nice-Sophia Antipolis, UMR 6543 CNRS, Parc Valrose, Nice cedex 2, 06108 France*

MARIA P. DE MIGUEL (25), *Department of Microbiology, Kimmel Cancer Center, Thomas Jefferson University, Philadelphia, Pennsylvania 19107*

PETER J. DONOVAN (25), *Kimmel Cancer Center, Thomas Jefferson University, Philadelphia, Pennsylvania 19107*

BILL EADES (13), *Department of Pathology & Immunology, Washington University School of Medicine, 660 S. Euclid Avenue, Box 8118, St. Louis, Missouri 63110*

KEVIN EGGAN (2), *Whitehead Institute for Biomedical Research, 9 Cambridge Center, Cambridge, Massachusetts 02142*

KOJI ETO (10), *Division of Vascular Biology, Department of Cell Biology, Scripps Research Institute, 10550 North Torrey Pines Road, VB-5, La Jolla, California 92037*

PAUL J. FAIRCHILD (12), *Sir William Dunn School of Pathology, University of Oxford, South Parks Road, Oxford, OX1 3RE, UK*

OLIVIER FERAUD (15), *Laboratoire Developpement et Vieillissement de L'endothelium, EMI INSERM 0219, DRDC/DVE, CEA Grenoble, 17 rue des martyrs, 38054 Grenoble cedex 09, France*

ADAM A. FILIPCZYK (30), *Monash Institute of Reproduction and Development, Monash University, 246 Clayton Road, Clayton, Victoria, 3168, Australia*

STUART T. FRASER (4), *Laboratory of Molecular Mouse Genetics, Institute for Toxicology, Johannes Gutenberg-University, Obere Zahlbacher Strasse 67, Mainz 55131, Germany*

LIOR GEPSTEIN (32), *Cardiovascular Research Laboratory, Bruce Rappaport Faculty of Medicine, Technion-Israel Institute of Technology, 2 Efron Street, POB 9649, Haifa 31096, Israel*

AMIRA GEPSTEIN (32), *Cardiovascular Research Laboratory, Bruce Rappaport Faculty of Medicine, Technion-Israel Institute of Technology, 2 Efron Street, POB 9649, Haifa 31096, Israel*

TAKASHI HAMAZAKI (20), *Department of Pathology University of Florida College of Medicine, P.O. Box 100275, Gainesville, Florida 32610 0275*

SUSAN M. HAWES (30) *Monash Institute of Reproduction and Development, Monash University, 246 Clayton Road, Clayton, Victoria, 3168, Australia*

SHIN-ICHI HAYASHI (7), *Division of Immunology, Department of Molecular and Cellular Biology, School of Life Science, Faculty of Medicine, Tottori University, 86 Nishi-Machi, Yonago, Tottori 683-8503, Japan*

CLAUDIA HEGERT (18), *Department of Medical Molecular Biology, Medical University of Lübeck, Ratzeburger Allee 160, Lübeck, D-23538, Germany*

IRIT HUBER (32), *Cardiovascular Research Laboratory, Department of Biophysics and Physiology, The Bruce Rappaport Faculty of Medicine, Technion-Israel Institute of Technology, 2 Efron Street, POB 9649, Haifa 31096, Israel*

KATSUHIKO ISHIZAKI (24), *Department of Tissue and Organ Development, Regeneration and Advanced Medical Science, Gifu University Graduate School of Medicine, Gifu 500-8705, Japan*

JOSEPH ITSKOVITZ-ELDOR (32), *Department of Physiology and Biophysics, Bruce Rappaport Faculty of Medicine, Technion-Israel Institute of Technology, 2 Efron Street, POB 9649, Haifa 31096, Israel*

RUDOLF JAENISCH (2), *Whitehead Institute for Biomedical Research, 9 Cambridge Center, Cambridge, Massachusetts 02142*

L. MARTIN JAKT (4), *RIKEN Centre for Developmental Biology, 2-2-3 Minatojima-minamimachi, Chuo-ku, Kobe, Japan 650-0047*

GABRIELA KANIA (21), *In Vitro Differentiation Group, Institute of Plant Genetics and Crop Plant Research, Corrensstr. 3, Gatersleben, D-06466, Germany*

JOSEPH B. KEARNEY (6), *Program in Genetics and Molecular Biology, University of North*

Carolina at Chapel Hill CB#3280, Chapel Hill, North Carolina 27599

IZHAK KEHAT (32), *Cardiovascular Research Laboratory, Bruce Rappaport Faculty of Medicine, Technion-Israel Institute of Technology, 2 Efron Street, POB 9649 Haifa 31096, Israel*

GORDON M. KELLER (3,9), *Carl C. Icahn Center for Gene Therapy and Molecular Medicine, Mount Sinai School of Medicine, 1425 Madison Avenue, Box 1496, New York, New York 10029-6574*

MARION KENNEDY (3), *Carl C. Icahn Institute for Gene Therapy and Molecular Medicine, Mount Sinai School of Medicine, 1425 Madison Avenue, Box 1496, New York, New York 10029-6574*

JONG-HOON KIM (22), *Laboratory of Molecular Biology, NINDS, National Institutes of Health, Bethesda, Maryland 20892-4092*

KENJI KITAJIMA (5), *Department of Molecular Cell Biology, Research Institute for Microbial Diseases, Osaka University, 3-1 Yamadaoka, Suita Osaka, Osaka 565-0871, Japan*

RAJA KITTAPPA (22), *Laboratory of Molecular Biology, NINDS, National Institutes of Health, Bethesda, Maryland 20892-4092*

ÅSA KOLTERUD (14), *Umeå Centre for Molecular Medicine, Umeå University, 901 87 Umeå, Sweden*

JAN KRAMER (18), *Department of Internal Medicine I, University of Lübeck, Lübeck, D-23538, Germany*

TAKAHIRO KUNISADA (24), *Department of Tissue and Organ Development, Regeneration and Advanced Medical Science, Gifu University Graduate School of Medicine, Gifu 500-8705, Japan*

MICHAEL KYBA (8), *Center for Developmental Biology, UT Southwestern Medical Center, Dallas, Texas 75390-9133*

ANDREW L. LASLETT (30), *Monash Institute of Reproduction and Development, Monash University, 246 Clayton Road, Clayton, Victoria, 3168, Australia*

ANDREW L. LEAVITT (10), *Departments of Laboratory Medicine and Internal Medicine, University of California at San Francisco, San Francisco, California 94143*

JONATHAN G. LIEBER (9), *Department of Medicine, National Jewish Medical and Research Center, 1400 Jackson Street, Denver, Colorado 80206*

TERRY MAGNUSON (28), *Department of Genetics, University of North Carolina at Chapel Hill, Chapel Hill, North Carolina 27599*

RON MCKAY (22), *Laboratory of Molecular Biology, NINDS, National Institutes of Health, Bethesda, Maryland 20892-4092*

TSUTOMU MOTOHASHI (24), *Department of Tissue and Organ Development, Regeneration and Advanced Medical Science, Gifu University Graduate School of Medicine, Gifu 500-8705, Japan*

TORU NAKANO (5,10), *Department of Molecular Cell Biology, Research Institute for Microbial Diseases, Osaka University, 3-1 Yamadaoka, Suita Osaka, Osaka 565-0871, Japan*

NORIO NAKATSUJI (29), *Department of Development and Differentiation, Institute for Frontier Medical Sciences, Kyoto University, Sakyo-ku, Kyoto, 606-8507, Japan*

ANNE NAVARRETE-SANTOS (21), *Department of Anatomy and Cell Biology, Martin Luther University, Halle-Wittenberg, Halle (Saale), D-06108, Germany*

SATOMI NISHIKAWA (4), *RIKEN Centre for Developmental Biology, 2-2-3 Minatojima-minamimachi, Chuo-ku, Kobe 650-0047, Japan*

SHIN-ICHI NISHIKAWA (4), *RIKEN Centre for Developmental Biology, 2-2-3 Minatojima-minamimachi, Chuo-ku, Kobe, Japan 650-0047*

KATHLEEN F. NOLAN (12), *Sir William Dunn School of Pathology, University of Oxford, South Parks Road, Oxford, OX1 3RE, UK*

MINETARO OGAWA (4), *Department of Cell Differentiation, Institute of Molecular Embryology and Genetics, Kumamoto University, Japan*

MITSUHIRO OKADA (4), *Department of Cell Differentitation, Institute of Molecular Embryology and Genetics, Kumamoto University, Kumamoto, Japan 650-0047*

HIROMI OKUYAMA (7), *Biotechnology Research Laboratories, Takara Bio Inc., Otsu, Siga 520-2193, Japan*

DAVID PANCHISION (22), *Laboratory of Molecular Biology, NINDS, National Institutes of Health, Bethesda, Maryland 20892-4092*

MARTIN F. PERA (30), *Monash Institute of Reproduction and Development, Monash University, 246 Clayton Road, Clayton, Victoria 3168, Australia*

RITA C. R. PERLINGEIRO (8), *ViaCell Inc, 26 Landsdowne St, Cambridge, MA 2139*

PERPÉTUA PINTO DO Ó (14), *Umeå Centre for Molecular Medicine, Umeå University, 901 87 Umeå, Sweden*

MARIE-HÉLÈNE PRANDINI (15), *Laboratoire Developpement et Vieillissement de L'endothelium, EMI INSERM 0219, DRDC/DVE, CEA Grenoble, 17 rue des martyrs, 38054 Grenoble cedex 09, France*

JOY RATHJEN (1), *Department of Molecular Biosciences and ARC Special Research Centre for the Molecular Genetics of Development, Adelaide University, Molecular Life Sciences/335, Adelaide, 5005, Australia*

PETER D. RATHJEN (1), *Department of Molecular Biosciences and ARC Special Research Centre for the Molecular Genetics of Development, Adelaide University, Molecular Life Sciences/335, Adelaide, 5005, Australia*

JÜRGEN ROHWEDEL (18). *Department of Medical Molecular Biology, Medical University of Lübeck, Ratzeburger Allee 160, Lübeck, D-23538, Germany*

EIKO SAKAI-OGAWA (5), *Department of Molecular Cell Biology, Research Institute for Microbial Diseases, Osaka University, 3-1 Yamadaoka, Suita Osaka, Osaka 565-0871, Japan*

MAYA SCHULDINER (31), *Department of Genetics, The Silberman Institute of Life Sciences, The Hebrew University, Jerusalem 91904, Israel*

SANFORD J. SHATTIL (10), *Division of Vascular Biology, Department of Cell Biology, The Scripps Research Institute, 10550 North Torrey Pines Road, VB-5, La Jolla, California 92037*

AUSTIN G. SMITH (23), *Institute for Stem Cell Research, University of Edinburgh, King's Buildings, West Mains Road, Edinburgh, Scotland EH9 3JQ, UK*

PHILIPPE SORIANO (26), *Program in Developmental Biology, Division of Basic Sciences, Fred Hutchinson Cancer Research Center, 1100 Fairview Avenue N, A2-025, P.O. Box 19024, Seattle, Washington 98109-1024*

HEIDI STUHLMANN (27), *Department of Cell Biology, Division of Vascular Biology, Scripps Research Institute, Mail CVN-26, 10550 North Torrey Pines Road, La Jolla, California 92037*

HIROFUMI SUEMORI (29), *Stem Cell Research Center, Institute for Frontier Medical Sciences, Kyoto University, Sakyo-ku, Kyoto, 606-8507, Japan*

MAKOTO TANAKA (5), *Department of Molecular Cell Biology, Research Institute for Microbial Diseases, Osaka University, 3-1 Yamadaoka, Suita Osaka, Osaka 565-0871 Japan*

NAOHIRO TERADA (20), *Department of Pathology, University of Florida College of Medicine, P.O. Box 100275, Gainesville, Florida 32610-0275*

MOTOKAZU TSUNETO (7), *Division of Immunology, Department of Molecular and*

Cellular Biology, School of Life Science, Faculty of Medicine, Tottori University, 86 Nishi-Machi, Yonago, Tottori 683-8503, Japan

PHI VILLAGEOIS (19), *Centre de Biochimie, Institut de Recherches Signalisation, Biologie du Developpement et Cancer, Université de Nice-Sophia Antipolis, UMR 6543 CNRS, Parc Valrose, Nice cedex 2, 06108 France*

DANIEL VITTET (15), *Laboratoire Developpement et Vieillissement de L'endothelium, EMI INSERM 0219, DRDC/DVE, CEA Grenoble, 17 rue des martyrs, 38054 Grenoble cedex 09, France*

JAY L. VIVIAN (28), *Department of Genetics, University of North Carolina at Chapel Hill, Chapel Hill, North Carolina 27599*

HERMAN WALDMANN (12), *Sir William Dunn School of Pathology, University of Oxford, South Parks Road, Oxford, OX1 3RE, UK*

EWA WANDZIOCH (14), *Umeå Centre for Molecular Medicine, Umeå University, 901 87 Umeå, Sweden*

BRIGITTE WDZIEKONSKI (19), *Centre de Biochimie, Institut de Recherches Signalisation, Biologie du Developpement et Cancer, Université de Nice-Sophia Antipolis, UMR 6543 CNRS, Parc Valrose, Nice cedex 2, 06108 France*

ANNA M. WOBUS (21), *In Vitro Differentiation Group, Institute of Plant Genetics and Crop Plant Research, Corrensstr. 3, Gatersleben, D-06466, Germany*

G. SCOTT WORTHEN (9), *Department of Medicine, National Jewish Medical and Research Center, 1400 Jackson Street, Denver, Colorado 80206*

TOSHIYUKI YAMANE (7,24), *Department of Pathology, Stanford University School of Medicine, B259 Beckman Center, Stanford, California 94305*

JUN YAMASHITA (4), *RIKEN Centre for Developmental Biology, 2-2-3 Minatojima-minamimachi, Chuo-ku, Kobe 650-0047, Japan*

HIDETOSHI YAMAZAKI (7), *Division of Regenerative Medicine and Therapeutics, Department of Genetic Medicine and Regenerative Therapeutics, Institute of Regenerative Medicine and Biofunction, Tottori University Graduate School of Medical Science, and Division of Immunology, Department of Molecular and Cellular Biology, School of Life Science, Faculty of Medicine, Tottori University, 86 Nishi-Machi, Yonago, Tottori 683-8503, Japan*

QI-LONG YING (23), *Institute for Stem Cell Research, University of Edinburgh, King's Buildings, West Mains Road, Edinburgh, Scotland, EH9 3JQ, UK*

NAOKO YOSHIMURA (24), *Department of Tissue and Organ Development, Regeneration and Advanced Medical Science, Gifu University Graduate School of Medicine, Gifu 500-8705, Japan*

WEN JIE ZHANG (13), *Department of Pathology & Immunology, Washington University School of Medicine, 660 S. Euclid Avenue, Box 8118, St. Louis, Missouri 63110*

JIE ZHENG (5), *Department of Molecular Cell Biology, Research Institute for Microbial Diseases, Osaka University, 3-1 Yamadaoka, Suita Osaka, Osaka 565-0871, Japan*

JUAN CARLOS ZÚÑIGA-PFLÜCKER (11), *Department of Immunology, University of Toronto, Toronto, Ontario M5S 1A8, Canada*

Preface

Since their isolation over twenty years ago, embryonic stem (ES) cells have come to play a prominent role in many different fields in biomedical research. While used extensively for gene targeting studies for the generation of "knock-out mice," it is their capacity to differentiate into a wide array of lineages in culture that has most recently captured the attention of basic scientists, clinical researchers, and the lay public. The capacity of an ES cell to differentiate into almost any cell type in a culture dish offers unprecedented opportunities for studies in lineage commitment and development, gene function, and cancer. In addition, the development of human ES cells in 1998 has expanded the potential uses of the *in vitro* differentiation approach to include the generation of specific cell types for cell replacement therapy and regenerative medicine.

It has been known for some time that mouse ES cells can differentiate in culture and generate derivatives of the three primary germ cell layers: ectoderm, endoderm, and mesoderm. In many of these early studies, however, differentiation was not well controlled and the cultures often consisted of mixtures of tissues. Significant advances in the methodologies of ES cell differentiation have been made in recent years, and it is now possible to generate relatively pure populations of cells from a number of different lineages in a reproducible fashion. These new approaches have enabled investigators to begin to use the system to probe the molecular events regulating early lineage commitment, as well as to generate populations appropriate for cell replacement therapy in preclinical models. With the establishment of human ES cells, the challenge for those in the field of stem cell and developmental biology is to now translate the information from the mouse to the human system.

In this volume, we have brought together a comprehensive collection of the most up-to-date methods for the differentiation of both mouse and human ES cells into a broad spectrum of lineages. Together with the differentiation protocols, we have solicited a select set of protocols that highlight approaches for gene discovery and lineage selection using ES cells. The chapters are written by leaders in the field from the international community and provide technical details that will enable the reader to establish and maintain ES cell cultures and induce their differentiation into a large number of different lineages.

We hope that this volume on *Differentiation of Embryonic Stem Cells* will find its way into many laboratories and proves to be useful at the bench for investigators world-wide. We are extremely grateful to the many authors for their excellent contributions to this volume and their patience in dealing with publication schedules. Finally, we extend our appreciation to Shirley Light at Academic Press who organized the assembly of this and hundreds of other volumes of *Methods in Enzymology* over several decades.

Paul M. Wassarman
Gordon M. Keller

METHODS IN ENZYMOLOGY

VOLUME XIII. Citric Acid Cycle
Edited by J. M. LOWENSTEIN

VOLUME XIV. Lipids
Edited by J. M. LOWENSTEIN

VOLUME XV. Steroids and Terpenoids
Edited by RAYMOND B. CLAYTON

VOLUME XVI. Fast Reactions
Edited by KENNETH KUSTIN

VOLUME XVII. Metabolism of Amino Acids and Amines (Parts A and B)
Edited by HERBERT TABOR AND CELIA WHITE TABOR

VOLUME XVIII. Vitamins and Coenzymes (Parts A, B, and C)
Edited by DONALD B. MCCORMICK, AND LEMUEL D. WRIGHT

VOLUME XIX. Proteolytic Enzymes
Edited by GERTRUDE E. PERLMANN AND LASZLO LORAND

VOLUME XX. Nucleic Acids and Protein Synthesis (Part C)
Edited by KIVIE MOLDAVE AND LAWRENCE GROSSMAN

VOLUME XXI. Nucleic Acids (Part D)
Edited by LAWRENCE GROSSMAN AND KIVIE MOLDAVE

VOLUME XXII. Enzyme Purification and Related Techniques
Edited by WILLIAM B. JAKOBY

VOLUME XXIII. Photosynthesis (Part A)
Edited by ANTHONY SAN PIETRO

VOLUME XXIV. Photosynthesis and Nitrogen Fixation (Part B)
Edited by ANTHONY SAN PIETRO

VOLUME XXV. Enzyme Structure (Part B)
Edited by C. H. W. HIRS AND SERGE N. TIMASHEFF

VOLUME XXVI. Enzyme Structure (Part C)
Edited by C. H. W. HIRS AND SERGE N. TIMASHEFF

VOLUME XLI. Carbohydrate Metabolism (Part B)
Edited by W. A. WOOD

VOLUME XLII. Carbohydrate Metabolism (Part C)
Edited by W. A. WOOD

VOLUME XLIII. Antibiotics
Edited by JOHN H. HASH

VOLUME XLIV. Immobilized Enzymes
Edited by KLAUS MOSBACH

VOLUME XLV Proteolytic Enzymes (Part B)
Edited by LASZLO LORAND

VOLUME XLVI. Affinity Labeling
Edited by WILLIAM B. JAKOBY AND MEIR WILCHEK

VOLUME XLVII. Enzyme Structure (Part E)
Edited by C. H. W. HIRS AND SERGE N. TIMASHEFF

VOLUME XLVIII. Enzyme Structure (Part F)
Edited by C. H. W. HIRS AND SERGE N. TIMASHEFF

VOLUME XLIX. Enzyme Structure (Part G)
Edited by C. H. W. HIRS AND SERGE N. TIMASHEFF

VOLUME L. Complex Carbohydrates (Part C)
Edited by VICTOR GINSBURG

VOLUME LI. Purine and Pyrimidine Nucleotide Metabolism
Edited by PATRICIA A. HOFFEE AND MARY ELLEN JONES

VOLUME LII. Biomembranes (Part C: Biological Oxidations)
Edited by SIDNEY FLEISCHER AND LESTER PACKER

VOLUME LIII. Biomembranes (Part D: Biological Oxidations)
Edited by SIDNEY FLEISCHER AND LESTER PACKER

VOLUME LIV. Biomembranes (Part E: Biological Oxidations)
Edited by SIDNEY FLEISCHER AND LESTER PACKER

VOLUME LV. Biomembranes (Part F: Bioenergetics)
Edited by SIDNEY FLEISCHER AND LESTER PACKER

VOLUME LVI. Biomembranes (Part G: Bioenergetics)
Edited by SIDNEY FLEISCHER AND LESTER PACKER

VOLUME LVII. Bioluminescence and Chemiluminescence
Edited by MARLENE A. DELUCA

VOLUME LVIII. Cell Culture
Edited by WILLIAM B. JAKOBY AND IRA PASTAN

VOLUME LIX. Nucleic Acids and Protein Synthesis (Part G)
Edited by KIVIE MOLDAVE AND LAWRENCE GROSSMAN

VOLUME LX. Nucleic Acids and Protein Synthesis (Part H)
Edited by KIVIE MOLDAVE AND LAWRENCE GROSSMAN

VOLUME 61. Enzyme Structure (Part H)
Edited by C. H. W. HIRS AND SERGE N. TIMASHEFF

VOLUME 62. Vitamins and Coenzymes (Part D)
Edited by DONALD B. MCCORMICK AND LEMUEL D. WRIGHT

VOLUME 63. Enzyme Kinetics and Mechanism (Part A: Initial Rate and Inhibitor Methods)
Edited by DANIEL L. PURICH

VOLUME 64. Enzyme Kinetics and Mechanism (Part B: Isotopic Probes and Complex Enzyme Systems)
Edited by DANIEL L. PURICH

VOLUME 65. Nucleic Acids (Part I)
Edited by LAWRENCE GROSSMAN AND KIVIE MOLDAVE

VOLUME 66. Vitamins and Coenzymes (Part E)
Edited by DONALD B. MCCORMICK AND LEMUEL D. WRIGHT

VOLUME 67. Vitamins and Coenzymes (Part F)
Edited by DONALD B. MCCORMICK AND LEMUEL D. WRIGHT

VOLUME 81. Biomembranes (Part H: Visual Pigments and Purple Membranes, I)
Edited by LESTER PACKER

VOLUME 82. Structural and Contractile Proteins (Part A: Extracellular Matrix)
Edited by LEON W. CUNNINGHAM AND DIXIE W. FREDERIKSEN

VOLUME 83. Complex Carbohydrates (Part D)
Edited by VICTOR GINSBURG

VOLUME 84. Immunochemical Techniques (Part D: Selected Immunoassays)
Edited by JOHN J. LANGONE AND HELEN VAN VUNAKIS

VOLUME 85. Structural and Contractile Proteins (Part B: The Contractile Apparatus and the Cytoskeleton)
Edited by DIXIE W. FREDERIKSEN AND LEON W. CUNNINGHAM

VOLUME 86. Prostaglandins and Arachidonate Metabolites
Edited by WILLIAM E. M. LANDS AND WILLIAM L. SMITH

VOLUME 87. Enzyme Kinetics and Mechanism (Part C: Intermediates, Stereochemistry, and Rate Studies)
Edited by DANIEL L. PURICH

VOLUME 88. Biomembranes (Part I: Visual Pigments and Purple Membranes, II)
Edited by LESTER PACKER

VOLUME 89. Carbohydrate Metabolism (Part D)
Edited by WILLIS A. WOOD

VOLUME 90. Carbohydrate Metabolism (Part E)
Edited by WILLIS A. WOOD

VOLUME 91. Enzyme Structure (Part I)
Edited by C. H. W. HIRS AND SERGE N. TIMASHEFF

VOLUME 92. Immunochemical Techniques (Part E: Monoclonal Antibodies and General Immunoassay Methods)
Edited by JOHN J. LANGONE AND HELEN VAN VUNAKIS

VOLUME 130 Enzyme Structure (Part K)
Edited by C. H. W. HIRS AND SERGE N. TIMASHEFF

VOLUME 131. Enzyme Structure (Part L)
Edited by C H W HIRS AND SERGE N. TIMASHEFF

VOLUME 132. Immunochemical Techniques (Part J: Phagocytosis and Cell-Mediated Cytotoxicity)
Edited by GIOVANNI DI SABATO AND JOHANNES EVERSE

VOLUME 133. Bioluminescence and Chemiluminescence (Part B)
Edited by MARLENE DELUCA AND WILLIAM D. MCELROY

VOLUME 134. Structural and Contractile Proteins (Part C: The Contractile Apparatus and the Cytoskeleton)
Edited by RICHARD B. VALLEE

VOLUME 135. Immobilized Enzymes and Cells (Part B)
Edited by KLAUS MOSBACH

VOLUME 136. Immobilized Enzymes and Cells (Part C)
Edited by KLAUS MOSBACH

VOLUME 137. Immobilized Enzymes and Cells (Part D)
Edited by KLAUS MOSBACH

VOLUME 138. Complex Carbohydrates (Part E)
Edited by VICTOR GINSBURG

VOLUME 139. Cellular Regulators (Part A: Calcium- and Calmodulin-Binding Proteins)
Edited by ANTHONY R. MEANS AND P. MICHAEL CONN

VOLUME 140. Cumulative Subject Index Volumes 102–119, 121–134

VOLUME 141. Cellular Regulators (Part B: Calcium and Lipids)
Edited by P. MICHAEL CONN AND ANTHONY R. MEANS

VOLUME 142. Metabolism of Aromatic Amino Acids and Amines
Edited by SEYMOUR KAUFMAN

VOLUME 143. Sulfur and Sulfur Amino Acids
Edited by WILLIAM B. JAKOBY AND OWEN GRIFFITH

VOLUME 144. Structural and Contractile Proteins (Part D: Extracellular Matrix)
Edited by LEON W. CUNNINGHAM

VOLUME 145. Structural and Contractile Proteins (Part E: Extracellular Matrix)
Edited by LEON W. CUNNINGHAM

VOLUME 146. Peptide Growth Factors (Part A)
Edited by DAVID BARNES AND DAVID A. SIRBASKU

VOLUME 147. Peptide Growth Factors (Part B)
Edited by DAVID BARNES AND DAVID A. SIRBASKU

VOLUME 148. Plant Cell Membranes
Edited by LESTER PACKER AND ROLAND DOUCE

VOLUME 149. Drug and Enzyme Targeting (Part B)
Edited by RALPH GREEN AND KENNETH J. WIDDER

VOLUME 150. Immunochemical Techniques (Part K: *In Vitro* Models of B and T Cell Functions and Lymphoid Cell Receptors)
Edited by GIOVANNI DI SABATO

VOLUME 151. Molecular Genetics of Mammalian Cells
Edited by MICHAEL M. GOTTESMAN

VOLUME 152. Guide to Molecular Cloning Techniques
Edited by SHELBY L. BERGER AND ALAN R. KIMMEL

VOLUME 153. Recombinant DNA (Part D)
Edited by RAY WU AND LAWRENCE GROSSMAN

VOLUME 154. Recombinant DNA (Part E)
Edited by RAY WU AND LAWRENCE GROSSMAN

VOLUME 155. Recombinant DNA (Part F)
Edited by RAY WU

VOLUME 156. Biomembranes (Part P: ATP-Driven Pumps and Related Transport: The Na, K-Pump)
Edited by SIDNEY FLEISCHER AND BECCA FLEISCHER

VOLUME 157. Biomembranes (Part Q: ATP-Driven Pumps and Related Transport: Calcium, Proton, and Potassium Pumps)
Edited by SIDNEY FLEISCHER AND BECCA FLEISCHER

VOLUME 158. Metalloproteins (Part A)
Edited by JAMES F. RIORDAN AND BERT L. VALLEE

VOLUME 159. Initiation and Termination of Cyclic Nucleotide Action
Edited by JACKIE D. CORBIN AND ROGER A. JOHNSON

VOLUME 160. Biomass (Part A: Cellulose and Hemicellulose)
Edited by WILLIS A. WOOD AND SCOTT T. KELLOGG

VOLUME 161. Biomass (Part B: Lignin, Pectin, and Chitin)
Edited by WILLIS A. WOOD AND SCOTT T. KELLOGG

VOLUME 162. Immunochemical Techniques (Part L: Chemotaxis and Inflammation)
Edited by GIOVANNI DI SABATO

VOLUME 163. Immunochemical Techniques (Part M: Chemotaxis and Inflammation)
Edited by GIOVANNI DI SABATO

VOLUME 164. Ribosomes
Edited by HARRY F. NOLLER, JR., AND KIVIE MOLDAVE

VOLUME 165. Microbial Toxins: Tools for Enzymology
Edited by SIDNEY HARSHMAN

VOLUME 166. Branched-Chain Amino Acids
Edited by ROBERT HARRIS AND JOHN R. SOKATCH

VOLUME 167. Cyanobacteria
Edited by LESTER PACKER AND ALEXANDER N. GLAZER

VOLUME 168. Hormone Action (Part K: Neuroendocrine Peptides)
Edited by P. MICHAEL CONN

VOLUME 169. Platelets: Receptors, Adhesion, Secretion (Part A)
Edited by JACEK HAWIGER

VOLUME 194. Guide to Yeast Genetics and Molecular Biology
Edited by CHRISTINE GUTHRIE AND GERALD R. FINK

VOLUME 195. Adenylyl Cyclase, G Proteins, and Guanylyl Cyclase
Edited by ROGER A. JOHNSON AND JACKIE D. CORBIN

VOLUME 196. Molecular Motors and the Cytoskeleton
Edited by RICHARD B. VALLEE

VOLUME 197. Phospholipases
Edited by EDWARD A. DENNIS

VOLUME 198. Peptide Growth Factors (Part C)
Edited by DAVID BARNES, J. P. MATHER, AND GORDON H. SATO

VOLUME 199. Cumulative Subject Index Volumes 168–174, 176–194

VOLUME 200. Protein Phosphorylation (Part A: Protein Kinases: Assays, Purification, Antibodies, Functional Analysis, Cloning, and Expression)
Edited by TONY HUNTER AND BARTHOLOMEW M. SEFTON

VOLUME 201. Protein Phosphorylation (Part B: Analysis of Protein Phosphorylation, Protein Kinase Inhibitors, and Protein Phosphatases)
Edited by TONY HUNTER AND BARTHOLOMEW M. SEFTON

VOLUME 202. Molecular Design and Modeling: Concepts and Applications (Part A: Proteins, Peptides, and Enzymes)
Edited by JOHN J. LANGONE

VOLUME 203. Molecular Design and Modeling: Concepts and Applications (Part B: Antibodies and Antigens, Nucleic Acids, Polysaccharides, and Drugs)
Edited by JOHN J. LANGONE

VOLUME 204. Bacterial Genetic Systems
Edited by JEFFREY H. MILLER

VOLUME 205. Metallobiochemistry (Part B: Metallothionein and Related Molecules)
Edited by JAMES F. RIORDAN AND BERT L. VALLEE

VOLUME 206. Cytochrome P450
Edited by MICHAEL R. WATERMAN AND ERIC F. JOHNSON

VOLUME 220. Membrane Fusion Techniques (Part A)
Edited by NEJAT DÜZGÜNES

VOLUME 221. Membrane Fusion Techniques (Part B)
Edited by NEJAT DÜZGÜNES

VOLUME 222. Proteolytic Enzymes in Coagulation, Fibrinolysis, and Complement Activation (Part A: Mammalian Blood Coagulation Factors and Inhibitors)
Edited by LASZLO LORAND AND KENNETH G. MANN

VOLUME 223. Proteolytic Enzymes in Coagulation, Fibrinolysis, and Complement Activation (Part B: Complement Activation, Fibrinolysis, and Nonmammalian Blood Coagulation Factors)
Edited by LASZLO LORAND AND KENNETH G. MANN

VOLUME 224. Molecular Evolution: Producing the Biochemical Data
Edited by ELIZABETH ANNE ZIMMER, THOMAS J. WHITE, REBECCA L. CANN, AND ALLAN C. WILSON

VOLUME 225. Guide to Techniques in Mouse Development
Edited by PAUL M. WASSARMAN AND MELVIN L. DEPAMPHILIS

VOLUME 226. Metallobiochemistry (Part C: Spectroscopic and Physical Methods for Probing Metal Ion Environments in Metalloenzymes and Metalloproteins)
Edited by JAMES F. RIORDAN AND BERT L. VALLEE

VOLUME 227. Metallobiochemistry (Part D: Physical and Spectroscopic Methods for Probing Metal Ion Environments in Metalloproteins)
Edited by JAMES F. RIORDAN AND BERT L. VALLEE

VOLUME 228. Aqueous Two-Phase Systems
Edited by HARRY WALTER AND GÖTE JOHANSSON

VOLUME 229. Cumulative Subject Index Volumes 195–198, 200–227

VOLUME 230. Guide to Techniques in Glycobiology
Edited by WILLIAM J. LENNARZ AND GERALD W. HART

VOLUME 231. Hemoglobins (Part B: Biochemical and Analytical Methods)
Edited by JOHANNES EVERSE, KIM D. VANDEGRIFF, AND ROBERT M. WINSLOW

VOLUME 336. Microbial Growth in Biofilms (Part A: Developmental and Molecular Biological Aspects)
Edited by RON J. DOYLE

VOLUME 337. Microbial Growth in Biofilms (Part B: Special Environments and Physicochemical Aspects)
Edited by RON J. DOYLE

VOLUME 338. Nuclear Magnetic Resonance of Biological Macromolecules (Part A)
Edited by THOMAS L. JAMES, VOLKER DÖTSCH, AND ULI SCHMITZ

VOLUME 339. Nuclear Magnetic Resonance of Biological Macromolecules (Part B)
Edited by THOMAS L. JAMES, VOLKER DÖTSCH, AND ULI SCHMITZ

VOLUME 340. Drug–Nucleic Acid Interactions
Edited by JONATHAN B. CHAIRES AND MICHAEL J. WARING

VOLUME 341. Ribonucleases (Part A)
Edited by ALLEN W. NICHOLSON

VOLUME 342. Ribonucleases (Part B)
Edited by ALLEN W. NICHOLSON

VOLUME 343. G Protein Pathways (Part A: Receptors)
Edited by RAVI IYENGAR AND JOHN D. HILDEBRANDT

VOLUME 344. G Protein Pathways (Part B: G Proteins and Their Regulators)
Edited by RAVI IYENGAR AND JOHN D. HILDEBRANDT

VOLUME 345. G Protein Pathways (Part C: Effector Mechanisms)
Edited by RAVI IYENGAR AND JOHN D. HILDEBRANDT

VOLUME 346. Gene Therapy Methods
Edited by M. IAN PHILLIPS

VOLUME 347. Protein Sensors and Reactive Oxygen Species (Part A: Selenoproteins and Thioredoxin)
Edited by HELMUT SIES AND LESTER PACKER

VOLUME 348. Protein Sensors and Reactive Oxygen Species (Part B: Thiol Enzymes and Proteins)
Edited by HELMUT SIES AND LESTER PACKER

VOLUME 349. Superoxide Dismutase
Edited by LESTER PACKER

VOLUME 350. Guide to Yeast Genetics and Molecular and Cell Biology (Part B)
Edited by CHRISTINE GUTHRIE AND GERALD R. FINK

VOLUME 351. Guide to Yeast Genetics and Molecular and Cell Biology (Part C)
Edited by CHRISTINE GUTHRIE AND GERALD R. FINK

VOLUME 352. Redox Cell Biology and Genetics (Part A)
Edited by CHANDAN K. SEN AND LESTER PACKER

VOLUME 353. Redox Cell Biology and Genetics (Part B)
Edited by CHANDAN K. SEN AND LESTER PACKER

VOLUME 354. Enzyme Kinetics and Mechanisms (Part F: Detection and Characterization of Enzyme Reaction Intermediates) (in preparation)
Edited by DANIEL L. PURICH

VOLUME 355. Cumulative Subject Index Volumes 321–354
(in preparation)

VOLUME 356. Laser Capture Microscopy and Microdissection
(in preparation)
Edited by P. MICHAEL CONN

VOLUME 357. Cytochrome P450, Part C (in preparation)
Edited by ERIC F. JOHNSON AND MICHAEL R. WATERMAN

VOLUME 358. Bacterial Pathogenesis (Part C: Identification, Regulation, and Function of Virulence Factors)
Edited by VIRGINIA L. CLARK AND PATRIK M. BAVOIL

VOLUME 359. Nitric Oxide (Part D: Nitric Oxide Detection, Mitochondria and Cell Functions, and Peroxynitrite Reactions)
Edited by ENRIQUE CADENAS AND LESTER PACKER

VOLUME 360. Biophotonics (Part A)
Edited by GERARD MARRIOTT AND IAN PARKER

VOLUME 361. Biophotonics (Part B)
Edited by GERARD MARRIOTT AND IAN PARKER

VOLUME 362. Recognition of Carbohydrates in Biological Systems (Part A: General Procedures)
Edited by YUON C. LEE AND REIKO T. LEE

VOLUME 363. Recognition of Carbohydrates in Biological Systems (Part B: Specific Applications)
Edited by YUON C. LEE AND REIKO T. LEE

VOLUME 364. Nuclear Receptors
Edited by DAVID W. RUSSELL AND DAVID J. MANGELSDORF

VOLUME 365. Differentiation of Embryonic Stem Cells
Edited by PAUL M. WASSARMAN AND GORDON M. KELLER

VOLUME 366. Protein Phosphatases
Edited by SUSANNE KLUMPP AND JOSEF KRIEGLSTEIN

VOLUME 367. Liposomes (Part A) (in preparation)
Edited by NEJAT DUZGUNES

VOLUME 368. Macromolecular Crystallography (Part C) (in preparation)
Edited by CHARLES W. CARTER AND ROBERT M. SWEET

VOLUME 369. Combinatorial Chemistry (Part B) (in preparation)
Edited by GUILLERMO A. MORALES AND BARRY A. BUNIN

VOLUME 370. RNA Polymerases and Associated Factors (Part C) (in preparation)
Edited by SANKAR L. ADHYA AND SUSAN GARGES

VOLUME 371. RNA Polymerases and Associated Factors (Part D) (in preparation)
Edited by SANKAR L. ADHYA AND SUSAN GARGES

VOLUME 372. Liposomes (Part B) (in preparation)
Edited by NEJAT DUZGUNES

VOLUME 373. Liposomes (Part C) (in preparation)
Edited by NEJAT DUZGUNES

VOLUME 374. Macromolecular Crystallography (Part D) (in preparation)
Edited by CHARLES W. CARTER AND ROBERT M. SWEET

VOLUME 375. Chromatin and Chromatin Remodeling Enzymes (Part A) (in preparation)
Edited by CARL WU AND C. DAVID ALLIS

VOLUME 376. Chromatin and Chromatin Remodeling Enzymes (Part B) (in preparation)
Edited by CARL WU AND C. DAVID ALLIS

VOLUME 377. Chromatin and Chromatin Remodeling Enzymes (Part C) (in preparation)
Edited by CARL WU AND C. DAVID ALLIS

VOLUME 378. Quinones and Quinone Enzymes (Part A) (in preparation)
Edited by HELMUT SIES AND LESTER PACKER

VOLUME 379. Energetics and Biological Macromolecules (Part D) (in preparation)
Edited by JO M. HOLT, MICHAEL L. JOHNSON, AND GARY K. ACKERS

VOLUME 380. Energetics and Biological Macromolecules (Part E) (in preparation)
Edited by JO M. HOLT, MICHAEL L. JOHNSON, AND GARY K. ACKERS

VOLUME 381. Oxygen Sensing (in preparation)
Edited by CHANDAN K. SEN AND GREGG L. SEMENZA

VOLUME 382. Quinones and Quinone Enzymes (Part B) (in preparation)
Edited by HELMUT SIES AND LESTER PACKER

Section I

Differentiation of Mouse Embryonic Stem Cells

[1] Lineage Specific Differentiation of Mouse ES Cells: Formation and Differentiation of Early Primitive Ectoderm-like (EPL) Cells

By JOY RATHJEN and PETER D. RATHJEN

Introduction

The development of methods for controlled, regulated and reproducible differentiation of embryonic stem cells *in vitro* will be required before the projected basic and therapeutic applications of pluripotent cell differentiation capabilities can be fully realized.[1] There is particular merit in differentiation regimes that recapitulate lineage establishment during normal embryogenesis. These are anticipated to provide a nontransformed model system for the investigation of cell fate choice, identification, characterization and production of transient differentiation intermediates, and identification of signaling pathways that regulate cell identity and acquisition of positional information. In conjunction with the extraordinary experimental malleability of the ES cell genome, this enables sophisticated analysis of gene function at the cellular level in a system that is not restricted by limitations associated with maintenance of a viable embryo. Directed, lineage specific differentiation protocols are also likely to find application in the production of cell populations suited for therapeutic transplantation to alleviate diseases caused by cell loss, damage or dysfunction.

The most commonly used protocols for ES cell differentiation rely on differentiation within complex cellular aggregates known as embryoid bodies (EBs).[1,2] ES cell EBs undergo a developmental program that recapitulates many of the events of early mammalian embryogenesis. Initially, outer cells of the aggregate differentiate to establish the extra-embryonic endodermal lineage, followed by formation from the inner cells of a second pluripotent cell population equivalent to primitive ectoderm. After about 4 days in culture, inner cells begin to lose pluripotence and

[1] J. Rathjen and P. D. Rathjen, Mouse ES cells: experimental exploitation of pluripotent cell differentiation potential, *Curr. Opin. Gen. Dev.* **11**, 589–596 (2001).

[2] T. C. Doetschman, H. Eistetter, M. Katz, W. Schmidt, and R. Kemler, The in vitro development of blastocyst-derived embryonic stem cell lines: formation of visceral yolk sac, blood islands and myocardium, *J. Embryol. Exp. Morph.* **87**, 27–45 (1985).

0076-6879/2003 $35.00

differentiate into cells representative of the primary germ layers, ectoderm, endoderm, and mesoderm. Further differentiation results in formation of many of the cell populations of the embryo and adult including beating cardiomyocytes, blood, hepatocytes, neurons, epidermis, and gut endothelium. The major difference from development *in vivo* is the lack of organizational cues associated with the body axes. The consequent disorganization can result in temporal and spatial misalignment of tissue populations within the EB and potential exposure of cells to inappropriate environmental and inductive signaling. Similarly, the effectiveness of this differentiation system is compromised by the presence within the environment of cell populations, such as visceral endoderm, known to be sources of inductive signals that regulate pluripotent cell maintenance and differentiation *in vivo*. The addition of signaling molecules such as growth factors to the differentiation environment is therefore of only limited utility because of the heterogeneity of the responsive population and the presence of competing cytokine signals.

ES cells share a gene expression, differentiation potential and cytokine responsiveness with their source cell population, the pluripotent cells of the ICM.[3–5] The immediate developmental fate of ICM en route to formation of the embryo proper, differentiation to primitive ectoderm, can be recapitulated *in vitro* by formation of *e*arly *p*rimitive ectoderm-*l*ike (EPL) cells.[3] Unlike ES cells, when differentiated *in vitro* EPL cells do not form visceral endoderm,[4,6] providing a technology for pluripotent cell differentiation that is devoid of extraembryonic signals. This enables control over the formation of somatic lineages from pluripotent cells by manipulation of the differentiation environment and thereby overcomes many of the inherent problems associated with ES cell EBs.

The purpose of this chapter is to discuss, at a technical level, the formation and use of EPL cells, including methodologies for further differentiation of EPL cells to the mesodermal or ectodermal lineage, and identification of alternative pluripotent and somatic cell populations.

[3] J. Rathjen, J.-A. Lake, M. D. Bettess, J. M. Washington, G. Chapman, and P. D. Rathjen, Formation of a primitive ectoderm like cell population, EPL cells, from ES cells in response to biologically derived factors, *J. Cell Sci.* **112**, 601–612 (1999).

[4] J.-A. Lake, J. Rathjen, J. Remiszewski, and P. D. Rathjen, Reversible programming of pluripotent cell differentiation, *J. Cell Sci.* **113**, 555–566 (2000).

[5] T. A. Pelton, S. Sharma, T. C. Schulz, J. Rathjen, and P. D. Rathjen, Transient pluripotent cell populations during primitive ectoderm formation: Correlation of in vivo and in vitro pluripotent cell development, *J. Cell Sci.* **115,** 329–339 (2001).

[6] J. Rathjen, B. H. Haines, K. Hudson, A. Nesci, S. Dunn, and P. D. Rathjen, Directed differentiation of pluripotent cells to neural lineages: homogeneous formation and differentiation of a neurectoderm population, *Development* **129**, 2649–2661 (2002).

Cell Culture Requirements for ES and EPL Cells

EPL cells are formed by culture of ES cells in the presence of medium conditioned by exposure to the hepatocellularcarcinoma cell line HepG2. As a pluripotent cell population equivalent to primitive ectoderm, which exists transiently in the embryo and is fated to differentiate in response to signals regulating gastrulation, EPL cells are extremely sensitive to inductive signals or trace contaminants in the culture environment. Accordingly the formation and maintenance of homogeneous EPL cell populations requires rigorous quality control of cell culture components. Procedures that yield satisfactory outcomes for the selection of key tissue culture reagents are outlined below.

General Consumables

1. Tissue culture plasticware is sourced from Falcon or Nunc, bacterial grade dishes from Technoplas.
2. Trypsin is obtained as a 0.5% stock solution in 5.3 m*M* EDTA, 145 m*M* NaCl (Gibco-BRL) and diluted 1 : 10 in PBS before use. Trypsin is not prewarmed to 37°C before use but used cool, between 4°C and room temperature.
3. 1× PBS, pH 7.3, is made from analytical grade salts in Milli-Q treated H_2O. 8.00 g/liter NaCl, 0.20 g/liter KCl, 0.20 g/liter KH_2PO_4, 1.15 g/liter Na_2HPO_4.

Wherever possible disposable plastic consumables, such as plastic pipettes, are used for the culture of pluripotent cells.

Foetal Calf Serum (FCS)

Foetal calf serum is tested for the ability to sustain proliferation of undifferentiated ES cells in culture, for toxicity to ES cell growth, and for the ability to support induction and culture of EPL cells. Sample lots of serum are sourced from multiple companies and batches sufficient for 2 years use are reserved. Serum analysis takes approximately 4 weeks, so testing is commenced 2–3 months before existing supplies are depleted.

1. 1×10^3 low passage (passages 14–16) feeder independent D3 ES cells[2] are seeded into gelatinized 2 ml tissue culture wells in 1 ml of ES cell medium (Table I) made with FCS of the samples to be tested. ES cell colony forming ability (plating efficiency) and integrity are compared in six duplicate wells for each FCS sample with medium containing the FCS in use at the time of assay. After 5 days culture at 37°C and 10% CO_2, colonies are assessed by microscopic inspection of live cells, and then stained for alkaline phosphatase

TABLE I
MEDIUM FORMULATIONS

ES cell medium:	DMEM[a] 10% foetal calf serum (FCS) 40 μg/ml gentamycin (Pharmacia and Upjohn) 1 m*M* L-glutamine[b] (Life Technologies) 0.1 m*M* β-mercaptoethanol[c] (β-ME; Sigma Aldrich) LIF[d], approximately 1000 units
HepG2 medium:	DMEM[a] 10% FCS
50% MEDII:	50% DMEM[a] supplemented with 10% FCS 50% MEDII[e] 40 μg/ml gentamycin 1 m*M* L-glutamine[b] 0.1 m*M* β-ME[c]
EB/differentiation medium:	DMEM[a] 10% FCS 40 μg/ml gentamycin (Pharmacia and Upjohn) 1 m*M* L-glutamine[b] 0.1 m*M* β-ME[c]
Chemically defined medium:	50% DMEM[a] 50% Hams F12 (Gibco-BRL # 11765-054) 1× insulin-transferrin-sodium selenite (ITSS; Roche) 1 m*M* L-glutamine

[a]DMEM with high glucose and no HEPES (Life Technologies # 11995-065).
[b]100 μl β-ME diluted in 14.2 ml of 1× PBS gives a 1000× stock solution. Fresh stocks should be made every 14 days.
[c]L-glutamine is added to medium only if the medium stock has been open and exposed to light for greater than 1 month.
[d]LIF is sourced as a conditioned medium from transfected Cos-1 cells (Smith, 1991).[6a]
[e]Conditioned medium from the human hepatocellularcarcinoma cell line, HepG2.
Cell culture medium is made in small aliquots that are renewed every passage.

activity with Sigma kit 86-R using the modifications described in Rathjen *et al.*[3]

2. Samples are assessed by three criteria. These allow an assessment of cell morphology and quantitation for each FCS sample of both colony forming ability and the degree of differentiation.
   - *Maintenance of morphological integrity:* ES cells should grow in round, domed colonies of approximately equal size, with smooth edges and a high refractive index (for example, see Fig. 1D and E). Colonies should be compact, such that individual cell–cell

[6a] A. G. Smith, *J. Tiss. Cult. Meth.* **13,** 89–94 (1991).

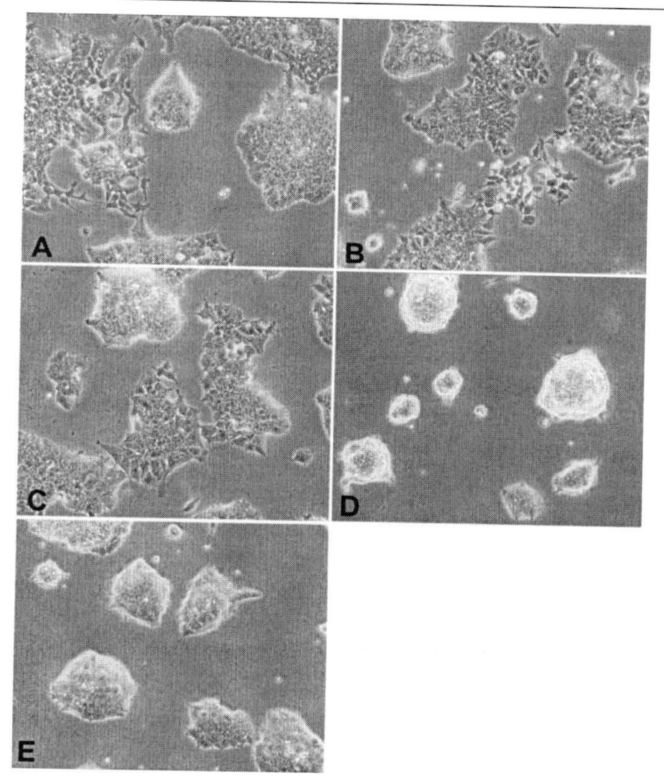

FIG. 1. Morphology of ES cells grown on gelatin samples. ES cells grown on (A, B) Type A gelatin derived from acid-cured porcine skin, 300 bloom, obtained from Sigma Aldrich (G1890; lot # 69H1227 and lot # 117H0527, respectively), (C) a type B gelatin from lime-cured bovine skin, 225 bloom, obtained from Sigma Aldrich (G9391; lot # 68H024), (D) a food grade gelatin from ICN (901771; lot # 8453B) and compared to gelatin free tissue culture plastic (E). Each 3 ml well was treated with 0.1% gelatin solution overnight at room temperature before seeding with 1×10^5 D3 ES cells in ES cell medium. Cells were cultured for 3 days. Images were taken using phase contrast at 20× magnification.

junctions cannot be easily seen. FCS samples that fail to support greater than 90% of the colonies in this morphology are abandoned.

- *Plating efficiency:* The number of colonies in each well is counted and expressed as a percentage of the total number of cells seeded. The plating efficiency should be between 25 and 50%.
- *Degree of differentiation:* Each colony is scored for the presence or absence of differentiated, alkaline phosphatase negative cells. In

mixed colonies these cells normally surround the colony. Less than 10% of the ES cell colonies should contain any differentiated cells.

3. FCS samples that fulfill all three criteria in the initial screen are assessed for toxicity at higher concentrations. ES cells are seeded as above in medium containing varying concentrations of FCS, from 5 to 25%, before assessment as described earlier. Those FCS samples in which differentiation is increased or plating densities are decreased with increasing serum concentration are not pursued.
4. The final screening procedure is use of FCS samples for the production of MEDII and formation of EPL cells in adherent culture. The methods for production of MEDII and formation of EPL cells are given elsewhere in this review. For FCS quality testing, 1×10^3 low passage (passages 14–16) feeder independent D3 ES cells are seeded into six duplicate gelatinized 2 ml tissue culture wells in 1 ml of 50% MEDII (Table I) made with FCS of the samples to be tested. After 5 days culture at 37°C and 10% CO_2, colonies are assessed by microscopic inspection of live cells. EPL cell cultures are inspected for the homogeneous formation of EPL cell colonies (Fig. 2) and the loss of all colonies of ES cell morphology. Cells can be stained for alkaline phosphatase activity with Sigma kit 86-R using the modifications described in Rathjen *et al.*[3] and the degree of differentiation within the cultures assessed as for ES cells. The levels of differentiation should be comparable or less than in the FCS in use.
5. Long-term storage of FCS stocks is carried out at −20°C. Before use, FCS is heat inactivated at 56°C for 30 min, aliquoted and frozen at −20°C. When required, FCS aliquots are thawed at 4°C overnight and kept at 4°C for 1–2 months. Medium is made in small batches that are used rapidly, usually within 2 days, and repeated warming/cooling cycles of serum or medium are avoided.

Gelatin

The potential adverse effects of gelatin, an impure mixture of matrix proteins derived from porcine or bovine skin, on the growth of pluripotent cells are often overlooked. Gelatin can increase the differentiation of EPL cells when compared to other, purified matrix components,[7] and variations in ES cell morphology cultured on individual batches of gelatin (Fig. 1) suggest that it cannot be viewed as an inert component of the system.

[7] J. Rathjen, J. M. Washington, M. D. Bettess, and P. D. Rathjen, Identification of a biological activity that supports primitive ectoderm maintenance and proliferation, *Biol. Reprod.*, in press.

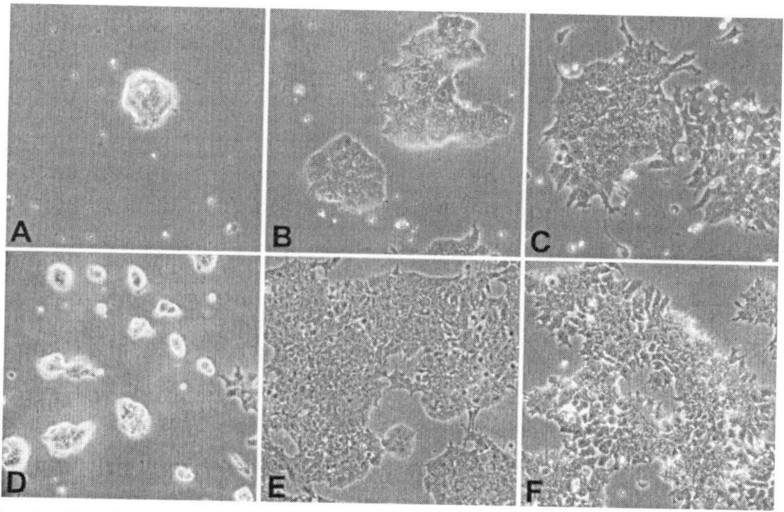

FIG. 2. Morphology of ES and EPL cells in culture. (A–C) ES cells were seeded into 1 ml gelatinized tissue culture wells at low density (1000 cells/well) and cultured for 5 days in (A) ES cell medium, (B) 50% MEDII supplemented with 1000 units/ml LIF, and (C) 50% MEDII without addition of exogenous LIF. Cells in B and C have differentiated to form EPL cells. (D–F) 1×10^6 ES cells were seeded into a 10 cm tissue culture plate pretreated with 0.1% gelatin (high density culture) and cultured for 2 days in (D) ES cell medium, (E) 50% MEDII supplemented with 1000 units/ml LIF, and (F) 50% MEDII without addition of exogenous LIF. Note the more open appearance of EPL cells grown without LIF (C, F) in comparison to those grown with LIF supplementation (B, E).

Wherever possible, ES cell lines are cultured on nongelatinized tissue culture plastic. However, gelatin is a convenient matrix for the formation and maintenance of EPL cells that grow poorly in the absence of exogenous ECM. Gelatin samples are tested for the ability to maintain ES cell morphology.

- 0.1% gelatin solutions in 1× PBS are prepared and the gelatin is dissolved by autoclaving. Duplicate 3 cm tissue culture wells are pretreated with 0.1% gelatin overnight at room temperature.
- Soluble gelatin is aspirated from the well and 1×10^5 ES cells are seeded in 2.5 ml of ES cell medium and cultured for 3 days at 37°C in 10% CO_2.
- Cells are examined microscopically for key indicators of ES cell morphology. Approaching 100% of the colonies should be round, three-dimensional and compact, with a high refractive index, as demonstrated in Fig. 1D and E. Gelatin samples in which the

morphology does not conform to this description, as illustrated in Fig. 1A–C, are not pursued.

Maintenance of ES Cells

ES cells are cultured in ES cell medium (Table I) on nongelatinized tissue culture dishes in 10% CO_2 in humidified incubators. We routinely use D3 ES cells (ATCC catalogue # CRL 1934)[2] between passage 14 and 35, but have used equivalent conditions successfully with a number of feeder independent, 129-derived mouse ES cells. The methods below are for ES cells cultured in a 10 cm diameter plate (we use 10 cm Falcon tissue culture dishes that have an actual diameter of 8.5 cm).

1. Near confluent ES cells are washed twice with 1× PBS and 1 ml of trypsin is added and swirled around the plate. The plate is left at room temperature. ES cells cultured on nongelatinized tissue culture plastic are more likely to detach from the substratum than cells grown on gelatin so that great care must be used during experimental manipulation.
2. When the boundaries of the cells become apparent within the ES cell colonies and individual cells appear loosely adherent (approximately 30–60 sec after addition of trypsin), 1 ml of DMEM supplemented with 10% FCS, 0.1 m*M* β-ME is added and the cells are dislodged by pipetting trypsin : medium repeatedly over the surface of the plate with a 1 ml Gilson pipetteman. This medium is not stored beyond 1 week. Cells are transferred to a 30 ml centrifuge tube and pelleted at 1200 rpm at room temperature for 2–3 min.
3. Medium is aspirated from the cells, the tube is tapped 3–4 times to loosen the pellet, and the pellet is resuspended in 1 ml DMEM with 10% FCS, 0.1 m*M* β-ME by pipetting twice with a Gilson 1 ml pipetteman. Cell number is determined by counting on a hemocytometer and cells are seeded in 10–15 ml of ES cell medium in 10 cm diameter dishes at a range of concentrations, usually between 1.5×10^6 and 5×10^5 cells/dish (3×10^4 to 1×10^4 cells/cm^2). These cell densities should necessitate passaging in 2–4 days with medium being replenished every 2 days. A 10 cm diameter plate will yield $1–2\times10^7$ ES cells at near confluence.

ES Cell Morphology

ES cell colonies grown under these conditions should be smooth, round and domed, with a compact appearance such that the individual cells are barely discernible (Figs. 1D and E, and 2A and D).

This morphology clearly distinguishes ES from EPL cell colonies. Plates with a high proportion (>10%) of colonies not conforming to this morphology and/or with a high proportion of differentiated cells are not pursued. Cells can either be repassaged, which enriches for pluripotent cells because differentiated cells within the cultures adhere and grow poorly after passage, or, if not improving, reinitiated from frozen stocks.

Preparation of MEDII

MEDII is a conditioned medium taken from the human hepatocellular carcinoma cell line, HepG2.[8] HepG2 cells can be obtained from ATCC (catalogue # HB-8065). We find it most convenient to culture cells and condition medium in 175 cm^2 flasks.

1. HepG2 cells are passaged when near confluence.
2. Cells are washed twice thoroughly with 1× PBS. Two to 3 ml trypsin is added to the flask and incubated with the cells at 37°C for 5–7 min, until detachment of the cells from each other and the substratum is observed microscopically. The cells are detached from the plastic by hitting the side of the flask forcefully. Five milliliter of HepG2 medium (Table I) is added and the cell suspension is pipetted vigorously up and down a 10 ml pipette before transfer to a 30 ml centrifuge tube. Cells are pelleted by centrifugation at 1200 rpm for 5 min at room temperature, the medium is aspirated and the pellet is loosened by tapping the side of the tube. Cells are resuspended in 1–3 ml of DMEM supplemented with 10% FCS by pipetting repeatedly with a 1 ml Gilson pipetteman.
3. For production of MEDII conditioned medium, HepG2 cells are seeded at a density sufficient to give a near confluent layer of cells after 4 days in culture. In our hands this approximates to 5×10^4 cells/cm^2. Cells are seeded at a ratio of approximately 1.75×10^5 cells/ml, or 50 ml of medium in a 175 cm^2 flask, and maintained in 5% CO_2 in a humidified incubator. MEDII medium is collected at the next cell passage (day 4). For production of active MEDII, cells must be reseeded in accordance with the procedure described above. Replenishment of medium without passage of cells cannot be guaranteed to give repeatable results.

[8] B. B. Knowles, C. C. Howe, and D. P. Aden, Human hepatocellular carcinoma cell lines secrete the major plasma proteins and hepatitis B surface antigen, *Science* **209**, 497–499 (1980).

4. Before use, conditioned medium is clarified by centrifugation at 3000 rpm and sterilized through a 0.22 μm filter. MEDII is kept for up to 4 weeks at 4°C. The medium can be frozen but in some assays[7] the activity appears to deteriorate following freezing. Note that MEDII does not contain β-ME.
5. Approaches and protocols for the quality control of MEDII are outlined later in this review.

Notes. 1. It can be extremely difficult to reduce HepG2 cells to a single cell suspension, hence all cell counts are approximate. If care is not taken to achieve a suspension as near as practically possible to single cells, the cultures will grow in distinct clumps, not spread over the plastic substrate, and become very difficult to handle.

2. Fresh HepG2 cells from frozen stocks are initiated at least every 3 months and whenever the cells appear stressed or unhappy on visual inspection. In culture HepG2 cells grow as a monolayer with an irregular and vacuolated appearance and a high refractive index. Cells are localized in distinct colonies that have a lumpy and irregular appearance, and spontaneously form rounded, three-dimensional clones within the colonies. When the three-dimensional clumps become extensive and dominate the culture, new cultures are initiated from frozen stocks.

Formation of EPL Cells

EPL cells are formed from ES cells by culture in medium supplemented with MEDII conditioned medium in the presence or absence of exogenous LIF. EPL cell formation can be achieved both in adherent culture,[3] from ES cells cultured on gelatinized tissue culture plastic, or in suspension culture,[6] from ES cells aggregated and cultured in MEDII. EPL cell formation can be identified by alterations in morphology, gene expression profile and differentiation potential.[3,4,6]

EPL cells are formed and maintained in 50% MEDII (Table I). If LIF is required it is used at approximately 1000 units/ml, as for ES cell medium. EPL cells are cultured at 37°C in 10% CO_2 in humidified incubators.

Adherent Culture

- ES cells, 1–1.5$\times 10^6$ (2×10^4 to 3×10^4 cells/cm^2), are seeded in 10–15 ml of 50% MEDII (Table I) in a gelatinized 10 cm dish. Morphological indications of EPL cell formation, a flattening of the colony morphology and the ability to discern individual cells within the colonies, should become apparent by 24 hr, and after 48 hr all colonies should have adopted an EPL cell phenotype (Fig. 2). Addition of LIF to the culture medium results in a subtle alteration of EPL cell morphology (Fig. 2) and a delay in the establishment of EPL cell gene expression by approximately 2 days, such that cells cultured for 2 days in 50% MEDII are similar in gene expression to those cultured for 4 days in 50% MEDII + LIF.[4,9] EPL cells in adherent culture are passaged every 2 days as these cells proliferate more rapidly than ES cells. Trypsinization and passaging conditions for EPL cells are as for ES cells.

Notes. 1. Long-term maintenance of EPL cells in adherent culture is technically difficult as the cells are inherently less stable and more prone to differentiation than their ES cell counterparts. In particular, the cells can appear to undergo a "crisis" after 8 days in culture indicated by high levels of differentiation within the culture and poor seeding efficiencies after passage. Increasing the seeding densities during these passages to account for the increase in differentiated cells should allow continued maintenance of EPL cells.

2. For assay of bioactive factors such as growth factors, cytokines or fractionated medium samples, it can be advantageous to form and culture EPL cells at low density where effects on differentiation can be more easily assessed. In these cases the above methodologies are used except that ES cells are seeded at a density of $2.5–5 \times 10^2$ cells/cm^2.

Suspension Culture

ES cells are seeded into bacterial plates in 50% MEDII (Table I; 10 ml/10 cm dish) at a density of 5×10^4 to 1×10^5 cells/ml. On day 2, the aggregates (*E*mbryoid *B*odies in *M*EDII; EBMs) are transferred to a 30 ml centrifuge tube and allowed to settle. The medium is aspirated, and EBMs are resuspended in 10 ml of 50% MEDII. Depending on projected usage, EBMs are divided between 2 (1 : 2) and 4 (1 : 4) new bacterial plates and 50% MEDII is added to a total volume of 10 ml/plate. Aggregates will form within 24 hr and subtle differences in morphology in comparison to ES cell EBs will be apparent from day 3, most notably a smoother, rounder appearance of EBMs. Changes in gene expression associated with EPL

[9] M. D. Bettess, Purification, Identification and Characterisation of Signals Directing Embryonic Stem (ES) cell differentiation. Department of Molecular Biosciences, Adelaide University, Adelaide, South Australia, 2001.

cells, for example downregulation of *Rex1* and upregulation of *Fgf5*, are established by day 2 and persist until day 4/5.

Notes. LIF is a potent cytokine and even small amounts within the culture medium can alter the pluripotent cell differentiation described in this work. For culture of EPL cells in the absence of exogenous LIF, care should be taken to minimize potential cytokine carryover from ES cell culture. In particular, all ES cell manipulations during passaging should be carried out in medium without LIF.

EPL Cell Morphology

The morphology of EPL cells is most easily seen in adherent culture. These cells are similar in size and morphology to ES cells, small cells with little cytoplasm and 2–3 clearly discernible nucleoli. In contrast to the three-dimensional colony structure of ES cells, EPL cells form strictly two-dimensional colonies, or monolayers, in which individual cells can be discerned (Fig. 2). Cells on the edges of the colonies appear to be loosely attached to the colony, a phenomenon enhanced in cells that have been cultured in the absence of exogenous LIF.

Gene Expression in EPL Cells

Formation of EPL cells is accompanied by alterations in gene expression that can be used for identification of EPL cells and discrimination from ES cells. Marker genes that can be used for pluripotent cell identification fall into three classes (Table II):

- genes expressed in ES and EPL cell populations and indicative of pluripotence;
- genes expressed in ES cells/ICM and downregulated on formation of EPL cells/primitive ectoderm;
- genes upregulated on the formation of EPL cells/primitive ectoderm.

For identification of EPL cells we routinely use a combination of the pluripotent cell marker *Oct4*,[10–12] the ICM marker *Rex1*[19] and the primitive

[10] M. H. Rosner, A. Vigano, K. Ozato, P. M. Timmons, F. Poirier, P. W. J. Rigby, and L. M. Staudt, A POU-domain transcription factor in early stem cells and germ cells of the mammalian embryo, *Nature* **345**, 686–692 (1990).

[11] H. R. Schöler, G. R. Dressler, R. Balling, H. Rohdewohld, and P. Gruss, *Oct-4*: a germline-specific transcription factor mapping to the mouse t-complex, *EMBO J.* **9**, 2185–2195 (1990).

[12] Y. I. Yeom, H.-S. Ha, R. Balling, H. Schöler, and K. Artzt, Structure, expression and chromosomal location of the *Oct-4* gene, *Mech. Dev.* **35**, 171–179 (1991).

TABLE II
GENE EXPRESSION MARKERS USED FOR THE IDENTIFICATION OF EPL CELLS

	ES	ICM	EPL	Primitive ectoderm
Pluripotent cell markers				
Oct4[10–12]	+	+	+	+
Alkaline phosphatase[13,14]	+	+	+	+
SSEA1[15,16]	+	+	+	+
ES/ICM markers				
CRTR-1[5]	+	+	−	−
Gbx2[17,18]	+	+	−	−
Psc1[5]	+	+	−	−
Rex1[19]	+	+	low	−
EPL/primitive ectoderm markers				
Fgf5[20,21]	−	−	+	+
PRCE[5]	−	−	+	+[a]

[a]PRCE expression *in vivo* is limited to early primitive ectoderm.

[13] A. C. Hahnel, D. A. Rappolee, J. L. Millan, T. Manes, C. A. Ziomek, N. G. Theodosiou, Z. Werb, R. A. Pederson, and G. A. Schultz, Two alkaline phosphatase genes are expressed during early development in the mouse embryo, *Development* **110**, 555–564 (1990).

[14] S. Pease, P. Braghetta, D. Gearing, D. Grail, and R. L. Williams, Isolation of embryonic stem (ES) cells in media supplemented with recombinant leukemia inhibitory factor (LIF), *Dev. Biol.* **141**, 344–352 (1990).

[15] D. Solter and B. B. Knowles, Monoclonal antibody defining stage-specific mouse embryonic antigen (SSEA-1), *Proc. Natl. Acad. Sci. USA* **75**, 5565–5569 (1978).

[16] A. G. Smith, Mouse embryo stem cells: their identification, propagation and manipulation, *Semin. Cell Biol.* **3**, 385–399 (1992).

[17] A. Bulfone, L. Puelles, M. H. Porteus, M. A. Frohman, G. R. Martin, and J. L. R. Rubenstein, Spatially restricted expression of *Dlx-1*, *Dlx-2* (*Tes-1*), *Gbx-2*, and *Wnt-3* in the embryonic day 12.5 mouse forebrain defines potential transverse and longitudinal segmental boundaries, *J. Neurosci.* **13**, 3155–3172 (1993).

[18] G. Chapman, J. L. Remiszewski, G. C. Webb, T. C. Schulz, C. D. K. Bottema, and P. D. Rathjen, The mouse homeobox gene, *Gbx2*: genomic organization and expression in pluripotent cells *in vitro* and *in vivo*, *Genomics* **46**, 223–233 (1997).

[19] M. B. Rogers, B. A. Hosler, and L. J. Gudas, Specific expression of a retinoic acid-regulated, zinc-finger gene, *Rex-1*, in preimplantation embryos, trophoblast and spermatocytes, *Development* **113**, 815–824 (1991).

[20] O. Haub and M. Goldfarb, Expression of the fibroblast growth factor-5 gene in the mouse embryo, *Development* **112**, 397–406 (1991).

[21] J. M. Hébert, M. Boyle, and G. M. Martin, mRNA localization studies suggest that murine FGF-5 plays a role in gastrulation, *Development* **112**, 407–415 (1991).

ectoderm marker *Fgf5*.[20,21] In comparison to ES cells, EPL cells express approximately equivalent levels of *Oct4*, low levels of *Rex1* and high levels of *Fgf5*.[3] The formation of EPL cells is dynamic and gene expression is acquired over time, a phenomenon particularly apparent in the upregulation of *Fgf5* expression.[3]

For analysis by Northern blot, 10 cm plates of ES and EPL cells at days 2, 4, and/or 6 are trypsinized to a single cell suspension and pelleted at 1200 rpm for 3 min in a 30 ml centrifuge tube. The pellet is washed once in PBS and the cells are stored at −80°C. Total RNA is extracted from the cells using RNAzol B (Tel-Test Inc., Texas) according to the manufacturer's instructions. Gene expression is routinely assessed by Northern blot using techniques and probes described in Rathjen *et al.*[3] Qualitative approaches such as wholemount *in situ* analysis are complicated by low level expression of some transcripts, such as *CRTR-1* and *Fgf5* (in early EPL cells), and persistent low level expression of others such as *Rex1*. We have not explored the possibility of using RT-PCR for EPL cell identification.

Differentiation of EPL Cells: Formation of Nascent and Differentiated Mesoderm

Differentiation of EPL cell aggregates in suspension allows the formation of populations of cells highly enriched/homogeneous for specific germ lineages. This contrasts with differentiation of ES cells within EBs that results in uncontrolled formation of all three germ layers. This difference has been hypothesized to result from insufficiencies in the extraembryonic endodermal lineage formed during differentiation of EPL cell aggregates, most notably a lack of visceral endoderm.[4,22]

Differentiation of EPL cells as EBs (EPLEBs) results in the formation of a population enriched in mesoderm progenitors in the absence of ectoderm or visceral endoderm.

1. EPL cells are formed from ES cells for 2 days in adherent culture without exogenous LIF or 3–4 days in suspension culture, at which time the cells will have downregulated *Rex1* and upregulated *Fgf5*.
2. Cells are trypsinized to a single cell suspension as described previously and seeded at a density of 1–1.2 $\times 10^5$ cells/ml in bacterial dishes in EB/differentiation medium (Table I). Medium is replenished

[22] J. Rathjen, S. Dunn, M. D. Bettess, and P. D. Rathjen, Lineage specific differentiation of pluripotent cells *in vitro*: a role for extraembryonic cell types, *Reprod. Fertil. Dev.* **13**, 15–22 (2001).

every 2 days and the aggregates are divided 1 : 2 on day 2 of culture. The day of seeding is denoted as day 0. Cellular aggregates will form during the initial 24–36 hr of suspension culture.

3. The presence of nascent mesoderm and contaminating lineages is assessed by the expression of lineage-restricted marker genes at days 2, 3, and 4 of culture. Different techniques, including wholemount *in situ* hybridization analysis and quantitative determination of RNA expression are suited to different marker genes.
   - *Quantitative determination of RNA levels:* Northern blot analysis is used to measure expression of the nascent mesoderm marker *brachyury*[23] on days 2, 3, and 4 of EPLEB differentiation, and RNase protection is used to measure the ectoderm specific marker *Sox1*[24] on days 5–9 of culture. For RNA preparation, aggregates (30 ml days 2, 3; 20 ml days 4–6; 10 ml days 7–9) are pelleted by centrifugation at 300 rpm, resuspended gently in 1× PBS and repelleted at 1200 rpm, and frozen at −80°C. Total RNA is extracted using RNAzol B (Tel-Test Inc., Texas) and analyzed using the probes and protocols described in Lake *et al.*[4] Successful differentiation of EPL cells to nascent mesoderm is indicated by upregulation of *brachyury* expression on days 2 and 3. This correlates with expression of *goosecoid*[4,25] and differs from *brachyury* expression in ES cell EBs, which occurs between days 4 and 6. Additionally, the level of *brachyury* expression within EPLEBs should be several fold higher than in ES cell EBs. Expression of *Sox1* should not be detected in EPLEBs, in contrast to ES cell EBs where this gene is expressed from day 6 to 10.
   - *Wholemount* in situ *hybridization analysis:* Wholemount *in situ* hybridization is used for quantitative assessment of the proportion of cells/EBs expressing *brachyury*[23] and *Oct4*[10–12] on days 2, 3, and 4, *AFP*[26] on day 4 and *Sox1* on day 9. Sufficient EPLEBs are collected for several *in situ* hybridization experiments, usually 10–20 ml/timepoint. Aggregates are pelleted

[23] B. G. Herrmann, Expression pattern of the *Brachyury* gene in whole-mount T^{wis}/T^{wis} mutant embryos, *Development* **113**, 913–917 (1991).

[24] L. H. Pevny, S. Sockanathan, M. Placzek, and R. Lovell-Badge, A role for SOX1 in neural determination, *Development* **125**, 1967–1978 (1998).

[25] M. Blum, S. J. Gaunt, K. W. Y. Cho, H. Steinbeisser, B. Blumberg, D. Bittner, and E. M. De Robertis, Gastrulation in the mouse: the role of the homeobox gene goosecoid, *Cell* **69**, 1097–1106 (1992).

[26] M. Dziadek and E. Adamson, Localization and synthesis of alphafetoprotein in post-implantation mouse embryos, *J. Embryol. Exp. Morphol.* **43**, 289–313 (1978).

by centrifugation at 300 rpm, resuspended in 1× PBS and repelleted at 1200 rpm, then fixed by resuspension in 4% paraformaldehyde : PBS at room temperature for 30–60 min. Wholemount *in situ* hybridization is performed using the method of Rosen and Beddington[27] with modifications and probes described in Lake *et al.*[4] and Rathjen *et al.*[6]

By wholemount *in situ* hybridization, a peak in *brachyury* expression is detected on days 2 and 3 of EPLEB culture, with approaching 100% of the aggregates containing *brachyury*-expressing cells. In contrast, ES cell EBs will contain no *brachyury*-expressing cells at day 2, and low levels at day 4. *Oct4* will be expressed uniformly across EPLEBs on day 2 but become patchy on days 3 and 4, indicating the loss of pluripotence that accompanies differentiation. *AFP* expression indicative of visceral endoderm is detected in the outer layer of cells in ES cell EBs by day 4 of culture. Equivalent expression of *AFP* in outer cells of EPLEBs on day 4 is not detected, although expression in internal cells, representative of other cell populations, can sometimes be seen. The sites of expression can be discriminated clearly by sectioning stained aggregates. *Sox1* expressing cells should not be detected in EPLEBs.

4. The efficiency of EPLEB differentiation to mesodermal progenitors is also assessed by analysis of two differentiated populations formed by further differentiation of the EPLEBs, the mesodermal derivative cardiomyocytes and the ectodermal population, neurons.

For analysis of terminally differentiated cell populations individual EPLEBs are seeded into 2 ml gelatinized tissue culture wells in 1 ml EB/ differentiation medium on or between days 4 and 6 of culture. These are monitored microscopically for the formation of beating cardiomyocytes. While the absolute proportion of EPLEBs forming contracting cardiomyocytes can be variable, foci of contracting cells should begin to appear on day 6, approximately 2 days earlier than in ES cell EBs. These cultures are also scored for formation of neurons between days 10 and 14 of culture. By comparison with ES cell EBs, EPLEBs should exhibit a dramatic reduction or absence of neurons.

Notes. 1. Trypsinization conditions vary between laboratories. We routinely trypsinize ES and EPL cells for 30–60 sec at room temperature. Over-trypsinization, particularly of EPL cells, can result in inefficient formation of aggregates.

[27] B. Rosen and R. S. Beddington, Whole-mount in situ hybridisation in the mouse embryo: gene expression in three dimensions, *Trends Genet.* **9**, 162–167 (1993).

2. A common problem encountered in ES cell EB and EPLEB formation is adherence of aggregates to the bacterial plastic. In our experience, this can be attributed to variations in bacterial grade plastic dishes which can be overcome by testing plates from alternative manufacturers, or to incomplete heat inactivation of FCS.

3. *Brachyury* is expressed transiently during mesodermal differentiation.[23] Determination of the absolute number of cells adopting a mesodermal fate in EBs by wholemount *in situ* hybridization is therefore not yet possible.

4. Differences in the kinetics or magnitude of gene expression and differentiation from those described above, or in the formation of neurons is observed in EPLEB, could be attributed to poor quality MEDII or ES cells.

Differentiation of EPL Cells: Formation of Nascent and Differentiated Ectoderm

Manipulation of the differentiation environment of EPL cells has allowed the development of a differentiation protocol that yields a homogeneous population of ectodermal and neurectodermal progenitors, without the formation of either the mesodermal or endodermal lineages.[6] EPL cell-derived neural progenitors have been characterized by the expression of multiple neural markers and by differentiation potential, with the ability to form neurons, glia and neural crest.[6]

1. EPL cells are produced in suspension as EBM, formed by seeding ES cells into bacterial plates in 50% MEDII as previously described. For ectoderm differentiation it is advisable to divide the aggregates 1:4 on day 2. EBM are timed from the seeding of ES cells into MEDII, denoted as day 0.
2. On day 4, EBM are further divided 1:2 and the medium is replenished with 50% MEDII.
3. Aggregates are maintained in 50% MEDII until day 7 with daily replenishment of medium. On day 7 EBM are transferred to a chemically defined medium (Table I) supplemented with 10 ng/ml FGF2 (Sigma Aldrich) and maintained with daily medium replenishment until day 9.

   The timing and progression of this pathway can be monitored by morphological features of the aggregates (Fig. 3). Early aggregates are smooth and round, and cellular aggregation is very efficient. By day 4, the aggregates remain smooth, with no outer layer of endoderm, but appear slightly "angular" as the surface of the aggregate develops

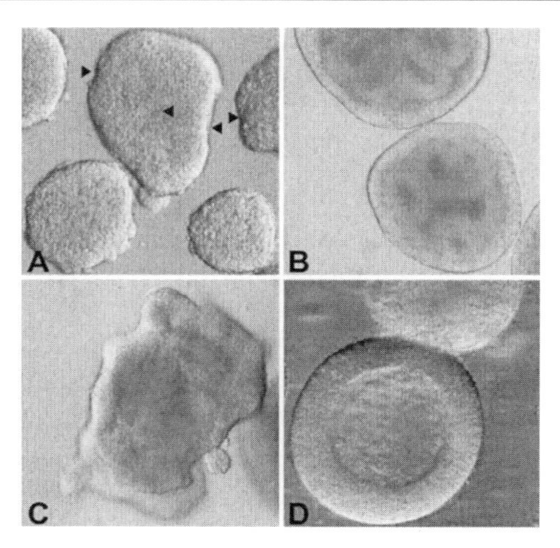

FIG. 3. Morphology of EBM during differentiation of EPL cells to the ectodermal lineage. (A) EBM on day 4 of culture showing the slightly angular appearance of the aggregates. Arrows denote the dips and craters that form on the surface of EBM. (B) EBM on day 7 of culture, showing the typical and distinct layered morphology of the psuedostratified epithelium that comprises EBM on days 5–7. (C, D) EBM on day 9 of culture, showing the stratified columnar epithelium. Note in (C) the contortion of the cell layers. Images (A) and (D) were taken using Hoffman interference contrast and 40× magnification, (B) and (C) were taken using phase contrast and 20× magnification.

dips and craters (Fig. 3A). Larger bodies will have a single darkened area in the center indicative of the single focus of cell death and cavity formation;[6] in smaller bodies the internal cavity is not as apparent. Between days 5 and 7 EBMs are comprised entirely of cells arranged within morphologically equivalent epithelial layers. The morphology of the aggregates is distinctive, a convoluted and folded internal arrangement of the cell layer(s) within a smooth outer layer (Fig. 3B). It is not clear from morphological investigation if the aggregates comprise a single or multiple cell layers. Again, no layers consistent with the presence of extraembryonic endoderm can be seen associated with the bodies. On days 5 through 7 the cell layers are not highly stratified and the cells have not adopted a strictly columnar appearance. Days 8 and 9 are characterized by transformation of the layers into a highly ordered stratified epithelial layer (Fig. 3C–D). The spherical regularity of the bodies is lost and they assume a variety of structures as the cell layers contort into new and wonderful shapes (Fig. 3C).

4. As was the case with confirmation of mesoderm formation within EPLEBs, the presence of neurectoderm and contaminating lineages is

assessed by the expression of lineage-restricted marker genes. Unlike mesoderm, the earliest marker (*Sox1*) is associated with formation of neural progenitors and is distinct from loss of pluripotence and formation of the ectodermal germ layer. However, the persistent expression of *Sox1* does allow quantitation of the number of cells adopting a neural cell fate.

- *Quantitative determination of RNA:* The progression of differentiation within EBM is measured by analysis of the pluripotent cell marker *Oct4*,[10–12] mesodermal marker *brachyury*[23] and neural specific marker *Sox1*.[24] Ten to 20 ml of EBMs are collected on days 4–9 and total RNA is isolated as for EPLEB. Gene expression is analyzed using the probes and protocols described in Lake *et al.*[4]

Loss of pluripotence in EBM occurs about day 5 and is marked by an approximate four-fold decrease in the expression of *Oct4* to levels consistent with expression of *Oct4* in nascent neurectoderm in the developing embryo.[6] *Sox1* expression, consistent with the formation of neurectoderm is upregulated after day 7. Expression of *brachyury* should not be observed.

- *Wholemount* in situ *hybridization analysis:* Wholemount *in situ* hybridization is used for quantitative assessment of the proportion of cells expressing *brachyury*[23] on days 6 and 7, *Sox1*[24] on day 9 and *AFP*[25] on day 4 or 9. Ten to 20 ml/timepoint EBMs are collected and prepared for wholemount *in situ* hybridization as described for EPLEBs. Wholemount *in situ* hybridization is performed using the method of Rosen and Beddington[27] with modifications described in Lake *et al.*,[4] using probes as described therein and in Rathjen *et al.*[6] Consistent with the formation of a homogeneous population of neurectoderm, cells expressing *brachyury* and *AFP* within the aggregates should not be detected. Uniform expression of *Sox1* should be detected throughout each EBM.

5. Quantitative assessment of the homogeneity of neural precursor formation can be achieved by FACS analysis of a single cell suspension stained by immunocytochemistry for the expression of the neural adhesion molecule N-CAM.EBM on day 9–10 of differentiation are transferred to a 30 ml centrifuge tube and pelleted by centrifugation at 1200 rpm.

EBM are washed once with PBS and resuspended and incubated in 1 ml 0.5 m*M* EDTA/PBS for 5 min at room temperature. The aggregates are dissociated to a single cell suspension by vigorous pipetting with a 1 ml Gilson pipetteman. Cells are pelleted by centrifugation at 1200 rpm and washed three times in PBS, before resuspension and fixation in 4% PFA/PBS for 30 min at room temperature. Fixed cells are pelleted by centrifugation at 1200 rpm, washed with 1% BSA/PBS, and resuspended at 1×10^6 cells/ml in 1% BSA/PBS containing an antibody directed against N-CAM (Santa Cruz Biotech, SC-1507) at a dilution of 1 : 2. The cells are incubated with antibody for 1 hr. The cells are washed twice with 1% BSA/PBS before incubation with a preadsorbed (1 hr in 1% BSA/PBS) FITC conjugated goat anti-mouse IgM (μ-specific: Sigma) used at a concentration of 1 : 100 in 1% BSA/PBS. Cells are washed in PBS and fixed in 1% PFA for 30 min before being subjected to FACS analysis using standard protocols. Within a cell suspension generated from EBM greater than 95% of the cells should bind the N-CAM antibody, and, in comparison to N-CAM binding cells within ES cell EBs the fluorescent intensity of the cells should be greater.

6. The formation of EPL cell-derived neural progenitors and the absence of mesodermal differentiation within the aggregates is also assessed by analysis of two differentiated populations, neurons and cardiomyocytes, within differentiated populations derived from EBM.

   Individual aggregates are seeded into 2 ml gelatinized tissue culture wells in 500 μl of 50% MEDII on day 7. After adherence (approximately 16 hr) the medium is changed to 1 ml of chemically defined medium (Table I) without FGF2. Bodies are scored for beating cardiomyocytes and neurons on days 8, 10, and 12. Approaching 100% of the aggregates should contain neurons by day 12, with none or few of the bodies (<2%) containing foci of beating muscle.

Notes. 1. Detection of mesodermal expression markers and cell populations indicates that the differentiation procedure is not optimal. "Breakthrough" can be attributed to poor quality MEDII or ES cells, or to aggregation conditions that result in aggregates growing from 1 or 2 ES cells which have a greater tendency to give rise to mesoderm regardless of the culture conditions.

2. Although this differentiation pathway is relatively robust and reproducible, occasionally aggregates seeded on day 7 die rather than differentiate. Although the explanation for this has not been ascertained, it correlates with accelerated morphological progression through the pathway and may reflect a response of later, more differentiated cell populations to serum within the medium.

Reversion of EPL Cells to ES Cells

ES and EPL cells represent distinct but interchangeable pluripotent cell states.[3,4] Passage of EPL cells into medium supplemented with LIF but not

MEDII, ES cell medium, results in formation of ES cell colonies. Reversion of EPL cells can be monitored by:

- Morphology, with the cells adopting a three-dimensional colony morphology consistent with formation of ES cells.
- Gene expression, demonstrated by the maintenance of *Oct4* expression, upregulation of *Rex1* expression and downregulation of *Fgf5* expression.
- Differentiation potential within EBs, verified by expression of *brachyury* on day 4 of development, the later onset of contracting cardiomyocytes and the formation of neurons within the aggregates.

The experimental procedures for the assessment of parameters are discussed above.

Quality Control of MEDII

Key to the formation and further differentiation of EPL cells is a consistent and high quality supply of MEDII. MEDII is tested routinely to ensure consistency and reproducibility. Medium testing is incorporated into ongoing experiments. Poor quality or inconsistent MEDII can alter the kinetics of EPL cell formation and further differentiation, reduce the homogeneity of EPL cells and their differentiated progeny, and adversely affect the reproducibility of the system. Two parameters are measured to ensure MEDII quality.

- *Pluripotent cell morphology:* High quality MEDII yields 100% of colonies with a homogeneous EPL cell morphology after 2 days in culture in 50% MEDII (Table I) but without LIF. Poor quality MEDII yields cell populations that lack homogeneity, indicated by incomplete transition and persistence of colonies with an ES cell morphology.
- *Pluripotent cell differentiation:* The differentiation potential of EPL cells is perhaps the most rigorous assessment of MEDII quality. We routinely use the precocious and extensive formation of mesodermal progenitors and differentiated mesodermal populations in EPLEB[4] for assessment of EPL cell differentiation potential as an indicator of MEDII quality.

The timing and extent of mesoderm formation in EPLEBs, a reproducible measure of the formation of EPL cells, is detected by the kinetics and magnitude of *brachyury* expression during EPLEB differentiation and/or the timing of onset of cardiomyocyte contractions. Expression of *brachyury*[23] can be detected by wholemount *in situ* hybridization or

Northern blot analysis. High quality MEDII induces the formation of EPL cells that reproducibly express *brachyury* on days 2 and 3 of differentiation in approaching 100% of EPLEBs, and the onset of contractions within cardiomyocytes on day 6.

Suboptimal MEDII will result in EPL cells that do not conform to this description of differentiation, with patterns of *brachyury* expression on Northern blots that appear intermediate between EPL and ES cells. These include broader patterns of *brachyury* expression initiating on day 2 but extending beyond day 3, and later onset of expression, with the expression of *brachyury* on days 3 and 4 rather than days 2 and 3. Similarly, a later, more ES cell-like onset in the formation of contracting cardiomyocytes indicates deficiencies in the quality of MEDII.

Concluding Remarks

By analogy with the primitive ectoderm of the early mammalian embryo, EPL cells represent an obligatory intermediary in the differentiation of ES/ICM cells to the primary germ lineages. Manipulation of their differentiation environment has led to the development of protocols for directed differentiation to homogeneous/highly enriched populations representative of the mesodermal or ectodermal lineages. The success of these protocols is proposed to arise from a lack within the differentiation environment of the extraembryonic endodermal derivative, visceral endoderm, a known source of signaling molecules regulating cell fate choice during pluripotent cell differentiation. The lack of endogenous signaling enables controlled alterations to the differentiation environment and effective direction of pluripotent cell differentiation by exogenously added factors. Cells produced from EPL cells appear representative of those produced within the embryo, with analogous gene expression profiles and differentiation potential, and EPL cell differentiation recapitulates the temporal progression of lineage formation observed during embryonic development. Further exploitation of EPL cell differentiation will:

- allow characterization at a molecular and cellular level of the processes of cell fate choice during gastrulation;
- serve as a technology for the production of normal, nontransformed cell populations for further investigation;
- provide a resource for the identification and characterization of intermediate populations arising during cell differentiation;
- provide a technological basis for production of human cell populations from ES cells that can be applied to the treatment of human disease.

Acknowledgment

The authors would like to acknowledge members of the Rathjen laboratory, past and present, who have contributed to the development and troubleshooting of the methods and markers presented here. This work was supported by the Australian Research Council and the National Health and Medial Research Council.

[2] Differentiation of F1 Embryonic Stem Cells into Viable Male and Female Mice by Tetraploid Embryo Complementation

By KEVIN EGGAN and RUDOLF JAENISCH

Introduction

Production of embryonic stem (ES) cell-tetraploid embryo chimeras, or tetraploid embryo complementation, has proven to be a powerful method in mammalian developmental biology. Tetraploid embryo complementation involves the production of tetraploid embryos, generally by electrofusion at the two-cell stage, followed by injection of tetraploid blastocysts with ES cells and finally transfer of these composite embryos into a suitable surrogate mother. In chimeric embryos generated by these means, the extraembryonic lineages are derived from the tetraploid host embryo, while the embryonic lineages are dominated by derivatives of the ES cells.[1] The use of tetraploid embryo complementation has become particularly widespread for establishing the relative importance of a gene's function in embryonic or extraembryonic development.[2] By aggregating mutant embryos with tetraploid embryos or by injecting mutant ES cells into tetraploid blastocysts, it can be tested whether a developmental phenotype observed in a mutant can be rescued by the normal trophoblast tissues derived from the tetraploid component of the conceptus. Additionally, as the contribution of tetraploid cells to the embryonic lineages is either nonexistent or vanishingly small, tetraploid embryo complementation provides an important means for creating completely ES cell-derived embryos and mice.[1]

Mice completely derived from ES cells by tetraploid embryo complementation (ES cell-tetraploid) survive to birth at a high frequency,

[1] A. Nagy *et al.*, *Development* **110**, 815 (1990).
[2] J. Rossant and J. C. Cross, *Nat. Rev. Genet.* **2**, 538 (2001).

0076-6879/2003 $35.00

formally demonstrating that ES cells are sufficiently pluripotent to give rise to all of the embryonic lineages.[1] However, it was observed that neonatal ES cell-tetraploid mice died due to respiratory failure with an extremely high penetrance.[1] This respiratory distress prevented the production of ES cell-derived adults and therefore greatly limited the utility of this procedure. A number of ES cell lines have subsequently been identified that allow the production of adult ES cell-tetraploid mice.[3,4] However, the properties of these ES cell lines that permitted neonatal survival remained unclear. Furthermore, the efficiency by which viable mice could be produced from these lines, particularly after long-term *in vitro* culture, was still limited. Thus, although normal and wild-type mutant mice could be produced by tetraploid embryo complementation, the use of this method for the production of mice from targeted ES cell lines did not become wide-spread due to its inefficient and unpredictable nature.

Recently, we have demonstrated that hybrid vigor plays a critical role in the survival of ES cell-tetraploid animals.[5] Animals derived from ES cells with an inbred genetic background die of respiratory failure, as previously observed.[1,5] However, mice generated from a number of ES cell lines with an F1 genetic background develop to term and survive to maturity with a high frequency.[5] Remarkably, several of these lines retain the potency to efficiently generate viable mice even after as many as 5 serial rounds of *in vitro* gene-targeting.[6] Thus, tetraploid embryo complementation with F1 ES cell lines represents a reasonable, reliable and easily adopted methodology for the production of mice from genetically engineered ES cells.

This chapter briefly describes methods currently used in our laboratory for the production of viable ES cell-derived mice by tetraploid embryo complementation. Our aim is to provide sufficient information concerning experimental methodologies and strategies such that the reader will be able to use information garnered here to accelerate the production of mutant mice in their laboratory. With this goal in mind, these protocols are written with the assumption that the reader has prior knowledge of methodologies required for the standard production of mutant and transgenic mice. These methods include targeted mutagenesis of ES cell by homologous recombination, injection of ES cells into diploid blastocysts for creation of chimeric offspring and creation of transgenic mice by pronuclear injection. For those not well versed in these arts and as a companion to this

[3] A. Nagy, J. Rossant, R. Nagy, W. Abramow-Newerly, and J. C. Roder, *Proc. Natl. Acad. Sci. USA* **90**, 8424 (1993).

[4] Z. Q. Wang, F. Kiefer, P. Urbanek, and E. F. Wagner, *Mech. Dev.* **62**, 137 (1997).

[5] K. Eggan *et al.*, *Proc. Natl. Acad. Sci. USA* **98**, 6209 (2001).

[6] K. Eggan *et al.*, *Nat. Biotech.* **20**, 455 (2002).

chapter, we recommend the excellent text by Hogan *et al.*[7] Additional reviews and information concerning tetraploid embryo complementation and the production of tetraploid embryos can also be found elsewhere.[2,8,9]

Overall Strategy

The conventional production of genetically engineered mice is highly time consuming and involves the generation of mutant alleles in ES cells by homologous recombination, the production and breeding of chimeric founder mice and finally, crosses between the resulting mouse strains carrying the desired allele to produce homozygous mutant offspring (Fig. 1A). The primary advantage in using tetraploid embryo complementation in the production of mutant mice is that completely ES cell-derived mice can be produced directly from ES cells without a chimeric intermediate.[5] Thus, for instance, experimental mice carrying multiple mutations or transgenes may be directly produced from engineered ES cells without breeding (Fig. 1B).

The utility of ES cell-derived mouse production has been further increased by our recent observation that the Y chromosome is lost from male ES cells at a high frequency (2%).[6] Harnessing Y chromosome loss allows 39XO female ES cell-tetraploid mice to be produced from targeted male ES cell lines. These heterozygous XO females are fertile when mated with ES cell-tetraploid males derived from 40XY ES cell lines carrying the same mutation.[6,10] Thus, our strategy allows homozygous mutant mice to be produced in a single breeding cycle, thereby greatly expediting production of homozygous mutant offspring (Fig. 1C).

Importantly, 39XO subclones can be readily identified from male ES cell lines using established cell culture and molecular techniques, allowing our strategy to be utilized by any laboratory already carrying out gene-targeting experiments.[6] We find that the additional work required for identification of XO subclones is greatly compensated for by the shortening of breeding cycles needed to produce mice that are homozygous for a mutation or that carry multiple transgenes.

The conventional derivation of mice carrying more than one mutant allele of interest involves the independent production of targeted ES cell lines, then mutant mouse strains, followed by two cycles of mating to yield

[7] B. Hogan *et al.*, *in:* "Manipulating the Mouse Embryo." Cold Spring Harbor Laboratory Press, Cold Spring Harbor, NY, 1994.

[8] Nagy Lab website: http://www.mshri.on.ca/nagy/.

[9] K. J. Mclaughlin, *Methods Enzymol.* **225**, 919 (1993).

[10] B. Cattanach, *Genet. Res.* **3**, 487–490 (1962).

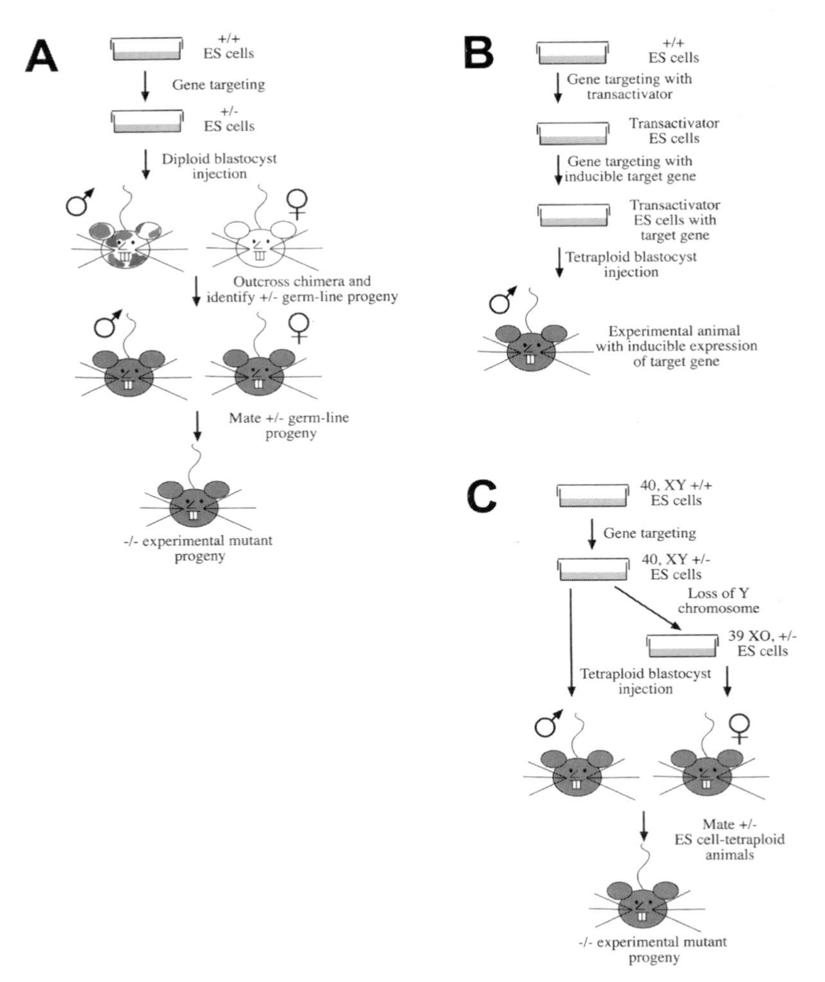

FIG. 1. Standard and accelerated production of mutant mice from ES cells. (A) Standard production of mutant mice from heterozygous ES cells requires generating chimeric founder animals by introducing targeted male ES cells into diploid blastocysts. Chimeric founders must then be outcrossed to fix the mutation in the male and female germ-line. These male and female heterozygous offspring are finally intercrossed to produce homozygous mutant progeny. (B) Animals carrying multiple transgenes can be rapidly generated using serial gene-targeting followed by the production of mice by tetraploid embryo complementation. For instance, tetracycline inducible gene expression requires two transgenes, a transactivator and an inducible target gene. Both of these components can be targeted to the F1 ES cells. Mice with tetracycline inducible gene expression can then be generated by tetraploid embryo complementation. (C) The accelerated production of homozygous mutant offspring can be achieved by isolating 39XO subclones from targeted 40XY ES cell lines and producing ES cell-derived males and females by tetraploid embryo complementation. These heterozygous mice can then be immediately intercrossed to produce homozygous mutant offspring, considerably shortening the time required to generate experimental animals.

compound heterozygous and finally compound homozygous mutant animals. Using tetraploid embryo complementation, it is possible to isolate 39XO derivatives of cell lines sequentially targeted multiple times and then to produce both male and female compound heterozygous mutant mice. Mating these compound heterozygotes allows production of mutant mice with all possible combinations of heterozygous and homozygous genotypes in a single mouse cross. This approach could be extremely beneficial for the genetic analysis of entire gene families or genetic pathways, allowing the rapid exploration of many potential phenotypes.

In addition to the expedited production of mutant mice, the production of female mice from male ES cell lines allows female germ-line transmission of targeted mutations or transgenes that either directly inhibit spermatogenesis[11,12] or cause developmental failure after paternal inheritance.[13]

An important requirement for the routine production of male and female ES cell tetraploid mice carrying multiple mutations is that ES cells retain their potency to generate mice by tetraploid embryo complementation after multiple rounds of gene-targeting. We have previously demonstrated that F1 ES cell lines subjected to one or two rounds of selection produce ES cell-tetraploid mice as efficiently, or more efficiently than the parental cell line.[5,6] Furthermore, even up to five consecutive rounds of genetic manipulation do not abolish the potency of the F1 ES cells to generate ES cell-tetraploid mice, establishing the efficacy of the strategy described here.[6]

However, before embarking on these methods it is important to consider that the offspring of F1 ES cell-tetraploid mutant mice are a genetically heterogeneous population of F2 animals. If mutant animals with an inbred genetic background are required for quantitative or qualitative analysis of a particular phenotype, substantial backcrossing would be necessary, abrogating many benefits of this approach.

ES Cell Culture and Genetic Manipulations

Derivation, Culture and in vitro *Gene-targeting of F1 ES Cells*

Isolation, culture and gene-targeting of F1 ES cells is carried out essentially as described.[7] ES cells are cultured in ES cell medium (DMEM with 15% fetal calf serum (Hyclone), 0.1 m*M* nonessential amino acids (Gibco), 2 m*M* L-glutamine, 50 IU Penicillin, 50 IU Streptomycin (Gibco)

[11] R. Al-Shawi *et al.*, *Mol. Cell. Biol.* **11**, 4207 (1991).

[12] P. J. Wang, J. R. McCarrey, F. Yang, and D. C. Page, *Nat. Genet.* **27**, 422 (2001).

[13] Y. Marahrens, B. Panning, J. Dausman, W. Strauss, and R. Jaenisch, *Genes. Dev.* **11**, 156 (1997).

and 0.1 m*M* Beta-mercaptoethanol (Sigma), 1000 U/ml LIF) on gelatinized tissue culture ware (Falcon) preplated with a mono-layer of gamma-irradiated primary mouse embryo fibroblasts (MEFs).

Subcloning of ES Cells to Identify 39XO Derivatives of Targeted Cell Lines

To isolate 39XO subclones of a targeted ES cell line we generally expand the ES cell line, trypsinize, count the cells and plate 5000 ES cells in a 10 cm dish preplated with irradiated MEFs. Approximately 8 days after plating, well defined ES colonies can be seen. Two hundred to 300 colonies are picked and expanded on female MEFs for freezing and DNA isolation. It is essential to expand the ES cell subclones on female MEFs as contaminating Y chromosome DNA from male MEFs may confound Y chromosome genotyping of the ES cell subclones. Y chromosome genotyping of subclones can be robustly carried out either by Southern hybridization or by PCR on crudely prepared genomic DNA.[14] After identification, frozen subclones can be expanded and female mutant mice can be produced by tetraploid embryo complementation.

Y Chromosome Genotyping by Southern Hybridization

For Y chromosome Southern analysis of ES cell subclones, DNA is digested with *Eco*RI, blotted and probed with a 720 bp *Mbo*I fragment of the plasmid pY2.[15] Alternatively, DNA can be digested with *Pst*I, blotted and probed with the 1.5 kb EcoRI fragment of the plasmid pY353.[16] Both of these probes hybridize to highly repetitive sequences located primarily on the Y chromosome, giving a strong signal at a range of molecular weights. It is advisable to always run a female DNA control on each blot so that the male-specific pattern of bands can be easily determined. 39XO subclones can then be recognized by the lack of these male-specific hybridization signals but remaining cross-reactivity with a small number of autosomal repeats which are also observed in the female control. 39XO subclones should be found at approximately a 1.5–2% frequency.[6]

Identification of 39XO Subclones by PCR

PCR genotyping for the Y chromosome can also rapidly be carried out using primers specific for the Y linked *Zfy* locus. *Zfy* primer 1: GAT AAG CTT ACA TAA TCA CAT GGA. *Zfy* primer 2: CCT ATG AAA TCC

[14] P. W. Laird *et al.*, *Nucl. Acids Res.* **19**, 429 (1991).
[15] E. E. Lamar and E. Palmer, *Cell* **37**, 171 (1984).
[16] C. E. Bishop and D. Hatat, *Nucl. Acids Res.* **15**, 2959 (1987).

TTT GCT GC. Also, primers specific for the Y chromosome EST *Tet35* have been successfully used to screen for loss of the Y chromosome. For *TET35* PCR, primer 1: CTCATGTAGACCAAGATGACC, primer 2: GGAATGAATGTGTTCCATGTCG. For these primer sets we perform PCR for 30–35 cycles, annealing at 60°C for 10–15 sec and extending for 30 sec at 72°C. These PCR products are then run on a 1% agarose gel to assay for the presence of the Y chromosome.

Karyotyping of ES Cell Lines During Serial Gene-Targeting

It has been observed that certain karyotypic abnormalities in ES cells may interfere with both production of ES cell-tetraploid mice and germ-line transmission of targeted mutations.[3,4,17] Surprisingly, we have found that chromosomal abnormalities present in most, if not all ES cell lines, did not interfere with their potency to generate adult mice by tetraploid embryo complementation.[6] Our chromosomal analyses have revealed that the abnormalities found in donor ES cell lines are not present in the animals derived from them. It may be that only a small subset of the injected ES cells, those with a normal karyotype, contribute to embryo formation while the cells with karyotypic abnormalities are selected against during development. However, any chromosomal abnormality resulting in a growth advantage *in vitro* can be expected to completely overgrow the population of karyotypically normal cells and eliminate its competence to generate mice.[17] Therefore, if multiple rounds of gene-targeting are to be performed before mutant mouse production, it may be prudent to perform chromosomal analysis at each round of targeting to exclude lines dominated by cells with abnormal karyotypes. If these precautions are taken, it may be that there is no limit to the number of times an F1 ES cell line can be genetically manipulated before mutant mouse production.

Production of Tetraploid Embryos

Tetraploid mouse embryos can be produced from diploid embryos by a number of methods including preventing cytokinesis with Cytochalasin B[18] and by chemically inducing cell–cell-fusion with PEG.[19] However, embryos for tetraploid embryo complementation are generally created by electrically fusing the two blastomeres of a two-cell embryo. The popularity of this

[17] X. Liu *et al.*, *Dev. Dyn.* **209**, 85–91 (1997).
[18] R. G. Edwards, *J. Exp. Zool.* **137**, 349 (1958).
[19] M. A. Eglitis, *J. Exp. Zool.* **213**, 309 (1980).

method stems from the simplicity by which a large number of tetraploid embryos can be generated and the extremely high efficiency by which these embryos complete preimplantation development. In short, diploid zygotes are isolated from superovulated females, cultured to the two-cell stage, when electrofusion is performed and then cultured to the blastocyst stage at which time ES cells are injected.

In our experience, one of the most critical determinants of successfully producing mice by tetraploid embryo complementation is the proper *in vitro* culture and maintenance of the preimplantion embryos. If special attention is not paid to proper *in vitro* culture technique, embryo health may be compromised rendering tetraploid blastocyst production inefficient and frustrating. In our hands, tetraploid embryos develop to the blastocysts stage with 90–95% efficiency. If this efficiency is not routinely achieved, postimplantation development may be less then optimal and efficient production of mice may be compromised. In order to routinely use tetraploid embryo complementation for the production of mutant mice, researchers should expect to produce and inject a minimum of 50 tetraploid blastocysts per experiment.

Isolation and in vitro *Culture of Preimplantation Embryos*

Fertilized zygotes used for the production of tetraploid embryos are isolated from superovulated B6D2F1 (NCI) females that have been mated with B6D2F1 males. Eight- to 12-week-old females are generally maintained in our colony for at least 2 weeks after arrival from the vendor before superovulation. This delay ensures time for the females to undergo day–night cycle adjustment. To induce superovulation, we first inject 5 IU of pregnant mares serum (PMS) (Calbiochem) followed 48 hr later by 5 IU of human chorianic gonadotropin (HCG) (Calbiochem). These injections are generally performed in the early afternoon. After HCG injection, females are individually housed with a B6D2F1 male overnight. Approximately 20–24 hr after HCG injection, fertilized zygotes are isolated from the oviduct. Superovulation of 8–10 females is generally sufficient for a single experiment, allowing isolation of around 200 fertilized zygotes.

All preimplantation embryo culture is carried out in microdrops on standard bacterial petri dishes (Falcon) under mineral oil (Squib). A detailed description of microdrop embryo culture and zygote isolation can be found elsewhere.[7] Briefly, after sacrificing the superovulated females, the oviducts are dissected and placed in a petri dish flooded with mineral oil, containing several 30 μl drops of M2 medium (Sigma) and several drops of M2 medium containing 0.1% w/v bovine testicular hyaluronidase (Sigma).

Under the dissecting microscope the ampula containing the fertilized zygotes should be easily observed. Using a sharp pair of forceps, the ampula can be torn releasing the zygotes, still surrounded by their cumulus masses, into the mineral oil. These cumulus masses can then be pushed into the drops of M2 media containing hyaluronidase. We find that this method allows for rapid quantitative isolation of zygotes with minimal hypertonic shock to the embryos. After several minutes in M2 with hyaluronidase the cumulus masses surrounding the zygotes will dissolve allowing the zygotes to be isolated. After cumulus mass dissociation, the zygotes should be washed through several drops of M2 media to both eliminate the hyaluronidase and to remove excess cumulus cells. Zygotes are then immediately transferred to microdrops of KSOM media (Specialty Medium) and placed at 37° with an atmosphere of 5% CO_2 in air for overnight culture.

Electrofusion of Two-cell Embryos

Application of a direct current (DC) electrical pulse across the cellular membranes separating the two blastomeres of a two-cell embryo leads to the opening of small holes in the membrane. Because of the direct juxtaposition of the two cells, these openings often resolve into small cytoplasmic bridges. As a spherical shape is the lowest energy equilibrium-state for the membrane, these bridges continue to widen until the two cells have completely fused to form one continuous cytoplasm containing two nuclei (Fig. 2A). Thus two diploid-cells become one tetraploid-cell. For most efficient fusion at the lowest current, the electrical field must be perpendicular to the membrane junction between the two cells of the embryo. The two-cell embryo can be aligned in this manner either manually

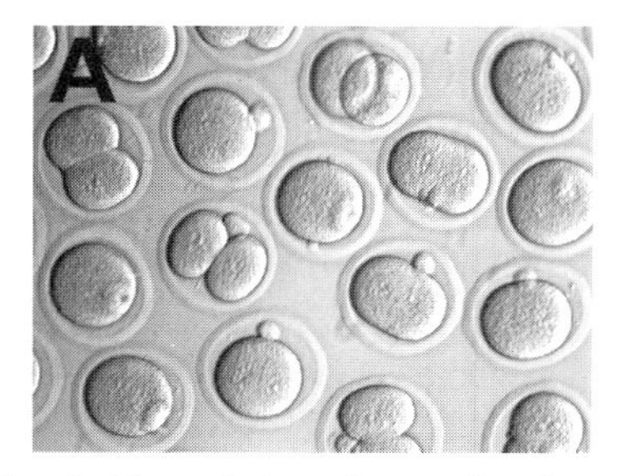

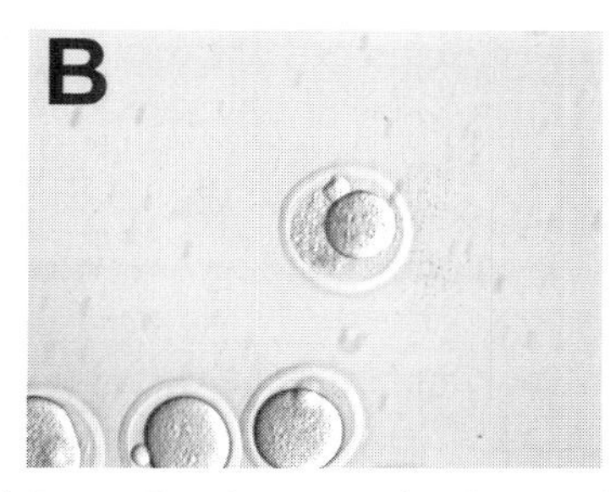

FIG. 2. Electrofusion of two-cell embryos. (A) Two-cell embryos 10 min after DC fusion. Note that the embryos are at various stages of cell-fusion. (B) A two-cell embryo in which one of the two blastomeres has died during the electrofusion procedure. These diploid embryos must be removed to ensure that they do not confound the results of the tetraploid embryo complementation experiment (see text).

via micromanipulation or electrically with the application of an alternating current (AC) in a nonelectrolyte buffer. Devices capable of applying sufficient AC and DC electrical fields are available from several sources. We recommend the CF-150B pulse generator BLS Ltd., Hungary.

Under the two sets of conditions described below it will be apparent in several minutes whether membrane fusion has begun, although it may take as long as an hour for all embryos to complete fusion. After 2 hr, tetraploid embryos having undergone cell-fusion must be separated from diploid embryos that have not fused. Using these methods we routinely observe fusion rates of 95% without embryo lysis. However, occasionally embryos in which one of the two blastomeres has lysed due to the electrical-treatment are observed (Fig. 2B). Lysis of one of the blastomeres results in a one-cell diploid embryo that can be easily confused with the tetraploid embryos. These diploid one-cell embryos often continue to develop into diploid blastocysts and can confound interpretation of tetraploid embryo complementation experiments if not eliminated. Exact field strengths and conditions may vary between pulse generators and should be experimentally determined. In general, if the rate of cell-fusion is low, then the DC voltage or pulse width should be increased. In contrast, if excessive embryo lysis or poor preimplantation development is observed, voltage or pulse width should be decreased.

Manual Alignment and DC Fusion of Two-cell Embryos

We perform manual alignment and DC fusion on an inverted microscope using the lid of a petri dish as a micromanipulation chamber. Platinum wires (200–300 μm in diameter) are used as both electrodes and micromanipulators to align two-cell embryos for fusion. A group of 15–25 two-cell embryos are placed on the stage in a 500 μl drop of M2 media (Sigma). Note that in this case, the M2 media is not covered in mineral oil, as mineral oil tends to coat the electrodes, causing substantial variability in electrical field strength and cell-fusion. As media not covered in mineral oil tends to evaporate quickly, changing the ionic strength of the media, it is critical to work quickly. In order to prevent hypertonic shock, operation on a single group should not take more than 5–10 min. For cell-fusion, embryos are aligned with the interface between their two blastomeres perpendicular to the electrical field and a single electrical pulse (1.5 kV/cm for 90 μs) is individually applied to each in turn. After electrofusion, embryos are washed through several drops of KSOM and returned to KSOM media at 37°. Each subsequent round of electrofusion should be carried out in a fresh drop of M2 medium.

Electrofusion by AC Alignment and DC Pulse

AC alignment of the embryos, followed by DC fusion allows many embryos to be simultaneously aligned and fused. When acting on a two-cell embryo, the AC field generates a membrane potential, polarizing both cells within the embryo. This cellular polarization has two effects. First, it causes adjacent membranes to become attracted to each other forming a tighter junction between the cells, thus helping to mediate DC cell-fusion. Second, the polarization causes the embryo to begin to rotate until the axis of the embryo is parallel to the electrical field. Thus, the interface of the two blastomeres is electrically aligned perpendicular to the subsequent DC pulse. With this method as many as 50–100 two-cell embryos can be simultaneously fused.

In order to obtain an AC field in the medium a nonelectrolyte salt (generally mannitol) is required. For AC alignment and DC fusion we use a fusion medium composed of 0.3 *M* mannitol, 0.1 m*M* $MgSO_4$, 50 μm $CaCl_2$, and 3% BSA at pH 7.4. The medium should be stored frozen and is generally good for several months. Before fusion, two-cell embryos must be equilibrated in fusion medium. Equilibration is accomplished by placing the embryos at the top of a large drop of fusion medium (1 ml) and allowing them to settle to the bottom. After equilibration, the embryos may be transferred into the fusion chamber.

For this fusion method we use a fusion chamber consisting of two parallel platinum wires (500 μm gap width) immobilized on a glass petri dish (such as GSS-500, BLS Ltd.). The chamber is flooded with electrofusion medium and the equilibrated embryos are placed midway between the two electrodes. The AC field should be initiated before the embryos settle to the bottom of the chamber in order to facilitate rotation of the embryos in the electrical field. We recommend trying an initial AC field strength of two AC volts, however, the field strength required to rotate the embryos may vary from chamber to chamber and is largely influenced by electrode gap distance. The minimal field strength required to rotate the embryos should be experimentally identified and used subsequently. Some embryos may not align in the field due to adherence to the chamber or other embryos and can be nudged with the end of a capillary, allowing them to align. Other embryos may never align due to asymmetry between the two blastomeres and will not undergo cell-fusion.

Once embryo alignment is deemed to be satisfactory, the DC pulse (1.5 kV for 90 μs) is applied to initiate cell-fusion. After the DC pulse the embryos should be removed from the chamber and washed through several large drops (200 μl) of KSOM before being returned to the incubator for long-term culture.

Culture of Tetraploid Embryos to the Blastocyst Stage

After electrofusion, the embryos are cultured to the blastocyst stage in KSOM. The tetraploid embryos should continue to develop with timing analogous to their diploid two-cell counterparts. About 24 hr after fusion, four-cell tetraploid embryos will undergo compaction (with similar timing to compaction of diploid eight-cell embryos). Forty-eight to 56 hr after fusion tetraploid embryos should start to inflate a blastocoel cavity and be ready for ES cell-injection. Substantial delays in either compaction or blastocyst development may indicate that cell-fusion parameters were not gentle enough or that the embryo culture medium has become too old. We generally find that new KSOM medium must be made every 2 weeks and that medium quality is one of the most important parameters for successful, rapid preimplantation development of tetraploid embryos.

Piezo-Micromanipulator Injection of Tetraploid Blastocysts with ES Cells

We perform blastocyst injections on an inverted microscope with Hoffman modulation contrast optics using the lid of a petri dish as micromanipulation chamber (Fig. 3). For microinjection, 10–15 blastocysts are placed in a drop of ES cell media without LIF under mineral oil. A flat tip microinjection-pipette with an internal diameter of 12–15 μm is used for ES cell-injection. It is critical that the internal diameter of the injection pipette is sufficiently large that the cells are not constricted inside the injection needle. If the pipette is too small the ES cells will be lysed during blastocyst injection. Fifty ES cells are generally picked up in the end of the injection pipette so that four or five blastocysts may be injected at a time. The blastocyst to be injected is held with a standard holding pipette. The injection pipette, containing the ES cells is then pressed against the zona-pallucida. A brief pulse of the Piezo-element (Primetech Pmm, Ibaraki, Japan) is applied and the injection needle is pushed through the zona and trophectoderm into the blastocoel cavity. About 10 ES cells are expelled from the injection pipette and pushed against the inner surface of the blastocyst. After injection of the entire group, blastocysts are returned to KSOM media and placed at 37° until transfer to recipient females.

Embryo Transfer to Recipient Females

After blastocyst injection, 10–12 tetraploid-ES cell embryos are transferred to each uterine horn of 2.5 days post coitum (dpc) pseudopregnant Swiss Webster recipients exactly as described.[7]

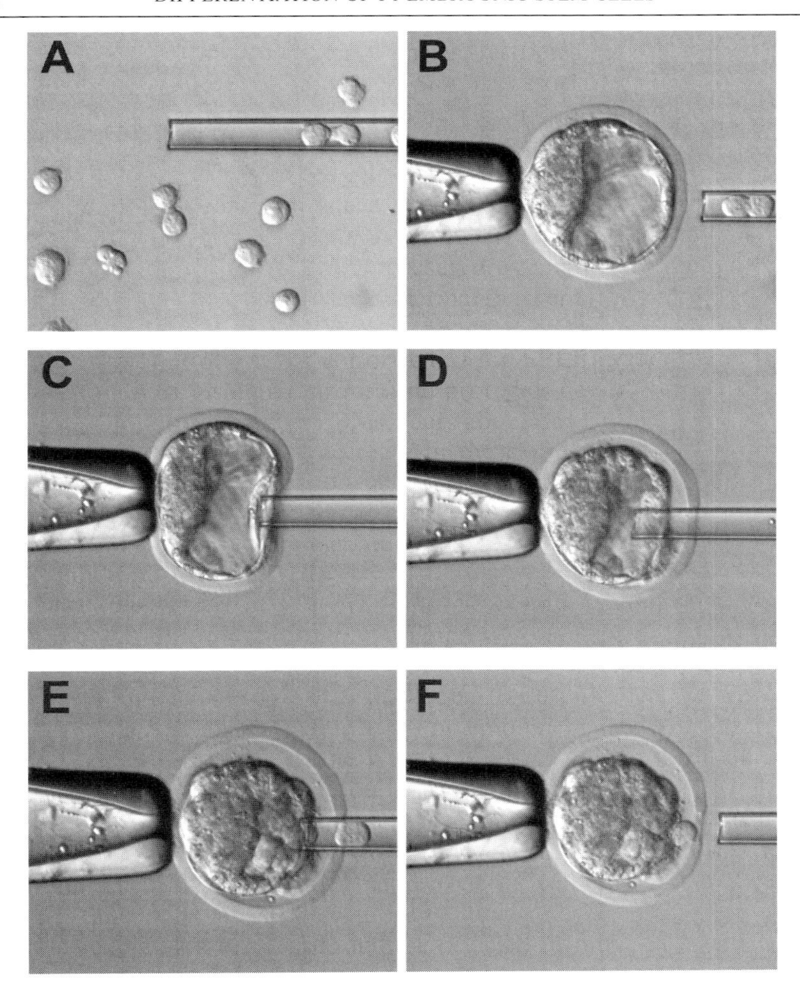

FIG. 3. Piezo-micromanipulator injection of ES cells into a blastocyst. (A) ES cells are collected in a flat tipped microinjection-pipette that is slightly larger than the ES cells. If the diameter of the needle is not larger than the ES cells, the ES cells may be killed during microinjection. After ES cell pick-up, the instruments are moved to the drop of media containing several blastocysts. (B) The blastocysts to be injected should be held adjacent to the inner cell mass (ICM) if present. However, the ICM will often be absent in tetraploid blastocysts. (C) The flat tipped injection pipette is pressed against the zona and (D) the piezo-micromanipulator is applied allowing the pipette to enter the blastocoel cavity. After ES cell deposition (E), the needle is removed (F) and the next blastocyst can be injected.

TABLE I
DEVELOPMENT AND SURVIVAL OF INBRED (A) AND F1 (B) ES CELL-TETRAPLOID MICE

ES cell line	Genotype	4N blasts injected	Pups alive at term (%Inj)	Pups respirating after C-section (%Alive)	Pups surviving to adulthood (%Alive)
(A) Survival of inbred ES cell-tetraploid pups					
J1	129/Sv	120	9 (7.5)	0	0
V18.6	129/Sv	48	5 (10)	1 (20)	0
V26.2	C57BL/6	72	3 (4)	1 (33)	0
V39.7	BALB/c	72	3 (4)	1 (33)	1 (33)
Total	Inbred	312	20 (6)	3 (15)	1 (5)
(B) Survival of F1 ES cell-tetraploid pups					
V6.5	C57BL/6 X 129/Sv	72	18 (25)	17 (94)	16 (89)
V6.5[a]	C57BL/6 X 129/Sv	60	11 (18)	9 (81)	9 (81)
V6.5[b]	C57BL/6 X 129/Sv	20	1 (15)	1 (100)	1 (100)
129B6	129/Sv x C57BL/6	48	2 (4)	1 (50)	1 (50)
F1.2-3	129/Sv x *M. Cast.*	48	4 (8)	3 (75)	3 (75)
V8.1	129/Sv x FVB	24	7 (30)	7 (100)	7 (100)
V17.2	BALB/c X 129/Sv	48	13 (27)	12 (92)	11 (85)
V30.11	C57BL/6 X BALB/c	24	4 (30)	4 (100)	3 (75)
Total	F1	344	60 (18)	54 (90)	51 (85)

[a]Three ES cell subclone targeted at the *Rosa26* locus.
[b]ES cell subclone serially targeted once at the *Rosa26* locus and once with a random insertion.

C-section and Cross-Fostering of ES Cell-Tetraploid Mice

Because of the high incidence of respiratory distress in ES cell-derived neonates,[1,5] we find it extremely beneficial to deliver pups by C-section on

More recent gene targeting studies, demonstrating that some of these genes are essential for the development of both the hematopoietic and endothelial lineages, is further evidence that they share a common ancestor.[17–20]

Over the past 10 years, extensive effort has been directed at establishing conditions for modeling the earliest stages of hematopoietic development in cultures of embryonic stem (ES) cells that have been induced to differentiate in culture. When ES cells differentiate *in vitro*, they generate colonies known as embryoid bodies (EBs) that contain developing cells of many lineages including those of the blood cell and vascular systems.[21,22] Most evidence to date suggests that the events leading to hematopoietic and endothelial development within the EBs recapitulates those that regulate the establishment of these lineages in the yolk sac, indicating that this system provides a valid model with which to study these commitment steps.[5,23]

The ES/EB system has been instrumental in identifying a progenitor with hemangioblast properties. When early EB-derived cells are cultured in the presence of vascular endothelial growth factor (VEGF), interleukin-6 and endothelial cell-conditioned medium, they generate blast cell colonies with endothelial and primitive and definitive hematopoietic potential.[24] The progenitor that gives rise to this colony, the blast colony-forming cell (BL-CFC) represents a transient population that is present within the EBs for approximately 24 hr, between days 2.5 and 3.5 of differentiation, preceding the onset of primitive erythropoiesis. The characteristics of the BL-CFC, namely its early development and its potential to generate primitive and/or definitive hematopoietic cells as well as endothelial progeny, suggests that it represents the *in vitro* equivalent of the yolk sac hemangioblast.

The ability to assay the early stages of hematopoietic and endothelial commitment within this model system provides a powerful approach to study the role of specific genes required for this process. We have taken

[17] F. Shalaby, J. Rossant, T. P. Yamaguchi, M. Gertsenstein, X. F. Wu, M. L. Breitman, and A. C. Schuh, *Nature* **376**, 62–66 (1995).
[18] R. Shivdasani, E. Mayer, and S. H. Orkin, *Nature* **373**, 432–434 (1995).
[19] L. Robb, I. Lyons, R. Li, L. Hartley, F. Kontgen, R. P. Harvey, D. Metcalf, and C. G. Begley, *Proc. Natl. Acad. Sci. USA* **92**, 7075–7079 (1995).
[20] M. C. Dickson, J. S. Martin, F. M. Cousins, A. B. Kulkarni, S. Karlsson, and R. J. Akhurst, *Development* **121**, 1845–1854 (1995).
[21] G. Keller, *Curr. Opin. Cell Biol.* **7**, 862–869 (1995).
[22] A. G. Smith, *Annu. Rev. Cell Dev. Biol.* **17**, 435–462 (2001).
[23] G. Keller, M. Kennedy, T. Papayannopoulou, and M. Wiles, *Mol. Cell. Biol.* **13**, 473–486 (1993).
[24] K. Choi, M. Kennedy, A. Kazarov, J. C. Papadimitriou, and G. Keller, *Development* **125**, 725–732 (1998).

advantage of this model to elucidate the role of two transcription factors, scl/tal-1 and AML-1/Runx1, in the establishment of the hematopoietic system. The findings from these studies demonstrated that scl/tal-1 is important for the establishment of the BL-CFC and/or its commitment to the hematopoietic lineages whereas AML-1/Runx1 functions in the commitment of the BL-CFC to the definitive hematopoietic program.[25,26] These findings are important as they highlight the power of the ES/EB system to allow studies to be performed that could not be done in the early embryo.

The protocol outlined in this chapter describes the growth of ES cells and their differentiation into two stages of development: the BL-CFC/hemangioblast stage and the primitive/definitive hematopoietic stage.

Materials and Methods

For successful, reproducible and efficient ES cell differentiation, it is essential to establish and maintain a stock of high quality, tested reagents. Below is a list of reagents used in our lab for ES cell differentiation to the hematopoietic lineages.

Materials and Reagents

Stock Reagent	Catalogue #
Dulbecco's Modified Eagle Medium (DMEM) powder	Gibco 12100-046
Tissue culture grade H_2O (TC-H_2O)	Cellgro MT 25-055-CV
Hepes 1 *M* solution	Gibco 15630-080
Penicillin/Streptomycin (P/S) 5000 u/ml	Gibco 15070-063
$NaHCO_3$	Sigma S-5761
Iscove's Modified Dulbecco's Medium (IMDM) prepared media	Cellgro MT 15-016-CV
L-Glutamine 200 m*M*	Gibco 25030-08
Monothioglycerol (MTG)	Sigma M-6145
Fetal bovine serum (FCS)	
(a) for ES cell growth and maintenance	Gemini Bioproducts Gem-cell 100–500

[25] S. M. Robertson, M. Kennedy, J. M. Shannon, and G. Keller, *Development* **127**, 2447–2459 (2000).

[26] G. Lacaud, L. Gore, M. Kennedy, V. Kouskoff, P. Kingsley, C. Hogan, L. Carlsson, N. Speck, J. Palis, and G. Keller, *Blood* **100**, 458–466 (2002).

(b) for EB generation: need to test batches from different suppliers
(c) for hematopoietic colony growth: Fetal Bovine Plasma-Derived Serum; Platelet Poor (PDS): Antech, Tyler, TX: Request Keller lab protocol serum

Gelatin	Sigma G-1890
Phosphate buffered saline 10X (PBS)	Cellgro MT 20-031-CV
Trypsin–EDTA (0.25%) prepared	Cellgro MT 25-053-CV
Trypsin 1 : 250 (powder)	Sigma T-4799
L-Ascorbic acid (AA)	Sigma A-4544
Transferrin 30 mg/ml	Roche 652-202
Protein-free hybridoma medium (PFHM-II)	Gibco 12040-093
Methylcellulose powder (MeC)	FLUKA 64630
IMDM powder	Gibco 12100-046
Matrigel (growth factor reduced)	BD biosciences 354230
Endothelial cell growth supplement (ECGS)	BD biosciences 356006

Recommended Culture Dishes

1. For EB generation: 60×15 mm petri dish (5 ml): Parker (through VWR 253-84-090), request ethylene oxide sterilization.
2. For hematopoietic colony assays: 35×10 mm petri dish: Falcon #1008.

Cytokines

Cytokines can be purchased from a number of different suppliers. We use the following from R&D systems:

G-CSF: #415-CS
GM-CSF: #415-ML
M-CSF: #416-ML
VEGF: #293-VE
IL-3: #403-ML
IL-6: #406-ML
IL-11: #418-ML
Thrombopoietin (Tpo): #488-TO
Erythropoietin (Epo): #959-ME
KL: #455-MC
LIF: #449-LR
bFGF: #233-FB
IGF-1: #791-MG

Prepared Reagents

Methylcellulose Stock (2.0%) is prepared from powder as follows:

1. Methylcellulose is prepared in a 1 liter volume in a 2 liter Erlenmeyer flask. Prior to starting, weigh the flask and then add 450 ml autoclaved TC-H_2O. Bring water to a boil on a hotplate. When the

water has started to boil, remove the flask from the hotplate and add 20 g of methylcellulose powder. Swirl to disperse the powder in the water. Reheat mixture and return to a boil. The mixture must be swirled repeatedly during this step as it will rise and spill over the flask as it boils. This heating and swirling step should be repeated 3–4 times, until the methylcellulose is dissolved and the mixture is homogeneous slurry. Note that boiling is required both for dissolving and sterilizing the methylcellulose powder.

2. After the final boil, cool the slurry for approximately 40 min until the flask is warm, but not hot, to the touch.
3. While cooling the mixture, prepare 500 ml of 2X IMDM and add 2X MTG (3.0×10^{-4} *M*). Sterile filter prior to use.
4. Add the 2X IMDM to the cooled methylcellulose mixture and swirl to mix.
5. Weigh the flask containing the methylcellulose/IMDM mixture. The final weight of the mixture minus the weight of the flask should be 1000 g. Adjust the weight with sterile water.
6. The methylcellulose mixture should be cooled overnight at 4°C. Swirl the mixture several times during the first hour of this cooling period to prevent separation of the methylcellulose and medium. The mixture will be considerably thicker at 4°C than at room temperature. The methylcellulose can now be aliquoted and stored at −20°C. Methylcellulose must be frozen before it can be used.

D4T Endothelial Cell-Conditioned Medium

D4T conditioned medium will enhance the growth of blast cell colonies when added together with VEGF and IL-6 to methylcellulose cultures of day 3 EB-derived cells.[24] D4T conditioned medium is produced as follows:

1. Grow the D4T cells to confluency on gelatinized flasks in IMDM supplemented with MTG (1.5×10^{-4} *M*), 10% FCS (selected for hemangioblast growth) and ECGS (50 ug/ml).
2. When the cells are confluent, replace the medium with new medium without ECGS and allow it to condition for 72 hr. Harvest the conditioned medium and add fresh medium to the cells to repeat the conditioning. This conditioning step can be repeated up to five times using the same cell population.
3. The medium from the different harvests should be pooled, filtered and tested for its ability to stimulate EB-derived blast cell

colonies in cultures containing VEGF and IL-6. Most batches of D4T conditioned medium display optimal activity between 15 and 25%. Tested conditioned medium should be aliquoted and stored frozen at −20°C.

Mouse Embryonic Feeder Cells (MEFs)

The protocol used for feeder cell generation is well established in the ES cell scientific community and is based on that of published protocols.[27] The feeders are expanded to confluency in the appropriate number of 100 mm tissue culture grade dishes. One dish yields approximately 1.0×10^7 cells. At this stage, the cells are harvested, irradiated at 3000R and frozen at 3×10^6 cells per vial. When thawed, this number of feeders is sufficient to cover the wells of 2×6-well plates.

Trypsin–EDTA

Trypsin 1 : 250: for 500 ml

Trypsin:	1.25 g
0.5 *M* EDTA:	1.08 ml
PBS:	500 ml

Warm to dissolve, filter sterilize, aliquot and store at −20°C.

Note. Trypsin can be prepared as described here or purchased as a prepared stock.

Gelatin

Prepare a 0.1% solution of gelatin in 1X PBS, dissolve and sterilize by autoclaving.

Gelatinized Flasks and Dishes

To gelatinize flasks or dishes, cover the surface of the vessel (e.g., 2.0 ml for a T25 flask) with the gelatin solution and incubate for 20 min at room temperature. It is possible to prepare dishes and flasks in advance and store them with the gelatin solution at 4°C for up to 1 week. Stored plates should be sealed with parafilm. Remove excess gelatin solution prior to use.

Ascorbic acid

Prepare a stock solution of 5 mg/ml in cold TC-H_2O, sterile filter, aliquot and store at −20°C. Use once and discard excess.

[27] E. J. Robertson, ed.), "Teratocarcinomas and Embryonic Stem Cells: A Practical Approach." IRL Press, Oxford, Washington DC, 1987.

Monothioglycerol (MTG)

While the amounts of MTG are indicated for the different media, it is important to test each new batch of MTG as there is variability between them. We aliquot MTG in 1.0 ml volumes that are stored frozen (−20°C). When aliquots are thawed, they can be used for several experiments and then discarded.

Matrigel-coated Wells

The stock bottle of Matrigel should be thawed slowly on ice, diluted 1 : 1 with IMDM, aliquoted (0.5 ml) and frozen (−20°C). Expansion of blast colony cell populations is carried out in microtiter wells pretreated with a thin layer of Matrigel. The wells are coated by first spreading 5 μl of diluted Matrigel over the surface with an Eppendorf pipette tip. The plate should be kept on ice during this procedure. When the required number of wells has been coated, incubate the plate on ice for 10–15 min. Following this incubation, remove excess Matrigel from each well and then incubate at 37°C for an additional 15 min. The plate is now ready to use.

Prepared Media

DMEM

We make our own DMEM from purchased powder, using the following recipe:

DMEM powder:	package for 1 liter
Penicillin/Streptomycin:	10 ml/liter
Hepes buffer 1 *M*:	25 ml/liter
$NaHCO_3$:	3.025 g/liter*
Tissue culture grade H_2O ($TC-H_2O$):	make up to 1 liter
Sterile filter	

*The amount of $NaHCO_3$ has been adjusted to that of IMDM.

IMDM Supplemented

For 500 ml:

Cellgro IMDM media:	490 ml
P/S:	5 ml
Glutamine:	5 ml

IMDM supplemented with P/S and glutamine is used for all the recipes detailed below.

Media for Growth and Maintenance of Undifferentiated ES Cells

DMEM-ES and IMDM-ES Medium

DMEM or IMDM supplemented:	85%
Fetal bovine serum (FCS)*:	15%
MTG:	1.5×10^{-4} *M*
LIF:	10 ng/ml

*FCS must be tested for ability to support ES cell growth.

IMDM–FCS Medium

IMDM supplemented:	85%
FCS:	15%
MTG:	1.5×10^{-4} *M*

EB Differentiation Medium for the Generation of Hemangioblasts

IMDM supplemented:	83%
FCS:	15%
Transferrin:	0.5–1%
Glutamine:	1%
MTG:	3.0×10^{-4} *M**
AA:	50 ng/ml

*The amount of MTG required will vary depending on the batch.

EB differentiation Medium for the Generation of Hematopoietic Progenitors

IMDM supplemented:	78%
FCS:	15%
Transferrin:	0.5–1%
Glutamine:	1%
MTG:	3.0×10^{-4} *M*
AA:	50 ng/ml
PFHM-II:	5%

Methylcellulose Mixture for the Growth of Hematopoietic Colonies from EBs

The BL-CFC as well as the committed hematopoietic progenitors have specific growth requirements and therefore the types of colonies that develop will depend on the factors and reagents included in the methylcellulose mixture. Selection of appropriate components enables one to generate cultures that contain predominantly one colony type or a broad mixture of colonies from the same EB cell population. The components of

TABLE I
COMPONENTS OF THE METHYLCELLULOSE MIXTURES USED FOR THE GENERATION OF DIFFERENT TYPES OF COLONIES

	Mix #1	Mix #2	Mix #3	Mix #4	Mix #5
Type of colony stimulated	Blast	Ery-P	Ery-D/Mega, Ery-D, Mega, Ery-P	Mac	Multilineage
Stock methylcellulose	55%	55%	55%	55%	55%
Serum	10% FCS	10% PDS	10% PDS	10% PDS or FCS	10% PDS or FCS
PFHM-II	–	5%	5%	5%	5%
MTG	3.0×10^{-4} *M*	3.0×10^{-4} *M*	3.0×10^{-4} *M*	3.0×10^{-4} *M*	3.0×10^{-4} *M*
Ascorbic acid	25 ng/ml	25 ng/ml	25 ng/ml	25 ng/ml	25 ng/ml
Glutamine	2 m*M*	2 m*M*	2 m*M*	2 m*M*	2 m*M*
Transferrin	300 μg/ml	300 μg/ml	300 μg/ml	300 μg/ml	300 μg/ml
D4T CM	25%	–	–	–	–
VEGF	5 ng/ml	–	–	–	–
KL	–	–	100 ng/ml	–	100 ng/ml
Epo	–	2 units/ml	2 units/ml	–	2 units/ml
Tpo	–	–	5 ng/ml	–	5 ng/ml
IL-3	–	–	–	1 ng/ml	1 ng/ml
IL-6	10 ng/ml	–	–	–	10 ng/ml
IL-11	–	–	5 ng/ml	–	5 ng/ml
GM-CSF	–	–	–	–	3 ng/ml
G-CSF	–	–	–	–	30 ng/ml
M-CSF	–	–	–	5 ng/ml	5 ng/ml
IMDM	to 100%	to 100%	to 100%	to 100%	to 100%

five different methylcellulose mixtures for the growth of different types of colonies are detailed in Table I. The generation of cultures with specific colony types is recommended when learning to recognize and score hematopoietic colonies.

Note. Methylcellulose is too viscous to be used with regular pipettes. The stock methylcellulose is handled with a 10 ml syringe without any needle. The final mixtures can be distributed with a 3 or 5 ml syringe with a 16-gauge needle.

Hemangioblast Expansion Medium

Hemangioblast medium is used for the expansion of the hematopoietic and endothelial cells from the blast cell colonies and contains the following components:

FCS:	10%
Horse serum:	10%
VEGF:	5 ng/ml

IGF-1:	10 ng/ml
bFGF:	10 ng/ml
Epo:	2 U/ml
KL:	100 ng/ml
IL-3:	1 ng/ml
MTG:	1.5×10^4 *M*
IMDM supplemented:	make to 100%

Endothelial Expansion Medium

IMDM supplemented:	80%
FCS:	20%
VEGF:	50 ng/ml
bFGF:	20 ng/ml
MTG:	1.5×10^4 *M*

Methods

The protocol for ES differentiation is presented as three separate steps as outlined in Fig. 1.

Step 1: Maintenance of ES Cells

To achieve efficient and reproducible differentiation, it is essential to begin with a healthy, undifferentiated and well-maintained stem cell culture. We maintain our undifferentiated ES cells on irradiated primary MEF cells

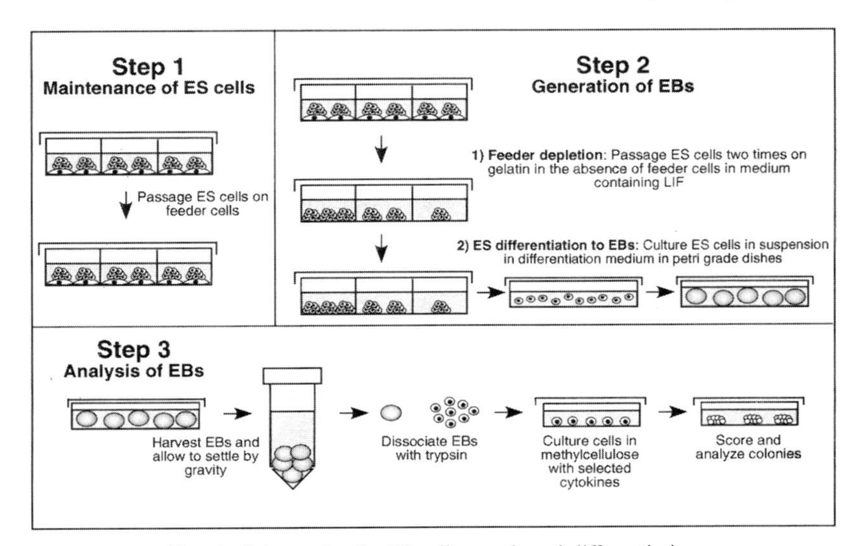

FIG. 1. Scheme for the ES cell growth and differentiation.

in 6-well plates in DMEM-ES medium supplemented with LIF. Other culture dishes or flasks can be used, however, we find that the 6-well plate format provides optimal flexibility for testing different cell concentrations. The conditions used for maintenance of ES cells are critical to allow proper *in vitro* differentiation. ES cells maintained on MEFs behave differently in differentiation cultures than ES cells maintained on STO fibroblasts or those grown in a feeder-independent fashion on gelatinized flasks. In our experience, the best results are obtained from ES cells maintained on MEFs. Proper techniques for passaging ES cells are essential for maintaining a healthy cell population. The following protocol works well with all ES cell lines tested.

1. Twenty-four hours prior to passaging or thawing ES cells, plate irradiated feeders into a 6-well plate. Approximately 1.5×10^6 cells are sufficient for 1 plate.
2. On the day of passage, remove the medium from the ES cell culture, add 1 ml of trypsin per well and return the plate to the 37°C incubator for 2–3 min. Following this short incubation step, disperse the cells by pipetting 3–4 times with a P1000 pipette. Following dispersion, transfer the 1 ml of cells in trypsin to a tube containing 9 ml of DMEM supplemented with 15% FCS. Centrifuge cells at 1000 rpm for 5 min, remove the supernatant and resuspend the pellet in 1 ml of DMEM-ES medium using a P1000 pipette. It is important to disperse the ES cells into a single cell suspension at this stage as passaging undissociated aggregates will result in the growth of large ES cell colonies that undergo spontaneous differentiation in the maintenance cultures. If properly trypsinized, the ES cells disperse easily using this protocol. If the cells do not disperse after 3 min of trypsinization, the trypsin solution is suboptimal. It is advisable to replace the trypsin stock as extending the time of trypsinization can damage the ES cells.
3. Seed several dilutions of the cells onto the new feeders in DMEM-ES medium (final volume of 3 ml per well). Most ES cell lines can be diluted 5- to 10-fold at each passage and have to be passaged every 48 hr. However, it is always a good idea to passage at several different dilutions to ensure that proper cell density is maintained during the 48-hr growth period. This is particularly important when working with an unfamiliar cell line. For the most reproducible differentiation results, it is best not to maintain the undifferentiated ES cells in culture over extended periods of time. We routinely thaw a new vial of cells for every experiment.

Step 2: Generation of Embryoid Bodies

1. Prior to initiating EB development, it is important to deplete the ES culture of feeder cells, as they will alter the kinetics of differentiation if present in the EB cultures. Feeder depletion is accomplished by passaging the cells two times on gelatin-coated wells in the presence of LIF. This is a critical step as most feeder-dependent ES cells will differentiate when moved to a gelatin-coated surface. To minimize the extent of differentiation, the ES cell density should be kept high when passaging without feeders. To ensure appropriate cell density, we use 2–3 different dilutions (ranging from two- to six-fold) and select the well with the appropriate concentration for the next passage. The appropriate concentration will contain a high density of undifferentiated ES cell colonies that have not yet acidified the medium. The ES cell colonies should retain their typical three-dimensional structure with well-defined smooth perimeters. During this feeder depletion step, the cells are passaged at 24-hr intervals, given that they are maintained at high density. For hematopoietic differentiation of the ES cells, the first feeder depletion passage is done in DMEM-ES medium, whereas the second is done in IMDM-ES medium. In our experience, carrying out the second passage in IMDM-ES medium provides the most efficient and reproducible differentiation pattern. Cells passaged twice in the absence of feeders are used for the generation of EBs
2. To establish EB cultures, trypsinize the feeder depleted ES cells and wash two times with IMDM–FCS (no LIF) to remove residual LIF from the medium. EBs are generated in nonadherent petri grade dishes. If tissue culture grade dishes are used, the ES cells will differentiate and form a complex adherent cell population that does not generate hematopoietic progeny in a reproducible fashion. We routinely use 60 mm dishes for EB generation. Other size dishes can be used, however, it is essential to test different brands to find those that are least adhesive. The number of ES cells plated in the differentiation cultures will depend on the cell line used and the progenitor population to be assayed. Table II summarizes the range of cell numbers routinely used for the generation of EBs for hemangioblasts and hematopoietic progenitors from three different ES cells lines. The numbers provided in the table are estimates. It is important to test different cell numbers for the particular ES cell line being used as too few or too many cells will result in suboptimal differentiation. Optimal conditions should yield 1.5×10^6 to 3.0×10^6 cells per 60 mm dish. In addition to cell yield, differentiation can be

TABLE II
CELL NUMBER REQUIRED FOR OPTIMAL GENERATION OF EBs FROM THREE DIFFERENT ES CELL LINES

ES cell line	Cell number (per ml) for day 3 EBs (hemangioblast stage)	Cell number (per ml) for day 6 EBs (hematopoietic stage)
CCE	1×10^4	1×10^3
E14	3.0×10^4	2×10^3
J1	1.5×10^4	1×10^3

The input number indicated should yield a total of between 1.5×10^6 and 3.0×10^6 cells on days 3 and 6 of differentiation.

monitored by levels of expression of the receptor Flk-1[13,28] and/or the number of blast colonies at the hemangioblast stage of development (days 2.5–4.0) (Fig. 2). Hematopoietic stage EBs (day 6–7 of differentiation) can be assayed for hematopoietic progenitor potential. It is important to note that the cell numbers and the components of the medium used for the generation of the two stages of EB development are different.

Step 3: Analysis of EBs

Harvesting of EBs

1. Harvest EBs from the differentiation cultures and place in a 50 ml tube. Allow the EBs to settle into a loose pellet by gravity for approximately 5–10 min. A maximum of eight dishes can be pooled into a single 50 ml tube.
2. Remove the supernatant, add 3 ml of trypsin to the EBs and place in a 37°C water bath for 3 min. At the end of the 3-min incubation period, add 1 ml of FCS to the trypsin and vortex the tube to initiate dissociation of the EBs. To complete the dissociation process, pass the cells through a 20-gauge needle 1–2 times, using a 5 ml syringe and transfer to a 14 ml tube containing 6 ml of IMDM–FCS. Centrifuge, and resuspend the pellet in 1-2 ml IMDM–FCS, using a P1000 pipette. The EB cells can be counted and analyzed for hemangioblast or hematopoietic potential.

[28] P. Faloon, E. Arentson, A. Kazarov, C. X. Deng, C. Porcher, S. Orkin, and K. Choi, *Development* **127**, 1931–1941 (2000).

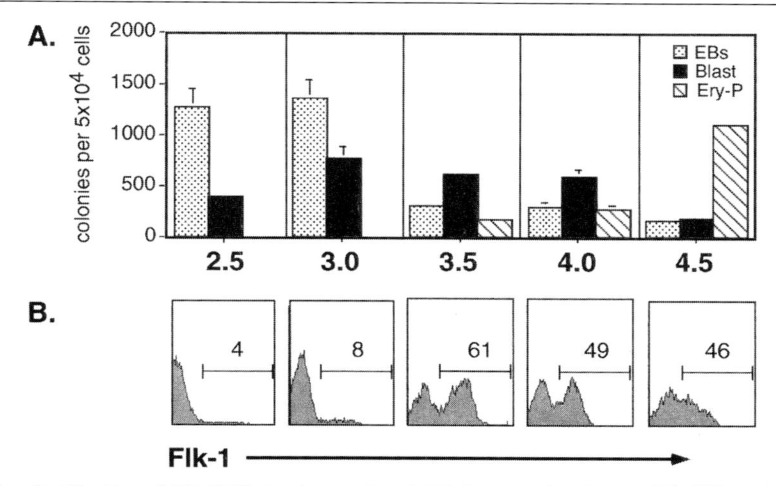

FIG. 2. Kinetics of BL-CFC development and Flk-1 expression during EB differentiation. (A) Number of secondary EBs, blast colonies and primitive erythroid (Ery-P) colonies generated from EBs differentiated for 2.5–4.5 days. Numbers below the histogram indicate the day of EB differentiation. (B) Expression of Flk-1 within EB populations harvested at the indicated days of differentiation. Numbers above the bar represent the % Flk-1 positive cells.

Analysis of Hemangioblast and Hematopoietic Potential of EBs

Hemangioblast stage. The hemangioblast stage of development precedes the onset of hematopoiesis within the EBs and is found between days 2.5 and 4.0 of differentiation for most ES cell lines. The BL-CFC represents a transient population that is found in the EBs for only 18–36 hr. Consequently, any changes in culture conditions that alter the kinetics of development by as little as 6 hr, can lead to inconsistencies between experiments. Figure 2A shows a typical kinetic analysis demonstrating the appearance of BL-CFCs between days 2.5 and 4.0 of differentiation during which time the number of secondary EBs is declining. Secondary EBs develop from residual ES cells that have not yet differentiated in the primary EBs. Beyond day 4.0 of differentiation, the number of BL-CFC declines with the commitment to the hematopoietic program as indicated by the appearance of significant numbers of primitive erythroid progenitors. Given that the BL-CFC expresses the receptor tyrosine kinase Flk-1,[28] its development is associated with the upregulation of receptor expression within the EBs, as shown in Fig. 2B. These dynamic changes in Flk-1 expression are the expected pattern and indicative of efficient and appropriately timed differentiation. To minimize any changes in the kinetics of differentiation, it is important to adhere strictly to the protocol, using the same reagents for each experiment. Even with identical reagents, however, one can experience slight changes in the kinetics of BL-CFC development. Thus, it is advisable to assay EBs from several time points when assessing BL-CFC potential. Based on findings from a number of different ES cell lines, we analyze day 3.0, 3.5 and 4.0 EBs (dependent on the levels of Flk-1 expression). While the presence of Flk-1 within an EB population does not guarantee the presence of large numbers of BL-CFCs, the lack of significant Flk-1 expression does indicate that the population has not yet progressed to the hemangioblast stage of development. Expression of Flk-1 can be used as an initial screen to define the appropriate day of EB development for BL-CFC analysis in subsequent experiments.

1. For FACS analysis, antibody staining is carried out as follows: 2×10^5 cells are resuspended in 100 μl of PBS containing 10% FCS and 0.02% sodium azide. An appropriate amount of antibody is added and the cells are incubated on ice for 20 min. Following the staining step, the cells are washed two times with the same media. It is important to minimize the number of centrifugation steps, as EB cells are fragile.
2. To assay for the BL-CFC, take an aliquot of the suspension with the appropriate number of cells (see below) for 3.5 ml media and place it in a 14 ml snap cap tube. The volume of the cell suspension should not exceed 10% of the total mixture. Larger volumes will reduce the viscosity of the methylcellulose mixture. Add 3.5 ml of the hemangioblast colony methylcellulose mixture (#1, Table I) to the cells using a syringe with a 16-gauge needle. It is important to have more methylcellulose (approximately 0.5 ml) than is actually needed as it is impossible to recover all of the mixture from the tube. Vortex the mixture with the cells and then allow it to settle for 5 min to permit the air bubbles to rise. Aliquot 1 ml of the mixture into each of 3×35 mm petri grade dishes. Gently swirl the dishes to disperse the methylcellulose mix evenly over the bottom surface. Place these dishes in a larger dish, together with a 35 mm open dish of water for humidity. Culture at 37°C in an environment of 5% CO_2 in air. For most ES cell lines, we culture between 3×10^4 and 1×10^5 cells per ml for BL-CFC analysis. The frequency of the BL-CFC ranges from 0.2 to 2.0% of the total EB population.
3. Blast colonies develop within 3–4 days of culture and can be recognized as clusters of cells that are easily distinguished from secondary EBs that develop from residual undifferentiated ES cells (Fig. 3). Blast colonies and secondary EBs are the predominant type of colonies present in these cultures.
4. To analyze the hematopoietic and endothelial potential of the blast colonies, individual colonies are picked from the methylcellulose and

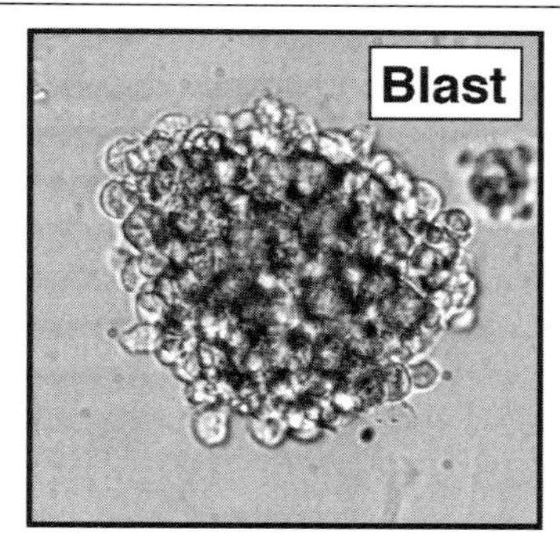

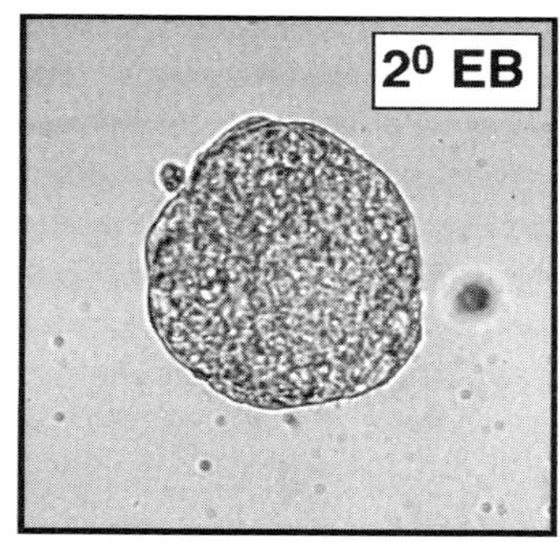

FIG. 3. Photograph of a blast cell colony and secondary EB. These colonies were generated from EBs differentiated for 3.5 days.

cultured further in hemangioblast expansion medium on Matrigel-coated microtiter wells. The hemangioblast expansion medium contains cytokines that promote the growth of both adherent (endothelial) and nonadherent (hematopoietic) cell populations. Following 2–3 days of culture, the nonadherent cells can be harvested by gentle pipetting and assayed for hematopoietic progenitor potential in methylcellulose cultures using the multi-lineage mix (mix #5) that will support the growth of multiple progenitors. The remaining adherent population is cultured for an additional 4 days in endothelial expansion medium. At this point, the adherent population can be lysed directly in the well and subjected to RT-PCR for the analysis of expression of genes associated with endothelial development.

Hematopoietic Stage. Shortly following the peak of the hemangioblast stage of development, committed hematopoietic progenitors can be detected within the EBs. Both primitive erythroid and definitive hematopoietic progenitors appear as early as day 4 of differentiation. Following the initiation of primitive erythropoiesis between days 3.5 and 4.0 of differentiation, this progenitor population increases in size dramatically, reaching a peak by day 6. Primitive erythroid progenitor numbers drop sharply to undetectable levels over the next 48 to 72 hr. The number of definitive progenitors increases at this stage and reaches a plateau between days 6 and 7 of differentiation. While definitive progenitor numbers decline beyond this point, the decrease is not as dramatic as observed with the primitive erythroid lineage. It is worth stressing that these patterns are based on EBs differentiated in medium containing PFHM-II and serum preselected for optimal differentiation to the hematopoietic lineages. PFHM-II is included in these cultures as we have found that it increases the efficiency of EB differentiation to hematopoiesis. It is not included in the EB hemangioblast medium, however, as it reduces the number of detectable BL-CFC, possibly due to acceleration of the kinetics of differentiation at this stage. All wild type ES cell lines that we have tested follow this pattern of development. To evaluate the primitive and definitive hematopoietic potential of a given cell line, we routinely analyze EBs differentiated for 6 days. For this analysis, the EBs are harvested and processed as described above. The EB-derived cells are plated in methylcellulose media containing different hematopoietic cytokines. The selection of the media will depend on the type of progenitor to be analyzed as described below. While the large mix of cytokines present in mix #5 will stimulate progenitors of all the lineages present within the day 6 EBs, at times it is desirable and advantageous to evaluate the growth of a select subset of hematopoietic lineages.

Primitive Erythroid Colonies. Most primitive erythroid progenitors require only Epo for growth and generate relatively small (20–100 cells), tight, bright red colonies within 3–5 days of culture in methycellulose mix #2. The colonies are homogeneous in size, suggesting that this progenitor population is generated within the EBs in a single synchronous wave. The primitive erythroid progenitors are sensitive to components found in most regular FCS and thus require platelet poor PDS for growth. In addition, we have found that the growth and development of these colonies is significantly enhanced by the addition of the protein-free medium, PFMH-II (5%) to the methylcellulose culture. All other hematopoietic progenitors found in day 6 EBs require cytokines in addition to Epo for growth. Thus, the vast majority of colonies in a methylcellulose culture of day 6 EB cells stimulated with only Epo are of the primitive erythroid lineage. In addition to the characteristic colony morphology, primitive erythroid colonies can be identified by morphological assessment of the cells following May-Grunwald/Giemsa staining and by analysis of the globin expression patterns. Primitive erythrocytes are large, remain nucleated and express embryonic globins such as βH1. Under optimal conditions, one should expect between 1000 and 5000 colonies per 1×10^5 day 6 EB cells plated. The numbers will vary between different cell lines and serum.

Definitive Hematopoietic Colonies.

Definitive Erythroid and Megakaryocyte Colonies. The definitive erythroid and megakaryocyte lineages are often found within the same colony when assaying day 6 EB cells, suggesting that they arise from a common progenitor at this stage of development. It is difficult to score megakaryocytes in cultures containing macrophage colonies as some macrophages grow to a large size in methylcellulose. To be able to assess the megakaryocyte potential of a population, it is best to culture the cells in a methylcellulose mix (mix #3) devoid of macrophage-stimulating cytokines. When cultured in these conditions, day 6 EB-derived cells will give rise to the following types of colonies: primitive erythroid, bilineage definitive erythroid/megakaryocyte, pure definitive erythroid and pure megakaryocyte. The primitive erythroid colonies will be the most abundant. Bilineage definitive erythroid/megakaryocyte colonies will be the largest in the cultures and can be recognized by their distinct morphology, consisting of hemoglobinized cells in the center surrounded by large megakaryocytes at the periphery. Pure definitive erythroid colonies can be distinguished from the primitive erythroid colonies by the fact that they are often larger, less red and contain smaller cells. Cells in the definitive erythroid colonies express βmajor but no βH1 globin. Pure megakaryocyte colonies consist of relatively few (10–50) cells. Most of these colonies contain cells of quite different sizes, representing different stages of megakaryocyte development. These cells tend to have few or no visible granules in the cytoplasm and have a defined, smooth, dark membrane. Megakaryocyte colonies should be scored between days 4 and 6 of culture, whereas the definitive erythroid and bilineage megakaryocyte/definitive erythroid are best scored between days 7 and 9 of culture.

Macrophage Colonies. The second most abundant progenitor within the day 6 EBs is that of the macrophage lineage. Macrophage progenitors can be evaluated in cultures containing a broad mix of cytokines or in conditions that specifically promote their growth. When cultured in the presence of methylcellulose mix #4, day 6 EB-derived cells will form colonies consisting almost exclusively of macrophages. As with other hematopoietic colonies, the pure macrophage colonies display a distinct morphology. Macrophage colonies stimulated with M-CSF and IL-3 can be recognized as tightly packed colonies of large cells with cytoplasmic vacuoles. The cells can be confirmed as belonging to the macrophage lineage by morphological analysis and expression of the *c-fms* gene that encodes the receptor for M-CSF. Macrophage colonies grown under these conditions can be scored between days 5 and 8 of culture.

Multilineage Colonies. In addition to lineage-restricted progenitors, day 6 EBs also contain multipotential cells that generate mixed lineage colonies. These colonies require a broad mix of cytokines to develop. Two types of mixed colonies develop following the culture of day 6 EB cells in methylcellulose mixture #5. The first are large colonies that consist predominantly of macrophages with a small component of definitive erythroid cells and neutrophils. These colonies appear initially as large macrophage colonies and subsequently develop the erythroid and neutrophil component. The second type of mixed colony is somewhat smaller in size, and is more erythroid in composition, but also contains megakaryocytes, macrophages, mast cells and neutrophils. Both types of mixed colonies should be scored between days 8 and 12 of culture. The composition of the mixed colonies can be assessed by morphology. In addition to the mixed multilineage colonies, this mixture of cytokines will also stimulate the growth of the lineage-restricted colonies. Given the mixture of colonies, these cultures are the most challenging to score.

Selection of Serum for ES Cell Growth and Differentiation. As serum is the most essential component of the ES cell differentiation cultures, it is important to select appropriate batches that will promote the most efficient and reproducible development of hematopoietic cells within the EBs.

Serum for ES Cell Growth (Growth Serum). Serum tested for the growth and maintenance of ES cells is commercially available. However, given that the pretested batches are more expensive than regular FCS, we routinely select our own serum for ES growth. For this selection, ES cells are passaged 4–5 times in the different test lots and the cultures monitored for cell viability, growth and maintenance of cells with an undifferentiated morphology. It is easiest to test serum on feeder cell-depleted CCE ES cells.

Serum for the Generation of EBs (Differentiation Serum). The serum selected for the growth and maintenance of ES cells in an undifferentiated state is most likely not going to be the best lot for the generation of well-differentiated EBs. Consequently, a different lot of serum will need to be identified for this purpose. Differentiation serum should be selected for its ability to support the efficient development of EBs that contain optimal numbers of hematopoietic progenitors. Efficiency of EB development is determined by counting the total number of cells in the differentiation cultures at days 3 and 6. For a good serum, the ratio of cell input and output should be similar to that outlined in Table II. The potential to support hematopoietic differentiation is assessed by assaying EBs at different stages of development for the presence of desired progenitor populations. To select serum optimal for hemangioblast development, EBs differentiated in different lots of serum for 2.5–4.0 days are analyzed for Flk-1 expression and the presence of BL-CFC. For the selection of serum supporting later stage hematopoietic development, day 6 EBs differentiated in the different serum lots are analyzed for hematopoietic progenitor cell content. The selection of serum for the hemangioblast stage of development is the most challenging, given the transient nature of this population. Sera that support the efficient development of the hemangioblast will often support the development of primitive and definitive precursors in later stage EBs. However, the reverse is not always true. Therefore, the serum test should be set up based on the types of progenitors to be studied.

SERUM TO SUPPORT THE GROWTH OF BLAST CELL AND HEMATOPOIETIC COLONIES (MATURATION SERUM). For the growth and maturation of colonies in methylcellulose, sera should be selected for its ability to support the growth of large numbers of colonies as well as for its ability to support the growth and maturation of cells within the colonies. For maturation, the most sensitive cells are those of the primitive erythroid lineage. We have found that the majority of batches of regular FCS do not efficiently support the growth and differentiation of the primitive erythroid colonies. In contrast, almost all batches of PDS are supportive of the development of these colonies. As PDS also supports the growth of other colonies we use it routinely for all our hematopoietic colony assays. Selected lots of normal FCS will also support the development of colonies from all definitive lineages and may be used in place of PDS. Pretested lots of either PDS or FCS can be used for the growth of the hemangioblast-derived colonies. The appropriate serum should be selected based on numbers of blast cell colonies that develop and on the potential of these colonies to generate both hematopoietic and endothelial progeny in the expansion cultures.

Conclusions

The protocols outlined in this chapter will enable one to assay two stages of hematopoietic development from most mouse ES cell lines. With good reagents and well-maintained ES cell stocks, commitment to the hematopoietic program is efficient and highly reproducible. The majority of problems encountered with ES cell differentiation are related to poor quality ES cells, suboptimal reagents or deviation from established protocols.

[4] *In Vitro* Differentiation of Mouse Embryonic Stem Cells: Hematopoietic and Vascular Cell Types

By STUART T. FRASER, JUN YAMASHITA, L. MARTIN JAKT, MITSUHIRO OKADA, MINETARO OGAWA, SATOMI NISHIKAWA, and SHIN-ICHI NISHIKAWA

Introduction

Summary

Embryonic stem cells have been posited as sources of differentiated cell types for regenerative medicine. One of the most enticing cell types is the recently described endothelial progenitor cell (EPC) which can contribute to blood vessels and is a candidate for therapy against vascular

0076-6879/2003 $35.00

diseases.[1] Here, we describe an *in vitro* differentiation system which results in the generation of endothelial, smooth muscle and hematopoietic cells from ES cells. We have generated an ES-derived equivalent of the lateral plate mesoderm (LPM), a critical source of vascular and hematopoietic cells during embryonic development.[2] This *in vitro* differentiation system is useful for the analysis of developmental pathways and is highly amenable for the analysis of gene expression at distinct stages of the development of the vascular and hematopoietic lineages from Flk1$^+$ precursors.

The Lateral Plate Mesoderm and Its Derivatives

Shortly after gastrulation, the mesoderm forms as a distinct layer between the endoderm and ectoderm. Expression of Vascular Endothelial Growth Factor Receptor-2, also known as Flk1 in the mouse, is a clear indicator of the development of the LPM.[3] Recently, Flk1 expression has been utilized to isolate putative LPM cells from 7.5 dpc mouse embryos which were found to contain endothelial and hematopoietic potential.[4] Flk1 expression has also been used to isolate a population capable of generating complex three-dimensional vascular structures consisting of endothelial cells and smooth muscle cells with hematopoietic cells appearing in the lumen of these tubes.[5–7]

[1] T. Asahara, T. Murohara, A. Sullivan, M. Silver, R. van der Zee, T. Li, B. Witzenbichler, G. Schatteman, and J. M. Isner, Isolation of putative progenitor endothelial cells for angiogenesis, *Science* **275**, 964–967 (1997).

[2] T. P. Yamaguchi, D. J. Dumont, R. A. Conlon, M. L. Breitman, and J. Rossant, flk-1, an flt-related receptor tyrosine kinase is an early marker for endothelial cell precursors, *Development* **118**, 489–498 (1993).

[3] N. Kabrun, H. J. Buhring, K. Choi, A. Ullrich, W. Risau, and G. Keller, Flk-1 expression defines a population of early embryonic hematopoietic precursors, *Development* **124**, 2039–2048 (1997).

[4] S. T. Fraser, M. Ogawa, R. T. Yu, S. Nishikawa, M. C. Yoder, and S.-I. Nishikawa, Definitive hematopoietic commitment within the embryonic VE-cadherin$^+$ population, *Exp. Hematol.* **30**, 1070–1078 (2002).

[5] M. Hirashima, H. Kataoka, S. Nishikawa, and N. Matsuyoshi, Maturation of embryonic stem cells into endothelial cells in an in vitro model of vasculogenesis, *Blood* **93**, 1253–1263 (1999).

[6] J. Yamashita, H. Itoh, M. Hirashima, M. Ogawa, S. Nishikawa, T. Yurugi, M. Naito, K. Nakao, and S. Nishikawa, Flk1-positive cells derived from embryonic stem cells serve as vascular progenitors, *Nature* **408**, 92–96 (2000).

[7] S. I. Nishikawa, S. Nishikawa, M. Hirashima, N. Matsuyoshi, and H. Kodama, Progressive lineage analysis by cell sorting and culture identifies FLK1 + VE-cadherin$^+$ cells at a diverging point of endothelial and hemopoietic lineages, *Development* **125**, 1747–1757 (1998).

Differentiation of Mesoderm and Mesodermal Derivatives from ES Cells

In Vitro *Generation of Lateral Plate Mesoderm*

ES cells have been used to generate various cell types by allowing differentiation into clusters termed embryoid bodies.[8–16] To improve the uniformity of culture conditions, we have devised a system which allows the differentiation of ES cells into mesodermal cells without embryoid body formation.[5] In our previously described system, we cultured ES cells as a monolayer on collagen type IV-coated dishes in the absence of LIF for 4 days.[5] At this stage, presumptive Flk1$^+$ LPM cells were isolated using flow cytometry to enrich vascular and hematopoietic precursors. Flk1$^+$ cells delaminate as part of the mesoderm during embryogenesis. During this developmental stage, upregulation of Flk1 occurs as E-cadherin which stabilizes epithelial structures is downregulated.[17] In devising our strategy to isolate vascular and hematopoietic precursors from ES we found that Flk1$^+$ E-cadherin$^-$ cells serve this function well. In contrast to the complex nature of embryoid bodies, our system allow for easy monitoring or manipulation, such as addition of doxycycline for conditional gene ablation or activation. This system is also amenable for the analysis of gene expression during differentiation.

[8] W. Risau, H. Sariola, H. G. Zerwes, J. Sasse, P. Ekblom, R. Kemler, and T. Doetschman, Vasculogenesis and angiogenesis in embryonic-stem-cell-derived embryoid bodies, *Development* **102**, 471–478 (1988).

[9] D. Vittet, M. H. Prandini, R. Berthier, A. Schweitzer, H. Martin-Sisteron, G. Uzan, and E. Dejana, Embryonic stem cells differentiate in vitro to endothelial cells through successive maturation steps, *Blood* **88**, 3424–3431 (1996).

[10] R. M. Schmitt, E. Bruyns, and H. R. Snodgrass, Hematopoietic development of embryonic stem cells in vitro: cytokine and receptor gene expression, *Genes Dev.* **5**, 728–740 (1991).

[11] M. V. Wiles and G. Keller, Multiple hematopoietic lineages develop from embryonic stem (ES) cells in culture, *Development* **111**, 259–267 (1991).

[12] U. Burkert, T. von Ruden, and E. F. Wagner, Early fetal hematopoietic development from in vitro differentiated embryonic stem cells, *New Biol.* **13**, 698–708 (1991).

[13] G. Keller, M. Kennedy, T. Papayannopoulou, and M. V. Wiles, Hematopoietic commitment during embryonic stem cell differentiation in culture, *Mol. Cell Biol.* **13**, 473–486 (1993).

[14] T. Nakano, H. Kodama, and T. Honjo, Generation of lymphohematopoietic cells from embryonic stem cells in culture, *Science* **265**, 1098–1101 (1994).

[15] T. Nakano, H. Kodama, and T. Honjo, In vitro development of primitive and definitive erythrocytes from different precursors, *Science* **272**, 722–724 (1996).

[16] T. Fujimoto, M. Ogawa, N. Minegishi, H. Yoshida, T. Yokomizo, M. Yamamoto, and S. Nishikawa, Step-wise divergence of primitive and definitive haematopoietic and endothelial cell lineages during embryonic stem cell differentiation, *Genes Cells* **6**, 1113–1127 (2001).

[17] C. A. Burdsal, C. H. Damsky, and R. A. Pedersen, The role of E-cadherin and integrins in mesoderm differentiation and migration at the mammalian primitive streak, *Development* **118**, 829–844 (1993).

Vascular Progenitor Cells Arising from ES-derived Mesodermal Cells

Developing EBs contain endothelial cells often forming blood vessels.[8,9] Endothelial cells can also develop from Flk1$^+$ cells generated as a monolayer rather than as a complex cellular cluster.[5] This culture system is therefore easy to manipulate by the addition of growth factors or ablation of cell adhesion by blocking antibodies. Endothelial cells can form on collagen type IV-coated dishes and as sheets or cord-like structures when cultured on OP9 stromal cells. These cells express a variety of endothelial markers such as vascular endothelial-cadherin (VE-cadherin), PECAM-1 and CD34.[5] When Flk1$^+$ cells were cultured as a monolayer in the presence of platelet derived growth factor (PDGF)-BB, adherent cells expressing a variety of pericyte markers including smooth muscle actin, desmin and smooth muscle tropomyosin developed.[6] These results clearly indicate that Flk1$^+$ cells can give rise to both of the major cell types forming complex vascular structures, namely endothelial cells and pericytes. When pelleted and grown in semi-solid collagen matrix, Flk1$^+$ cells can form clusters from which tubes consisting of endothelial cells creating a luminal space and surrounded by pericytes, which is typical of blood vessels.[6] These findings clearly show that Flk1$^+$ LPM cells derived from ES cells can form the vasculature and function as vascular progenitor cells.

Generation of Hematopoietic Cells from ES Cells

The hematopoietic potential of ES cells either from embryoid bodies or from cocultures on OP9 stroma has been previously demonstrated.[10–16] Primitive erythroid cells are the first lineage to appear during development. We have found that a subset of Flk1$^+$ cells express the transcription factor GATA-1. Flk1$^+$ expressing GATA-1$^+$ cells are committed to the primitive erythroid lineage.[16] We have also shown that endothelial cells developed from Flk1$^+$ cells and distinguished by expression of VE-cadherin can generate definitive hematopoietic cells.[7] VE-cadherin$^+$ precursors have served to generate definitive erythroid, myeloid, megakaryocytic and lymphoid lineages suggesting that they are capable of generating all the major hematopoietic lineages.[7] A detailed description of the generation of hematopoietic cells from ES cells has been published.[18]

Assessing Gene Expression During In Vitro *Differentiation of ES Cells*

The developmental pathways taken by ES cells differentiating into vascular and hematopoietic cell types are becoming increasingly well

[18] S. T. Fraser, M. Ogawa, S. Nishikawa, and S. Nishikawa, Embryonic stem cell differentiation as a model to study hematopoietic and endothelial cell development, *Methods Mol. Biol.* **185**, 71–81 (2002).

characterized. The ability to manipulate this culture system to a fine degree, such as using serum-free culture, as well as the ease in isolating intermediate differentiating cell types, makes this system ideal for assessing gene expression during the development of vascular and hematopoietic cell types. Microarray analysis allows for the large-scale analysis of gene expression at specific points of development. The analysis of the vast amount of data generated by comparing gene expression in a range of cell types from undifferentiated ES through to mature hematopoietic, endothelial or smooth muscle cells presents several problematic issues, of which some are discussed here.

Materials

ES Cells Lines

Currently, numerous ES lines are available to researchers. The potential to generate different mature populations may vary between lines. However, in our experience CCE, D1 and other 129 mouse strain-derived ES lines have all generated FLK1$^+$ cells, endothelial cells and hematopoietic cells. The line chosen must be examined by the researcher for its ability to generate these cell types. The medium chosen to maintain ES cells in an undifferentiated state may also differ between lines.

Reagents

Culture Medium

1. CCE ES maintenance medium: Knockout DMEM (GIBCO, 10829-018) supplemented with MEM nonessential amino acids solution (NEAA) (GIBCO, 11140-050), Sodium Pyruvate (Sigma, USA, cat. no. S8636), L-glutamine (Gibco BRL. cat. no., 25030-081), 5×10^{-5} *M* final concentration 2-mercaptoethanol (2-ME) (Merck, Germany), 50 U/ml penicillin, 50 μg/ml streptomycin (Pen/Strep) and 15% FCS (JRH Biosciences, USA, cat. no. 8G4011).
2. E14 tg2a-derived ES line maintenance medium: Glasgow MEM (Gibco BRL, cat. no. 11710-035), NEAA, Sodium Pyruvate, 2-ME, Pen/Strep, 1% FCS, 10% knockout serum replacement (GIBCO, 10828-028).
3. ES differentiation medium: Alpha modified Eagle's Medium (αMEM) containing 2-ME, Pen/Strep and 10% FCS (Gibco BRL, USA).
4. SFO3 serum-free cell culture medium (Sanko Junyaku, Co. Ltd., Tokyo, Japan) supplemented with 2-ME and Pen/Strep.

5. OP9 stromal cell culture medium: αMEM supplemented with Pen/Strep and 20% FCS (JRH Biosciences, USA, cat. no. 12003-78P).

Isolating FLK1$^+$ cells Derived from In Vitro *Differentiating ES cells*

1. Collagen type-IV coated 6-well plates, Biocoat (Becton Dickinson, USA, cat no. 254428).
2. Cell dissociation buffer (CDB) (Gibco BRL, cat. no. 13150-016).
3. Phosphate buffered saline lacking Ca^{2+} and Mg^{2+} (PBS^-).
4. Normal mouse serum (OEM Co. Ltd., USA, cat. no. 175-85340).
5. Fluorescently labeled monoclonal antibodies (mAbs) recognizing FLK1[18] and E-cadherin with distinct fluorochromes.
6. Hank's buffered saline solution (HBSS) containing 1% bovine serum albumin (HBSS/BSA) and HBSS/BSA containing 5 μg/ml propidium iodide (Sigma Co., USA cat. no. P-4170) (HBSS/BSA + PI).

Differentiation of Endothelial Precursors Isolated from Differentiating ES Cultures

1. Fluorescently conjugated anti-mouse VE-cadherin mAb.[19]
2. HBSS/BSA and HBSS/BSA + PI.
3. Collagen type-IV coated 6-well plates (Biocoat, Becton Dickinson, MA, USA, cat. no. 354438).
4. Differentiation medium.

Generation of Smooth Muscle Cells and Vascular Tubes from ES cells

1. Differentiation medium.
2. Recombinant human PDGF-BB (Gibco BRL, USA).
3. Recombinant human Vascular Endothelial Growth Factor (VEGF) (R&D Systems, USA).
4. Collagen type-IV coated 6-well plates (Biocoat, Becton Dickinson, USA).
5. Type I collagen gel (3 mg/ml) (Nitta gelatin, Niigata, Japan).
6. Anti-smooth muscle actin monoclonal antibody (mAb) (Clone 1A4; Sigma Co., USA).
7. Petri dishes.

[19] H. Kataoka, N. Takakura, S. Nishikawa, K. Tsuchida, H. Kodama, T. Kunisada, W. Risau, T. Kita, and S. I. Nishikawa, Expressions of PDGF receptor alpha, c-Kit and Flk1 genes clustering in mouse chromosome 5 define distinct subsets of nascent mesodermal cells, *Dev. Growth Differ.* **39**, 729–740 (1997).

Primitive and Definitive Hematopoietic Cultures from ES-derived Progenitors

1. Mouse hematopoietic growth factors: Erythropoietin (Epo), Flt3 ligand (Flt3lig), Interleukin-3 (IL-3), Interleukin-7 (IL-7), stem cell factor (SCF) (R&D System, USA).
2. Six-well culture dish containing confluent OP9 stromal cells.[20]
3. Fluorescently conjugated monoclonal antibodies recognize hematopoietic differentiation markers such as CD45, Ter119, Gr-1, Mac-1, c-kit, CD19, B220.[21] In general, conjugated mAbs were purchased from Pharmingen, USA. Anti-embryonic hemoglobin pAb.

Assessing Gene Expression During In Vitro *Differentiation of ES Cells*

mRNA Isolation and Reverse Transcription

1. TRIzol reagent (Invitrogen Corp., CA, USA; cat. no. 15596-018).
2. Prep Micro mRNA purification Kit (Amersham Biosciences, UK; cat. no. 27-9255-01).
3. Superscript Choice System (Invitrogen Corp., CA, USA; cat. no. 18090-019).
4. Reverse transcription primer T7 (Amersham Biosciences; cat. no. 72-0591-01).
5. Superscript II reverse transcriptase (200 units/ul) (Invitrogen; cat. no. 18064-014).
6. T4 DNA polymerase (1–8 units/ul) (Invitrogen; cat. no. 18005-017).
7. BioArray RNA Transcript labeling Kit (Affymetrix, Inc., Santa Clara, CA, USA; cat. no. 900182).
8. RNAeasy spin columns (Qiagen GmbH, Germany; cat. no. 74103).

Microarray Assay

1. Murine genome U74 GeneChip array (Affymetrix).
2. Hybridization solution of 1 *M* NaCl, 10 m*M* Tris (pH 7.6), 0.005% Triton X-100, 50 p*M* control oligonucleotide B2 (5′ bioGTCAAGA TGCTACCGTTCAG3′) (Affymetrix).

[20] N. Matsuyoshi, K. Toda, Y. Horiguchi, T. Tanaka, S. Nakagawa, M. Takeichi, and S. Imamura, In vivo evidence of the critical role of cadherin-5 in murine vascular integrity, *Proc. Assoc. Am. Physicians* **109**, 362–371 (1997).

[21] H. Kodama, M. Nose, Y. Yamaguchi, J. Tsunoda, T. Suda, S. Nishikawa, and S. Nishikawa, In vitro proliferation of primitive hemopoietic stem cells supported by stromal cells: evidence for the presence of a mechanism(s) other than that involving c-kit receptor and its ligand, *J. Exp. Med.* **176**, 351–361 (1992).

3. Control cRNA (Bio B[150 p*M*], Bio C[500 p*M*], Bio D [2.5 n*M*], and Cre X [10 n*M*]) (American Type Tissue Collection, Manassas, VA, and Lofstrand Labs, Gaithersburg, MD, USA).
4. A, B, C murine genome arrays (Affymetrix).
5. Rotisserie hybridization oven (model 320; Affymetrix).
6. Streptavidin phycoerythrin (Molecular Probes, Inc., Eugene, OR, USA).
7. GeneArray Scanner (Hewlett-Packard, Santa Clara, CA, USA).
8. GeneChip 3.1 software (Affymetrix).

Analysis of Gene Expression Data

1. GIN (gene information), .CDF (probe set layouts), .SIF (probe set sequences) library files specific for each Affymetrix oligonucleotide array (Affymetrix).
2. Mouse genome sequence (version 5.3) Ensembl (http://www.ensembl.org/).
3. Blast programme, NCBI (http://www.ncbi.nlm.nih.gov/).

Methods

Induction and Purification of Lateral Plate Mesoderm from ES Cells

1. Split undifferentiated ES cells by brief trypsinization.
2. Add 1×10^4 undifferentiated ES cells to each well of a 6-well collagen-coated dish (see section "Mesoderm Induction").
3. Add 3 ml differentiation medium to each well.
4. Culture is undisturbed for 4 days in 5% CO_2 37°C environment.
5. Harvest differentiated ES cells by aspirating medium, washing cultures with PBS^-, add 3 ml cell dissociation buffer to each well and incubate for 10–20 min until the differentiated cells begin to dislodge from the bottom of the well.
6. Collect and centrifuge harvested cells.
7. Incubate with NMS for 10–20 min on ice to block non-F(Ab) specific interactions.
8. Add fluorescently conjugated rat anti-mouse Flk1 and rat anti-mouse E-cadherin mAbs and incubate on ice for 20 min.
9. Wash cells twice with PBS^-.
10. Resuspend cells in HBSS/BSA/PI.
11. Perform flow cytometric isolation of $Flk1^+$ E-$cadherin^-$ population.

Differentiation and Isolation of ES-derived Endothelial Cells

1. Sort $Flk1^+$ cells as described in section "Induction and Purification of LPM from ES Cells."

2. Add 1×10^4 cells/well of a 6-well culture dish containing confluent OP9 cells.
3. Add 3 ml differentiation medium.
4. Culture undisturbed for 3 days.
5. Harvest cells and prepare for antibody staining as described in section "Induction and Purification of LPM from ES Cells."
6. Incubate cells with fluorescently labeled anti-mouse VE-cadherin mAb for 20 min on ice.
7. Wash and prepare for cell sorting as described in section "Induction and Purification of LPM from ES Cells."
8. Sort VE-cadherin$^+$ cells using flow cytometer.

In Vitro *Generation of Vascular Smooth Muscle Cells*

1. Sort Flk1$^+$ cells as described in section "Induction and Purification of LPM from ES Cells."
2. Culture Flk1$^+$ cells on collagen IV coated dishes for 2–3 days with differentiation medium. For serum-free cultures replace differentiation medium with SFO3 medium supplemented with PDGF-BB (10 ng/ml) (see section "*In Vitro* Differentiation of Smooth Muscle Cells," Notes 1 and 2).
3. Fix and stain dishes with anti-smooth muscle actin Ab (see Fig. 1 and section "*In Vitro* Differentiation of Smooth Muscle Cells," Note 3).

In Vitro *Formation of Vascular Tubes*

An outline of this methodology is presented in Fig. 2.

1. Incubate purified Flk1$^+$ cells on a petri dish to induce the formation of cell aggregates. LPM cells are cultivated at a concentration of

FIG. 1. Generation of smooth muscle actin$^+$ cells following 3 days of culture of Flk1$^+$ cells on collagen type-IV coated dishes in the presence of PDGF-BB.

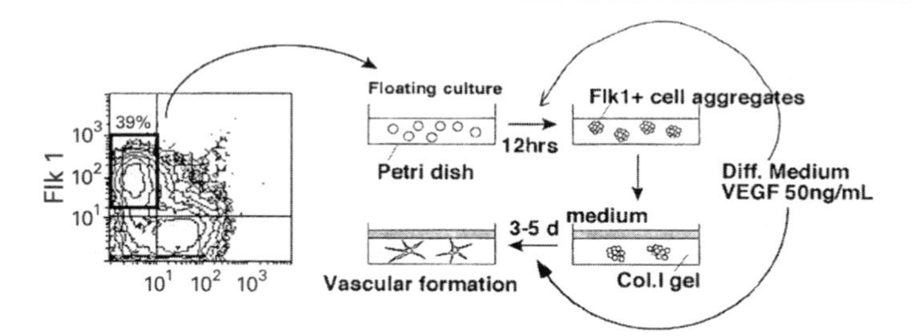

FIG. 2. *In vitro* generation of complex vascular structures from Flk1$^+$ mesodermal cells.

4×10^5 cells/ml in a differentiation medium supplemented with 50 ng/ml of VEGF for 12 hr.

2. Transfer medium including aggregates to a 15-ml Falcon tube with blue tip, the tip of which has been cut, to avoid destruction of aggregates by shear stress. Incubate for 15–30 min.
3. Aspirate supernatant gently leaving the cell pellet in approximately 100 μl of medium.
4. Resuspend aggregates with 2× differentiation medium at a final cell concentration of approximately 5×10^5 cells/ml. Perform following steps on ice.
5. Add an equal volume of Type I collagen gel (3 mg/ml) to cell suspension and mix by pipetting gently and apply to 24 well dish immediately using 300 μl/well.
6. Incubate cultures in CO_2 incubator for 15–30 min to allow gel to polymerize.
7. Add 500–700 μl/well warm differentiation media supplemented with 50 ng/ml VEGF. Volume includes both medium and gel. Change medium every second day.
8. Observe tube formation after 5–7 days (see Fig. 3).

Primitive and Definitive Hematopoietic Cultures from ES-derived Progenitors

Primitive Erythroid Cultures

1. Sort Flk1$^+$ cells as described in section "Induction and Purification of LPM from ES Cells."
2. Culture on confluent OP9 stromal cells in differentiation medium containing 2 U/ml Epo.
3. Harvest hematopoietic cells after 4 days culture and analyse by immunostaining with anti-embryonic Hemoglobin Ab.

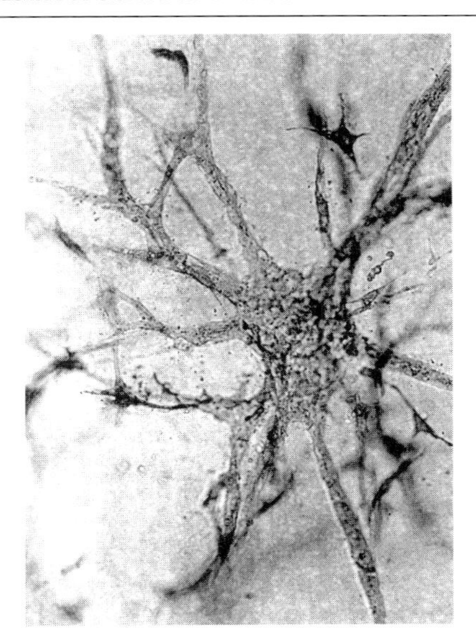

FIG. 3. Smooth muscle actin immunostaining of tube structures generated in collagen gel from Flk1^{+} precursors after 5 days culture.

Definitive Hematopoietic Cultures

1. Sort VE-cadherin^{+} endothelial cells as described in section "Differentiation and Isolation of ES-derived Endothelial Cells."
2. Add endothelial cells to confluent cultures of OP9 stromal cells.
3. Add differentiation medium containing hematopoietic growth factors. To encourage erythroid–myeloid lineages add SCF, Epo and IL-3. B lymphoid cells can be generated by culture with differentiation medium containing SCF, Flt3 ligand and IL-7.
4. Change medium every 3–4 days.
5. Harvest hematopoietic cells and analyse by flow cytometry using fluorescently conjugated mAbs recognizing lineage-specific markers. Erythropoiesis occurs within 3–4 days. Myeloid cells can be detected from 4 to 7 days while B lymphoid cells require approximately 2 weeks of culture.

Assessing Gene Expression During In Vitro *Differentiation of ES Cells*

mRNA Isolation and Reverse Transcription

1. Using differentiated ES cell populations as a source of RNA, prepare total RNA with TRIzol reagent according to the manufacturer's instructions.

2. Prepare Poly(A)+RNA from total RNA using the Quick Prep Micro mRNA purification Kit method according to the manufacturer's instructions.
3. Synthesize double-stranded cDNA using the Superscript Choice System and reverse transcription primer T7 according to the manufacturer's instructions.

Microarray Assay

1. Synthesize biotin-labeled cRNA by *in vitro* transcription using the BioArray RNA Transcript labeling Kit (Affymetrix) according to the manufacturer's instructions.
2. Hybridize the biotin-labeled and fragmented cRNA to the murine genome U74 GeneChip array according to the manufacturer's instructions.
3. Wash the GeneChip arrays using automated fluidics station protocol.
4. Stain with streptavidin phycoerythrin.
5. Wash and Scan with a GeneArray Scanner using the GeneChip 3.1 software.

Analysis of Gene Expression Data

1. Use the Affymetrix GeneChip 3.1 software to create .CHP and .CEL files from the .DAT file obtained from the scan of the array. The .CEL file contains the intensity value for each probe on the chip ordered by their X–Y position, whereas the .CHP file contains a set of aggregate values for each probe set based on the probe pair values. Base subsequent analyses on the .CEL file values as these give access to the individual probe pair intensity values.
2. Map the intensity values in the .CEL file to their specific probe sets using the information provided in the .CDF files (supplied with the chip).
3. Normalize the intensity values obtained from the .CEL file to compensate for the amount of probe loaded on the chip (see "Analysis of Gene Expression Data," Note 1).
4. Enter the normalized probe set intensities into your database cross-referenced to the annotation (see "Analysis of Gene Expression Data," Note 2).
5. To gauge the quality, calculate the ANOVA score for variation across the experimental conditions for the normalized probe pair profiles for each probe set. This score is a good indicator of the covariation of individual probe pair profiles and shows a bimodal distribution indicating reasonable threshold levels. Use this score to

sort genes selected by database queries and to select high-quality data for analysis, such as clustering.

6. Compare the profiles according to Euclidean distances of prior normalized expression profiles.

Notes

Mesoderm Induction

1. The generation of useful quantities of Flk1$^+$ cells is dependent on the seeding concentration of undifferentiated ES cells. While 1×10^4 cells/well is appropriate for CCE ES cells, each distinct ES cell line should be examined for an appropriate seeding density, which should result in confluency but not over-crowding or cell death after 4 days of culture.

In Vitro *Differentiation of Smooth Muscle Cells*

1. If VEGF is absent, Flk1 expression on Flk1$^+$ cells will be downregulated within 12 hr of culture.
2. When Flk1$^+$ cells are cultured on collagen IV only with differentiation medium, >95% of cells express smooth muscle actin (SMA$^+$) with few PECAM-1$^+$ cells. When 50 ng/ml VEGF is added to the culture, approximately 50% become SMA$^+$ cells and 50% become endothelial cells. When Flk1$^+$ cells are cultivated with SFO3 alone, few SMA$^+$ cells are observed. If PDGF-BB is added to the culture, SMA$^+$ cells proliferate and no endothelial cell sheets are observed. VEGF treatment induces predominant appearance of PECAM-1$^+$ EC while simultaneous administration of VEGF and PDGF-BB induce both cell types.
3. Induction of SMA$^+$ cells from Flk1$^+$ cells occurs from 1.5 to 2.5 days after reculture of Flk1$^+$ cells.

Analysis of Gene Expression Data

1. In our experience, various normalization methods which have been suggested to overcome nonlinear effects have yielded no obvious differences to the final analyses. We use a minimum/median normalization based on the mismatch values as these should represent nonspecific hybridization related to the amount of probe present, where all values have been normalized by subtracting the minimum mismatch value followed by division by the median of the mismatch values. We recommend users to try a range of normalization methods.
2. Affymetrix provides annotation for the probe sets on the chip in the .GIN and the .SIF sequence library files supplied with the arrays.

References to Genbank entries are frequently not useful as they indicate entries in Genbank for lacking further annotation or have been retired. Resulting gene loci can be correlated with the predicted gene locations in the Ensembl database.

Gauge of Confidence and Biological Meaning: Each gene on the Affymetrix chips is represented by 16–20 individual probe pairs. The mean difference is assumed to be proportional to specific mRNA concentration. Commonly, a set threshold of mean difference (two- or three-fold) between samples is used to distinguish meaningful differences. As each probe set contains 16 individual probe-pair profiles, there are a variety of ways to compare expression profiles. These include, comparison of probe sets based on the mean of the Euclidean distances of an all against all probe pairs, comparison of probe sets against a user-defined expression profile or comparison based on aggregate expression profiles rather than individual probe-pair profiles.

Acknowledgment

The culture systems presented have been developed by many members of the Department of Molecular Genetics, Kyoto University. We would like to thank Drs. Masanori Hirashima, Tetsuhiro Fujimoto and Takami Yurugi for useful comments and suggestions. This work was supported by grant from the Ministry of Education, Science, Sports and Culture, the Japan Society for the Promotion of Science and the Ministry of Health and Welfare of Japan.

[5] *In Vitro* Differentiation of Mouse Embryonic Stem Cells to Hematopoietic Cells on an OP9 Stromal Cell Monolayer

By KENJI KITAJIMA, MAKOTO TANAKA, JIE ZHENG, EIKO SAKAI-OGAWA, and TORU NAKANO

Introduction

Embryonic stem cells are totipotent cell lines. That is, these cells can participate in normal developmental processes and can differentiate into cells of all lineages, including germ cells, following reintroduction into blastocysts by microinjection. Although the *in vitro* differentiation capacity of the cells is rather limited, as compared to their *in vivo* totipotency, they can be readily induced to differentiate into hematopoietic lineage cells, and constitute a powerful tool for investigating hematopoietic development and differentiation. Embryoid body (EB) formation is the conventional method

0076-6879/2003 $35.00

most frequently used for *in vitro* differentiation from ES cells.[1] However, coculture with some stromal cell lines constitutes a very powerful tool for the induction of differentiation to some specific cell lineages. Induction of differentiation into cells of the lympho-hematopoietic lineage and dopamine neurons can be induced efficiently by coculture with stromal cell lines.[2–5] In this chapter, we describe the method established for induction of differentiation, the OP9 system. The OP9 system is a unique method that uses the macrophage colony stimulating factor (M-CSF)-deficient stromal cell line OP9. The OP9 system can be used for hematopoietic development, hematopoietic differentiation, B cell formation, and megakaryocyte formation, and for analysis of gene function using the tetracycline (Tet)-regulatory gene expression system. In addition, we describe here the merits and demerits of the OP9 system in comparison with the conventional EB formation method.

OP9 Stromal Cells

Stromal cell lines are cells that can support the proliferation and differentiation of hematopoietic cells. The vast majority of the stromal cell lines available are derived from mesenchymal cells of mouse bone marrow. Various stromal cell lines have facilitated the study of hematopoiesis and B lymphopoiesis. However, the proliferation of macrophages interferes with analysis of lympho-hematopoiesis, as the macrophages spread and prevent the observation of other cell lineages. This is probably due to M-CSF, as almost all stromal cell lines established from normal mice produce M-CSF constitutively. To avoid this macrophage proliferation, Kodama *et al.* established a cell line from mutant *op/op* mice.[6] These mice suffer from "osteopetrosis," i.e., ivory bone disease, and their bones lack the bone marrow cavity due to a mutation in the coding region of the M-CSF gene.[7] The *op/op* mice are defective in osteoclast formation, and consequently their bones lack the bone marrow cavity, as M-CSF is an essential factor for the differentiation of osteoclasts. The stromal cell line OP9 was established from

[1] M. V. Wiles, *Methods Enzymol.* **225**, 900–918 (1993).
[2] T. Nakano, H. Kodama, and T. Honjo, *Science* **265**, 1098–1101 (1994).
[3] T. Nakano, *Semin. Immunol.* **7**, 197–203 (1995).
[4] H. Kawasaki, K. Mizuseki, and Y. Sasai, *Methods Mol. Biol.* **185**, 217–227 (2002).
[5] H. Kawasaki, K. Mizuseki, S. Nishikawa, S. Kaneko, Y. Kuwana, S. Nakanishi, S. I. Nishikawa, and Y. Sasai, *Neuron* **28**, 31–40 (2000).
[6] H. Kodama, M. Nose, S. Niida, and S. Nishikawa, *Exp. Hematol.* **22**, 979–984 (1994).
[7] H. Yoshida, S. Hayashi, T. Kunisada, M. Ogawa, S. Nishikawa, H. Okamura, T. Sudo, and L. D. Shultz, *Nature* **345**, 442–444 (1990).

the calvaria of newborn *op/op* mice. This cell line can support differentiation of not only myeloid lineage cells but also B lymphoid lineage cells.

When mouse ES cells are cocultured on stromal cell lines established from normal mice, macrophages emerge almost exclusively and no hematopoietic cells of other lineages can be detected. Therefore, we investigated the use of the stromal cell line OP9. Several groups have reported that M-CSF has some inhibitory effects on the process of hematopoiesis.[8,9] However, it is unclear whether M-CSF has a direct effect on hematopoietic cells or some biologically active factors secreted from macrophages induce the observed effects indirectly. Although we believe that the lack of M-CSF is a very important characteristic of OP9 cells for the efficient induction of differentiation of mouse ES cells, it is not yet clear whether additional factors, such as other unidentified factors or a combination of factors expressed in OP9 cells, are also involved in the observed effects of this cell line.

Maintenance of OP9 Cells

One of the most important points for the induction of differentiation is appropriate maintenance of OP9 stromal cells. As is common for most stromal cell lines, OP9 cells can easily lose the ability to maintain lymphohematopoiesis. Generally, it is necessary to carefully select a serum lot that can maintain OP9 stromal cells well. However, in our experience, most serum lots are suitable for the maintenance. In contrast, the quality of the medium is very important. Medium purchased in solution is not appropriate for maintenance of the cells. Freshly made α-modified Eagle's medium (α-MEM) with ribonucleosides and deoxyribonucleosides (GIBCO/BRL, cat no. 12571-063) produced with ultrapure water (Millipore, Bedford, MA, or equivalent) is required, and the medium should be used within one month of preparation.

OP9 cells can be obtained from our laboratory, various other laboratories worldwide, or ATCC (#CRL-2749FL), on request. The cells require a relatively high concentration of FCS, and the recommended culture medium is α-MEM supplemented with 20% FCS, 1% L-glutamine (GIBCO/BRL), and 1% nonessential amino acids (GIBCO/BRL). Overgrowth of the cells has a deleterious effect for the maintenance and

[8] Y. Mukouyama, T. Hara, M. Xu, K. Tamura, P. J. Donovan, H. Kim, H. Kogo, K. Tsuji, T. Nakahata, and A. Miyajima, *Immunity* **8**, 105–114 (1998).

[9] K. Minehata, Y. S. Mukouyama, T. Sekiguchi, T. Hara, and A. Miyajima, *Blood* **99**, 2360–2368 (2002).

hematopoiesis-supporting activity of the cells. The cells are usually maintained in dishes 10 cm in diameter (Nunc), but of course, the size of the culture is optional. Cultures should be split just before they reach confluence by washing the cells twice with standard Ca/Mg-free phosphate buffered saline (PBS), followed by addition of 1 ml of 0.1% trypsin–EDTA, and incubation in a CO_2 incubator for 5 min. After incubation, the cells are suspended in 8–10 ml of culture medium, and centrifuged for 5 min at 1500 rpm. Sometimes, cell clumps remain, but this is of no consequence. The cells should be seeded from one dish to four dishes and spread evenly. As OP9 cells grow relatively slowly and do not move well, an even distribution at seeding is necessary for optimal maintenance. Six-well plates are used for the induction of differentiation, but the size of the dishes can be varied. For this purpose, cells are split from one 10 cm dish to four 6-well plates. Medium should be changed every 2 days to keep the cells healthy.

Maintenance of Mouse ES Cells

All of the mouse ES cell lines tested showed induction of differentiation on OP9 cells. We found some differences among the ES cell lines in proliferation during the first 5 days of differentiation induction. However, this was of no consequence as the number of cells can be optimized at the initial seeding. ES cell lines should be cultured according to the standard procedure. Recently, we use the E14tg2a line derived from 129/la strain, as the cells can be maintained without feeder cells.[10]

Induction of Differentiation into Multipotential Hematopoietic Progenitors

During ontogeny, hematopoiesis can be divided into two phases. The first of these is primitive hematopoiesis, which occurs in yolk sac blood islands, and produces primitive erythrocytes almost exclusively. Definitive hematopoiesis begins in the fetal liver, and transfers to the bone marrow around the time of birth. The differentiation of various types of hematopoietic cells, including definitive erythrocytes, occurs during definitive hematopoiesis. Coculture of ES cells on OP9 cells can give rise to primitive erythrocytes and definitive hematopoietic progenitors.[2,3]

Mouse ES cells can be maintained in the immature state by the cytokine leukemia inhibitory factor (LIF). Then, the initial step of the induction of differentiation is the removal of LIF from the culture medium. In some

[10] H. Niwa, T. Burdon, I. Chambers, and A. Smith, *Genes Dev.* **12**, 2048–2060 (1998).

cases, embryonic fibroblasts are used for the maintenance of ES cells. It is not necessary to remove the fibroblasts at the induction of differentiation, as the fibroblast cell number is small compared to that of ES cells, and the cells are mitotically inactive.

A confluent OP9 cell layer is used for the induction of differentiation. The cells can be used until 5–6 days after reaching confluence. However, the medium must be changed every 2 days for maintenance. ES cells under logarithmic growth should be used for induction. For the induction of differentiation, culture medium used for the maintenance of OP9 cells is used. No particular cytokines are necessary for basic induction of differentiation. Single-cell suspensions of ES cells in the differentiation induction media are obtained at the passage of ES cells. Aliquots of 10^4 ES cells are seeded into each well of 6-well plates. The number of cells can be varied to some extent, depending on the cell line used.

The most critical period for the induction of differentiation is the first 5 days. On day 5, two types of the colonies can be observed. One is the colony resembling immature ES cells and the other is the colony differentiating into mesoderm. The more the mesodermal colonies appear, the better the induction (Fig. 1). The initial induction is perfect when the percentage of mesodermal colonies is over 90%. At least five times more mesodermal colonies than undifferentiated colonies are required for good differentiation induction. Another index for good induction of differentiation during the first 5 days is the proliferation of the cells; more than 10^6 cells can be harvested when 10^4 ES cells are seeded, and proper induction of differentiation occurs.

On day 5, the cells with hematopoietic potential emerge in mesodermal colonies. However, the frequency of these cells is not high, with only 1/500–1/1000 cells possessing hematopoietic activity. To select these hematopoietic cells, trypsinization and reseeding are necessary on day 5. On day 5 of induction, the induced cells are washed twice with PBS, and 1 ml of 0.25% trypsin–EDTA is added. After a 5-min incubation in a CO_2 incubator, 5 ml of culture medium is added, and the induced cells are harvested with the stromal cells. Vigorous pipetting, similar to that required to prepare a single-cell suspension of ES cells, is necessary, because OP9 cells tend to produce large clumps. Usually, visible cell clumps remain in suspension. To remove the clumps, the tubes should be allowed to stand at room temperature for 1 min. Only the supernatant containing single cells is harvested, and centrifuged at 1500 rpm for 5 min. Then, the cells are reseeded onto freshly prepared OP9 cells. Usually, 10^5 day-5 induced cells are plated into each well of 6-well plates. Seeding of more cells is not recommended. However, fewer cells can be seeded for some purposes, such as clonal analysis.

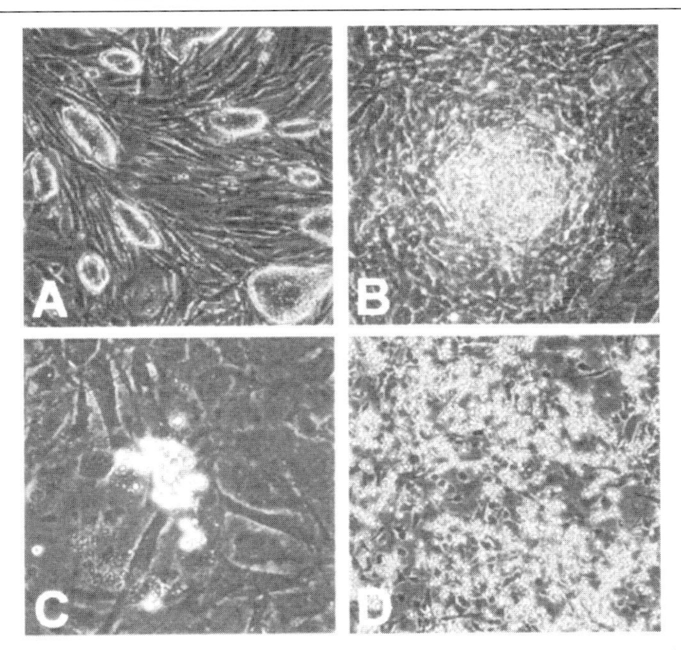

FIG. 1. Morphological changes during induction of differentiation by OP9. (A) ES cell colonies; (B) day 5 mesodermal colonies; (C) day 8 hematopoietic cell cluster; (D) day 14 differentiated blood cells.

After passage on day 5, small and round cell clusters of hematopoietic cells appear. These clusters are of clonal origin, and can give rise to various types of blood cells, such as erythrocytes, macrophages, granulocytes, megakaryocytes, mast cells, and B cells. Adherent cells of unknown origin appear on some occasions. These cells are disadvantageous for the maintenance of hematopoietic cells, because they cover the OP9 layer and inhibit hematopoiesis. To avoid this deleterious effect, the induced hematopoietic cells should be passaged again between days 8 and 10. At this second passage, no trypsinization is required. Hematopoietic cells on or beneath the stromal cells are harvested by vigorous pipetting. Then, the cells are seeded onto a fresh OP9 cell layer.

The cells continue to grow until day 14 of induction. Around this time, the vast majority of the blood cells detach from OP9 cells, and only a small number of cells, such as macrophages, mast cells and pre-B cells, remain. Induction of differentiation by OP9 does not give rise to self-renewing hematopoietic stem cells, as in the EB formation method. Notably, the induction of differentiation occurs only once, and the differentiation proceeds in a highly synchronous manner.

Differentiation into Primitive Erythrocytes

During mouse embryogenesis, the production of "primitive" erythrocytes (EryP) precedes the production of "definitive" erythrocytes (EryD) in parallel with transition of the site of hematopoiesis from the yolk sac to the fetal liver. In the process of induction of differentiation in the OP9 system, EryP and EryD emergence occur with time courses similar to those seen during ontogeny. Induction of differentiation should be carried out as described in the above section.[11] To obtain maximum numbers of primitive erythrocytes, erythropoietin (EPO) should be added from day 3 of the induction of differentiation (1 U/ml). If EPO is not added, the vast majority of EryP undergo apoptotic cell death on day 7 of induction. The time courses of the emergence of primitive and definitive hematopoiesis in the induction of differentiation by OP9 are very similar to those seen during mouse ontogeny.

Induction of Differentiation to Megakaryocytes, Osteoclasts, and B Lineage Cells

Induction of differentiation into megakaryocytes and B lymphoid lineage cells can be achieved using the OP9 system. In contrast, it is difficult to induce differentiation of these cells with the EB formation method. In the induction of megakaryocyte differentiation, not only megakaryocytes but also platelet shedding can be observed.[12,13] Although the addition of TPO alone can induce efficient megakaryocyte differentiation, the optimal conditions for induction of megakaryocytic differentiation are reported by Eto *et al.*, in Chapter 10 of this volume. This method will facilitate elucidation of the functions of various genes in megakaryocytes and platelets. Osteoclasts are the progeny of hematopoietic stem cells. The development of these cells *in vitro* can also be achieved using OP9 cells,[14] as described by Hayashi (Tsuneto *et al.*) in Chapter 7 of this volume.

Addition of Flt-3 ligand during the induction of differentiation facilitates the development of B lineage cells.[15] B lineage cells generated *in vitro* from ES cells are functionally analogous to normal fetal liver-

[11] T. Nakano, H. Kodama, and T. Honjo, *Science* **272**, 722–724 (1996).

[12] T. Era, T. Takahashi, K. Sakai, K. Kawamura, and T. Nakano, *Blood* **89**, 1207–1213 (1997).

[13] K. Eto, R. Murphy, S. W. Kerrigan, A. Bertoni, H. Stuhlmann, T. Nakano, A. D. Leavitt, and S. J. Shattil, *Proc. Natl. Acad. Sci. USA* **99**, 12819–12824 (2002).

[14] T. Yamane, T. Kunisada, H. Yamazaki, T. Era, T. Nakano, and S. I. Hayashi, *Blood* **90**, 3516–3523 (1997).

[15] S. K. Cho, T. D. Webber, J. R. Carlyle, T. Nakano, S. M. Lewis, and J. C. Zuniga-Pflucker, *Proc. Natl. Acad. Sci. USA* **96**, 9797–9802 (1999).

derived, or bone marrow-derived, B lineage cells at three important developmental stages: first, they respond to Flt-3 ligand during the early lymphopoietic progenitor stage; second, they become targets for Abelson murine leukemia virus (A-MuLV) infection at the pre-B cell stage; third, they secrete Ig upon stimulation with lipopolysaccharide at the mature, mitogen-responsive stage. A detailed protocol for induction of differentiation to B lineage cells is presented in Chapter 11 by Zúñicka-Pflücker (Cho *et al.*) in this volume.

Lineage-specific Gene Expression Analysis Using OP9 Cells

The continuous generation of mature blood cells from hematopoietic progenitor cells requires a highly complex series of molecular events. To examine lineage-specific gene expression during the differentiation process, we developed a novel method that combines LacZ reporter gene analysis with the OP9 system.[16] As a model system using this method, we chose the erythroid and megakaryocytic differentiation pathways. Although erythroid and megakaryocytic cells possess distinct functional and morphological features, these two lineages originate from bipotential erythro-megakaryocytic progenitors, and share common lineage-restricted transcription factors. A portion of the 5′ flanking region of the human glycoprotein IIb (αIIb) integrin gene, extending from base −598 to base +33, was examined in detail.

First, stable transformants are produced by electroporation with the reporter gene and neomycin resistance gene at a molar ratio of 20 : 1, followed by G418 selection. Preliminary experiments should be performed to determine how many electroporated cells should be seeded onto each culture plate. As pools of 50–100 G418 resistant colonies are used for analysis, 50–100 G418-resistant ES cell colonies should develop in each culture plate. Stably transformed ES cell colonies in each culture dish are harvested altogether, and aliquots of 10^4 of the cells are subjected to induction of differentiation as described above. We confirmed that the results of such pool analysis are consistent with the mean results of clonal analysis. To analyze the gene expression in individual lineages, appropriate cytokines are added from day 3–5 of induction. EPO (1 U/ml), TPO (20 ng/ml), and M-CSF (200 ng/ml) are added for analysis of erythroid, megakaryocyte, and macrophage lineages, respectively. EryP, EryD, megakaryocytes, and macrophages are analyzed on day 8, 12, 10, and 9,

[16] T. Era, T. Takagi, T. Takahashi, J. C. Bories, and T. Nakano, *Blood* **95**, 870–878 (2000).

respectively. The percentages of LacZ-positive cells are correlated well with the promoter activity.

Our data clearly showed that an approximately 200-bp enhancer region, extending from position −598 to −400, was sufficient for megakaryocyte-specific gene expression. This experimental system has advantages over those using erythro-megakaryocytic cell lines because the OP9 system corresponds closely to normal hematopoietic cell development and differentiation. Furthermore, this system is more efficient than transgenic analysis, and can be used to easily examine gene expression with null mutations of specific genes. We are continuing to improve this method by using EGFP instead of LacZ (data unpublished).[16] EGFP is very useful as its expression can be examined with monoclonal antibodies against lineage-specific surface markers. Furthermore, bacterial artificial chromosome (BAC) constructs spanning the ~100 kb promoter region can be analyzed similarly. For analysis of BAC clones, Lipofectamine 2000 (GIBCO/BRL) gave the best results for gene introduction, which should be carried out in accordance with the manufacturer's recommendations. The gene introduction efficiency of BAC clones is much lower than that of small plasmids. In this instance, clonal analysis is carried out instead of pool analysis in the BAC clone experiment.

Conditional Gene Expression by the Tet System During the Induction of Differentiation

To investigate the effects of various genes during hematopoietic differentiation, a method consisting of a combination of the OP9 system and tetracycline (Tet)-regulated expression has been developed.[17–19] To avoid the toxicity of tetracycline transactivator (tTA), a modified version of the transactivator was developed. This Tet-OP9 system constitutes a very powerful tool for analysis of genes of interest. To analyze the functions of genes involved in hematopoiesis, it is theoretically possible for induction of differentiation by OP9 to be carried out on the ES cell clones expressing the gene. However, in many cases, important functional genes show various effects in ES cells. Thus, it is often very difficult to analyze the gene function using the constitutively active promoter.

Using this method, Era and Witte analyzed the function of the *Bcr-abl* gene, which is a fusion gene responsible for human chronic myeloid

[17] M. Gossen and H. Bujard, *Proc. Natl. Acad. Sci. USA* **89**, 5547–5551 (1992).

[18] T. Era, S. Wong, and O. N. Witte, *Methods Mol. Biol.* **185**, 83–95 (2002).

[19] T. Era and O. N. Witte, *Proc. Natl. Acad. Sci. USA* **97**, 1737–1742 (2000).

leukemia.[19] We analyzed the function of GATA-2, an immature hematopoietic cell-specific zinc finger type transcription factor, and compared its functions with those of its fusion protein with the estrogen receptor ligand-binding domain, GATA-2/ER.[20] The results of these experiments clearly demonstrate that the function of GATA-2 is the opposite of that of GATA-2/ER. This Tet-OP9 system will facilitate elucidation of the functions of many genes involved in hematopoietic differentiation.

OP9 System vs the EB Method

Technically, the OP9 system is more complicated than the EB formation method. However, the OP9 system has several benefits, as follows: (1) During the induction of differentiation by OP9, the process of hematopoiesis can be observed easily. EB is essentially a "black box," and it is difficult to observe what is going on inside. (2) EB formation can induce hematopoietic progenitors, while the OP9 system can induce not only hematopoietic progenitors but also fully mature blood cells. (3) The OP9 system follows the same time course as hematopoiesis during development *in vivo*. (4) The OP9 system can induce efficient differentiation to megakaryocytes, while the EB method cannot. (5) The OP9 system can induce efficient differentiation to B lineage cells, while the EB method cannot. (6) It is difficult to produce EB from some ES cell lines, and consequently the EB formation method cannot be used with these cells. However, the OP9 system is applicable to all ES cell lines examined to date. On the other hand, serum-free culture conditions cannot be used in the OP9 system, while EB formation can be achieved under serum-free conditions.

Generally, the OP9 system and the EB method provide similar results in the induction of differentiation of double-knockout ES cells. However, in one case, flk-1 knockout ES cells, the two results showed a marked discrepancy.[21] Genetic studies in mice demonstrated an intrinsic requirement for the VEGF receptor Flk-1 in the early development of both the hematopoietic and endothelial cell lineages. ES cells homozygous for a targeted null mutation in flk-1 (flk-1$^{-/-}$) were examined for their hematopoietic potential *in vitro* during EB formation or when cultured on monolayers of the stromal cell line OP9. Surprisingly, in EB cultures flk-1$^{-/-}$ ES cells were able to differentiate into all myeloid–erythroid

[20] K. Kitajima, M. Masuhara, T. Era, T. Enver, and T. Nakano, *EMBO J.* **21**, 3060–3069 (2002).

[21] M. Hidaka, W. L. Stanford, and A. Bernstein, *Proc. Natl. Acad. Sci. USA* **96**, 7370–7375 (1999).

lineages, albeit at half the frequency of heterozygous lines. In contrast, although flk-1$^{-/-}$ ES cells formed mesodermal-like colonies on OP9 monolayers, they failed to generate hematopoietic clusters, even in the presence of exogenous cytokines. However, flk-1$^{-/-}$ OP9 cultures did contain myeloid precursors, albeit at greatly reduced percentages. This defect was rescued by first allowing flk-1$^{-/-}$ ES cells to differentiate into EB, and then passaging these cells onto the OP9 stromal cell monolayer. Thus, the *in vivo* phenotype of flk-1$^{-/-}$ is similar to the *in vitro* phenotype of the OP9 system, rather than to EB formation. The same group also reported that the OP9 system was useful for gene trap experiments.[22,23]

It has long been controversial whether EryP and EryD are derived from common progenitors.[11] The OP9 system gives rise to EryP and EryD sequentially, with a time course similar to that seen in murine ontogeny. From the different growth factor requirements and the results of limiting dilution analysis of precursor frequencies, we concluded that EryP and EryD developed from different precursors *via* distinct differentiation pathways.

In contrast, a study using the EB formation method drew contradictory conclusions.[24] They reported that primitive erythrocytes and other hematopoietic lineages arise from a common multipotential precursor that develops within EB generated from differentiated ES cells. In response to vascular endothelial growth factor (VEGF) and c-kit ligand, these precursors give rise to colonies containing immature cells (blasts) expressing marker genes characteristic of hematopoietic precursors. They demonstrated by kinetic study that the blast colony-forming cells represent a transient population, preceding the establishment of the primitive erythroid and other lineage-restricted precursors.

Keller *et al.* suggested that this discrepancy was because our results were obtained at a later stage of the induction of differentiation. However, this surmise is incorrect because we carefully examined the time course of the appearance of the progenitor cells. We propose that Keller's transient blast colony-forming cells were not common "hematopoietic" progenitors, but were common "endothelial-hematopoietic" progenitors, i.e., so-called "hemangioblasts." In fact, they themselves demonstrated that blast colony-forming cells have the potential to differentiate into both endothelial cells and blood cells. Taken together, these observations suggest that there

[22] M. Hidaka, G. Caruana, W. L. Stanford, M. Sam, P. H. Correll, and A. Bernstein, *Mech. Dev.* **90**, 3–15 (2000).

[23] W. L. Stanford, G. Caruana, K. A. Vallis, M. Inamdar, M. Hidaka, V. L. Bautch, and A. Bernstein, *Blood* **92**, 4622–4631 (1998).

[24] M. Kennedy, M. Firpo, K. Choi, C. Wall, S. Robertson, N. Kabrun, and G. Keller, *Nature* **386**, 488–493 (1997).

are common progenitors of EryP, EryD, and endothelial cells (i.e., hemangioblasts), but that there are no common "hematopoietic" progenitors, which can differentiate into EryP and EryD but not endothelial cells. A recent study analyzing RUNX-1 function seemed to support this suggestion.[25]

Conclusions

The induction of differentiation by OP9 has several advantages over the conventional EB formation method. However, the two methods, i.e., the OP9 system and EB method, are not mutually exclusive. Kyba *et al.* demonstrated the induction of differentiation from ES cells to transplantable hematopoietic stem cells[26] using a combination of both EB formation and the OP9 system. We are confident that the OP9 system can achieve everything that can be done by the EB method, as well as having some additional capabilities not available using any other method, as discussed above. However, the OP9 system is more difficult to use than the EB method. Thus, it is necessary to determine on a case-by-case basis which method is appropriate for any given application.

[25] G. Lacaud, L. Gore, M. Kennedy, V. Kouskoff, P. Kingsley, C. Hogan, L. Carlsson, N. Speck, J. Palis, and G. Keller, *Blood* **100**, 458–466 (2002).
[26] M. Kyba, R. C. Perlingeiro, and G. Q. Daley, *Cell* **109**, 29–37 (2002).

[6] *In Vitro* Differentiation of Mouse ES Cells: Hematopoietic and Vascular Development

By JOSEPH B. KEARNEY and VICTORIA L. BAUTCH

Introduction

The recent excitement surrounding potential stem-cell therapies has led to increased interest in the ability of embryonic stem (ES) cells to differentiate into various cell types *in vitro*. Mouse ES cells have been differentiated *in vitro* since the mid-1980s,[1,2] and vasculogenesis and

[1] T. C. Doetschman, H. Eistetter, M. Katz, W. Schmidt, and R. Kemler, *J. Embryol. Exp. Morph.* **87**, 27 (1985).
[2] W. Risau, H. Sariola, H. G. Zerwes, J. Sasse, P. Ekblom, R. Kemler, and T. Doetschman, *Development* **102**, 471 (1988).

0076-6879/2003 $35.00

hematopoiesis were among the earliest developmental processes studied using this model. ES cells are totipotent, or capable of giving rise to all embryonic lineages, when placed back into the embryo. However, when placed in basal medium with added serum, they undergo a programmed differentiation that leads to successive restriction of cell fate, and the elaboration of a limited number of cell types.[3] Most cell types that are prevalent under these conditions are those found in the embryonic yolk sac—visceral endoderm, vascular endothelial cells, and hematopoietic cells such as erythrocytes and macrophages.[4–8] Additionally, hematopoietic precursors of other lineages are present and can be scored by secondary plating for colony identification. Cardiac mesoderm is not normally found in the yolk sac, but it also forms and begins to beat in individual areas. Moreover, the vascular endothelial cells that differentiate during ES cell differentiation form primitive blood vessels, analogous to the first vessels that form in the yolk sac.

Vascular development during mouse ES cell differentiation involves two processes also used *in vivo*: vasculogenesis, the assembly of mesodermal precursor cells called angioblasts and their differentiation into endothelial cells, and angiogenesis, the expansion of vessels by migration and cell division of endothelial cells in vessels.[9,10] Primitive erythrocytes, and later embryonic macrophages, initially mature in ES cell-derived blood islands surrounded by endothelial cells. At later times the embryonic macrophages extravasate from the blood islands into the surrounding nonvascular area, as do embryonic macrophages *in vivo*.[11] Thus this model is ideal for molecular, cellular, and pharmacological studies of early vascular and hematopoietic development. Because vascular and hematopoietic development occurs in the context of other cell types that are normally associated with these developmental processes in the embryo, we feel that it faithfully recapitulates blood vessel formation and hematopoiesis *in vivo*, but this complexity provides challenges to analysis. Thus we have focused efforts on developing protocols for *in situ* localization of vascular and hematopoietic

[3] G. M. Keller, *Curr. Opin. Cell Biol.* **7**, 862 (1995).
[4] M. V. Wiles and G. Keller, *Development* **111**, 259 (1991).
[5] R. M. Schmitt, E. Bruyns, and H. R. Snodgrass, *Genes Dev.* **5**, 728 (1991).
[6] R. Wang, R. Clark, and V. L. Bautch, *Development* **114**, 303 (1992).
[7] A. Leahy, J.-W. Xiong, F. Kuhnert, and H. Stuhlmann, *J. Exp. Zool.* **284**, 67 (1999).
[8] G. Keller, M. Kennedy, T. Papayannopoulou, and M. Wiles, *Mol. Cell. Biol.* **13**, 473 (1993).
[9] T. J. Poole and D. Coffin, *J. Exp. Zool.* **251**, 224 (1989).
[10] W. Risau, *Nature* **386**, 671 (1997).
[11] M. Inamdar, T. Koch, R. Rapoport, J. T. Dixon, J. A. Probolus, E. Cram, and V. L. Bautch, *Dev. Dyn.* **210**, 487 (1997).

cells in this model, and ways to accurately quantitate vascular development in this model. These assays are described in this chapter.

Another feature of ES cell differentiation is the ability to analyze mutant ES cells that lack specific genes. Since all mouse "knock-out" mutational analysis *in vivo* requires gene targeting in ES cells to eliminate one copy of the gene, there is a ready-made source of thousands of specific mutations that can potentially be analyzed *in vitro*. Protocols are available for generating homozygosity at a targeted locus by incubating ES cells in higher concentrations of drug to select for cross-overs,[12] and these protocols work reasonably well in our hands. For example, we used the ES cell model to analyze the role of several mutations in the VEGF signaling pathway.[13,14] Moreover, we and others[15–17] have developed approaches for random gene targeting in ES cells using gene trap vectors, and several interesting mutations have been isolated using this approach. Protocols for chemical mutagenesis of ES cells have also been developed.[18] Recent reports describing the construction of deletion chromosomes and mitotic cross-overs in ES cells[19–21] suggest the exciting possibility that soon we may have genetic manipulability approaching that of *Drosophila* in the ES cell model. In any case, these genetic tools and advances make it imperative to develop analytical cell biological tools applicable to ES cell differentiation.

Finally, it occurred to us several years ago that vascular development during ES cell differentiation was uniquely suited for the new imaging technologies being developed around green fluorescent protein (GFP) and a new generation of confocal microscopes. Success has been achieved in imaging developmental processes in optically transparent organisms such as *C. elegans* and zebrafish, but imaging mouse embryos is difficult for several reasons. However, when partially differentiated ES cell-derived embryoid bodies (EBs) are attached to a permissive surface, they adhere and spread

[12] R. M. Mortensen, D. A. Conner, S. Chao, A. A. Geister-Lowrance, and J. G. Seidman, *Mol. Cell. Biol.* **12**, 2391 (1992).

[13] V. L. Bautch, S. D. Redick, A. Scalia, M. Harmaty, P. Carmeliet, and R. Rapoport, *Blood* **95**, 1979 (2000).

[14] J. B. Kearney, C. A. Ambler, K. A. Monaco, N. Johnson, R. G. Rapoport, and V. L. Bautch, *Blood* **99**, 2397 (2002).

[15] W. L. Stanford, G. Caruana, K. A. Vallis, M. Inamdar, M. Hidaka, V. L. Bautch, and A. Bernstein, *Blood* **92**, 4622 (1998).

[16] P. Friedrich and P. Soriano, *Genes Dev.* **5**, 1513 (1991).

[17] J.-W. Xiong, R. Battaglino, A. Leahy, and H. Stuhlmann, *Dev. Dyn.* **212**, 181 (1998).

[18] Y. Chen, J. Schimenti, and T. Magnuson, *Mamm. Genome* **11**, 598 (2000).

[19] Y. Yu and A. Bradley, *Nat. Genet. Rev.* **2**, 780 (2001).

[20] L. Lefebvre, N. Dionne, J. Karaskova, J. A. Squire, and A. Nagy, *Nat. Genet.* **27**, 257 (2001).

[21] P. Liu, N. A. Jenkins, and N. G. Copeland, *Nat. Genet.* **30**, 66 (2001).

while continuing their program of differentiation.[22] Thus a primitive vascular plexus develops in a structure that physically resembles a "pancake" but nevertheless constitutes a valid biological context. The relative two-dimensionality of this culture model allows us to actually watch the process of mammalian vascular development! It also provides approaches to further understanding processes such as vessel assembly and vessel expansion. Thus we conclude this chapter with a set of protocols that we have developed for real-time imaging of mammalian vascular development.

ES Cell Maintenance

Since numerous protocols for ES cell maintenance have been published, we focus here on the unique aspects of maintaining ES cells for *in vitro* differentiation. ES cells are traditionally maintained on a feeder layer of mouse embryo fibroblasts or STO cells to provide signals that prevent differentiation, chief among them being LIF (Leukemia Inhibitory Factor). ES cells must be "weaned" off of feeder layers prior to initiating *in vitro* differentiation or they will remain as ES cells. This is done by removing the feeder layer and providing LIF from another source. Sources of LIF include: (1) recombinant LIF that is commercially available (i.e., StemCell Technologies, Sigma-Aldrich); (2) transient transfection of COS cells with LIF-expressing plasmids and use of the medium to supplement ES cell growth medium; and (3) harvesting medium conditioned by the 5637 human bladder cancer cell line (ATCC #HTB9) and using it as a supplement in ES cell growth medium. We prefer the 5637 cell conditioned medium because in our hands it preserves the undifferentiated morphology of ES cells better than the first two options. Perhaps additional factors in the conditioned medium also act in concert with LIF to prevent differentiation.

In any case, ES cell colonies that develop on tissue culture plastic with supplements generally look flatter than colonies grown on feeder layers, and they have cells around the edges that have begun to differentiate. However, if properly cared for they provide reproducible *in vitro* differentiation from 1 to 2 passages off of feeder layers until they reach passages 40–50, when they are discarded. Alternatively, some protocols call for maintenance on feeder layers, with weaning off of feeder layers for 1–2 passages just prior to differentiation. We find this approach very cumbersome for routine and constant *in vitro* differentiation experiments.

ES cells must also be passed frequently to prevent differentiation, whether on or off feeder layers. We generally pass ES cells every 2–4 days.

[22] V. L. Bautch, W. L. Stanford, R. Rapoport, S. Russell, R. S. Byrum, and T. A. Futch, *Dev. Dyn.* **205**, 1 (1996).

Standard protocols for trypsinization and passage are followed.[23] ES cells are maintained in 17% fetal calf serum (FCS), and it is important to use serum that has been screened on ES cells. Prescreened serum is commercially available (i.e., Invitrogen-Gibco, Hyclone, StemCell Technologies), but we have had good success screening serum lots in our own laboratory. In general, 50–75% of lots approved for general tissue culture use are suitable for ES cell maintenance.

In Vitro Differentiation

There are numerous protocols for ES cell differentiation, some of which are described in this volume. Many protocols are designed to take advantage of the initial totipotency of ES cells, and they provide supplements and specific factors that promote the differentiation of a specific embryological lineage, such as neuronal precursor cells.[24,25] However, as described in the Introduction, the conditions that we and others use initiates a programmed differentiation that leads to the formation of several cell types typically found in the mouse yolk sac. Since this approach allows for differentiation of a primitive vasculature and primitive hematopoietic cells in a biological context, it is a useful model to study developmental processes.

There are several ways to prepare ES cells for programmed *in vitro* differentiation. The important parameters are that: (1) ES cells be set up in small groups of associated cells; and (2) they be prevented from attaching to a surface until they have established cell–cell contacts and begun a differentiation program. These requirements are met by two protocols that we use routinely—a method using Dispase to provide small clumps of ES cells, and a hanging drop method that promotes reassociation of single ES cells by close association in a small volume of medium (Fig. 1). The Dispase method has the advantage of providing almost unlimited amounts of material for differentiation, which is sometimes needed for biochemical, molecular biological, and cell sorting experiments. The hanging drop method is less amenable to scale-up, but it provides better synchronization of the differentiation program and a simpler start-up procedure. In the following sections, we briefly describe each method, then conclude this section with a section describing common protocols for differentiation once the EBs are generated.

[23] M. L. Roach and J. D. McNeish, *in* "Embryonic Stem Cells: Methods and Protocols" (K. Turksen, ed.), p. 1. Humana Press, Totowa, New Jersey, 2002.

[24] M. Li, *in* "Embryonic Stem Cells: Methods and Protocols" (K. Turksen, ed.), p. 205. Totowa, New Jersey, Humana Press, 2002.

[25] T. Mujtaba and M. S. Rao, *in* "Embryonic Stem Cells: Methods and Protocols" (K. Turksen, ed.), p. 189. Totowa, New Jersey, Humana Press, 2002.

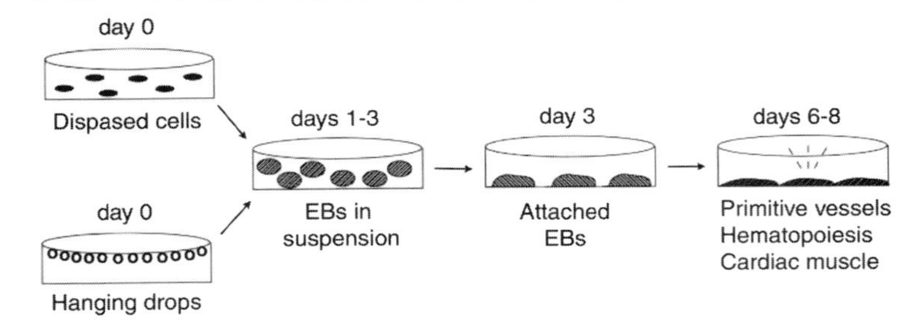

FIG. 1. Scheme for programmed differentiation of ES cells. ES cells are dispersed in suspension by either of two protocols (far left) on day 0. They remain in suspension for 3 days, then are attached to tissue culture plastic dishes or wells. At days 6–8 vascular development and hematopoiesis (erythropoiesis) peak, while macrophage development peaks at days 10–12 (not shown).

Embryoid Body Generation—Method 1 (Dispase)

Detailed protocols for EB generation using Dispase digestion of ES cell colonies have been published by ourselves and others (i.e., Ref. 26). To begin the process, ES cell colonies that are weaned off feeders are allowed to "age" for an additional 5–6 days after the normal day of passage without feeding or disturbance. It is thought that this period begins the differentiation process and promotes the formation of cell–cell junctions that are resistant to Dispase digestion. After this period, the enlarged, partially differentiated colonies are incubated with cold Dispase (Roche Diagnostics) at 1.2 U/ml in PBS for 1–2 min. After rinsing in PBS, the ES cell clumps are resuspended in differentiation medium (described below) in bacteriological dishes. These dishes dramatically reduce the attachment of the ES cell clumps to the bottom of the dish during this period, which is crucial in obtaining a good differentiation.

Over a 2–3 day period the ES cell clumps round up and form EBs that contain endoderm and mesoderm, as well as precursors of both the vascular and hematopoietic lineage called hemangioblasts.[27] In our hands the most reproducible differentiation comes from placing day 3 EBs into tissue culture dishes or wells, although reattachment at any time from day 0 (day of Dispase treatment) to day 4 provides some differentiated vascular and hematopoietic cells (Ref. 28 and V. L. B., unpublished results). After day 4

[26] V. L. Bautch, *in* "Embryonic Stem Cells: Methods and Protocols" (K. Turksen, ed.), p. 117. Totowa, New Jersey, Humana Press, 2002.

[27] K. Choi, M. Kennedy, A. Kazarov, J. C. Papadimitriou, and G. Keller, *Development* **125**, 725 (1998).

[28] S. D. Redick and V. L. Bautch, *Am. J. Pathol.* **154**, 1137 (1999).

the EBs reattach very inefficiently, probably because the cell–cell contacts have sufficient strength and numbers to remain intact. Of course it is possible to continue differentiation in suspension at this point and produce cystic embryoid bodies (CEBs),[1,6] but *in situ* localization in these structures is difficult. We find that the Dispase protocol produces EBs of various sizes, and that if we select EBs of a certain size group for reattachment (usually the mid-sized EBs) the programmed differentiation that ensues is reasonably synchronized and temporally reproducible (J. B. K., and V. L. B., unpublished results).

Embryoid Body Differentiation—Method 2 (Hanging Drop)

The hanging drop method of ES cell differentiation was first described by Wobus *et al.*[29] It basically consists of complete dissociation of slightly aged ES cell colonies (2–4 days after passage), then inoculating small drops of medium with the ES cells. The plate holding the drops is then covered and "flipped" so that the drops are suspended from the top of the dish. They are incubated this way for 2 days to promote cell–cell adhesion, then the plate is flipped again and flooded with medium. The resulting EBs are moved to a new plate and incubated in suspension for an additional day (3 days total) prior to seeding in tissue culture dishes or wells as described above.

To dissociate the ES cells, trypsin–EDTA and mechanical disruption are used. The dissociated cells are rinsed and resuspended in differentiation medium at a concentration of 83,000 cells/ml. This solution is used to place 30 μl drops on a bacteriological dish. Dishes sterilized with ethylene oxide (currently difficult to find) work best in preventing premature attachment of the ES cells to the plate. The flipping is also important to prevent cell attachment to the surface, but it can be difficult to do without running the drops together. Placing the drops 1 cm apart is recommended, and using a single smooth motion when turning the plate helps prevent running. If the plate is held right side up at waist level, then brought to head level and turned with one smooth arcing motion, the drops usually stay in place.

Attachment and Differentiation of Embryoid Bodies

Whether EBs are generated by Dispase digestion or the hanging drop method, they are generally plated into tissue culture dishes or wells on day 3

[29] A. M. Wobus, G. Wallukat, and J. Hescheler, *Differentiation* **48**, 173 (1991).

of differentiation (Fig. 1). When seeded at densities of 10–20 EBs/well of a 24-well dish, they will spread to cover most of the bottom surface by day 8 of differentiation. Even dispersal of EBs prior to attachment is important for good coverage. Differentiation medium consists of DMEM-H, 20% lot-selected FCS, 150 μM monothioglycerol, and 50 μg/ml gentamicin. EBs usually attach within 2–4 hr.

It is critical to change the medium frequently to obtain a good differentiation. With normal seeding and growth, feeding the cultures every 48 hr is sufficient, but we monitor the pH with phenol red indicator, and change medium prior to 48 hr if we note acidity in the medium. This sometimes occurs with very heavy seeding of EBs, or with regular seedings at later time points of differentiation (day 10–12) used to score macrophage development. This regimen is continued until the cultures are fixed for end-point analyses. The phenol red is sometimes omitted, for example when GFP-expressing vessels are imaged using a confocal microscope (see below).

Visualization and Quantitation of Blood Vessels

Fortunately, it is relatively easy to visualize primitive ES cell-derived mouse blood vessels using several markers and protocols. We and others have used *in situ* hybridization of RNA probes to PECAM-1, flk-1 and flt-1 to visualize blood vessels in ES cell cultures (i.e., Ref. 13). However, it is much easier to visualize vessels using antibodies that recognize proteins expressed by mouse endothelial cells. A number of commercially available antibodies provide good and reasonably cell-type selective immunolocalization reagents. We and others have successfully used antibodies to flk-1 (Ly-73, B-D Pharmingen), CD-34 (RAM34, B-D Pharmingen), and VE-cadherin (11D4.1, B-D Pharmingen).[13,28,30] However, our workhorses are antibodies against PECAM-1 (Mec 13.3, B-D Pharmingen) and ICAM-2 (3C4, B-D Pharmingen) (Fig. 2A–B). Several years ago we and others showed that PECAM-1 is expressed on ES cells, a very limited number of hematopoietic precursor cells, and on mouse angioblasts and endothelial cells.[28,31,32] In contrast, we showed that ICAM-2 is only expressed on mouse endothelial cells that have formed a primitive vasculature and a limited number of hematopoietic cells.[13] The hematopoietic cells that are PECAM+ and/or ICAM-2+ are largely confined to the interior of the

[30] D. Vittet, M. H. Prandini, R. Berthier, A. Schweitzer, H. Martin-Sisteron, G. Uzan, and E. Dejana, *Blood* **88**, 3424 (1996).

[31] L. Piali, S. M. Albelda, H. S. Baldwin, P. Hammel, R. H. Ginsler, and B. A. Imhof, *Eur. J. Immunol.* **23**, 2464 (1993).

[32] P. Robson, P. Stein, B. Zhou, R. M. Schultz, and H. S. Baldwin, *Dev. Biol.* **234**, 317 (2001).

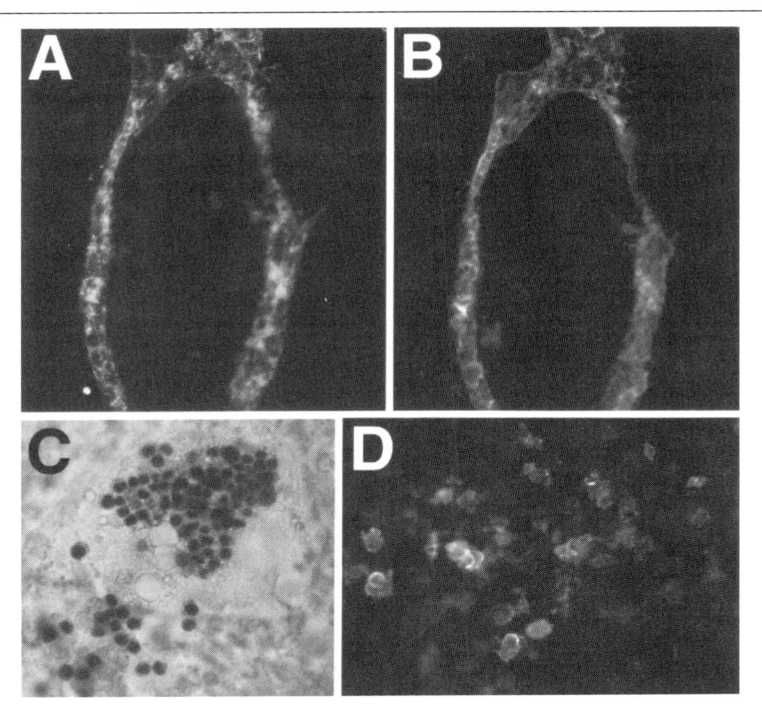

FIG. 2. Vascular and hematopoietic development in ES cell differentiation cultures. ES cells were broken up using Dispase, then differentiated as described after attachment on day 3. (A–B) Day 8 ES cell culture double stained for PECAM-1 (A) and ICAM-2 (B) using different secondary antibodies. Different fluorescent filters were used to visualize each staining pattern. Note the almost complete overlap of the two signals. (C) Day 8 ES cell culture reacted with benzidine to visualize the primitive erythrocytes, seen as dark circles. (D) Day 10 ES cell culture stained with Mac-1 to visualize the embryonic macrophages.

primitive vessels, and they have a very different shape from endothelial cells. Therefore, these markers together constitute a very selective assay for primitive mouse blood vessels and their precursor angioblasts.

We use both of these markers in conjunction with a quantitative imaging technique that we have developed to assay the relative amount of vasculature in attached ES cell differentiation cultures. A detailed protocol has been published.[26] Briefly, cultures are fixed in fresh cold MeOH : acetone (1:1) for 5 min, then rinsed in PBS. Cultures are blocked, then incubated with primary and secondary antibodies according to standard protocols. Rat anti-mouse PECAM-1 is used at a 1:1000 dilution, and rat anti-mouse ICAM-2 is used at a 1 : 500 dilution. Secondary antibodies can be conjugated to enzymes for colorimetric detection or to fluorescent molecules for immunofluorescence. For quantitation we use fluorescent detection, and

B-phycoerythrin (BPE) is linked to the secondary antibody for image quantitation.

Individual wells of a 24-well dish are set up for image analysis. Six to seven nonoverlapping fields are imaged or photographed using low magnification (i.e., 5× to 10× total magnification). It is important that the fields be chosen randomly and have complete cell coverage when viewed with phase contrast optics. This is done by analyzing sequential nonoverlapping fields. Photography is done using black and white film, and these images are then digitized by scanning. Alternatively, digital image capture can be used. Once images are digitized, they are manipulated with Adobe Photoshop and a plug-in package (Image Processing Tool Kit, Rev. 2.1, Reindeer Games, Asheville, NC) to remove areas of obvious background. Background is often associated with the attachment site of the original EB, where the cells remain rounded and more three-dimensional. This diffuse signal is easy to distinguish from the bona fide vascular staining, which is strong and crisp and cell border-associated for both PECAM-1 and ICAM-2. A plug-in is then used to change the images to the binary mode, and the computer can then easily determine the percentage of the area that is stained. It should also be possible to carry out these manipulations using NIH Image.

The percentages derived from nonoverlapping images of a given well are averaged for that well. We then average these percentages from 2 to 4 duplicate wells, and these numbers are used to calculate standard error. We realize that we are quantitating in two dimensions what is essentially a flattened but nevertheless three-dimensional structure. However, the percent stained area should be proportional to the amount of vasculature. Thus these numbers are used in comparative assays—for example, we always compare the percent area stained between wild-type controls and a given mutant background, and the fold difference is used to describe the effect of the mutation. We have also compared this imaging protocol with other quantitative assays such as RNAse protection analysis and FACS analysis,[13] and the different assays all give comparable results. We prefer this assay to FACS for vascular structures because: (1) we cannot distinguish hematopoietic precursors from endothelial cells once they are dissociated, and (2) the PECAM-1 epitope is trypsin-sensitive in our hands, and it is sometimes difficult to obtain good single cell suspensions from ES cell differentiation cultures using collagenase.

Visualization and Quantitation of Hematopoietic Cells

The two hematopoietic lineages that differentiate past the precursor stage during programmed ES cell differentiation are primitive erythrocytes and embryonic macrophages. The protocols and timing for visualizing each of these lineages differ and are described below. It should be noted that when we test lots of FCS in differentiation assays, we find that hematopoietic differentiation is more sensitive to lot variation than vascular development, so it is important to test for these lineages in addition to the vascular lineage.

Primitive Erythrocyte Differentiation

The erythrocytes that form during ES cell differentiation are "primitive," and analogous to the first erythrocytes that form during embryonic development.[1,5] In the mouse, yolk sac derived erythrocytes are nucleated, and they express a different subset of β-globin genes than do mature erythrocytes that are produced by the fetal liver and later the bone marrow. These nucleated erythrocytes begin to accumulate hemoglobin from days 6 to 12 of ES cell differentiation.[22] Thus they are easily visualized by a histochemical stain for benzidine, which reacts with the hemoglobin to produce a colored stain (Fig. 2C). There are several protocols available for benzidine staining. One caveat is that the most sensitive method in our hands[33] does not produce a stable reaction product. Thus the cultures must be imaged in a timely manner, and it is not possible to rigorously quantitate erythrocyte development using the benzidine method. However, visual inspection and comparisons of wells can yield semi-quantitative information. We mix 1 part benzidine stock (3% *p,p′*-diaminobiphenyl {Sigma #B3503} in 90% HAc, 10% H_2O), 1 part fresh 30% H_2O_2, and 5 parts water, then immediately add the mixture to the medium in a well at a 1:10 dilution. The benzidine positive cells turn a deep dark blue within 1–2 min, and the stain is stable for 5–15 min. We rinse quickly in PBS prior to photographing the benzidine stained erythrocytes using standard protocols.

We have not been successful in identifying an antibody that selectively stains all hematopoietic cells (or even erythrocytes), but not vascular cells. Others have successfully used TER-119 in FACS analysis for primitive erythrocytes,[34] but this antibody does not work for immunolocalization in our hands. We have used anti-CD45 to stain a subset of the hematopoietic cells, but mature primitive erythrocytes do not express detectable levels of CD45 (R. Rapoport and V. L. B., unpublished results). However, there are several genes whose promoters are selectively expressed in hematopoietic cells. One example is SCL/Tal, which initially is expressed in all

[33] B. R. O'Brien, *Exp. Cell Res.* **21**, 226 (1960).
[34] M. C. Olson, E. W. Scott, A. A. Hack, G. H. Su, D. G. Tenen, H. Singh, and M. C. Simon, *Immunity* **3**, 703 (1995).

hematovascular cells, but quickly becomes restricted to the hematopoietic lineage.[35] We and others are working to link such promoters to GFP to provide both quantitation and imaging capacity to hematopoietic development in ES cell cultures.

Embryonic Macrophage Differentiation

Embryonic macrophages can be visualized using an antibody to Mac-1 (MI/70, B-D Pharmingen) (Fig. 2D), which allows for quantitative analysis by FACS. Embryonic macrophages are first detectable at days 7–8 of ES cell differentiation, but their numbers peak at days 10–12, in contrast to vascular cells and erythrocytes that peak around day 8 of differentiation. Interestingly, embryonic macrophages initially form inside blood islands surrounded by endothelial cells, then apparently extravasate from the vessels into the surrounding cellular area, as do embryonic macrophages that form in the yolk sac *in vivo*.[11]

The best detection with the Mac-1 antibody involves a modification of the fixation protocol for ES cell differentiation. The cultures are fixed in fresh 4% paraformaldehyde for 6 min, then rinsed in PBS. They are then subjected to incubation with a very dilute trypsin solution (0.005%) for 60 sec. After this preparation, the blocking, staining, and visualization is straightforward. We generally use biotinylated Mac-1 and strepavidin-PE. This provides a strong signal and also allows for double staining with vascular markers such as PECAM-1 or ICAM-2 (see Ref. 11 for details).

One way to quantitate embryonic macrophages is by FACS.[36] Since macrophages are individual migratory cells, they do not have the extensive cell–cell junctions of endothelial cells and are easier to dissociate into single cell suspension (R. Rapoport and V. L. B., unpublished results). ES cultures are dissociated in 0.2% collagenase, then incubated in primary and secondary antibodies as described above. The cell suspensions are then fixed in 1% PFA at 4°C and analyzed with a Becton-Dickenson FACSCAN. We find that wild-type cultures have 3–6% Mac-1 positive cells at days 10–12 of differentiation, and that mutations predicted to impair embryonic macrophage development reduce this percentage to less than 1%.[36]

Time-lapse Imaging to Visualize Vascular Development

Attached ES cell differentiation cultures constitute a mammalian developmental system for imaging that has significantly reduced tissue

[35] C. J. Drake and P. A. Fleming, *Blood* **95**, 1671 (2000).

[36] R. Biggs, K. A. Monaco, C. E. Blickarz, M. S. Inamdar, and V. L. Bautch, *Oncogene*, submitted.

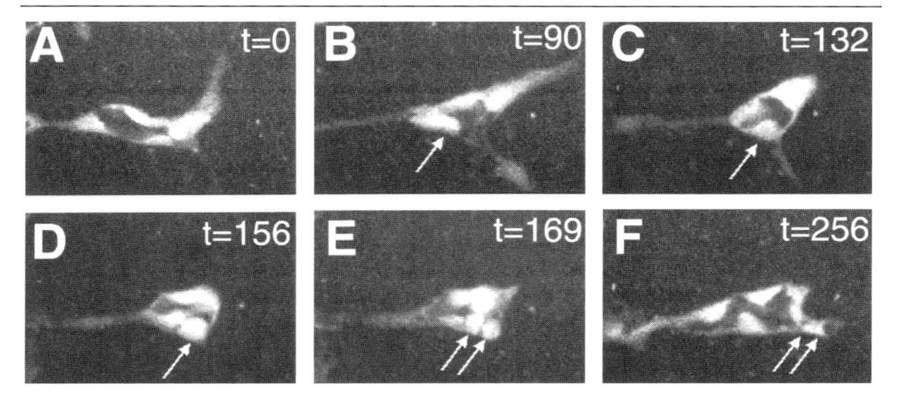

FIG. 3. Time-lapse imaging of ES cell differentiation cultures. An ES cell line stably transfected with a construct in which GFP is expressed using the PECAM promoter was differentiated to day 8. It was imaged on a Perkin Elmer spinning disk confocal microscope. Arrows point to a single cell that migrates to the distal tip of a vessel (B–D), undergoes a cell division (E), and forms two daughter cells that migrate to form a vascular extension (F). Time is in minutes at the top right corner of each panel.

thickness and low auto-fluorescence compared to intact embryos. ES cells can be stably transfected with fluorescent reporter genes by electroporation, and there are several tissue-specific promoters available. The Tie-2 promoter and the PECAM promoter have been successfully used to express fluorescent proteins in angioblasts and endothelial cells during ES cell differentiation (Ref. 37, P. Robson and H. S. Baldwin, personal communication; J. B. K., C. Ellerstrom, and V. L. B., unpublished results).

We have generated stably transfected ES cell lines that exhibit vascular-specific GFP expression during days 6–8 of differentiation using both the PECAM and the Tie-2 promoters. Day 6 ES cultures have GFP positive clumps of cells that resemble vasculogenic aggregates of angioblasts. On days 7–8 the GFP positive cells acquire endothelial cell morphology and exhibit angiogenic growth. For example, they migrate extensively and form sprouts, and they fuse and divide to form a primitive vascular plexus. Figure 3 shows time-lapse confocal microscope images of blood vessel development in differentiated ES cell cultures at day 8 of differentiation.

[37] T. M. Schlaeger, S. Bartunkova, J. A. Lawitts, G. Teichmann, W. Risau, U. Deutsch, and T. N. Sato, *Proc. Natl. Acad. Sci. USA* **94**, 3058 (1997).

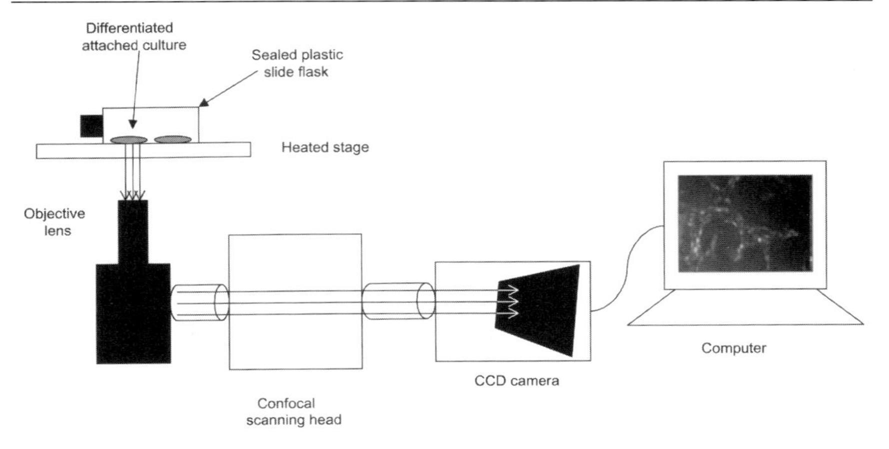

FIG. 4. Schematic representation of a confocal time-lapse imaging system for *in vitro* differentiated ES cultures. A sealed flask containing the culture is placed on a heated stage. The laser light excites the GFP, and the emitted light goes to the objective lens, through the confocal scanning head, and to the CCD camera. The image is projected onto the computer screen and can be saved as a TIFF file. The compilation of images is played as a time-lapse "movie" after collection.

The setup used to collect fluorescent image data from biological samples generally consists of a microscope with appropriate image capture equipment, and equipment to preserve the sample while filming. Figure 4 depicts the setup that we use for visualizing vascular development in differentiated ES cell cultures. A plastic slide flask (Nunc) containing attached ES cell differentiation cultures is first equilibrated in an incubator at 37°C and 5% CO_2, then tightly sealed. The flask is then placed on a heated stage (37°C) attached to the stage of an inverted microscope. Typically, ES cell differentiation cultures set up in this manner can be imaged for up to 12 hr without loss of integrity. Moreover, the same flask can be imaged over multiple days without adverse effects on differentiation or tissue integrity, provided that it is put back in the conventional incubator and unsealed between imaging sessions. Our inverted microscope (Nikon Eclipse TE300) is outfitted with a spinning disk confocal unit (Perkin–Elmer Cetus Ultraview Confocal Scanner) and a CCD camera for digital image capture. The entire apparatus is interfaced with a computer (Dell, Pentium III 600 MHz processor, 20 GB hard drive, 255 MB RAM).

Using this setup to image GFP-expressing vascular cells, we have found that a 2× air objective provides an adequate field of view and allows for single cell resolution. Generally, imaging in a single focal plane with the 20× objective provides enough focal depth to image ES cell-derived blood vessels without loss of information. However, it is important to monitor automated image capture every 30 min, and refocus the sample if areas of interest move out of the plane of focus or the field of view. As an alternative to refocusing, confocal microscopes can be outfitted with a *Z*-axis step controller that images multiple focal planes. In our hands, setting the laser and camera for a 2 sec exposure every 60 sec provides sufficient GFP signal to reconstruct the vascular cell biological events that have occurred. The laser is shuttered between exposures to prevent specimen decay, and we see little or no photobleaching even after 12 hr of image capture. It should be noted, however, that image capture parameters must be optimized for each transgene reporter, confocal microscope setup, CCD camera, and imaging software package that is used.

Confocal microscopy typically reduces background and provides clearer fluorescent images when compared with conventional fluorescence microscopy. Spinning disk, laser scanning, and multi-photon confocal microscope models are all adequate for imaging of ES cell differentiation cultures. Imaging software packages such as Metamorph (Version 4.6, Universal Imaging Corp.) allow for the automation of image capture. In the case of Metamorph, drivers are used to coordinate the laser shutter, the CCD camera, and transfer of the digital image to the computer. Parameters such as duration of imaging, time-lapse between frames, *X*-, *Y*-, and/or *Z*-axis movements, and excitation/emission wavelengths are programmable. Many software packages store each exposure as an individual TIFF image file, which is convenient for time-lapse editing purposes. The ability to delete or enhance frames is critical to producing a quality, finished product. NIH Image is a convenient movie editing software program that allows one to generate movies from stacks of individual TIFF image files. It is currently free and publicly available through the NIH Website.

Concluding Remarks

The described protocols and assays together provide a set of tools for quantitative analysis of vascular and hematopoietic parameters during ES cell differentiation. The *in situ* localization techniques can be adapted to measure other cell biological parameters, such as the number of cytoplasmic extensions emanating from vascular tips. Moreover, we have only started to utilize the imaging capability of the ES cell differentiation model to answer questions about dynamic aspects of vascular and hematopoietic development. We anticipate that these approaches will be combined with genetic and molecular biological tools (i.e., microarray analysis) in the future to provide new insights into vascular and hematopoietic development.

Acknowledgment

We thank members of the Bautch Lab for discussion and for input on the protocols presented here. We thank Susan Whitfield for artwork and photography.

Funding sources: This work was supported by a grant from the NIH (HL43174) to V. L. B. V. L. B. was partially supported by an NIH Career Development Award (HL02908), and J. B. K. was supported by a predoctoral fellowship from the Department of Defense (DAMD 17-00-1-0379).

[7] *In Vitro* Differentiation of Mouse ES Cells into Hematopoietic, Endothelial, and Osteoblastic Cell Lineages: The Possibility of *In Vitro* Organogenesis

By Motokazu Tsuneto, Toshiyuki Yamane, Hiromi Okuyama, Hidetoshi Yamazaki, and Shin-Ichi Hayashi

Introduction

Embryonic stem (ES) cells are pluripotent cells derived from the inner cell mass of embryonic day (E) 3.5 blastocysts. Since ES cells can differentiate into all kinds of cells, including germ cells *in vivo*, any cell lineage should be inducible if appropriate conditions can be determined *in vitro*. Mainly, in a given study, one tries to induce only one cell lineage from ES cells. Here, we focus on the nature of ES cell differentiation and attempt to construct organized cell groupings, in other words, tissues and organs *in vitro* for the purpose of investigating in detail the processes of organogenesis.

As a model of hematopoietic cell differentiation, we are studying osteoclastogenesis. Osteoclasts are derived from hematopoietic stem cells, and function in bone remodeling by resorbing the bone matrix. They are present close to osteoblasts in the bone marrow, and there osteoblasts supply them with two essential factors for osteoclastogenesis, namely, macrophage colony-stimulating factor (M-CSF) and receptor activator of nuclear factor κB ligand (RANKL).[1-3]

[1] S. I. Hayashi, T. Yamane, A. Miyamoto, H. Hemmi, H. Tagaya, Y. Tanio, H. Kanda, H. Yamazaki, and T. Kunisada, *Biochem. Cell. Biol.* **76**, 911 (1998).

[2] L. E. Theill, W. J. Boyle, and J. M. Penninger, *Annu. Rev. Immunol.* **20**, 795 (2002).

[3] H. Yoshida, S. I. Hayashi, T. Kunisada, M. Ogawa, S. Nishikawa, H. Okamura, T. Sudo, L. D. Shults, and S. I. Nishikawa, *Nature* **345**, 442 (1990).

0076-6879/2003 $35.00

We have established three alternative systems for osteoclastogenesis from undifferentiated ES cells utilizing coculturing with a stromal cell line.[4,5] This system is very useful in investigating the function of critical factors and genes involved in all the osteoclastogenesis processes.

In the first system, the hematopoietic cell lineage is induced from ES cells by means of coculturing with OP9 cells,[6] which is a stromal cell line derived from the fetal calvaria of an M-CSF-deficient *op/op* mouse. After induction of the hematopoietic cell lineage, including osteoclast precursors, the cells are seeded on ST2 stromal cells to induce the formation of mature osteoclasts. ST2 cells are derived from bone marrow cells and have the potential to support osteoclastogenesis from hematopoietic cells in the presence of 1α,25-dihydroxyvitaminD_3 [$1\alpha,25(OH)_2D_3$] and dexamethasone. They constitutively produce M-CSF, and are induced by $1\alpha,25(OH)_2D_3$ to produce RANKL. After more than 6 days of culturing on ST2, osteoclasts can be detected in the cultures. This is referred to as "2- or 3-step culture" (Figs. 1 and 2). In the first step on OP9 cells, ES cells differentiate into hematopoietic cells containing osteoclast precursors, in the second step on OP9 cells, they selectively propagate, and in the final step on ST2 cells, immature osteoclast precursors terminally differentiate. This culture system is useful in investigating and enriching cells at each differential stage.

In the second system, undifferentiated ES cells are directly seeded on ST2 cells and cultivated in the presence of $1\alpha,25(OH)_2D_3$ and dexamethasone for 11 days. This is referred to as the "single-step culture" (Fig. 1), and this culture system enables us to investigate the complete osteoclastogenesis process without any manipulations except for regular medium changes. In this culture system, a single ES cell forms a colony. Osteoclasts (tartrate-resistance acid phosphatase: TRAP-expressing cells) appear on the edge of the colony, and osteoblasts (alkaline phosphatase-expressing cells) are observed close to the osteoclasts[7,8] (Fig. 3A, C). Furthermore, endothelial cells are also present radially in the colony[9] (Fig. 3B).

[4] T. Yamane, T. Kunisada, H. Yamazaki, T. Era, T. Nakano, and S. I. Hayashi, *Blood* **90**, 3516 (1997).

[5] T. Yamane, T. Kunisada, H. Yamazaki, S. H. Orkin, and S. I. Hayashi, *Exp. Hematol.* **28**, 833 (2000).

[6] T. Nakano, H. Kodama, and T. Honjo, *Science* **265**, 1098 (1994).

[7] H. Okuyama, H. Yamazaki, T. Yamane, and S. I. Hayashi, *Res. Adv. Blood* **1**, 75 (2001).

[8] Hiromi Okuyama, 2003. Manuscript in preparation.

[9] H. Hemmi, H. Okuyama, T. Yamane, S. Nishikawa, T. Nakano, H. Yamazaki, T. Kunisada, and S. I. Hayashi, *Biochem. Biophys. Res. Commun.* **280**, 526 (2001).

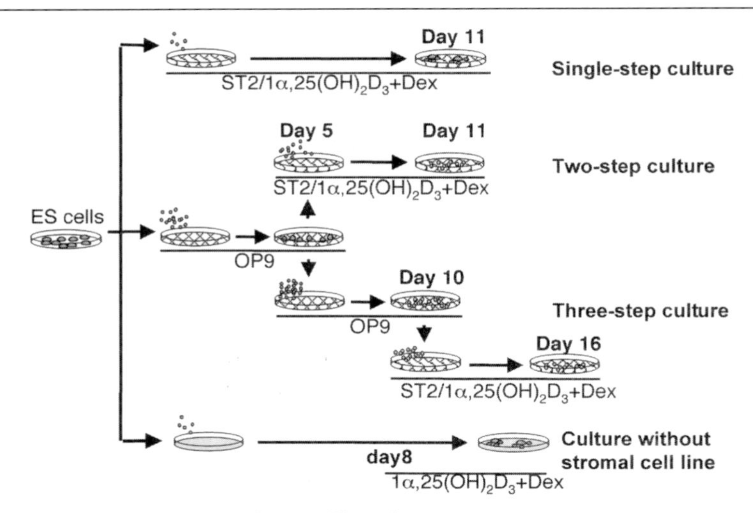

FIG. 1. The culture systems.

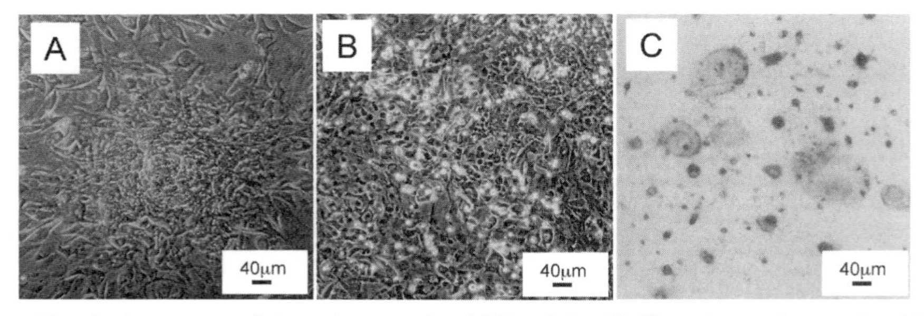

FIG. 2. Appearance of step cultures at day 5 (A) and day 10 (B), and osteoclasts at day 16 (C) with TRAP staining. (See Color Insert.)

In the third system, both osteoclasts and their supporting cells are induced from ES cells without stromal cell lines.[10] These systems will enable us to construct cell communities derived from ES cells alone *in vitro*.

Materials

0.25% trypsin in phosphate-buffered saline (PBS) + 0.5 m*M* EDTA
0.2% trypsin/1 m*M* EDTA

[10] Motokazu Tsuneto, 2003. Manuscript in preparation.

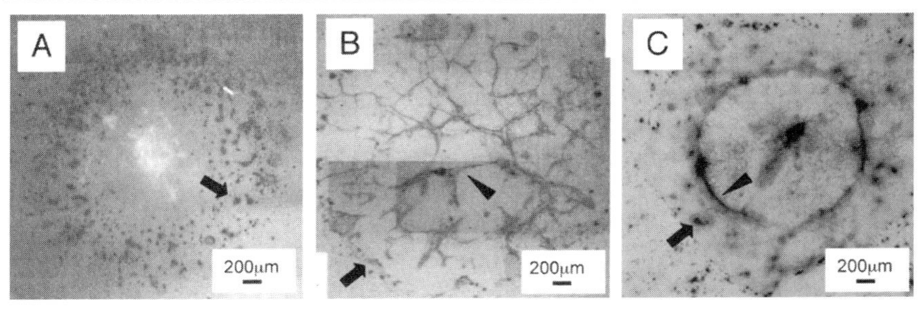

FIG. 3. Appearance of a single colony with various types of staining in the single-step cultures. (A) TRAP staining: Arrow: TRAP$^+$ cells. (B) CD31 and TRAP staining: CD31$^+$ cells (arrowhead) form networks and TRAP$^+$ cells (arrow) appear at the edge of the colony. (C) ALP and TRAP staining: ALP$^+$ cells (arrowhead) are round shaped and TRAP$^+$ cells (arrow) are present around them.

0.1% trypsin/0.5 m*M* EDTA
0.05% trypsin/0.53 m*M* EDTA (Life Technologies, Grand Island, NY, cat. no. 25300-062).

Commercial 2.5% trypsin (Life Technologies, cat. no. 15090-046) is divided into small aliquots in 15-ml tubes and stored at −20°C. When an aliquot is used, it is thawed, diluted with 1× PBS and supplemented with EDTA. After dilution, it is stored at 4°C.

Dulbecco's modified Eagle's medium (DMEM) (Life Technologies, cat. no. 12800-058).
RPMI medium 1640 (RPMI-1640) (Life Technologies, cat. no. 31800-014).
Minimum essential medium alpha medium (αMEM) (Life Technologies, cat. no. 11900-016).
Penicillin (Meiji Chemical, Tokyo, Japan).
Streptomycin (Meiji Chemical).

All media are supplemented with 50 μg/ml streptomycin and 50 U/ml penicillin.

Heat-inactivated fetal bovine serum (FBS) (Notes 1, 2).
200 m*M* L-glutamine (Life Technologies, cat. no. 25030-081).
100×MEM nonessential amino acids (Life Technologies, cat. no. 11140-050).
Mouse leukemia inhibitory factor (mLIF).

Mouse LIF is derived from medium conditioned by Chinese hamster ovary (CHO) cells transfected with LIF expression vector. Commercial mLIF (ESGRO™, Life Technologies, cat. no. 13275-011) can be substituted for this conditioned medium at more than 1000 U/ml for maintaining the ES cells.

2-Mercaptoethanol (2-ME) (Wako Pure Chemical Industries, Osaka, Japan, cat. no. 137-06862)
Gelatin, Type A: from porcine skin (Sigma, cat. no. G2500)
Knockout SR (KSR) (Life Technologies, cat. no. 10828-028)
$1\alpha,25(OH)_2D_3$ (Biomol Research Laboratories, Inc., Plymouth Meeting, PA, cat. no. DM-200)

Since $1\alpha,25(OH)_2D_3$ is unstable, it is diluted to 10^{-4} *M* with ethanol and divided into small aliquots to cryostat tubes. The tubes are stored in the dark at −80°C.

Dexamethasone (Sigma, St. Louis, MO, cat. no. D4902)
Dexamethasone is stored at 10^{-3} *M* in ethanol at 4°C.
Sodium tartrate (dihydrate) (Wako, cat. no. 190-03455)
Sodium acetate (trihydrate) (Wako, cat. no. 198-01055)
Naphthol AS-MX phosphate (Sigma, cat. no. N-5000)
Fast red violet LB salt (Sigma, cat. no. F-3381)
Formalin (37% formaldehyde solution: Wako, cat. no. 064-00406)
Acetone (Wako, cat. no. 012-00343)
Naphthol AS-BI phosphate (Sigma, cat. no. N-2250)
N,*N*-Dimethylformamide dehydrated (Wako, cat. no. 041-25473)
Fast blue BB salt (Sigma, cat. no. F3378)
DAB reagent set (Kirkegaard & Perry Laboratories, Inc., Gaithersburg, MD, cat. no. 54-10-00)
Polyoxyethylene (20) sorbitan monolaurate (a reagent equivalent to Tween 20) (Wako, cat. no. 167-11515)
Paraformaldehyde (Wako, cat. no. 160-00515)
Block Ace (Dainippon Seiyaku, Sapporo, Japan, cat. no. UK-B25)

Block Ace can be substituted for normal goat serum (KPL, cat. no. 71-00-27)

Monoclonal rat anti-mouse CD31 IgG (clone 390, Immunotech, Marseille, France, cat. no. 2081)
Horseradish peroxidase (HRP)-conjugated goat anti-rat IgG (ICN Pharmaceuticals, Inc., Aurora, Ohio, cat. no. 55778)
1 *M* Tris–HCl (pH 8.5)
1× PBS supplemented with 0.05% Tween 20 [Polyoxyethylene (20) sorbitan monolaurate] (PBS-T)

PBS-T is stored at 4°C.

Medium for maintenance of ES cells with embryonic fibroblasts (EF) (referred to as ESM) (Note 3).

		Final concentration
DMEM	196 ml	
Heat-inactivated FBS (Notes 1, 2)	36 ml	15%
200 mM L-glutamine	2.4 ml	2 mM
100×MEM nonessential amino acids	2.4 ml	0.1 mM
CHO-LIF	2.4 ml	1000 U/ml
$10^{-2}M$ 2-ME in DMEM	2.4 ml	10^{-4} M
Total	241.6 ml	

Medium for maintenance of ES cells without EF [EF(–)ESM] (Note 3).

		Final concentration
DMEM	206 ml	
Knockout SR	24 ml	10%
Heat-inactivated FBS (Notes 1, 2)	2.4 ml	1%
200 mM L-glutamine	2.4 ml	2 mM
100×MEM nonessential amino acids	2.4 ml	0.1 mM
CHO-LIF	2.4 ml	1000 U/ml
10^{-2} M 2-ME in DMEM	2.4 ml	10^{-4} M
Total	242 ml	

0.1% gelatin

Distilled water (DW)	500 ml
Gelatin, Type A: from porcine skin	0.5 g

Autoclave 0.1% gelatin solution at 120°C for 20 min and store it at RT.

10% formalin (3.7% formaldehyde) solution in 1×PBS (v/v)
Ethanol/acetone solution (50/50, v/v. Freshly prepared)
TRAP staining solution (500 ml)

To 450 ml of distilled water add:

		Final concentration
Sodium tartrate (dihydrate)	5.75 g	59.3 M
Sodium acetate (trihydrate)	6.8 g	165.7 M
Naphthol AS-MX phosphate	250 mg	

Stir TRAP solution until the solutes have dissolved. Adjust the pH to 5.0 with acetic acid and the volume to 500 ml with distilled water. Store at 4°C in the dark.

Just before performing TRAP staining, the TRAP solution should be freshly prepared as described below. Dissolve fast red violet LB salt in TRAP staining solution at a final concentration of 0.5 mg/ml.

Alkaline phosphatase (ALP) staining solution (180 ml)

Naphthol-AS-BI phosphate	18 mg
N,*N*-Dimethylformamide dehydrated	1.8 ml

Mix until the solutes have dissolved and add

DW	162 ml
1 *M* Tris–HCl (pH 8.5)	18 ml

Stir until the solutes have dissolved. Store at 4°C in the dark.

Just before performing ALP staining, dissolve fast blue BB salt in ALP staining solution at a final concentration of 0.16 mg/ml.

4% paraformaldehyde (PFA) solution (200 ml)

1×PBS	200 ml
Paraformaldehyde	8 g

Mix and heat to 70°C. Do not boil. Add 400 μl of 4 N NaOH and mix. Then add 700 μl of 2 N HCl and mix to dissolve the solutes completely. Confirm that the pH is 7–8. If not, adjust pH with NaOH or HCl.

Methods

Medium is prewarmed to 37°C and all cultures are maintained at 37°C in 95% air/5% CO_2 in a humidified incubator.

Thawing Frozen Cells

1. Quickly thaw frozen cells in a 37°C water bath until just a small bit of frozen solution remains in the tube.
2. Add 1 ml of medium drop by drop gently to the cell suspension, transfer the suspension to a 15-ml centrifuge tube, and gently add 7 ml of medium.
3. Centrifuge at 200*g* for 5 min at 4°C [for ES cells, at 20°C].
4. Aspirate the supernatant and suspend the cell pellet in the appropriate medium.
5. Seed cells in culture dishes or plates.

Gelatin Coating of Culture Dishes

Cover the bottom of culture dishes with 0.1% gelatin solution and incubate them at 37°C for 10 min. After removing the gelatin solution completely by aspiration, use the dishes for cell culturing.

Preparation of EF

1. Prepare E13.5–15.5 embryos and transfer them to freshly prepared 1× PBS in a petri dish to wash out blood.
2. Remove their heads and internal organs. Wash the embryos in fresh 1× PBS twice and transfer them into 10 ml of freshly prepared 1× PBS in a petri dish.
3. Cut them into small pieces and add 10 ml of 0.2% trypsin/EDTA.
4. After mixing the tissue fragments by gentle pipetting, incubate them at 37°C for 10 min.
5. Suspend them by vigorous pipetting and incubate at 37°C for 5 more minutes.
6. Suspend them again by vigorous pipetting and remove tissue debris.
7. Add 20 ml of DMEM/10%FBS to the mixture and mix by pipetting. After letting it stand for 5 min to remove debris, transfer the supernatant to a 50-ml centrifuge tube.
8. After letting the mixture stand at 25°C for 5 min, pass the supernatant through a mesh to remove debris, and centrifuge it at 400*g* for 5 min at 4°C.
9. Aspirate the supernatant, suspend pellets in DMEM/10%FBS and count the cell number.
10. Seed 8×10^6 cells per 150-mm culture dish and cultivate them at 37°C.
11. To remove debris, 4 hr after seeding, aspirate the culture medium and wash the cell layer with 1× PBS once. Add 25 ml of fresh medium and cultivate the cells at 37°C.
12. Freeze a portion of the cells after suspending them in [DMEM/10%FBS]/10% DMSO.
13. After the rest of the cells grow to 100% confluency, aspirate the medium, wash with 1× PBS once, and add 2 ml of 0.05% trypsin/EDTA. Incubate at 37°C for 5 min.
14. Add 12 ml of DMEM/10% FBS and suspend the cells by vigorous pipetting. Transfer them to a 50-ml centrifuge tube and centrifuge at 400*g* for 5 min at 4°C.
15. Aspirate the supernatant and suspend them in DMEM/10% FBS. Split them 1 : 3 into fresh 150-mm culture dishes.
16. After the cells grow to confluency, passage them as described in steps 13–15.
17. When third passage is carried out, count the cell number and seed more than 3×10^6 cells per 150-mm dish.
18. Treat them with mitomycin C as described in the following subsection.

Treatment of EF with Mitomycin C and EF Stock Preparation

1. After EF grow to 100% confluency in the 150-mm dishes, aspirate 10 ml of the medium and add 10 μg/ml of mitomycin C. Incubate the cells at 37°C for 2.5 hr.
2. Wash the cells gently with 1×PBS twice and add 2 ml of 0.05% trypsin/EDTA. Incubate them at 37°C for 5 min.
3. Add 12 ml of DMEM/10%FBS and suspend them by vigorous pipetting. Transfer them to a 50-ml centrifuge tube and centrifuge at 400*g* for 5 min at 4°C.
4. Aspirate the supernatant and suspend the cells in DMEM/10%FBS. Count the number of cells and store them as a mitomycin C-treated EF(MMC-EF) stock at 6×10^6 cells/tube in liquid N_2.

Maintenance of Undifferentiated ES Cells (Note 4)

For coculturing with the stromal cell line, undifferentiated ES cell lines are maintained in ESM on MMC-EF. ES cells are usually cultivated in 60-mm cell culture dishes.

1. After coating the culture dishes with gelatin, thaw EF and then seed them on gelatin-coated dishes.
2. More than 6 hr after seeding MMC-EF, thaw frozen stock ES cells, wash with ESM and sow them on precultured MMC-EF.
3. After ES cells grow to 70% confluency, wash them with 1×PBS three times, add 0.25% trypsin/EDTA and incubate them at 37°C for 5 min. Then add ESM and dissociate the cultures to single cells by vigorous pipetting. Wash and seed 2.5×10^6 cells/60-mm dish on freshly prepared MMC-EF (Table I).

For culturing ES without a stromal cell line, the undifferentiated ES cell line is maintained in EF(−)ESM on gelatin-coated dishes. These ES cells are referred to as EF(−)ES cells (Note 5)

1. After coating culture dishes with gelatin, thaw EF(−) ES cells, wash them with ESM and seed them in gelatin-coated dishes (Table I).

TABLE I

OPTIMAL CONDITION FOR PASSAGING ES CELLS

2.5×10^6 cells/60-mm dish
5×10^6 cells/100-mm dish

After 24 hr, the ES cells reach 70% confluency.

2. When ES cells grow to 70% confluency, passage and seed them at 2.5×10^6 cells/60-mm dish.

Maintenance of ST2 Stromal Cell Line (Note 6)

Bone marrow-derived ST2 cells are maintained in RPMI-1640 medium supplemented with 5% FBS and 5×10^{-5} *M* 2ME. ST2 cells are usually cultivated in 100-mm cell culture dishes.

1. When ST2 cells reach 100% confluency, wash them with 1×PBS once, then add 1 ml of 0.05% trypsin/EDTA and incubate at 37°C for 2 min.
2. Add 7 ml of medium and dissociate the cells by vigorous pipetting. Wash and split them 1 : 4 into new culture dishes.

Maintenance of OP9 Stromal Cell Line (Note 7)

M-CSF-deficient fetal calvaria-derived OP9 cells are maintained in αMEM medium supplemented with 20% FBS in 100-mm dishes.

1 When OP9 cells reach 70% confluency (Fig. 4C), wash them with 1× PBS twice, then add 1 ml of 0.1% trypsin/EDTA and incubate at 37°C for 5 min.
2 Add 7 ml of medium and dissociate the cells by vigorous pipetting. Wash and split them 1 : 4 into new culture dishes.

Osteoclastogenesis from Undifferentiated ES Cells in 2- or 3-step Cultures (Notes 8, 9, 10, 11)

In these cultures, osteoclasts are specifically induced. ES cells form colonies containing hematopoietic cells and osteoclast precursors by day 5 on OP9. In the next step, when these cells are seeded on ST2 and cultivated

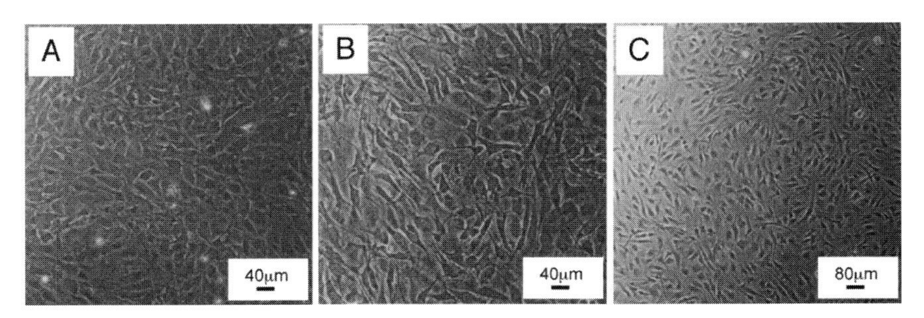

FIG. 4. Appearance of ST2 (A) and OP9 (B, C). (C) is 70% confluency. (See Color Insert.)

in the presence of 1α,25$(OH)_2D_3$ and dexamethasone, osteoclasts are induced (referred to as the 2-step culture).

After the initial 5 days of culturing on OP9, differentiated cells are passaged and reseeded on OP9 and cultivated for 5 more days. In this step, the number of osteoclast precursors is increased by 20-fold compared with that after the initial 5 days. After the second 5 days of culturing, the harvested cells are seeded on ST2 and cultivated in a medium supplemented with 1α,25$(OH)_2D_3$ and dexamethasone for 6 days, during which time osteoclasts are induced (referred to as the 3-step culture).

This culturing is performed in 6- and 24-well culture plates.

1. Prepare OP9 cells in a 6-well culture plate. After OP9 cells reach 100% confluency in this plate, they are used for this culture.
2. Grow ES cells to 70% confluency, dissociate ES cells to single cells by 0.25% trypsin/EDTA treatment and washing. After aspirating the supernatant, suspend the ES cells in αMEM/20% FBS.
3. Seed 10^4 cells/well on OP9 cells.
4. On day 3, replace half of the medium with fresh medium.
5. On day 5, wash the cells three times with 1× PBS, trypsinize them with 0.5 ml of 0.25% trypsin/EDTA per well by incubating at 37°C for 5 min.
6. Add 3.5 ml/well of αMEM/20% FBS to the wells and dissociate the cells to single cells by vigorous pipetting. Transfer the suspension to a 15-ml tube and centrifuge at 250*g* for 5 min at 4°C. *When 2-step culturing is performed, go to step 12.*
7. Aspirate the supernatant and suspend the cells in αMEM/20% FBS.
8. Seed 10^5 cells/well on freshly prepared OP9 cells in 6-well culture plates.
9. On day 8, replace half of the medium with fresh medium.
10. On day 10, add 2 ml of αMEM/20% FBS, dissociate the cells to single cells by vigorous pipetting and transfer the suspension to a 15-ml centrifuge tube.
11. Allow to stand to remove debris for 5 min at 25°C. After the debris settles, transfer the supernatant to a fresh 15-ml tube. Centrifuge at 250*g* for 5 min at 4°C.
12. Aspirate the supernatant and suspend the cells in αMEM/10% FBS.
13. Prepare a suspension of 10^3 cells/ml cell supplemented with 10^{-7} *M* dexamethasone and 10^{-8} *M* 1α,25$(OH)_2D_3$. Seed 1 ml/well of this cell suspension on prepared ST2 cells in a 24-well culture plate.
14. Three days later, change the medium to fresh αMEM/10%FBS supplemented with 10^{-7} *M* dexamethasone and 10^{-8} *M* 1α,25 $(OH)_2D_3$.

15. On day 6 of the cultivation of ES on ST2, perform the TRAP staining.

Induction of Osteoclasts, Endothelial Cells and Osteoblasts from Undifferentiated ES Cells in Single-step Culture (Notes 9, 12)

It has been observed that single ES cells give rise to cell communities containing osteoclasts, endothelial cells and osteoblasts in this culture system. Osteoclasts and osteoblasts appear on the edge of the colony, and endothelial cells extend from its center to the edge (Fig. 3). Hematopoietic c-Kit$^+$ cells first appear at day 4 and TRAP$^+$ cells first appear at day 8. This timing is similar to that observed during *in vivo* development, suggesting that this system probably mimics *in vivo* development spatially and temporally.

In this culture system, αMEM/10%FBS is used to induce ES cells to differentiate to osteoclasts.

1. Prepare ST2 cells in a 24-well plate. After the ST2 cells reach 100% confluency in the plate, they are used in this culture system.
2. Grow ES cells to 70% confluency, and dissociate them to single cells by 0.25%trypsin/EDTA treatment and washing. After aspirating the supernatant, suspend the ES cells in αMEM/10%FBS.
3. Count the cell number and seed 50–100 cells/well in medium supplemented with 10^{-7} *M* dexamethasone and 10^{-8} *M* 1α,25$(OH)_2D_3$.
4. Change the medium supplemented with 10^{-7} *M* dexamethasone and 10^{-8} *M* 1α,25$(OH)_2D_3$ every 2 or 3 days.
5. On day 11, perform the desired staining.

Osteoclastogenesis of Undifferentiated ES Cells Without Stromal Cell Lines (Note 13)

In this culture system, ES cells differentiate into several lineages. Osteoclasts and their supporting cells are observed because supplementation with 1α,25$(OH)_2D_3$ and dexamethasone results in osteoclastogenesis, although the efficiency of induction is lower than in cocultures with a stromal cell line.

For the differentiation of these ES cells to osteoclasts, αMEM/15%FBS is used to culture the ES cells and their supporting cells in 24-well culture dishes (1.88 cm^2/well, Corning Incorporated, NY, USA, cat. no. 3526) (Notes 14, 15, see Table II).

1. Grow ES cells to 70% confluency, dissociate them to single cells by treating them with 0.25% trypsin/EDTA and suspend them in αMEM/15% FBS.

TABLE II
VOLUMES OF MEDIUM FOR CULTURES, AND CONDITIONS FOR TRYPSINIZATION

Culture vessel	Medium (ml)	Trypsin/EDTA (ml)	Medium for neutralization (ml)
150-mm dish	25	2	12
60-mm dish	5	0.5	4.5
100-mm dish	10	1	7
6-well plate	2	0.5	3.5
24-well plate	1		

2. Dissociate them by vigorous pipetting and seed them at 2×10^5 cells/well in a 24-well cell culture dish.
3. On day 2 and day 6, replace half of the medium and on day 4, replace the medium with fresh medium.
4. On day 8, replace the medium with fresh αMEM/15% FBS supplemented with 10^{-7} M dexamethasone and 10^{-8} M $1\alpha,25(OH)_2D_3$. On day 11, replace the medium with fresh medium supplemented with 10^{-7} M dexamethasone and 10^{-8} M $1\alpha,25(OH)_2D_3$.
5. On day 14, perform TRAP and ALP staining.

Tartrate-Resistance Acid Phosphatase (TRAP) staining

1. Aspirate the medium and add 1 ml/well of 10% formalin to cover the cultured cells. Incubate them at 25°C for 10 min. During this time, add fast red violet LB salt to TRAP staining solution at a final concentration of 0.5 mg/ml.
2. Aspirate the 10% formalin and fill each well with 1× PBS. Remove the 1× PBS by aspiration.
3. Cover the cells with 0.5 ml of ethanol/acetone solution and incubate them at 25°C for just 1 min, then fill each well with 1× PBS. Aspirate the 1× PBS and wash with 1× PBS once more.
4. Cover the cells with 250 μl/well of TRAP solution and let them stand for 5 min at 25°C.
5. Soak the culture dish in tap water for more than 30 min.

Alkaline Phosphatase Staining

1. Aspirate the medium and add 1 ml/well of 10% formalin to cover the cultured cells. Store them at 25°C for 10 min. During this time, add fast blue BB salt into the TRAP staining solution at a final concentration of 0.16 mg/ml.

TABLE III
RECOMMENDED CONDITIONS FOR FREEZING CELLS

Cell line	Condition	Medium for freezing	Optimal density
ST2	100% confluency	FBS/10%DMSO	100-mm dish/2 tubes
OP9	70% confluency	[αMEM/20%FBS]/10%DMSO	100-mm dish/2 tubes
ES cells	70% confluency	ESM/10%DMSO	5×10^6 cells/tube
EF(−)ES cells	70% confluency	EF(−)ESM/10%DMSO	5×10^6 cells/tube

2. Aspirate 10% formalin and fill each well with 1× PBS. Remove the 1× PBS by aspiration.
3. Cover the cells with 250 μl/well of ALP solution and let them stand for 5 min at 25°C.
4. Soak the culture dish in tap water for more than 30 min.

Immunohistochemical Staining of CD31

1. To fix the cultured cells, add 4% paraformaldehyde (PFA) to each well and keep the cells on ice for 15 min.
2. After aspiration of the PFA, add ice-cold 1×PBS to each well and let the dishes stand on ice for 5 min. Perform this washing three times.
3. After aspirating the 1× PBS, add 500 μl of 0.3% H_2O_2/methanol and let it stand on ice for 10 min.
4. After aspirating the 0.3% H_2O_2/methanol, add ice-cold 100% ethanol and let the dishes stand on ice for just 1 min.
5. Wash the cells twice, add Block Ace and let them stand on ice for 20 min.
6. Add 5 μg/ml of first antibody (rat anti-mouse CD31) diluted with Block Ace and let stand at 4°C overnight.
7. Aspirate the primary antibody solution, and wash twice with ice-cold PBS-T on ice for 5 min.
8. Add 35 μg/ml of HRP-conjugated secondary antibody (HRP-conjugated goat anti-rat IgG) diluted with Block Ace and let stand on ice for 1 hr.
9. Aspirate the secondary antibody solution, and wash twice with ice-cold PBS-T on ice for 5 min.
10. Use a DAB reagent set according to its instruction manual. Let stand at 25°C for 10 min, and then add distilled water to stop the reaction.

Future Prospects

In our single-step culture system, the first appearance of hematopoietic cells and endothelial cells is on day 4. Since ES cells are derived from day 3.5 blastocysts, this day 4 in culture is similar to E 7.5. This is the same timing at which yolk hematopoiesis and vasculogenesis appear for the first time *in vivo*. We used this system for osteoclastogenesis and tried to accelerate this timing by adding M-CSF and RANKL. The location of osteoclasts in the colonies could be manipulated, but temporal acceleration was impossible.[9] This may indicate that cell differentiation control by itself, or some other factor(s) can control the time course. If we could control this timing, it would not only be very beneficial for regenerative medicine, but also helpful is elucidating the mechanism of the temporal control of tissue development.

Notes

1. FBS that supports the growth of ES cells well and prevents them from differentiating should be used. Prepare ESM or EF(−)ESM containing several lots of FBS and check the growth rate of ES cells cultivated in these preparations of medium. To check whether ES cells are maintained in the undifferentiated state, cultivate ES cells in ESM or EF(−)ESM without mLIF and perform ALP staining. For cultivating ES cells, use of more FBS has shown to be appropriate according to these tests.
2. To inactivate complement, FBS is heated in a water bath at 56°C for 30 min.
3. To maintain undifferentiated ES cells, the medium used should be as fresh as possible. If medium prepared 2 weeks or longer ago is used, add 2 m*M* L-glutamine to the medium. Medium should not be used for more than a month after preparation.
4. The passage number should be kept as low as possible. We recommend preparing sufficient stocks of frozen cells.
5. To adapt undifferentiated ES cells to maintain without EF, at first, remove EF by incubating ES cells and EF on a gelatin-coated dish for 30 min at 37°C and then plate nonadherent cells on a freshly prepared gelatin-coated dish. After that, cultivate the ES cells for at least 5 days as described in *Maintenance of undifferentiated ES cells*, and then the ES cells can be maintained without EF.
6. If the shape of ST2 cells (Fig. 4A) becomes fibroblastic during long-term maintenance, discard them and prepare new ones.

7. OP9 cells (Fig. 4B) are sensitive to old medium. Fresh medium should be used within 1 month of preparation to maintain them. Also, OP9 cells should be passed when they reach 70% confluency (Fig. 4C) because at more than 70% confluency their ability to grow as single cells is disturbed. If the shape of OP9 cells becomes flattened, it is impossible for them to recover their initial state. Discard them and prepare fresh ones.
8. Similar results are obtained by using 3 ES lines: D3,[11] J1,[12] and CCE,[13] in our coculture system with stromal cell lines.
9. Seeding mixtures of ES cells and EF on OP9 and ST2 cells is not a problem, and EF need not be removed.
10. On day 5, differentiated (Fig. 2A) and undifferentiated colonies are observed.[6] The appearance of undifferentiated colonies is smooth, and the boundary of the cells is not clear. These colonies may comprise, at most, one-third of the total colonies.
11. During the culturing, OP9 cells sometimes differentiate to adipocytes, but this does not influence the experimental outcomes.
12. The efficiency of the formation of colonies by ES cells varies from 1 to 40% depending on the quantity of serum. We recommend seeding ES cells to form less than 20 colonies well in 24-well plates, because high colony density prevents ES cells from differentiating into osteoclasts. A preliminary experiment is performed to check the plating efficiency by seeding 10–1000 cells/well.
13. This method does not efficiently induce $TRAP^+$ cells. This efficiency increases when the period of culturing without $1\alpha,25(OH)_2D_3$ and dexamethasone is extended to 12 days.
14. The D3 ES cell line is used in this culture system. We have not yet determined the optimal conditions for the J1 ES cell line. Initial cell density seems to be important in this culture system.
15. Optimal density is responsible for the quantity of FBS. The optimal density for your FBS may be determined before you start the experiment.

[11] T. C. Doetschman, H. Eistetter, M. Katz, W. Schmidt, and R. Kemler, *J. Embryol. Exp. Morphol.* **87**, 27 (1985).

[12] E. Li, T. H. Bestor, and R. Jaenisch, *Cell* **69**, 915 (1992).

[13] E. Robertson, A. Bradley, M. Kuehn, and M. Evans, *Nature* **323**, 445 (1986).

Acknowledgment

We thank Drs. T. Nakano (Osaka University), T. Kunisada (Gifu University), T. Era (RIKEN) and H. Hemmi (Osaka University) for their collaboration. This study was supported by grants from the Special Coordination Funds for Promoting Science and Technology from the Ministry of Education, Culture, Sports, Science and Technology, the Japanese Government; and from the Molecular Medical Science Institute, Otsuka Pharmaceutical Co., Ltd.

[8] Development of Hematopoietic Repopulating Cells from Embryonic Stem Cells

By MICHAEL KYBA, RITA C. R. PERLINGEIRO, and GEORGE Q. DALEY

Introduction

Embryonic stem cells are defined as totipotent by virtue of their potential to chimerize any tissue of a developing embryo. From the standpoint of the development of cellular therapies, this virtue is problematic, because in general only a single or limited number of cell types is required for a given therapy. There are two possible solutions to this problem: one either identifies the critical genetic elements that control the lineage choices along the path toward a given cell type and then enforces these instructions on as much of the culture as possible, or one defines conditions which select out and amplify the cell type of interest from among its unwanted cousins. The latter approach has proven highly effective in our work towards the derivation of a repopulating hematopoietic stem cell (HSC) from embryonic stem cell cultures, particularly when used in combination with an element of genetic instruction.

The hematopoietic potential of mouse ES cells is easily demonstrated. ES cells grown under the appropriate conditions (suspension culture and the removal of LIF) will spontaneously form proliferating and differentiating cellular clusters referred to as embryoid bodies (EBs).[1] The EBs increase in both size and cellular variety with time, and within 8–10 days, superficial red islands containing hemoglobinized erythrocytes become

[1] G. R. Martin and M. J. Evans, Differentiation of clonal lines of teratocarcinoma cells: formation of embryoid bodies in vitro, *Proc. Natl. Acad. Sci. USA* **72**, 1441–1445 (1975).

0076-6879/2003 $35.00

apparent.[2,3] Hematopoiesis can be identified at earlier stages by disaggregating the embryoid bodies into single cells and plating these cells in suspension cultures with hematopoietic cytokines. Under these conditions, hematopoietic colonies will form containing differentiated cells of myeloid, erythroid, or mixed lineages.[4,5] The types of colony-forming cells (CFCs) that can be identified in this manner change over the course of time, but the pattern of change is highly reproducible: at early time points (5 days of differentiation) primitive erythroid precursors predominate. The primitive erythrocyte (Ery-P) is peculiar to early embryogenesis and is distinguished from the adult-type, or definitive erythrocyte (Ery-D) by the fact that it is nucleated, and therefore larger, and expresses embryonic globin isoforms (β-H1 in the mouse) with a higher affinity for oxygen than adult globin (β-major).[6,7] At later time points (8 days of differentiation and beyond) Ery-D, myeloid, and mixed colony types predominate.

Since hematopoietic CFCs are derived from a hematopoietic stem cell, one would predict that HSCs should be detectable prior to day 6 of EB differentiation. Many investigators, ourselves included, have looked for the presence of HSCs in early EBs by attempting to use their disaggregated cells as a substitute for bone marrow to rescue and reconstitute hematopoiesis in lethally irradiated adult mice. In our hands, neither rescue from the lethal effects of irradiation nor reconstitution of the hematopoietic system can be demonstrated. This would at first glance appear to present a paradox; however, there is a striking embryonic correlate to these results. In the early embryo, the first tissue to undergo hematopoietic differentiation is the extraembryonic mesoderm of the yolk sac. Like the EB, the yolk sac initially produces Ery-P, and at later time points myeloid and mixed CFC. However, prior to the onset of circulation, cells from the yolk sac will not reconstitute hematopoiesis in lethally irradiated adult hosts. Repopulating HSCs are detected in the yolk sac at around day 11, shortly after the establishment of

[2] T. C. Doetschman, H. Eistetter, M. Katz, W. Schmidt, and R. Kemler, The in vitro development of blastocyst-derived embryonic stem cell lines: formation of visceral yolk sac, blood islands and myocardium, *J. Embryol. Exp. Morphol.* **87**, 27–45 (1985).

[3] R. M. Schmidt, E. Bruyns, and H. R. Snodgrass, Hematopoietic development of embryonic stem cells in vitro: cytokine and receptor gene expression, *Genes Dev.* **5**, 718–740 (1991).

[4] M. V. Wiles and G. Keller, Multiple hematopoietic lineages develop from embryonic stem (ES) cells in culture, *Development* **111**, 259–267 (1991).

[5] G. Keller, M. Kennedy, T. Papayannopoulou, and M. V. Wiles, Hematopoietic commitment during embryonic stem cell differentiation in culture, *Mol. Cell. Biol.* **13**, 473–486 (1993).

[6] J. E. Barker, Development of the mouse hematopoietic system, *Dev. Biol.* **18**, 14–29 (1968).

[7] T. Brotherton, D. Chui, J. Gauldie, and M. Patterson, Hemoglobin ontogeny during normal mouse fetal development, *Proc. Natl. Acad. Sci. USA* **76**, 2853–2855 (1979).

circulation, however they are also detected in greater numbers at this time point at an intraembryonic site, the major vessels of the aorta-gonads-mesonephrous (AGM) region.[8] Because autonomous culture of AGM tissue explanted before the onset of circulation can give rise to adult-repopulating cells,[9] it is generally accepted that the AGM is the site of origin of the definitive HSC, while the yolk sac cells that initially give rise to Ery-P are specialized for that purpose and not capable of adult repopulation. Repopulating cells present in the yolk sac at day 11 are thus taken to have arrived there through the circulation. Although there are problems with this interpretation that are beyond the scope of this chapter, it should be apparent that hematopoiesis as described in EBs recapitulates the early stages of embryonic yolk sac hematopoiesis. It would therefore appear that the EB is incapable of producing cells corresponding to those of the AGM. The EB is certainly a smaller and much more chaotic structure than the embryo of equal developmental age.

Leukemic Engraftment

As a first step towards exploring the usefulness of EB culture systems for the generation of repopulating HSCs, we attempted to disconnect the repopulating activity of the putative HSC from its lymphoid–myeloid–erythroid differentiation potential. We reasoned that the ideal agent to accomplish this would be a transforming factor that acted either to block HSC apoptosis, or to drive HSC proliferation, or both, while at the same time leaving the natural ability of the HSC to differentiate along lymphoid, myeloid, and erythroid lineages unperturbed. Such a factor exists uniquely in the form of the oncogene, *Bcr/Abl.* This genetic fusion, the causative agent of chronic myeloid leukemia (CML), is encoded by the Philadelphia chromosome (a translocation between chromosomes 9 and 22) in CML patients. Bcr/Abl has the dual effect of driving HSC proliferation and blocking apoptosis.[10,11] The fact that Bcr/Abl-transformed HSC can still undergo normal differentiation is

[8] A. M. Muller, A. Medvinsky, J. Strouboulis, F. Grosveld, and E. Dzierzak, Development of hematopoietic stem cell activity in the mouse embryo, *Immunity* **1**, 291–301 (1994).

[9] A. Cumano, F. Dieterlen-Lièvre, and I. Godin, Lymphoid potential, probed before circulation in mouse, is restricted to caudal intraembryonic splanchnopleura, *Cell* **86,** 907–916 (1996).

[10] A. Bedi, B. A. Zehnbauer, J. P. Barber, S. J. Sharkis, and R. J. Jones, Inhibition of apoptosis by BCR-ABL in chronic myeloid leukemia, *Blood* **83**, 2038–2044 (1994).

[11] L. Puil, J. Liu, G. Gish, G. Mbamalu, D. Bowtell, P. G. Pelicci, R. Arlinghaus, and T. Pawson, Bcr-Abl oncoproteins bind directly to activators of the Ras signalling pathway, *EMBO J.* **13**, 764–773 (1994).

demonstrated by the presence of the Philadelphia chromosome in both lymphoid and myeloid cells of patients with CML. Indeed, the presence of this unique marker in all hematopoietic lineages of these patients is taken as classical evidence for the existence of an HSC.[12] Our strategy was to transduce Bcr/Abl into cells from day 5 EBs, the stage in which the stem cell for hematopoiesis is being specified as a derivative of the hemangioblast, the earliest progenitor defined for the hematopoietic lineage within EBs.[13]

Bcr/Abl-Induction of Hematopoietic Cultures

- Generate EBs by plating 10^4 ES cells per ml in *differentiation medium* {IMDM (Sigma) with 15 % fetal calf serum (FCS for differentiation; StemCell), 50 ug/ml ascorbic acid (Sigma), 200 ug/ml iron-saturated transferrin (Sigma), 4.5×10^{-4} *M* monothioglycerol (MTG; Sigma)} supplemented with 0.9% methylcellulose (M3120, StemCell Technologies) in 35 mm petri dishes (StemCell).
- Harvest on day 5 (120 hr) by diluting the methylcellulose with PBS and centrifuging the EBs. Wash once with PBS, and dissociate with 0.25% collagenase for 60 min at 37°C followed by repeated passage through a 23 G needle.
- Spin-infect (1300*g* for 90 min at 30°C in a Beckman GH-3.8 rotor) 3×10^5 cells with 10 ml of Bcr/Abl viral supernatant containing 4 μg/ml polybrene (Sigma) in three wells of a 6-well dish precoated with stromal cells (see below). Our viral vector expresses GFP from a downstream IRES, marking the transduced cells with green fluorescence.
- Culture on OP9 stroma at 37°C/5%CO_2. The growth conditions under which transduced populations are initiated are critical. We used a cytokine cocktail consisting of 0.5 ng/ml murine IL3 (interleukin 3; Peprotech), and 50 ng/ml each of human IL6 (interleukin 6; Peprotech), human SCF (stem cell factor; Peprotech), and human FL (FLT3 ligand; Peprotech), 50 μM β-mercaptoethanol, in IMDM/15% FCS, over OP9 stromal cells
- Once the colonies have become dense, passage in the absence of stroma in the same growth medium. Under these conditions, cultures become dominated by immature hematopoietic blast cells.

[12] P. J. Fialkow, R. J. Jacobson, and T. Papayannopoulou, Chronic myelocytic leukemia: clonal origin in a stem cell common to the granulocyte, erythrocyte, platelet and monocyte/macrophage, *Am. J. Med.* **63**, 125–130 (1977).

[13] K. Choi, M. Kennedy, A. Kazarov, J. C. Papadimitriou, and G. Keller, A common precursor for hematopoietic and endothelial cells, *Development* **125**, 725–732 (1998).

We favor colony induction on the stromal cell line OP9, which is genetically null for M-CSF,[14] based on our early experiments to optimize the system. Other stromal cell lines, including M2-10B4[15] and DAS104-4[16] gave an overabundance of macrophages and mast cells under these conditions, while OP9 promoted an expansion of hematopoietic blast cells (Fig. 1).

In addition to bulk cultures, clonal cell lines can be established by plating in methylcellulose suspension cultures and picking individual colonies. When expanded as above, the majority of our clonal cultures also consisted of blast cells. The *in vitro* differentiation potential of the clones was limited to producing Ery-P colonies or secondary blast cell colonies, however when we injected 4×10^6 cells into sublethally irradiated (500 Rad) 129Sv/Ev (Taconic) or NOD/SCID (Jackson) mice via lateral tail vein, mice succumbed to a donor-derived multilineage leukemia 5–9 weeks posttransplant. Because the leukemias were initiated from cloned cells, and because they contained lymphoid, myeloid, and erythroid elements, we concluded that Bcr/Abl had targeted a hematopoietic stem cell, albeit one with an intrinsic defect in adult engraftment that rendered these differentiation potentials

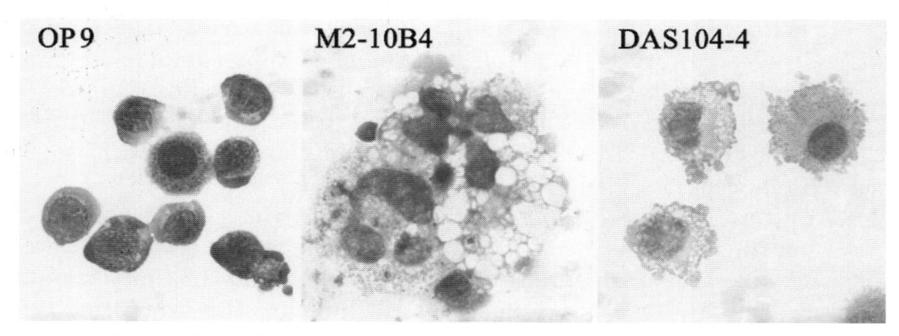

FIG. 1. Morphology of Bcr/Abl-transduced cells growing on different stromal cell lines. (See Color Insert.)

[14] T. Nakano, H. Kodama, and T. Honjo, Generation of lymphohematopoietic cells from embryonic stem cells in culture, *Science* **265**, 1098–1101 (1994).

[15] F. Lemoine, R. Humphries, S. Abraham, G. Krystal, and C. Eaves, Partial characterization of a novel stromal cell-derived pre-B-cell growth factor active on normal and immortalized pre-B cells, *Exp. Hematol.* **16**, 718–726 (1988).

[16] O. Ohneda, C. Fennie, Z. Zheng, C. Donahue, H. La, R. Villacorta, B. Cairns, and L. Lasky, Hematopoietic stem cell maintenance and differentiation are supported by embryonic aorta-gonad-mesonephros region-derived endothelium, *Blood* **92**, 908–919 (1998).

latent. We named this putative target cell the embryonic (or primitive) HSC in accordance with its primitive differentiation potential *in vitro*, and the primitive hematopoiesis to which it presumably would have contributed had it not been disaggregated from its EB environment and targeted with Bcr/Abl.[17]

Non-oncogenic Engraftment

With the formal demonstration that it was possible to target and expand repopulating hematopoietic blast cells from EBs, we turned to inducible expression systems with the aim of enabling nononcogenic engraftment. We began by investigating conditional regulation of signaling from the c-Mpl receptor. In addition to lineage-specific effects on megakaryocytes,[18,19] the cytokine thrombopoietin and its receptor, c-Mpl, had been shown to be required for proper HSC self-renewal,[20,21] and had been used to expand immature hematopoietic progenitors from bone marrow.[22] To enable inducible Mpl signaling in our model, CCE ES cells were infected with a retrovirus (provided by Anthony Blau, University of Washington, Seattle, WA) carrying the F36V mutant of the FKBP dimerization domain fused to the cytoplasmic domain of c-Mpl[23,24] allowing for activation of Mpl signaling in response to a chemical inducer of dimerization (CID), AP20187. The virus carried an IRES-GFP

[17] R. C. R. Perlingeiro, M. Kyba, and G. Q. Daley, Clonal analysis of differentiating embryonic stem cells reveals a hematopoietic progenitor with primitive erythroid and adult lymphoid-myeloid potential, *Development* **128**, 4597–4604 (2001).

[18] K. Kaushansky, S. Lok, R. D. Holly, V. C. Broudy, N. Lin, M. C. Bailey, J. W. Forstrom, M. M. Buddle, P. J. Oort, F. S. Hagen, G. Roth, T. Papayannopoulou, and D. Foster, Promotion of megakaryocyte progenitor expansion and differentiation by the c-Mpl ligand thrombopoietin, *Nature* **369**, 568–571 (1994).

[19] F. Wendling, E. Maraskovsky, N. Debili, C. Florindo, M. Teepe, M. Titeux, M. Methia, J. Breton-Gorius, D. Cosman, and W. Vainchenker, cMpl ligand is a humoral regulator of megakaryocytopoiesis, *Nature* **369**, 571–574 (1994).

[20] S. Kimura, A. W. Roberts, D. Metcalf, and W. S. Alexander, Hematopoietic stem cell deficiencies in mice lacking c-Mpl, the receptor for thrombopoietin, *Proc. Natl. Acad. Sci. USA* **95**, 1195–1200 (1998).

[21] G. P. Solar, W. G. Kerr, F. C. Zeigler, D. Hess, C. Donahue, F.J.d Sauvage, and D. L. Eaton, Role of c-mpl in early hematopoiesis, *Blood* **92**, 4–10 (1998).

[22] H. Ku, Y. Yonemura, K. Kaushansky, and M. Ogawa, Thrombopoietin, the ligand for the Mpl receptor, synergizes with steel factor and other early acting cytokines in supporting proliferation of primitive hematopoietic progenitors of mice, *Blood* **87**, 4544–4551 (1996).

[23] L. Jin, N. Siritanaratkul, D. W. Emery, R. E. Richard, K. Kaushansky, T. Papayannopoulou, and C. A. Blau, Targeted expansion of genetically modified bone marrow cells, *Proc. Natl. Acad. Sci. USA* **95**, 8093–8097 (1998).

[24] L. Jin, H. Zeng, S. Chien, K. G. Otto, R. E. Richard, D. W. Emery, and C. A. Blau, *In vivo* selection using a cell-growth switch, *Nat. Genet.* **26**, 64–66 (2000).

reporter, which allowed us to select high-level, stable expressers. The top 10% of GFP-expressing cells were sorted by FACS, expanded, and sorted a second time in the same way. In the third round of FACS, the brightest 1% of cells was selected for single cell cloning by sorting directly into 96-well dishes. After expansion to clonal cell lines, FKBPMpl-expressing ES cells were tested for their ability to give rise to hematopoietic blast cells on OP9 in the presence of the CID. Day 5 EBs were disaggregated, and 10^5 cells were plated in a single well of a 6-well dish of OP9, in the same medium used to expand Bcr/Abl-induced cells, with or without 100 n*M* AP20187. Although all clones expressed high levels of GFP as undifferentiated ES cells, the majority of clones silenced the provirus with differentiation. Silencing with differentiation is a problem that confounds many ES cell gene expression systems (see below). Nevertheless, several clones were able to generate a hematopoietic cell population whose expansion was now dependent on the presence of the CID (Fig. 2), indicating that in these clones, the hematopoietic lineage is competent for expression of the provirus, perhaps due to a fortuitous integration site. Like the Bcr/Abl-induced blast cells, the FKBPMpl cells grew semi-attached to the OP9 stroma, with a rounded morphology and relatively scant basophilic cytoplasm (Fig. 2). By FACS analysis (Table I) they were also superficially similar to the Bcr/Abl-induced cells. We injected nine sublethally conditioned (500 Rad) mice with 5×10^6 cells each, and treated five of these with CID (2 mg/kg per day AP20187, by intraperitoneal injection). Unlike the results we observed for Bcr/Abl, neither the CID-treated mice, nor the untreated mice, were repopulated with these cells. Although this model was not exhaustively exploited, this result demonstrates that continual activation of signaling pathways that are sufficient to generate a blast cell outgrowth *in vitro* is not necessarily sufficient to enable cellular survival and maintenance in adult recipients. It also suggests that in spite of superficial similarities, the cells generated by activation of different regulatory pathways may be intrinsically different.

In order to test regulators other than dimerizable receptors, we sought a more generic conditional expression system. An attractive candidate system developed by Wutz *et al.*[25] allowed for doxycycline-induced gene expression and coexpression of GFP from a bidirectional promoter. This system makes use of the reverse tetracycline transactivator (rtTA[26])

[25] A. Wutz, T. P. Rasmussen, and R. Jaenisch, Chromosomal silencing and localization are mediated by different domains of Xist RNA, *Nat. Genet.* **30**, 167–174 (2002).

[26] M. Gossen, S. Freundlieb, G. Bender, G. Muller, W. Hillen, and H. Bujard, Transcriptional activation by tetracyclines in mammalian cells, *Science* **268**, 1766–1769 (1995).

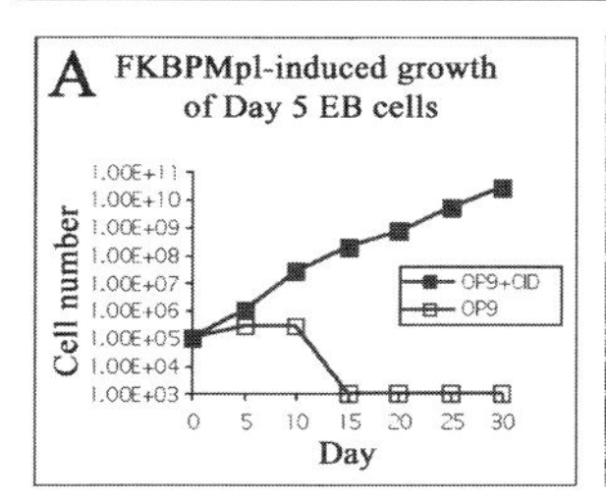

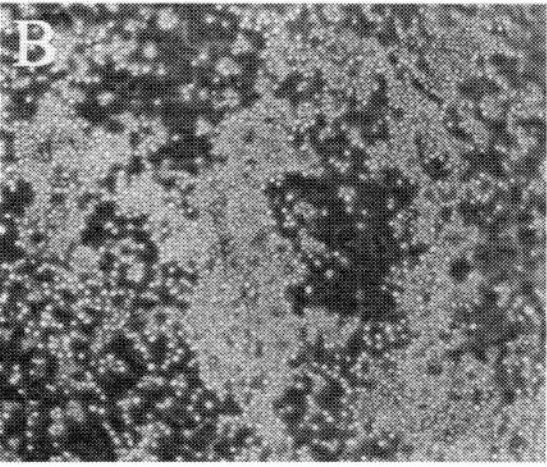

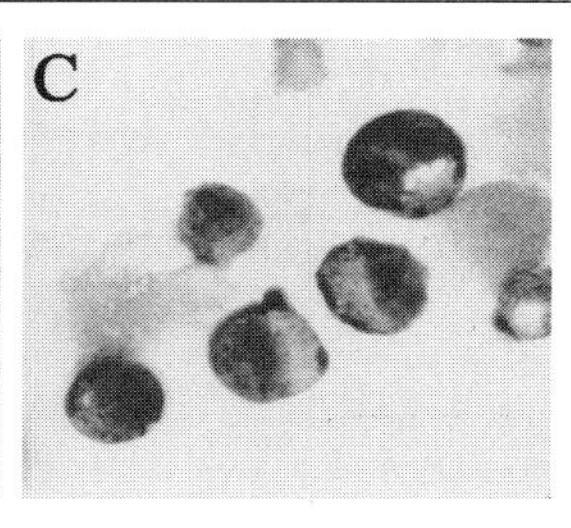

FIG. 2. FKPBMpl cells. (A) Cell growth, expressed as cumulative cell number, over time, in the presence and absence of CID. (B) Colony morphology on OP9 in the presence of CID. (C) Cellular morphology of CID-induced cells grown on OP9. (See Color Insert.)

TABLE I
IMMUNOPHENOTYPE OF EB-DERIVED CELLS *IN VITRO*

	Antigen	Bcr/Abl Cells	FKBPMpl Cells	HoxB4 Cells
Myeloid	Gr-1	0.25	0.1	5.6
	Mac-1	0.44	0.5	21.0
Erythroid	Ter119	0.7	6.4	0.7
Lymphoid	B220	0	2.1	0.6
	CD4	nd	1.9	0
	CD8	nd	0	0
Progenitor/megakaryocytic	CD41	nd	nd	47.8
Pan-hematopoietic	CD45	0.7	2.4	17.0
HSC	Sca-1	0.5	7.1	5.4
HSC/Progenitor	c-Kit	30.7	36	80.7
	AA4.1	0	1.4	0.7
HSC/Endothelial	CD31	0.0	9	78.0
	CD34	0.3	0.2	0.5
	Flk-1	0	0.2	0

A representative FACS analysis is shown for each of: a clone of Bcr/Abl-induced cells initiated on OP9 and expanded in liquid culture, FKBPMpl-induced cells, and HoxB4-induced cells, both initiated and expanded on OP9.

integrated into a constitutive locus, ROSA26,[27] allowing for transactivator expression in all tissues and cell types. The inducible locus is 5′ to the HPRT gene on the X-chromosome, and has a single loxP site, which allows for the

[27] B. P. Zambrowicz, A. Imamoto, S. Fiering, L. A. Herzenberg, W. G. Kerr, and P. Soriano, Disruption of overlapping transcripts in the ROSA beta geo 26 gene trap strain leads to widespread expression of beta-galactosidase in mouse embryos and hematopoietic cells, *Proc. Natl. Acad. Sci. USA* **94**, 3789–3794 (1997).

lox-in of a circular plasmid carrying any gene of interest. We tested this system with several candidate genes and found that although transgene expression was robust in ES cells, it was severely attenuated in differentiated EB cells. When a circular plasmid is integrated into a single lox-P site, the entire length of the plasmid separates the constitutive locus, HPRT, from the inducible locus (Fig. 3, original orientation). We reasoned that this extra distance may push the inducible locus out of the constitutively open transcriptional domain occupied by HPRT, subjecting the inducible locus to silencing with differentiation. We therefore reengineered the system, inverting its orientation such that the inducible locus stays proximal to HPRT, and eliminating the coinducible GFP reporter (Fig. 3, inverted orientation). This targeting ES cell line, which we named Ainv15, gave robust, inducible expression of transgenes even in highly differentiated cells.[28]

Lox-in to Derive Inducible ES Cell Lines from Ainv15 Targeting Cells

- Coelectroporate 20 μg each of the targeting plasmid carrying the inducible gene of interest and CRE-expression plasmid into 8×10^6 Ainv15 cells in 800 μl PBS. (We do this electroporation at room temperature on the BioRad gene pulser with capacitance extender, using the settings: 0.25 V, 500 μFD.)
- Plate the electroporated cells on 10 cm dishes with dense neo-resistant MEFs.
- Begin selection in G418 the next day (350 μg/ml) and maintain until colonies appear around day 10–14. In the first few days, cells will be very dense, feed twice per day. Thereafter, feed once per day.
- Pick colonies by flooding the dish with PBS, and extracting individual colonies with a P20 pipette. Transfer to 100 μl of 0.25% trypsin/EDTA and pipette to disrupt the colony. Incubate at 37°C for 2 min, and disrupt once again by pipetting. Add 900 μl of medium with 10% serum, and collect the cells of the colony by centrifugation. Replate the disrupted contents onto MEFs in 12-well dishes.
- Integration can be detected by using the following primers, which amplify across the loxP site to give band of approximately

[28] M. Kyba, R. C. R. Perlingeiro, and G. Q. Daley, HoxB4 confers definitive lymphoid-myeloid engraftment potential on embryonic stem cell and yolk sac hematopoietic progenitors, *Cell* **109**, 29–37 (2002).

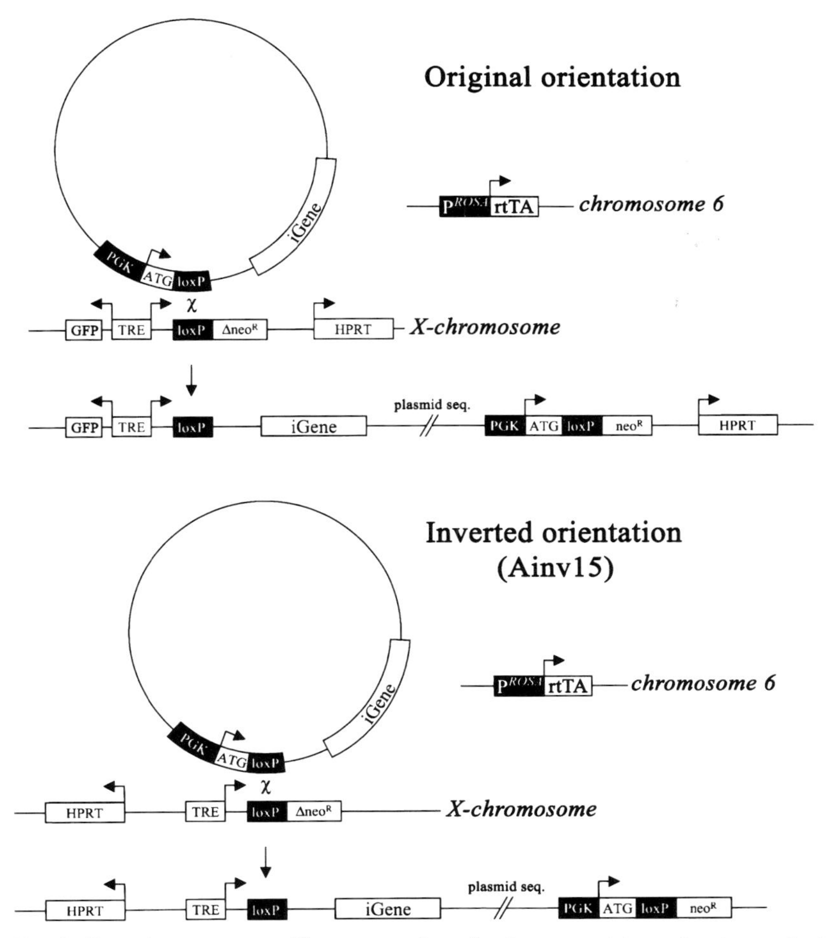

FIG. 3. Targeting constructs. Chromosome 6 carries the reverse tetracycline transactivator (rtTA) integrated into the ROSA26 locus. The X-chromosome carries modifications 5′ to the HPRT gene. After lox-in (shown by the symbol chi) of the targeting plasmid, the formerly nonfunctional neo gene acquires a start codon and a promoter/enhancer, making it functional and enabling selection for integration of the plasmid. In addition, the gene of interest, formerly carried by the plasmid, is now integrated downstream of the tetracycline responsive element (TRE), enabling its doxycycline-regulated expression. In the original configuration, the TRE also drives expression of a GFP reporter gene. After integration of the plasmid, the neo gene remains proximal to HPRT, while the inducible gene is located distally, separated by several kilobases of plasmid sequence. In this configuration, the inducible gene is subjected to silencing with differentiation. In the inverted configuration, the GFP reporter is eliminated, and the target locus is inverted relative to HPRT. Lox-in places the inducible gene proximal to HPRT and the neo gene distal, separated by several kilobases of plasmid sequence. In this configuration, the inducible gene is not subject to silencing with differentiation. PGK, phospho-glycero-kinase enhancer; P^{ROSA}, ROSA26 enhancer/promoter; iGene, inducible gene; loxP, recognition sequence for Cre recombinase, Δneo, deletion mutant of the neomycin (G418) resistance gene; ATG, start codon for the neo gene.

420 bp if integration has been successful: LoxinF: 5′-ctagatctcgaag-gatctggag-3′ LoxinR: 5′-atactttctcggcaggagca-3′ Cycle conditions: 45 sec at 95°C, 1 min at 60°C, 1 min at 72°C, repeat 29×, using Taq polymerase and Promega PCR buffer supplemented with 1.25 m*M* $MgCl_2$.

We selected HoxB4 as the first candidate on the basis of both information on its expression pattern (detected in the HSC but not in more differentiated hematopoietic progenitors,[29] nor in yolk sac[30]) and the competitive engraftment advantage that it affords HSC after bone marrow transplantation.[31] We introduced HoxB4 into the inducible locus and began testing its effects on the hematopoiesis of embryoid bodies. One of the great advantages of our inducible system is that it enables control over cells within their native EB environment, however our standard methodology for generating EBs, suspension of ES cells in methylcellulose, was not amenable to adding or removing an inducing agent at different time points in EB development. We therefore switched to a strictly liquid culture system where EBs were initiated in hanging drops (Fig. 4).

Hanging Drop EB Culturess

- Use an 8-well multichannel pipettor to plate approximately 300 drops per 15 cm non-tissue culture-treated dish, with each drop containing 100 ES cells in 10 μl of differentiation medium (see above, nb. no methylcellulose).
- Invert the dishes and incubate for 2 days at 37°C in 5% CO_2. ES cells quickly descend to the nadir of the drop, aggregate, and form a single EB per drop.
- Collect EBs after 48 hr, by flushing the dish with PBS, transferring to a 15 ml tube, and allowing the EBs to sediment by gravity for 3 min. Aspirate the liquid and collect the EBs in 10 ml of fresh differentiation medium. Transfer to 10 cm bacterial-grade dishes, and culture under

[29] G. Sauvageau, P. M. Landsdorp, C. J. Eaves, D. E. Hogge, W. H. Dragowska, D. S. Reid, C. Largman, H. J. Lawrence, and R. K. Humphries, Differential expression of homeobox genes in functionally distinct $CD34^+$ subpopulations of human bone marrow cells, *Proc. Natl. Acad. Sci. USA* **91**, 12223–12227 (1994).

[30] K. E. McGrath and J. Palis, Expression of homeobox genes, including an insulin promoting factor, in the Murine Yolk Sac at the time of hematopoietic initiation, *Mol. Reprod. Dev.* **48**, 145–153 (1997).

[31] G. Sauvageau, U. Thorsteinsdottir, C. J. Eaves, H. J. Lawrence, C. Largman, P. M. Lansdorp, and R. K. Humphries, Overexpression of HOXB4 in hematopoietic cells causes the selective expansion of more primitive populations in vitro and in vivo, *Genes Dev.* **9**, 1753–1765 (1995).

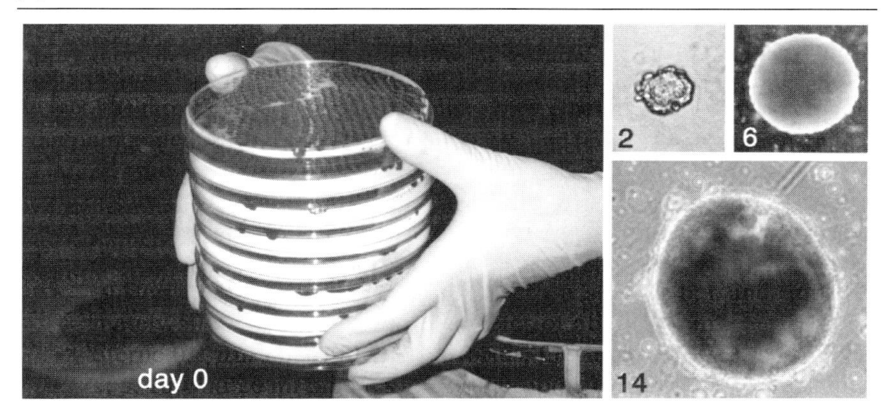

FIG. 4. Hanging drop EBs. Hanging drops are plated on day 0. The EBs that form are shown at days 2, 6, and 14, postplating. The first visual sign of hematopoiesis in the form of hemoglobinizing areas can be seen under dark-field at day 6. By day 14, well-hemoglobinized blood islands are apparent. (See Color Insert.)

slow swirling conditions on a rotating shaker (50 rpm) set up inside of a dedicated 37°C/5% CO_2 incubator.

- Feed the swirling cultures every two days by replacing half of the spent medium with fresh differentiation medium.

These techniques (hanging drops, use of bacterial-grade, non-tissue culture-treated dishes, and slow swirling) are all necessary to prevent attachment of the EBs to their dishes. We have observed that if EBs are allowed to attach, their complex structure quickly disintegrates, their cellular content spreads across the substrate, and they do not undergo hematopoietic differentiation.[32]

HoxB4-Induction of Hematopoietic Cultures

- Set up one dish of hanging drops using the iHoxB4 (inducible HoxB4) ES cells.
- Culture as described above, however induce HoxB4 expression at day 4 (96 hr) of EB development by adding doxycycline to the differentiation medium at a final concentration of 1 μg/ml.
- Allow the EBs to develop for two more days.

[32] S. M. Dang, M. Kyba, R. C. R. Perlingeiro, G. Q. Daley, and P. W. Zandstra, Efficiency of embryoid body formation and hematopoietic development from embryonic stem cells in different culture systems, *Biotechnol. Bioeng.* **78**, 442–453 (2002).

- On day 6, collect EBs by gravity, wash once with PBS, and collect again by gravity. Dissociate the EB cells by adding 0.5 ml of 0.25% trypsin and incubating 2 min at 37°C. Add 5 ml of IMDM/10% FCS and passage repeatedly through a 5 ml pipette until the EBs have disintegrated, and collect by centrifugation. Plate 10^5 cells per well onto semi-confluent 6-well dishes of OP9, in IMDM/10% FCS with 40 ng/ml each of murine VEGF (vascular endothelial growth factor; Peprotech), and human TPO (thrombopoietin; Peprotech), and 100 ng/ml of each of human SCF (stem cell factor; Peprotech), and human FL (FLT3 ligand; Peprotech). Maintain induction of HoxB4 by including doxycycline at 1 μg/ml. This cytokine cocktail was selected based on the developing understanding of the cytokine requirements of the definitive HSC as well as those of the EB-derived hemangioblast.
- After several days, colonies of semi-adherent cells arise on the OP9 stromal layer, and these can be passaged by trypsinization (collecting both adherent and nonadherent cells) onto fresh OP9 and expanded in the same medium.

HoxB4 cultures are rich in CFC, especially the CFU-GEMM (colony-forming unit-granulocyte, erythrocyte, macrophage, megakaryocyte) and by cytospin are dominated by immature blast cells similar in morphology to those produced by the Bcr/Abl experiment (Fig. 5). By antibody staining, the cells are prominently c-kit and PECAM-positive, however cultures contain both positive and negative cells for CD45, Sca-1, and CD41, and usually a small frequency of AA4.1-positive cells (Table I). The proportions

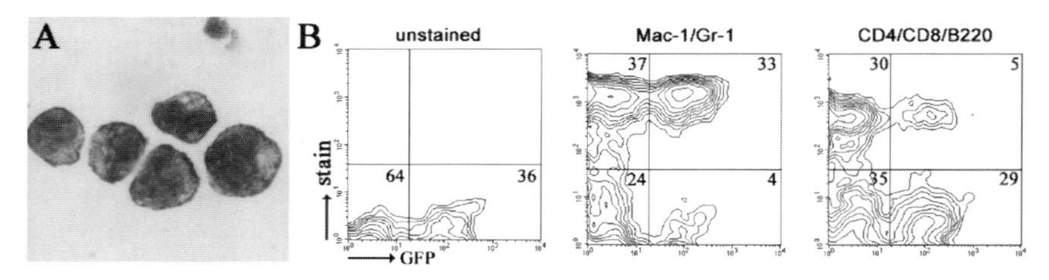

FIG. 5. HoxB4-induced cells. (A) Morphology of HoxB4-induced cells grown on OP9 stromal cells. (B) FACS analysis of peripheral blood of a recipient mouse three weeks posttransplant. GFP expression, which marks donor cells, is measured on the X-axis. Antibody staining is measured on the Y-axis. The percentage of cells falling into each quadrant is shown in the upper right-hand corner. In the first FACS, cells were stained with a nonspecific antibody, which does not recognize any blood cell type. In the second FACS, a cocktail of myeloid-specific antibodies was used (Gr-1 and Mac-1) to label circulating granulocytes and monocytes. In the third FACS, a cocktail of lymphoid-specific antibodies was used (B220, CD4, and CD8) to label circulating lymphocytes. (See Color Insert.)

of cells positive for these latter markers is somewhat variable from experiment to experiment, however we have observed that good adult engraftment potential correlates best with higher proportions of Sca-1 and AA4.1. Cells positive for lineage-specific markers are also often seen: cultures always contain some fraction Gr-1 and Mac-1 positive (myeloid-committed) cells, and may contain a smaller fraction of B220, CD4, or CD8 (lymphoid-committed) positive cells.

For the purposes of detecting donor contribution to adult recipients, cells were labeled by infection with a GFP-expressing retrovirus, sorted for GFP-positivity, and replated onto fresh OP9. After expansion, 2×10^6 cells were injected into lethally irradiated (2 doses of 500 Rad, separated by 4 hr) syngeneic mice (129 Ola, Harlan Laboratories). Cells home rapidly to the bone marrow, and we observe contribution to the peripheral blood almost immediately. Myeloid contribution is always more prevalent than lymphoid contribution, as shown in Fig. 5, and overall contributions vary from mouse to mouse, ranging, in the case of the iHoxB4 cells, from a few percent to a high of 50%, or higher with retroviral transduction of HoxB4 (see below). It is not necessary to maintain expression of HoxB4 in the recipients, although doing so (by adding doxycycline to the drinking water) correlates somewhat with higher levels of donor engraftment.

The hematopoietic progenitors of the early EB that we are targeting have a primitive embryonic hematopoietic fate specification. By causing ectopic expression of HoxB4 *in vitro*, this fate specification is perturbed, and certain aspects of definitive hematopoietic fate are conferred upon the cells, most importantly the ability to repopulate adult recipients. This reprogramming is also evident by comparing the expression of certain genes between the Bcr/Abl-induced and the HoxB4-induced *in vitro* populations. The HoxB4 cells downregulate primitive globin, β-H1, and upregulate definitive markers involved in bone marrow homing, CXCR4[33] and Tel,[34] whereas the Bcr/Abl cells do not downregulate β-H1, and have low levels of CXCR4. However, the HoxB4 cells' relative ineffectiveness at competing with endogenous bone marrow HSC (there is a very strong requirement for lethal donor conditioning, and even with the injection of 2×10^6 cells we still see fractional engraftment) as well as their propensity to contribute to myeloid over lymphoid lineages in transplanted recipients,

[33] A. Peled, I. Petit, O. Kollet, M. Magid, T. Ponomaryov, T. Byk, A. Nagler, H. Ben-Hur, A. Many, L. Shultz, O. Lider, R. Alon, D. Zipori, and T. Lapidot, Dependence of human stem cell engraftment and repopulation of NOD/SCID mice on CXCR4, *Science* **283**, 845–848 (1999).

[34] L. C. Wang, W. Swat, Y. Fujiwara, L. Davidson, J. Visvader, F. Kuo, F. W. Alt, D. G. Gilliland, T. R. Golub, and S. H. Orkin, The TEL/ETV6 gene is required specifically for hematopoiesis in the bone marrow, *Genes Dev.* **12**, 2392–2402 (1998).

suggests that they have not been completely respecified as bone marrow HSC. Significant additional work remains to understand the relationship between HoxB4-induced, ES-derived HSCs compared to their counterparts in fetal liver, cord blood, and adult bone marrow.

While the inducible system for expression of HoxB4 described above has many advantages over the retroviral transduction approach described for Bcr/Abl, we have also used retroviral expression of HoxB4 to the same end. In order to facilitate hematopoietic contribution from ES cells derived via therapeutic cloning, with a minimal number of additional genetic manipulations, we infected day 6 EB cells with an MSCV-HoxB4-iresGFP retrovirus and expanded these cells under the same conditions used to expand the inducible HoxB4 cells, however without doxycycline. HoxB4 efficiently targets the cell with the ability to form semi-adherent colonies on OP9, and rapidly expands it *in vitro*. Because the recipient mice in this experiment carried a null mutation for Rag2, the ability of HoxB4-induced EB-derived hematopoietic blast cells to produce lymphoid progeny could be monitored sensitively in the complete absence of host-derived lymphocytes. Under these circumstances we clearly observed the presence of donor-derived lymphocytes, by measuring genomic rearrangement of the IgH and TCR loci, the presence of circulating antibody, and the presence of some circulating IgM-positive donor cells, however the levels of lymphoid repopulation were much lower than the levels of myeloid.[35] While this may indicate that HoxB4 favors myeloid differentiation of EB-derived progenitors, it may also reflect an underlying differentiation bias of these formerly primitive hematopoietic progenitors.

Therapeutic Repopulation

Past work with genetically unmodified ES cells differentiated as EBs has shown that such cells are severely limited in their capacity to successfully chimerize the adult hematopoietic system.[36] Interestingly, the most successful attempts, in which some lymphoid repopulation was seen in Rag1$^{-/-}$ immunodeficient mice, made use of cells from 3-week-old EBs.[37] As discussed above, the origins of hematopoiesis in the EB occur in the first

[35] W. M. Rideout, K. Hochedlinger, M. Kyba, G. Q. Daley, and R. Jaenisch, Correction of a genetic defect by nuclear transplantation and combined cell and gene therapy, *Cell* **109**, 17–27 (2002).

[36] A. M. Müller and E. A. Dzierzak, ES cells have only a limited lymphopoietic potential after adoptive transfer into mouse recipients, *Development* **118**, 1343–1351 (1993).

[37] A. J. Potocnik, H. Kohler, and K. Eichmann, Hemato-lymphoid *in vivo* reconstitution potential of subpopulations derived from *in vitro* differentiated embryonic stem cells, *Proc. Natl. Acad. Sci. USA* **94**, 10295–10300 (1997).

week. This suggests that under some circumstances, the late stage EB may generate an environment conducive to a primitive-to-definitive respecification of hematopoietic progenitors, although as evidenced by the difficulty of repeating these results, the intrinsically variable nature of the differentiation process *in vitro* makes these circumstances elusive. We have targeted EB day 6 hematopoietic progenitors due both to their abundance and to the consistency with which they can be generated *in vitro*, however expression of HoxB4 at later times, when the environment may be more conducive to induction of definitive hematopoiesis, may give more completely respecified cells.

While genetic modification of murine ES cells has allowed us to model their differentiation to hematopoietic repopulating cells, less invasive methods will likely be needed before a hematopoietic therapy is available from human ES cells. In this regard, the challenge will be to identify environmental conditions that induce the same effects, to recapitulate *in vitro* the environment that the primitive HSC experiences in the embryo at the time of its definitive transition. This may involve the use of signaling molecules normally involved in these processes, as well as synthetic constructs such as transducable proteins, small molecule agonists or inhibitors, and transiently expressed nucleic acids. In any case, we envision that therapeutic differentiation protocols will ultimately combine elements of both instruction of lineage choice and selection and expansion of target cells.

[9] The *In Vitro* Differentiation of Mouse Embryonic Stem Cells into Neutrophils

By JONATHAN G. LIEBER, GORDON M. KELLER, and G. SCOTT WORTHEN

Abstract

A reliable and effective *in vitro* differentiation system is described for the production of functional neutrophils from mouse embryonic stem (ES) cells. A three-step culture method was developed that enables abundant production and effective harvesting of mature neutrophils at high purity without sorting. Utilization of the OP9 stromal cell line, which does not produce macrophage colony stimulating factor (M-CSF) was found to enhance the number, percentage and duration of neutrophils produced. Based on a number of criteria, morphologically and functionally mature

0076-6879/2003 $35.00

neutrophils can be produced using this method in approximately 16 days. This differentiation system provides a useful model system for studying neutrophil development and maturation *in vitro* and the many factors that regulate this process. Morphologically mature ES-derived neutrophils can be grown in culture that produce superoxide, flux calcium and directionally respond to the chemoattractant MIP-2. In addition, they express the granulocyte markers Gr-1 and the neutrophil specific antigen, as well as specific chloroacetate esterase.[1] Interestingly, during their development in culture, regional areas of apparent neutrophil production can be identified that recapitulate certain aspects of the marrow environment. As ES cells can be genetically modified, this system enables evaluation of the effects of specific genetic alterations on neutrophil differentiation and function.

Introduction

In man, approximately 120 billion neutrophils are produced daily, necessitating the need for dynamic control of their production and differentiation.[2] As neutrophils comprise a key component of the innate immune system, a better understanding of their development and regulation is necessary for appreciating their roles in normal homeostasis and disease.[3,4]

Recently, several studies have demonstrated that the *in vitro* differentiation of ES cells into a number of hematopoietic lineages can be used to study these processes in a controlled environment. Importantly, these studies have revealed that the *in vitro* differentiation of ES cells largely recapitulates the early stages of murine hematopoiesis with respect to the effects of cytokines on lineage development and the timing of specific hematopoiesis.[5–7]

As embryonic stem cells are pluripotent and readily genetically modifiable, they are well suited for studying the effects of genetic changes *in vitro* and *in vivo*. They also enable evaluation of the effects of specific

[1] J. G. Lieber, S. Webb, B. T. Suratt, S. K. Young, G. L. Johnson, G. M. Keller, and G. S. Worthen, submitted (2003).
[2] G. Cartwright, J. Athens, and M. Wintrobe, *Blood* **24**, 780 (1964).
[3] D. F. Bainton, *Br. J. Haematol.* **29**, 17 (1975).
[4] D. F. Bainton, *Prog. Clin. Biol. Res.* **13**, 1 (1977).
[5] M. V. Wiles and G. Keller, *Development* **111**, 259 (1991).
[6] T. Nakano, *Int. J. Hematol.* **65**, 1 (1996).
[7] G. Keller, M. Kennedy, T. Papayannopoulou, and M. V. Wiles, *Mol. Cell. Biol.* **13**, 473 (1993).

mutations *in vitro* that are embryonic lethal. However, effective *in vitro* differentiation of ES cells into neutrophils has only recently been reported.[1]

Stromal elements have previously been identified as being important for fetal as well as adult hematopoiesis.[8,9] Bone marrow-derived stroma has been used by several investigators to enhance *in vitro* hematopoiesis from ES cells but effective production of neutrophils has not been described.[8–12] Utilization of specific stroma for enhancing differentiation and support of particular cell types has not been adequately studied. As different stromal cells vary in their production of a wide variety of factors including cytokines, they are not equivalent in promoting and sustaining differentiation, suggesting the importance of utilizing specific stromal types for particular applications.

We have developed a three-step *in vitro* differentiation system for the differentiation of functional neutrophils from ES cells. In the first step, embryoid bodies (EBs) are formed to produce hematopoietic progenitors. In the Secondary Differentiation Mix, hematopoietic progenitors are expanded in a gp-130 based cytokine mix, while in the Tertiary Neutrophil Differentiation Mix, the neutrophil progenitors are further expanded, and differentiated into functional neutrophils. Importantly, in the secondary and tertiary differentiation steps, the differentiating cells are replated onto semiconfluent OP9 stromal cells, which were derived from the osteopetrotic mouse and hence lack functional M-CSF.[6,13,14] As M-CSF augments production and maintenance of macrophages and apparently impairs development of other hematopoietic lineages from ES cells,[15] we reasoned that utilization of the OP9 stromal cell line might be most effective for the *in vitro* production of neutrophils from ES cells.[6]

The methods described enable effective and reasonably sustained production of functional neutrophils from ES cells, an advance that in turn permits careful dissection of the process of neutrophil development and the factors that affects this. Furthermore, this system allows for the assessment of functions of particular genes on neutrophil differentiation, function and regulation. The described differentiation system simplifies handling and

[8] T. D. Allen, *Ciba Found. Symp.* **84**, 38 (1981).

[9] S. Rafii, R. Mohle, F. Shapiro, B. M. Frey, and M. A. Moore, *Leuk. Lymphoma* **27**, 375 (1997).

[10] J. C. Gutierrez-Ramos and R. Palacios, *Proc. Natl. Acad. Sci. USA* **89**, 9171 (1992).

[11] D. M. Haig, *J. Comp. Pathol.* **109**, 259 (1993).

[12] T. Nakano, *Semin. Immunol.* **7**, 197 (1995).

[13] H. Kodama, M. Nose, S. Niida, and S. Nishikawa, *Exp. Hematol.* **22**, 979 (1994).

[14] H. Yoshida, S. Hayashi, T. Kuooisada, M. Ogawa, S. Nishikawa, H. Okamura, T. Sudo, L. D. Shultz, and S. Nishikawa, *Nature* **345**, 442 (1990).

[15] T. Nakano, H. Kodama, and T. Honjo, *Science* **265**, 1098 (1994).

retrieval of cells compared to semi-solid methylcellulose based systems. In addition, at optimal times of neutrophil production, mature neutrophils generally account for 80–90% of the cell population in suspension, frequently permitting harvesting without the need for sorting.

Materials and Methods

Stock Media

1. Minimal Essential Medium Alpha (MEM-α), Gibco, Rockville, MD #32561-037.
2. Dulbecco's Modified Eagle Medium (DMEM), Gibco #11965-092.
3. Iscove's Modified Dulbecco's Medium (IMDM), Gibco, #12440-053.

Successful maintenance and differentiation of ES cells requires that all reagents, including media be of the highest quality and purity. Use of premade stock mediums such as DMEM, IMDM (Gibco #11965-092 and 12440-053, respectively) are readily available. Alternatively, these mediums can be made in house from packets of powdered medium (Gibco # 12100-103 and 12200-85, respectively). As described below, for our studies, we utilized premade stock mediums from Gibco due to their consistently high quality control and convenience. We routinely add additional (2 m*M*) L-glutamine to our differentiation mixes, as L-glutamine rapidly degrades.

ES Cell Growth and Transition Mediums

1. *DMEM-ES Medium:* Contains 83% by volume DMEM, 2 m*M* L-glutamine, 15% FBS (tested for ES cell growth), 1% Leukemia Inhibitory Factor (LIF) conditioned medium (a gift from Dr. Gordon Keller), 1.5×10^{-4} *M* monothioglycerol (MTG) and 100 U/ml penicillin and 100 μg/ml streptomycin.
2. *IMDM-ES Transition Medium:* Contains 83% by volume IMDM, 2 m*M* L-glutamine, 15% FBS (tested for ES cell growth), 1% Leukemia Inhibitory Factor (LIF) conditioned medium (a gift from Dr. Gordon Keller), 1.5×10^{-4} *M* monothioglycerol (MTG) and 100 U/ml penicillin and 100 μg/ml streptomycin.
3. *OP9 Growth Medium:* Contains 20% FBS, 80% by volume MEM-α and 0.75×10^{-4} *M* MTG.

Neutrophil Differentiation Mediums

1. *Primary Differentiation Medium:* Contains 15% heat inactivated FBS (tested for differentiation of ES cells), 2 m*M* L-glutamine, 4.5×10^{-4} *M*

MTG, 50 μg/ml of L-ascorbic acid, 5% protein-free hybridoma medium, 78% by volume IMDM and 100 U/ml penicillin and 100 μg/ml streptomycin.

2. *gp-130 Secondary Differentiation Medium:* Contains 10% pretested heat inactivated ES differentiation FBS, 10% horse serum, 5% protein-free hybridoma medium, 25 ng/ml of OSM, 10 ng/ml basic FGF, 5 ng/ml IL-6, 1% KL supernatant (conditioned medium from CHO cells transfected with an expression vector generously provided by the Genetics Institute, Cambridge MA), 5 ng/ml IL-11, and 1 ng/ml of rLIF (all recombinant cytokines from R&D) in 79% by volume IMDM containing 100 U/ml penicillin and 100 μg/ml streptomycin and 1.5×10^{-4} *M* MTG.
3. *Tertiary Neutrophil Differentiation Medium:* Contains 10% platelet depleted serum, 2 m*M* L-glutamine, 88% by volume IMDM, 100 U/ml penicillin, 100 μg/ml streptomycin, 1.5×10^{-4} *M* MTG, 60 ng/ml G-CSF (Amgen, Thousand Oaks), 3 ng/ml GM-CSF, and 5 ng/ml IL-6 (all recombinant cytokines except G-CSF from R&D).

Miscellaneous Materials and Reagents

1. Gelatin, 0.1% in water, StemCell Technologies, Vancouver, BC #07903 or 0.1% porcine gelatin Sigma, St Louis, #G-1890, in PBS
2. Ascorbic acid (AA), Sigma, St. Louis, #A4544
3. L-Glutamine, 200 m*M*, Gibco #25030-08
4. Penicillin/Streptomycin, Gibco, Grand Island, NY #15070-063
5. Protein-Free Hybridoma medium (PFHM-II), Gibco #12040
6. Phosphate Buffered Saline (PBS), Gibco #10010-023
7. Monothioglycerol (MTG), Sigma, #M-6145
8. Penicillin/Streptomycin, Gibco #15070-063
9. Syringe, 10 cc, B-D, Franklin Lakes, NJ, #309604
10. Syringe needles, Monoject Manufactured by Sherwood Medical, St. Louis, MO, #250115
11. Trypsin/EDTA, StemCell Technologies #07901.

Sera

1. FBS for growth and maintenance of undifferentiated ES cells. (Summit Biotech #S-100-05 or pretested FBS from StemCell Technologies, Vancouver, B.C. #06901.)

2. FBS for differentiation of ES cells. (Summit Biotechnology #FP-200-05, Ft. Collins, CO or pretested FBS from StemCell Technologies, Vancouver, B.C. #06900.)
3. FBS for growth and maintenance of OP9 stromal cells. (Summit Biotech, Ft. Collins, Co #S-100-05 or Gemini Products, Woodland CA #100-500.)
4. Fetal bovine plasma derived serum (PDS, platelet poor), Antech, Tyler TX, request Dr. Gordon Keller lab protocol serum.
5. Horse serum for differentiation of ES cells into neutrophils. (Untested horse serum can be obtained from Biocell Laboratories, #6402, Rancho Domingues, CA or if desired, horse serum specifically tested for myeloid differentiation can be obtained from StemCell Technologies, Vancouver, BC #06850.)

Different lots of Fetal Bovine Serum (FBS) vary in their ability to maintain ES cells in an undifferentiated growth-promoting state, as well as their ability to support *in vitro* differentiation of ES cells. Several vendors including StemCell Technologies, Vancouver, BC carry pretested FBS for the maintenance or differentiation of ES cells. Alternatively, one can evaluate different serum lots for efficacy. Wiles (1993) described several useful criteria for the selection of FBS.[16]

For production of neutrophils from ES cells, FBS was utilized that supported embryoid body (EB) formation by day 3, with distinct β-globin expression in 40–50% of day 8 EBs. Most importantly for our studies, this differentiation serum was able to support abundant neutrophil production. We have also successfully utilized tested ES cell growth promoting serum for routine culture of undifferentiated ES cells and differentiation supporting serum for effective neutrophil differentiation.

Sera requirements for growing OP9 stromal cells are not as fastidious as for ES cells and thus far, we have successfully utilized two different lots of FBS (one from Gemini Biosciences and the other from Summit Biotechnologies) for this purpose.

Heat inactivation of complement factors in completely thawed FBS should be done at 56°C for 45 min with mixing every 10 min. Following heat inactivation, serum should be aliquoted and stored at −20°C. After thawing, serum aliquots are used within one month to assure maintenance of serum containing growth factors.

Tested and untested horse serums were also used in the secondary differentiation mix. Horse serum contains significant levels of

[16] M. V. Wiles, *in* "Methods of Enzymology" (P.M.W.a.M.L. dePamphilis,, ed.), Vol. 225, p. 900. Academic Press, San Diego, 1993.

hydrocortisone among other factors, which has previously been shown to be important for formation of mature stroma in the Dexter long term bone marrow-derived culture system.[17] We have found that different lots of horse serum vary in their ability to enhance granulopoiesis in the ES/OP9 coculture system necessitating screening or using pretested horse serum to optimize production of neutrophils.

Recombinant Cytokines

The following recombinant cytokines can be obtained from R&D Systems, St. Louis, MN or in the case of G-CSF, Amgen, Thousand Oaks, CA.

1 G-CSF (Neupogen, Filgrastim, #NDC5513-530-01)
2 GM-CSF #415-ML
3 IL-6 #418-ML
4 IL-11 #418-ML
5 OSM #495-MO-025
6 Basic FGF #233-FB
7 LIF #449L.

Cytokines from Conditioned Medium

An excellent description of the preparation of conditioned medium for KL and LIF is provided by Keller *et al.* (2002).[18] For our studies, we utilized KL containing supernatant from conditioned medium from CHO cells transfected with an expression vector generously provided by the Genetics Institute, Cambridge, MA. LIF containing conditioned medium was obtained from CHO cells transfected with an expression vector expressing rLIF (a gift of Dr. Gordon Keller) that was constructed by Drs. Gordon Keller and Gary Johnson. Alternatively, 1000 U/ml of rLIF can be substituted.

Plasticware for Tissue Culture

For growing ES cells, flasks should be gelatin coated before use. For gelatin coating, use 2 or 6 mls of 0.1% gelatin per T-25 mm or T-75 mm

[17] T. M. Dexter, *J. Cell Physiol. Suppl.* **1**, 87 (1982).
[18] G. Keller, S. Webb, and M. Kennedy, *in* "Hematopoietic Development of ES Cells in Culture" (C.A.K.a.C.T. Jordan,, ed.), Vol. 63, p. 209. Humana Press, Totowa, 2002.

flask, respectively. The flasks should be coated for 20 min at room temperature before using and can be stored at 4°C for 2 weeks.

1 T-25 flasks (Greiner #690 160). Use for growth and expansion of undifferentiated ES cells. Gelatin coat flasks before use.
2 T-75 flasks (Greiner #658 170). Use for expansion of ES cells. Gelatin coat flasks before use.
3 Non-TC treated 60 mm×15 mm petri dishes (Fisher, Canada, #8-757-13-A). It is important to use the Canadian made 60 mm dishes, as the American made dishes with the same catalog number are not equivalent and the embryoid bodies adhere.
4 TC treated 6-well dishes (CorningCostar Corning, NY, #3516). Use for plating OP9 cells and ES-derived hematopoietic cells in the Secondary gp-130 and Tertiary Neutrophil Differentiation Mixes.

Reducing Agents

Reducing agents such as monothioglycerol (MTG) or dithiotreitol (DTT) are necessary for effective hematopoietic differentiation from ES and embryonal carcinoma cells.[5,19] We have utilized MTG at 1.5×10^{-4} *M* for ES cell differentiation and 0.75×10^{-4} *M* MTG for culturing OP9 stromal cells. Ascorbic acid was used at 50 μg/ml in the Primary Differentiation Mix. As there is variation between lots of MTG, it is important to test them to maximize effectiveness and to identify any lots that have toxic effects. This can be done by evaluating the effect of MTG on undifferentiated ES cell growth rates as well as formation of embryoid bodies and resulting hematopoietic progenitors.

Stock solutions (5 mg/ml) of L-ascorbic acid should be prepared using cold tissue culture grade water. Vortex the solution to adequately dissolve the ascorbic acid and 0.2 μm sterile filter before aliquoting into 1.5 ml tubes. Store aliquots at −20°C and do not reuse.

Routine culture of CCE-ES cells

CCE-ES cells[20,21] generally at passage 20 were used for these studies. For optimal consistency, CCE-ES cells from the same freeze down and passage should be used, although different passage numbers can be used if they are handled properly. For routine culture, ES cells were grown in DMEM-ES media on gelatinized plates and passaged every 2 or 3 days.

[19] R. Oshima, *Differentiation* **11**, 149 (1978).
[20] M. J. Evans and M. H. Kaufman, *Nature* **292**, 154 (1981).
[21] E. Robertson, A. Bradley, M. Kuehn, and M. Evans, *Nature* **323**, 445 (1986).

Two days prior to utilizing the ES cells for differentiation experiments, they were split into gelatinized T-25 flasks at a density of 0.5 to 1.0×10^5 cells/ml into IMDM-ES medium, which contained the same components as the DMEM-ES medium except IMDM was used.

OP9 Mouse Bone Marrow Stromal Cells

For culturing OP9 mouse bone marrow stromal cells, we utilize alpha MEM medium containing 20% FBS, 80% by volume MEM-α and 0.75×10^{-4} *M* MTG. The OP9 stromal cells should be passaged every 2–3 days at 1 : 6 to prevent excessive confluency and adipose conversion.

Utilization of an ES/OP9 Coculture Differentiation System for Efficient and Sustained Production of Mature Neutrophils from ES Cells

A three-step differentiation system was developed for effective production of neutrophils from mouse ES cells as outlined in Fig. 1. For reliable and optimal differentiation, mediums should be made within 24 hr of use.

Primary Differentiation for Production of Day 8 Embryoid Bodies

For induction of differentiation, CCE-ES cells should be plated at a density of 800–1000 cells/ml into a nontissue culture-treated petri dish containing 5 ml of Primary Differentiation Mix. If other ES lines are used, this concentration may have to be adjusted. Generally, ten to twenty 60 mm dishes of cells at 1000 cells/ml are set up to assure that an adequate number of neutrophils are produced for experimentation. After 6 days, the cultures should be fed with the same Primary Differentiation Mix.

gp-130 Secondary Differentiation Medium

After primary differentiation for 8 days, the EBs are harvested, settled for 15 min or centrifuged (225*g*, 5 min) and trypsinized using 5 ml of trypsin/45 ml volume of harvested EBs for 5 min at room temperature. After 3 min the EBs are disaggregated into a cell suspension using a 10 cc syringe with a 20 gauge needle. The cells are then washed in 20 ml of wash medium containing IMDM $+ 1.5 \times 10^{-4}$ *M* MTG containing 10% FBS, centrifuged (225*g*, 5 min) and resuspended in gp-130 Secondary Differentiation Medium and plated onto semi-confluent ($\sim$60–70% confluent) OP9 cells in tissue culture-treated 6-well plates. From twenty 60 mm dishes of day 8 EBs, approximately 1.6×10^7 cells are produced which can be replated onto 12 wells in two 6-well plates at about 1.4×10^6 cells/well containing the gp-130

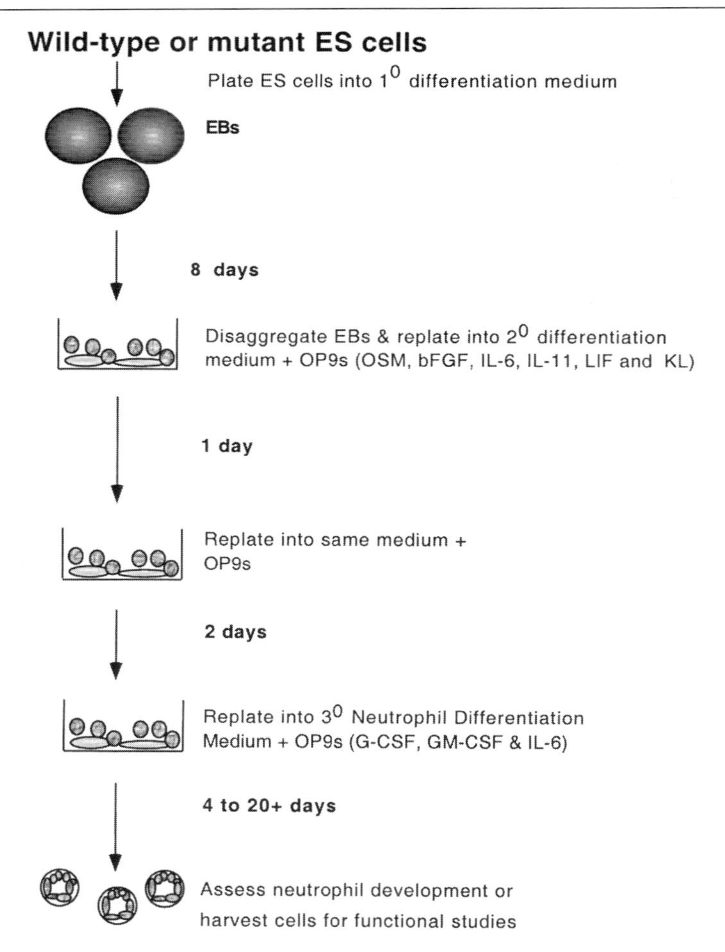

FIG. 1. Outline of the ES/OP9 differentiation strategy for production of neutrophils from ES cells.

Secondary Differentiation Mix. After 24 hr, the adherent cells associated with the monolayers are trypsinized and replated in the same medium into 12 wells containing new semi-confluent OP9 monolayers along with the cells in suspension.

Tertiary Neutrophil Differentiation Mix

After 3 days in the gp-130 Secondary Differentiation Mix, cells are transferred onto 12 wells of semi-confluent (~60–70% confluent) OP9 cells into two tissue culture-treated 6 well plates at a concentration of approximately $2\text{–}4 \times 10^5$ cells/ml ($\sim 0.8\text{–}1.6 \times 10^6$ cells) into a Tertiary Neutrophil Differentiation Mix. Cultures should be fed every 24–48 hr when the medium shows signs of acidification. Every 48–72 hr, remove and centrifuge the supernatants at 225*g* for 5 min at room temperature, resuspend the cell pellet in the Tertiary Neutrophil Differentiation Mix and add them back to the respective wells. After 4 days, the cells can be harvested for assays.

Harvesting Neutrophils from the ES/OP9 Coculture System

Retrieval and Staining of Differentiating Neutrophils for Histological Evaluation

To assess the overall number and percentage of neutrophils produced, the wells containing the differentiating neutrophils can be harvested, counted using a hemacytometer and cytospins prepared for staining with hematoxylin and eosin using the Hema 3 Staining Kit (Fisher Scientific, Pittsburgh, PA). The percentage of mature neutrophils can be morphologically determined from the hematoxylin and eosin stained cytospins.

Depending on the desired objective, cells can be harvested using four methods: (1) fine needle aspiration of the most superficially associated and generally most mature neutrophils over the neutrophil generating regions (NGRs), (2) gentle harvesting of the supernatants to enable continued granulopoiesis, (3) thorough harvesting to retrieve the majority of hematopoietic cells, and (4) gentle and thorough harvesting followed by trypsinization of the stroma to assess the entire cellular composition of the wells.

Neutrophil Production in the ES/OP9 Coculture System

Previous studies indicated that incorporation of the OP9 stromal cells at the gp-130 Secondary Differentiation Mix stage and throughout the Tertiary Neutrophil Differentiation Mix was most effective for sustained production of abundant neutrophils. In Fig. 2, the effects of OP9 cells on the number and duration of neutrophils produced in a single well (9.5 cm^2 growth area) of a 6-well plate over 23 days from a representative experiment are shown. As shown, OP9 stromal cells enhance neutrophil production from 5- to 10-fold compared to cultures without stroma. Generally, from 80,000 pluripotent ES cells, approximately 6×10^6 neutrophils can be obtained from twelve 9.5 cm^2 wells for 7–14 days using this approach. The results shown in Fig. 2 represent a conservative estimate of the production of neutrophils, as in some experiments peak neutrophil production approached

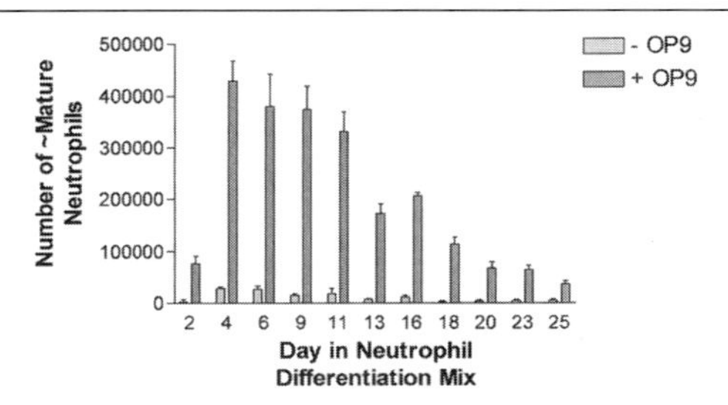

Fig. 2. OP9 stromal cells enhance the number and duration of neutrophils produced in the ES/OP9 coculture system. In this figure the effects of OP9 cells on neutrophil production in a single well of a six-well plate over 23 days from a representative experiment is shown. In the OP9 group, cells were cocultured with OP9 cells starting in the secondary gp-130 differentiation step. At the indicated times in the tertiary Neutrophil Differentiation Mix, 1 ml of gently harvested cells were retrieved and counted using a hemacytometer. Hematoxylin eosin stained (Hema 3 staining kit Fisher Scientific) cytospins were prepared and the percentage and number of mature neutrophils in multiple (8) wells were assayed in sequence to enable evaluation of each well every eight days, which should eliminate any effect of cell removal on the cell number.

8×10^5 cells/well. The numbers shown in Fig. 2 are approximately 80–90% of the total number of mature neutrophils, as the wells were gently flushed prior to harvest and many neutrophils associated with the stroma were not in suspension. In experiments where thorough harvesting of the wells was done the total number of bands to fully mature neutrophils recovered was typically 1.1–1.25 times the number when gentle harvesting was used (data not shown).

A High Percentage of Neutrophils can be Produced in the ES/OP9 Coculture System Starting at Day 4 in the Tertiary Neutrophil Differentiation Mix

As shown in Fig. 3, a high percentage of morphologically mature neutrophils can be produced in the ES/OP9 coculture system as early as day 4 in the Neutrophil Differentiation Mix. During periods of peak neutrophil production, approximately 75–95% of the gently harvested cells appear to be neutrophils at the band to polysegmented stage.

The Production of Neutrophils From Embryonic Stem Cells Occurs in Regional Areas

By day 3 in the tertiary Neutrophil Differentiation Mix, distinct areas of apparent neutrophil production can be observed, distinguished by clusters of neutrophils at multiple stages of maturation. As shown in

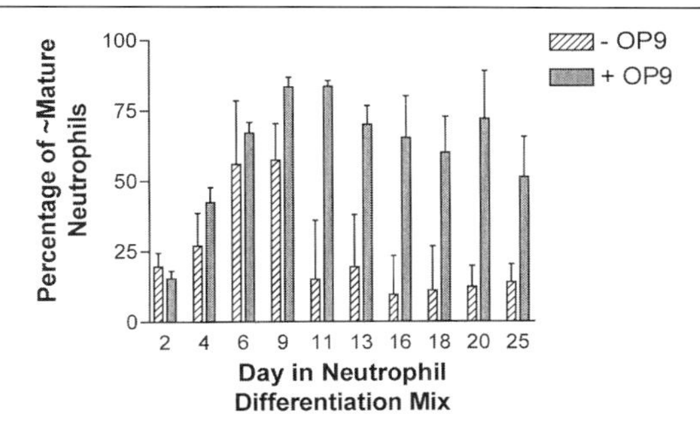

FIG. 3. OP9 stromal cells enhance the percentage of neutrophils produced. Cells were gently harvested as described in Fig. 2, cytospun and stained with the Hema 3 staining kit (Fisher Scientific, Pittsburgh, PA). The percentage of bands to polysegmented neutrophils was scored based on morphology.

Fig. 4a, by day 3 to 5 in the Tertiary Neutrophil Differentiation Mix, distinct areas associated with abundant neutrophil production are present. In Fig. 4a portions of two NGRs can be seen; one at the top and the other at the lower left. In such regions, multiple morphologically mature appearing neutrophils can be observed loosely associated with the surface of these regions (Fig. 4b). As the neutrophils mature they frequently appear to assume a more superficial position over these regions, with the most mature neutrophils being released into the medium. With increasing time in the tertiary Neutrophil Differentiation Mix, a greater percentage of poly-segmented PMNs are found in suspension, as shown in Fig. 4c.

Applications of the ES/OP9 Coculture System for Studying Neutrophil Differentiation, Function and for the *In Vitro* Expansion of Neutrophils

The ES/OP9 coculture system for production of neutrophils is well suited to studying neutrophil differentiation in a controlled environment. Information gained from such studies will be applicable to developing differentiation regimes for optimizing neutrophil expansion *in vitro*. This information could be used to reduce the period of neutropenia in cancer and bone marrow transplant patients, as well as understanding the roles of neutrophils in inflammation.

In addition, the ES/OP9 coculture system enables the *in vitro* differentiation of genetically modified ES cells into neutrophils. Preliminary studies using MEKK1 −/− ES cells have revealed phenotypic differences in neutrophils produced from these cells compared to their wild-type counterparts demonstrating the applicability of this system. Future studies will assess the effects of a wide variety of genetic manipulations,

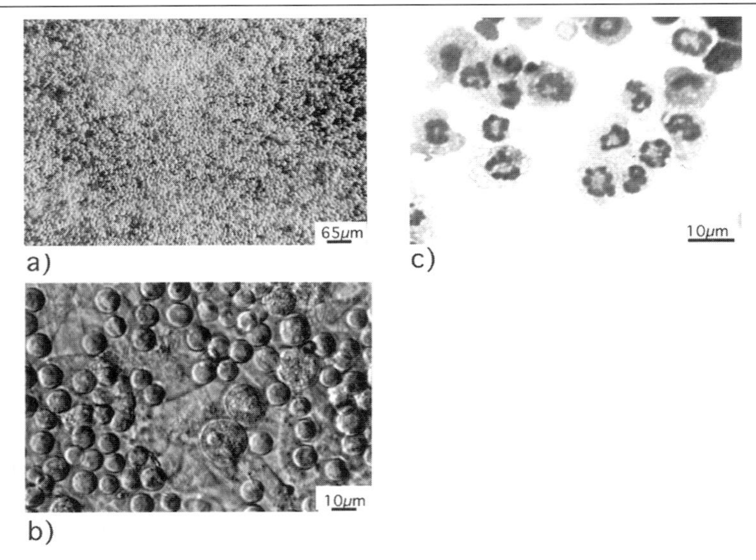

FIG. 4. Neutrophil generating regions contain differentiating neutrophils at all stages of maturation intimately associated with the OP9 cells. The composition of day 5 and 9 NGRs was evaluated by using fine needle aspiration to harvest loosely associated cells on the surface of these colonies. The aspirated cells were cytospun and stained with the Hema 3 staining kit (Fisher Scientific, Pittsburgh, PA). (a) Low power phase contrast photograph of portions of two NGRs from differentiating cells grown for 5 days in the Neutrophil Differentiation Mix. (b) At a higher magnification, differentiating neutrophils at multiple stages of maturation can be seen. (c) Photograph of hematoxylin and eosin stained neutrophils day 9 in the tertiary Neutrophil Differentiation Mix. (See Color Insert.)

including over and under expression of specific genes on neutrophil differentiation and function.

[10] Development and Analysis of Megakaryocytes from Murine Embryonic Stem Cells

By KOJI ETO, ANDREW L. LEAVITT, TORU NAKANO, and SANFORD J. SHATTIL

Introduction

Platelets are circulating anucleate cells whose adhesive and signaling functions are essential for normal hemostasis. Their aggregation within vascular wounds requires the regulated binding of adhesive ligands, such

0076-6879/2003 $35.00

12. Complete murine embryonic fibroblast (MEF) cell culture medium: DMEM, 10% FBS, 20 m*M* HEPES, 1× PSG.
13. 2× ES cell freezing medium: DMEM, 30% FBS, 40 m*M* HEPES, 20% dimethyl sulfoxide (DMSO). Make fresh for freezing cells.

Preparation of irradiated MEF cells

This method has been adapted from Ref. 20.

1. Isolate day 14 mouse embryos and place them in a dish containing PBS. Wash once in PBS.
2. Dissect individual embryos to remove the head and soft tissues (liver, intestine, kidneys, lung, heart), using watchmaker forceps. Wash carcasses once in PBS.
3. The following steps are performed inside a regular tissue culture hood. Place embryo carcasses into 10 cm dish and add 200–400 μl of 0.25% trypsin/EDTA. Mince into fine pieces with a scalpel. Add more trypsin/EDTA (e.g., 2 ml for 10 embryos), mix well with the embryo tissues and incubate for 30 min at 37°C.
4. Add 10 ml of complete MEF medium and transfer digested tissue into a 50 ml conical tube. Dissociate tissues further by vigorous pipetting. Allow large pieces to settle down, and transfer supernatant to a new tube. Add another 10 ml of complete MEF medium and repeat pipetting. Repeat this sequence three more times, resulting in a total sample of 40 ml.
5. Combine all supernatants in a 50 ml tube and plate cell suspension in T175 flasks (add the equivalent of cells from one embryo per flask). Supplement each flask with 30 ml of complete MEF medium.
6. Culture until confluency. Then trypsinize cells and resuspend in complete MEF medium containing 10% DMSO such that cells from each flask (one embryo) are placed into two freezer vials. Freeze at −70°C for 24 hr in a Cryo Freezing Container (NALGENE, catalogue no. 5100-0001). Then transfer vials to a liquid N_2 freezer.
7. Prior to use, thaw contents of one vial and culture and expand for 7–10 days to obtain approximately 10^9 cells. Then treat cells with a γ-irradiation source (1000 rad for 7.5 min). Freeze cells at $3–4 \times 10^6$ cells per vial and 10^7 cells per vial in complete MEF medium with 10% DMSO and store in liquid N_2.

[20] W. Wurst and A. L. Joyner, *in* "Gene Targeting: A Practical Approach" (A. L. Joyner,, ed.), pp. 37–39. IRL Press, New York, 1993.

Preparation of Flasks Containing MEF Cells

1. Add enough 0.7% gelatin solution to cover the surface of a T25 flask. Incubate for 20 min in the hood at room temperature, then aspirate.
2. Thaw one vial of frozen, irradiated MEF cells with gentle agitation in a 37°C water bath.
3. Transfer cell suspension to a 15-ml tube containing 10 ml of ES cell medium prewarmed to 37°C.
4. Sediment cells by centrifuging at 150*g* for 5 min at room temperature. Resuspend cell pellet in 1 ml of ES medium.
5. Seed MEF cells in the gelatin-coated T25 flask and supplement with 3 additional ml of ES medium.
6. After 2 hr under standard tissue culture conditions, cells should be approximately 80% confluent.

Thawing of ES Cells

1. ES cells are maintained in liquid N_2 at 2×10^6 per vial.
2. Thaw one vial in a 37°C water bath with gentle agitation.
3. Transfer cell suspension to a 15-m conical tube containing 9 ml of ES cell medium prewarmed to 37°C.
4. Sediment cells at 150*g* for 5 min at room temperature. Resuspend in 1 ml of complete ES cell medium. Eliminate cell aggregates by carefully pipetting 8–10 times using a 1 ml pipettor.
5. Seed 2×10^6 ES cells to the T25 flask containing the sub-confluent MEF cells and culture in a final vol of 5 ml by adding complete ES cell medium.
6. Change medium daily until the ES cells are approximately 80–90% confluent.

Passage and Splitting of ES Cells

1. Aspirate medium and wash cells twice with 5 ml of prewarmed PBS in a T25 flask.
2. Add 0.5 ml of 0.25% trypsin–EDTA to the flask. Incubate at 37°C for 1–2 min with periodic tapping to gently dislodge the ES cells.
3. Add 4.5 ml of ES cell medium to the trypsin-containing cell suspension and transfer to 15 ml conical tube.
4. Sediment cells at 150*g* for 5 min at room temperature. Resuspend to a single cell suspension in 1 ml of ES medium using a 1 ml pipettor. Establishing a single cell suspension is essential to prevent ES cell differentiation at this stage.

5. Seed one-third of the ES cells (e.g., $1–1.5\times10^6$ cells) to each of three T75 flasks containing sub-confluent MEF cells ($\sim10^7$ MEF cells per a T75) and 15 ml of ES medium. The gelatin-coated T75 flasks with a sub-confluent layer of irradiated MEF cells are generated as in section "Preparation of Flasks Containing MEF Cells" above.
6. Change medium daily. Cells should be split 1 : 10 every 2 to 3 days to avoid over-confluency.

Storage of ES Cells

1. Prepare fresh 2× freezing medium.
2. Remove cells from flask, sediment and resuspend to 8×10^6 cells per ml in complete ES cell medium. Maintain on ice.
3. Add an equal amount of 2× freezing medium.
4. Aliquot 0.5 ml (2×10^6 cells) into freezing vials. Freeze at −70°C for 24 hr in a Cryo Freezing Container (NALGENE, catalogue no. 5100-0001). Then transfer vials to a liquid N_2 freezer.

Maintenance of OP9 Stromal Cells

Rationale for Use of OP9 Cells

The OP9 stromal cell line was established from calvaria of newborn (C57BL/6xC3H) F2-osteopetrosis (op/op) mice, which lack functional macrophage colony-stimulating factor.[21] This cell line can support the expansion of multi-potential hematopoietic progenitors from murine ES cells, and in the presence of TPO, the development of megakaryocytes.[13,16,22]

Reagents

1. α-MEM medium with ribonucleosides, deoxyribonucleosides and L-glutamine (GIBCO/Invitrogen, catalogue no. 12571-063).
2. FBS (ES cell grade).
3. 0.10% (w/v) Trypsin–EDTA (0.25% Trypsin–EDTA is diluted with 1× Hanks' balanced salt solution, GIBCO/Invitrogen, catalogue no. 14170-120).
4. Penicillin–Streptomycin (PS), 100× (Sigma, catalogue no. P-0781).
5. OP9 cell culture medium: α-MEM, 20% FBS, PS.

[21] H. Yoshida, S. Hayashi, T. Kunisada, M. Ogawa, S. Nishikawa, H. Okamura, T. Sudo, and L. D. Shultz, *Nature* **345**, 442–444 (1990).

[22] T. Nakano, H. Kodama, and T. Honjo, *Science* **265**, 1098–1101 (1994).

Thawing of OP9 Cells

1. Thaw vial of frozen cells in a 37°C water bath with gentle agitation.
2. Transfer cells to a 15 ml centrifuge tube containing 10 ml of prewarmed OP9 cell culture medium.
3. Sediment at 1000 rpm for 5 min at room temperature. Resuspend cells in 5 ml of OP9 medium.
4. Seed 2×10^5 OP9 cells in a T25 flask.
5. Change medium every day until cells are 90% confluent (typically 2–3 days).

Passage of OP9 Cells

1. Aspirate medium and wash cells twice with PBS. Add 0.5 ml of 0.1% trypsin–EDTA and incubate at 37°C for 1–2 min.
2. Add 4.5 ml of OP9 medium.
3. Sediment at 150*g* for 5 min at room temperature. Resuspend cells in 5 ml of OP9 culture medium.
4. Seed $3–4 \times 10^5$ OP9 cells in a T75 flask containing a total of 15 ml OP9 medium.
5. Change medium every 2 days. Split cells 1 : 3 or 1 : 4 approximately every four days. OP9 cells should not be passaged for longer than 4 weeks. Do not use OP9 cells that are heavily vacuolated.

ES Cell Differentiation

Reagents

1. α-MEM.
2. FBS.
3. PBS.
4. PS.
5. 0.25% (w/v) Trypsin–EDTA.
6. Murine thrombopoietin (TPO) (kindly provided by KIRIN, Tokyo, Japan). Available commercially from R&D Systems (catalogue no. 488-TP).
7. Murine IL-6 (Biosource International, catalogue no. PMC-0065).
8. Human IL-11 (Biosource International, catalogue no. PHC-0115).
9. Differentiation culture media: Medium I, Day 0 to 5: αMEM, 20% FBS, PS; Medium II, Day 5 to 8: αMEM, 20% FBS, PS, 20 ng/ml TPO; Medium III, Day 8 to 12: αMEM, 20% FBS, PS, 10 ng/ml TPO, 10 ng/ml IL-6, 10 ng/ml IL-11.

Overall Differentiation Scheme

In the ES cell/OP9 cell coculture system, ES cells differentiate into "hemangioblasts" by day 4–5, and hematopoietic progenitors by day 5–7.[13,22] After the addition of TPO on day 5, megakaryocyte lineage cells proliferate and differentiate.[13,16] As a result, sufficient numbers of intermediate and large-sized megakaryocytes are available on days 12–15 to enable functional studies. After this, progressive cell death occurs, reducing the number of viable megakaryocytes for study (Fig. 1).

Day 0–5 of ES Cell Differentiation

1. Prepare 6-well plates (Falcon, catalogue no. 35-3046) containing confluent OP9 cells. For example, four days before the intended addition of ES cells, seed 6×10^4 OP9 cells in each well of the 6-well plate. Change OP9 culture medium two days before the addition of ES cells.
2. Remove confluent ES cells from a T75 flask as in section "Passage and Splitting of ES Cells."
3. Resuspend cells in 3 ml of prewarmed differentiation culture medium I using a 1 ml pipettor.
4. Seed 1 ml of ES cells into each of three 10 cm culture dishes containing 9 ml of differentiation medium I.
5. Incubate for 60–75 min in a CO_2 incubator to allow any contaminating feeder cells to adhere completely.
6. Collect supernatant with gentle washing of the plate to obtain both suspended and lightly attached ES cells. Count cells.

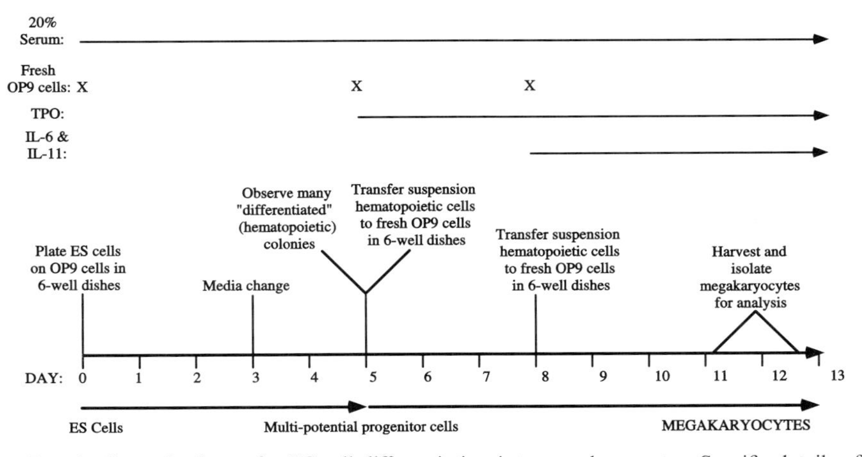

FIG. 1. General schema for ES cell differentiation into megakaryocytes. Specific details of the methodology are provided in the text.

7. After sedimentation at 150*g* for 5 min, seed 10^4 ES cells in 2 ml of differentiation medium I to each well of the 6-well plate containing fresh, confluent OP9 cells.
8. On day 3, change half of medium by aspirating 1 ml of medium from each well and adding 1 ml of new differentiation medium I.

Day 5–8 of ES Cell Differentiation

1. On day 5, check for the presence of "differentiated" colonies.[22] Collect the cells in suspension (sample A).
2. Wash each well twice with 2 ml PBS and add 0.5 ml of 0.25% trypsin–EDTA. Incubate at 37°C for 5–10 min.
3. Add 3.5 ml of differentiation medium I and pipet 10 times to dissociate the adherent cells, which will contain a mixture of loosely adherent hematopoietic cells and more firmly adherent OP9 cells.
4. Sediment at 150*g* for 5 min. Aspirate the supernatant and resuspend cell pellets from three wells of a 6-well plate in 10 ml of differentiation medium I.
5. Transfer the cell suspension to a new 10 cm dish and incubate in a CO_2 incubator for 30–35 min to allow any contaminating OP9 cells to attach.
6. Collect suspension cells (sample B) by pipetting gently 5–6 times, mix with sample A, and sediment as in step 4 to generate a pellet of hematopoietic cells that are depleted of OP9 cells.
7. Resuspend to 8×10^4/ml in differentiation medium II and add 2.5 ml aliquots to each well of a new 6-well plate containing fresh, confluent OP9 cells.

Day 8–12 of ES Cell Differentiation

1. Collect the supernatant cells (sample C) from a day 5 to 8 differentiation culture.
2. Wash the remaining adherent cells once with PBS and collect any cells that resuspend easily (sample D).
3. Add 0.5 ml of 0.1 % trypsin–EDTA to each well. Incubate at 37°C for 3–4 min.
4. Add 3.5 ml of differentiation medium II and pipet five times gently to dissociate adherent cells (sample E).
5. Sediment samples C and D at 100*g* for 10 min, resuspend cells to 10^5/ml in differentiation medium III, and pool. Seed 2×10^5 cells in 2 ml to each well of a new 6-well plate containing fresh, confluent OP9 cells. Culture until day 12. Different ES cell lines may have growth different kinetics, with some requiring culture until day 14–15.

6. Sediment sample E at 150*g* for 5 min. After resuspension to 10 ml, incubate in a 10 cm dish in a CO_2 incubator for 30–35 min to allow contaminating OP9 cells to adhere.
7. Collect suspension cells, sediment and resuspend to 10^5 cells/ml in differentiation medium III.
8. Seed 2 ml aliquots of Sample E to each well of a new 6-well plate containing fresh, confluent OP9 cells and culture up to day 12. For analysis of cells on day 12, this sample can now be pooled with those derived from samples C and D. On day 12 of differentiation, up to 80% of cells express the platelet markers, αIIbβ3 and GPIb-IX.[16]

Retrovirus-Mediated Transduction of Genes into Megakaryocytes

Reagents

1. DMEM.
2. PSG.
3. FBS (regular cell culture grade).
4. Virus culture medium: DMEM, 10% FBS, PSG.
5. 2 *M* $CaCl_2$ (29.2 g $CaCl_2$ in 100 ml H_2O; Filter sterilize).
6. 2× HEPES-buffered saline (16.4 g NaCl, 11.9 g HEPES acid, 0.21 g Na_2HPO_4, 800 ml H_2O. At room temperature, titrate with NaOH to a pH of exactly 7.1. Bring volume to 1 liter. Filter sterilize.
7. 10 m*M* chloroquine (51.6 mg chloroquine diphosphate (Sigma, catalogue no. C-6628) in 10 ml H_2O). Protect from light.
8. MSCV2.1 plasmid (derived from the MESV and LN retroviral vectors.[23,24] This plasmid includes a multi-cloning site for a gene of interest, followed by an internal ribosomal entry site (IRES) and the coding sequence for enhanced green fluorescence protein (GFP).
9. pCI-VSVG envelope plasmid.[25]
10. Packaging plasmid pHit-60.[26]
11. 0.45 μm syringe filter, surfactant free cellulose acetate (NALGENE, catalogue no. 190-2545).

[23] M. Grez, E. Akgun, F. Hilberg, and W. Ostertag, *Proc. Natl. Acad. Sci. USA* **87**, 9202–9206 (1990).

[24] A. D. Miller, and G. J. Rosman, *Biotechniques* **7**, 980–982, 84–6, 89–90 (1989).

[25] M. A. Curran, S. M. Kaiser, P. L. Achacoso, and G. P. Nolan, *Mol. Ther.* **1**, 31–38 (2000).

[26] Y. Soneoka, P. M. Cannon, E. E. Ramsdale, J. C. Griffiths, G. Romano, S. M. Kingsman, and A. J. Kingsman, *Nucl. Acids Res.* **23**, 628–633 (1995).

12. 1 mg/ml polybrene in H_2O (Sigma, catalogue no. H-9268).
13. 293T cells are used to produce retroviruses; NIH3T3 cells are used to establish retroviral titers.

Preparation of Retroviruses

Culture of 293T Cells

1. Thaw vial of frozen 293T cells in 37°C water bath with gentle agitation.
2. Seed 2×10^6 cells in 10 ml of virus culture medium to a 10 cm dish.
3. Split confluent cells 1 : 3 the day before transfection, and be sure they are evenly dispersed after plating.

Transfection of 293 Cells[27]

1. Mix DNA in a 15 ml conical polystyrene tube as follows: 10 μg MSCV2.1 plasmid + 10 μg packaging plasmid pHit-60p + 15 μg pCI-VSVG envelope plasmid + 33 μl of 10 m*M* chloroquine + 66 μl of 2 *M* $CaCl_2$. Add H_2O to 500 μl and incubate at room temperature for 3–5 min.
2. With air bubbling into the mixture through a 2 ml pipette, add 500 μl of 2× HEPES-buffered saline drop-wise using a 1 ml pipettor.
3. Add drop-wise 1 ml of this preparation to a 10 cm dish containing 293T cells at ~90% confluency. Add the mixture slowly while gently swirling the dish. Visible precipitates will appear.
4. Replace supernatant with fresh medium 8–10 hr later.
5. Forty-eight hours after step 4, collect viral supernatant, filter sterilize using a 0.45 μm syringe filter and store at −70°C. The time from media change in step #4 to virus harvest can vary, typically optimal at 48–72 hr.

Viral Titers

1. Seed a 10 cm dish with 4×10^6 NIH 3T3 cells in 10 ml of virus culture medium.
2. After confluency, remove cells, sediment and seed 10^5 cells in 2 ml of DMEM to each well of a 6-well plate. Incubate for 3–4 hr until cells have fully adhered and spread.
3. Thaw viral supernatant in 37°C water bath with gentle agitation. Then add 5, 50, and 500 μl of the supernatant in a total of 2 ml of virus culture medium to wells containing the 3T3 cells and culture for

[27] R. E. Kingston, *Curr. Protocols Mol. Biol.* 9.1.1–9.1.11 (1997).

12–18 hr in the presence of 4 μg/ml polybrene. Then aspirate supernatant and replace with fresh culture medium.

4. After another 48 hr of culture, monitor GFP expression by flow cytometry to determine viral titer.
5. Pick the two virus stock dilutions where the percentage of infected cells changes in accordance with the amount of input virus stock and take an average of the titers calculated according to the following formula:

Titer (infectious units (IU)/ml) = (% GFP+, infected cells) $\times$ (# of cells plated/well)/(volume of virus stock used as a fraction of an ml) $\times$ 100

Example: For a well in which 50 μl of virus stock yielded 50% infected cells, the titer would = $(50 \times 100{,}000)/((1/20) \times 100)$ = 1,000,000 IU/ml. Expected titers are > 4×10^6 IU/ml. It is important to realize that titer will vary depending on the cell type used for titer determination.

Retroviral Infection of ES Cell-Derived Megakaryocytes

1. On day 5 of the ES cell differentiation protocol (see section "Day 0–5 of ES Cell Differentiation"), resuspend 8×10^5 cells in 1 ml of differentiation medium II containing 8 μg/ml polybrene.
2. Adjust the viral stock with differentiation medium II and add 1 ml of this adjusted stock to 1 ml of cells such that the multiplicity of infection is 5.
3. Transfer 500 μl of the mixture to each well of a 12-well plate. Incubate for 18 hr at 37°C in a CO_2 incubator. There are no OP9 cells in the plate during the infection.
4. Wash cells once with α-MEM (without FBS) and add 4×10^5 infected cells in fresh differentiation medium II to each well of a 6-well plate containing fresh, confluent OP9 cells.
5. Culture up until day 12, as described in section "Day 5–8 of ES Cell Differentiation" and "Day 8–12 of ES Cell Differentiation."

Analysis of ES Cell-Derived Megakaryocytes

Preparation of ES Cell-Derived Megakaryocytes for Analysis

The culture system described above yields megakaryocyte progenitors and easily recognizable young and mature megakaryocytes by days 8–12. Under these conditions, about 10% of the mature megakaryocytes exhibit proplatelets. Typically, an initial culture of 1×10^4 ES cells yields 6×10^4

viable, αIIb-positive cells on day 12, of which two-thirds are intermediate and large-sized megakaryocytes. If desired, the more mature cells can be enriched further for biochemical studies by gravity sedimentation.[10] Alternatively as described below, integrin responses in the megakaryocytes, such as fibrinogen binding to αIIbβ3 and cell spreading on fibrinogen are easily monitored by flow cytometry and confocal microscopy, respectively.

1. For further biochemical and functional analyses, collect the cells in suspension derived from sample E in section "Day 8–12 of ES Cell Differentiation" (sample F).
2. Wash the adherent cells once with medium and collect any cells that resuspend with gentle pipetting (sample G).
3. Gently pipet the adherent cells again 5–8 times to collect any further cells that become resuspended (sample H).
4. In a 50 ml tube, pool samples C and D (from section "Day 8–12 of ES Cell Differentiation") with samples F and G. Pass sample H through a 75 μm filter (VWR no. 200 Sieve) to remove cell clumps and pool with the other cells. Incubate for 90 min in a CO_2 incubator.
5. Large megakaryocytes will sediment by gravity to the bottom of the tube during this interval. Gently collect them and resuspend to 10^5 cells/ml in 1% BSA-PBS solution, Tyrode's-HEPES buffer or Walsh buffer, depending on the analysis to be performed. The cells remaining in suspension will contain primarily progenitors and young megakaryocytes. If they are to be analyzed, sediment at 700 rpm for 10 min at room temperature and resuspend to 10^6 cells/ml in the desired buffer.

Immunocytochemistry

Reagents for Immunocytochemistry

1. Biotinylated antibodies (e.g., anti-CD41, anti-GPIb-IX and appropriate isotype controls).
2. Peroxidase Substrate kit (Vector Laboratories, catalogue no. SK-4100).
3. Vectastain Elite ABC kit (Vector Laboratories, catalogue no. PK-6100).
4. HEMA 3 Wright-Giemsa Staining Kit (Biochemical Sciences, Inc. or Fisher, catalogue no. CMS 122-911).
5. Microscope slides (Fisher, catalogue no. 12-544-7).
6. 5% bovine serum albumin (BSA) in PBS.
7. 0.3% H_2O_2 (Sigma, catalogue no. H-1009).

Methods for Immunocytochemistry

1. Cytospin 20,000–100,000 cells onto microscope slide.
2. Air dry for 5 min.
3. Stain cells with Wright-Giemsa according to the manufacturer's protocol.
4. For immunocytochemical staining,
   a. Fix with 70% methanol.
   b. Quench endogenous peroxidase activity by incubating in 0.3% H_2O_2 for 30 min at room temperature.
   c. Wash with PBS, and incubate with 5% BSA-PBS for 20 min.
   d. Remove excess liquid by blotting.
   e. Incubate with biotinylated primary antibodies for 30 min at room temperature, followed by a 5 min soak in PBS.
   f. Use peroxidase substrate and Vectastain ABC kits according to the manufacturers' protocols.

General Flow Cytometry Procedures

Reagents for Flow Cytometry

1. Rat anti-mouse Fc block (this and other commercial antibodies are from BD PharminGen; catalogue no. 553142).
2. FITC-conjugated rat anti-murine CD41 (catalogue no. 553848), Gr-1 (catalogue no. 553126), CD3 (catalogue no. 555274), B220 (catalogue no. 553087).
3. FITC-conjugated rat IgG_1 (catalogue no. 550617), IgG_{2a} (catalogue no. 553929), IgG_{2b} (catalogue no. 553988).
4. Phycoerythrin (PE)-conjugated rat anti-murine Sca-1 (catalogue no. 553336) and rat IgG_{2a} (catalogue no. 553930).
5. Normal rabbit serum (Vector Laboratories, catalogue no. S-5000).
6. FITC anti-rabbit IgG (H + L) (Vector Laboratories, catalogue no. FL-1000).

Cell size, surface antigen expression and DNA ploidy are assessed as described.[10,16]

Flow Cytometric Analysis of Fibrinogen Binding to Megakaryocytes

Reagents

1. Biotin-fibrinogen (1 mg/ml).[28]
2. Platelet agonists: 50 m*M* PAR4 thrombin receptor-activating peptide (AYPGKF) in H_2O; 10 m*M* ADP (Sigma, catalogue no. A-6521) in

[28] S. J. Shattil, M. Cunningham, and J. A. Hoxie, *Blood* **70**, 307–315 (1987).

H_2O; 10 mM epinephrine (Sigma, catalogue no. E-4375) in H_2O (prepare fresh).

3. PE-streptavidin (Molecular Probes, catalogue no. S-866).
4. Tyrode's-HEPES solution (137 mM NaCl, 2.9 mM KCl, 12 mM $NaHCO_3$, 0.2 mM $CaCl_2$, 0.2 mM $MgCl_2$, 0.1% BSA, 0.1% glucose, 5 mM HEPES, pH 7.4). Prepare fresh.

Fibrinogen Binding to Megakaryocytes

1. Resuspend 1–2×10^6 cells in 2 ml of Tyrode's-HEPES buffer.
2. Add 3–6×10^4 cells to a polystyrene flow cytometry tube (Falcon, catalogue no. 352008) containing 200 μg/ml biotin-fibrinogen and 10 μg/ml PE-streptavidin in a final vol of 50 μl.
3. Stimulate megakaryocytes with one or more platelet agonists by including 50 μM epinephrine, 50 μM ADP and/or 1 mM PAR4 (AYPGFK) in the tubes.
4. To assess nonspecific fibrinogen binding, include final concentrations of 10 mM of EDTA or 20 μg/ml of anti-αIIbβ3 antibody 1B5 in the tubes.[29]
5. Incubate tubes for 30 min at room temperature in the dark. Then add 500 μl of PBS containing 1 μg/ml propidium iodide and quantify fibrinogen binding in a flow cytometer. FITC-fibrinogen binding is monitored in the FL2 channel of the flow cytometer on the gated subset of viable cells (e.g., negative for propidium iodide, FL3) that have light scatter profiles consistent with megakaryocytes.[10] The FL1 channel can be used to monitor GFP fluorescence in order to identify virally transduced cells.[16]

Adhesion and Cytoskeletal Organization in Megakaryocytes

Reagents for Confocal Microscopy

1. Walsh buffer (137 mM NaCl, 2.7 mM KCl, 3.3 mM NaH_2PO_4, 1 mM $MgCl_2$, 0.1% BSA, 0.1% glucose, 3.8 mM HEPES, pH 7.4). Prepare fresh.
2. Microscope cover slips (Fisher, catalogue no. 12-546).
3. Microscope slides (Fisher, catalogue no. 12-544-7).
4. 2 mM phorbol myristate acetate (PMA, Sigma, catalogue no. P-8139) in ethanol. Store at −20°C.

[29] S. Lengweiler, S. S. Smyth, M. Jirouskova, L. E. Scudder, H. Park, T. Moran, and B. S. Coller, *Biochem. Biophys. Res. Commun.* **262**, 167–173 (1999).

5. 3.7% formaldehyde (Polysciences, Inc. catalogue no. 08018) in PBS. Store in dark at room temperature.
6. 0.2% Triton X-100 (Sigma catalogue no. T-9284) in PBS.
7. Normal goat serum (Vector Laboratories, catalogue no. S-1000).
8. Rhodamine-phalloidin (Molecular Probes, catalogue no. R-415).
9. Murine anti-vinculin monoclonal antibody (Sigma, catalogue no. V-4505).
10. FITC anti-mouse IgG (H + L) (Biosource, catalogue no. AMI 3408).
11. Coverslip mounting solution (ICN catalogue no. 622701).

Confocal Microscopy

1. Wash cover slips once with ethanol, let ethanol evaporate, and place one cover slip in each well of a 12-well plate.
2. Bring fibrinogen to 0.1 mg/ml in coating buffer (PBS, adjust to pH 8.0). Add 1 ml to each well and incubate at 37°C for 1 hr.
3. Aspirate excess solution, add 10 μg/ml of denatured BSA-PBS (heat-inactivated 90°C, 10 min), and incubate at room temperature for 1 hr. Wash once with PBS.
4. Place 10^5 cells in 1 ml of Walsh buffer onto each coverslip. If desired, add 100 nM PMA to promote cell spreading and incubate in a CO_2 incubator at 37°C for 45 min. Aspirate nonadherent cells and wash twice with PBS.
5. Add 1 ml of 3.7% formaldehyde-PBS. Incubate for 10 min at room temperature. Aspirate and wash twice with PBS.
6. Permeabilize fixed cells with 0.2% Triton X-100 in PBS for 5 min.
7. After aspiration, add 1 ml of 10% goat serum-PBS. Incubate at room temperature for 1 hr.
8. Wash twice with PBS and add 200 μl of a 1 : 60 dilution of rhodamine-phalloidin in 10% goat serum-PBS. Incubate at room temperature in the dark for 45 min.
9. Wash three times with PBS and incubate for 1 hr in the dark at room temperature with a 1 : 200 dilution of anti-vinculin antibody in 10% goat serum-PBS.
10. Wash three times with PBS, and incubate at room temperature in the dark for 45 min with 10 μg/ml of FITC-anti mouse IgG (H + L) in 10% goat serum-PBS.
11. Wash three times with PBS and mount cover slips onto microscope slide using 1 drop of mounting solution.
12. Analyze megakaryocytes by confocal microscopy.[10]

Perspective

These and related methods set the stage for using murine ES cells to study virtually any aspect of megakaryocyte development and function *in vitro*. To the extent that signaling pathways in mature megakaryocytes resemble their counterparts in platelets, the ES cell-derived megakaryocyte may also prove to be a useful model system for evaluating candidate genes involved in platelet function. These kinds of studies will be facilitated by utilizing homozygous knockout ES cell lines and conditional gene expression systems.[30,31] The potential to reconstitute or augment platelet production *in vivo* using ES cell-derived megakaryocyte lineage cells can also be explored. Studies in the murine system may presage similar ones starting with human ES cells, with implications for the prevention and treatment of bleeding disorders due to defects in platelet number or function.

[30] T. Era, S. Wong, and O. N. Witte, *Methods Mol. Biol.* **185**, 83–95 (2002).
[31] T. Era and O. N. Witte, *Proc. Natl. Acad. Sci. USA* **97**, 1737–1742 (2000).

[11] Development of Lymphoid Lineages from Embryonic Stem Cells *In Vitro*

By SARAH K. CHO and JUAN CARLOS ZÚÑIGA-PFLÜCKER

Abstract

Lymphocyte development proceeds through highly ordered and regulated stages, as multipotent hematopoietic progenitors differentiate into lineage-defined progenitors, and ultimately into mature effector cells. The ability to generate lymphocytes from embryonic stem (ES) cells *in vitro* should facilitate the study of these complex differentiation steps by providing a model system in which the effects of genetic and nongenetic manipulations can be examined in a controlled setting. These advances may also contribute to future therapeutic approaches. In this article, we describe the procedure for generating functional lymphocytes from mouse ES cells.

Introduction

Lymphocytes have been generated from ES cells by two main approaches. The first involves differentiation of ES cells into hematopoietic

0076-6879/2003 $35.00

progenitors *in vitro* followed by adoptive transfer of these progenitors into recipient mice.[1–3] This approach is useful for studying lymphocyte development in an *in vivo* context and, for example, can address questions about lymphocyte engraftment, homing, and migration. However, a complementary approach to study ES cells, as they differentiate into lymphocytes *in vitro*, is advantageous for detailed molecular studies under various culture conditions and for ease of manipulating the ES cells, subsequent progenitors, and lymphocytes throughout various stages of differentiation. Lymphocytes are generated *in vitro* by coculturing the ES cells on the stromal cell line, OP9,[4,5] which was derived from the bone marrow of *op/op* mice that are deficient for macrophage-colony stimulating factor (M-CSF).[6] Nakano *et al.* demonstrated that ES cells could differentiate into B lymphocytes *in vitro* when cocultured on a monolayer of OP9 cells (ES/OP9 coculture).[4] Stromal cells support hematopoietic differentiation in part by the cytokines they produce. M-CSF supports myeloid differentiation. In the ES/OP9 coculture system, ES cell differentiation and subsequent lymphopoietic induction was more efficient due to the absence of M-CSF, and exogenous addition of M-CSF resulted in extensive myelopoiesis and a decrease in lymphopoiesis.[4] Thus, the lack of M-CSF contributed to the ability of OP9 cells to support B cell differentiation from ES cells, albeit at a low frequency.[4] Cytokines that promote B cell differentiation include stem cell factor (SCF), Flt-3 ligand (Flt-3L), and interleukin-7 (IL-7).[7–12] Flt-3L in particular was shown to synergize with IL-7 to promote expansion of pro-B cells whereas SCF and IL-7 supported mostly myeloid differentiation of cultured Lin^{-} $CD117^{+}$

[1] J. C. Gutierrez-Ramos and R. Palacios, *Proc. Natl. Acad. Sci.* **89**, 9171–9175 (1992).
[2] A. J. Potocnik, P. J. Nielsen, and K. Eichmann, *EMBO J.* **13**, 5274–5283 (1994).
[3] A. J. Potocnik, H. Kohler, and K. Eichmann, *Proc. Natl. Acad. Sci. USA* **94**, 10295–10300 (1997).
[4] T. Nakano, H. Kodama, and T. Honjo, *Science* **265**, 1098–1101 (1994).
[5] T. Nakano, *Semin. Immunol.* **7**, 197–203 (1995).
[6] H. Yoshida, S. Hayashi, T. Kunisada, M. Ogawa, S. Nishikawa, H. Okamura, T. Sudo, and L. D. Shultz, *Nature* **345**, 442–444 (1990).
[7] O. P. Veiby, S. D. Lyman, and S. E. W. Jacobsen, *Blood* **88**, 1256–1265 (1996).
[8] S. E. Jacobsen, C. Okkenhaug, J. Myklebust, O. P. Veiby, and S. D. Lyman, *J. Exp. Med.* **181**, 1357–1363 (1995).
[9] S. D. Lyman and S. E. Jacobsen, *Blood* **91**, 1101–1134 (1998).
[10] B. E. Hunte, S. Hudak, D. Campbell, Y. Xu, and D. Rennick, *J. Immunol.* **156**, 489–496 (1996).
[11] S. Hudak, B. Hunte, J. Culpepper, S. Menon, C. Hannum, L. Thompson-Snipes, and D. Rennick, *Blood* **85**, 2747–2755 (1995).
[12] F. Hirayama, S. D. Lyman, S. C. Clark, and M. Ogawa, *Blood* **85**, 1762–1768 (1995).

Sca-1^{+} mouse bone marrow progenitors.[7] Analysis by reverse transcriptase polymerase chain reaction showed that OP9 cells express low levels of transcripts for Flt-3L relative to transcripts for SCF.[13] Comparison of ES/OP9 cocultures with or without the addition of exogenous Flt-3L revealed that B cell differentiation was dramatically enhanced in cocultures with exogenously-added Flt-3L, and substantial levels of B lymphopoiesis were consistently observed.[13] Thus, a practical protocol for efficient *in vitro* lymphopoiesis from ES cells became attainable.

The following method describes the procedure to generate B and natural killer (NK) lymphocytes from ES cells in the ES/OP9 coculture system (Fig. 1).

Materials

1. *Embryonic stem (ES) cells.* Lymphocytes have been generated *in vitro* from all ES cell lines tested (R1, D3, E14K cell lines derived from 129 mice; and ES cells derived from Balb/c and C57B1/6 mice). ES cells are maintained as undifferentiated adherent colonies on monolayers of mouse embryonic fibroblasts (EF).
2. *Embryonic fibroblast (EF) cells.* Undifferentiated ES cells are maintained on mouse primary EF cells. EF cells are mitotically inactivated by irradiation (20 Gy) or by treatment with mitomycin-C (Sigma M-0503). If treating EF cells with mitomycin-C, incubate confluent dishes with 10 ug/ml mitomycin-C for 2.5 hr at 37°C. Wash three times with 1X PBS, then add new media. For procedures to isolate primary EF cells and maintain feeder layers, see Refs. 14,15.
3. *OP9 cells.* OP9 cells are adherent cells cultured as monolayers. They can be obtained from the RIKEN cell depository (http://www.rtc.riken.go.jp/).
4. *Serum.* ES cell media and OP9 cell media require separate lots of fetal bovine serum (FBS). Prior to use, FBS should be heat-inactivated at 56°C for 30 min. Use ES cell-tested FBS to maintain undifferentiated ES cells. Use OP9 cell media to maintain OP9 cells and for ES/OP9 cocultures. For ES cell media, Hyclone offers prescreened characterized lots of FBS. For OP9 cell media, two or three FBS lots are tested in parallel in separate ES/OP9 cocultures.

[13] S. K. Cho, T. D. Webber, J. R. Carlyle, T. Nakano, S. M. Lewis, and J. C. Zúñiga-Pflücker, *Proc. Natl. Acad. Sci. USA* **96**, 9797–9802 (1999).

[14] E. J. Robertson, *Methods Mol. Biol.* **75**, 173–184 (1997).

[15] E. J. Robertson (ed.), "Teratocarcinomas and Embryonic Stem Cells: A Practical Approach." IRL Press, Oxford, UK, 1987.

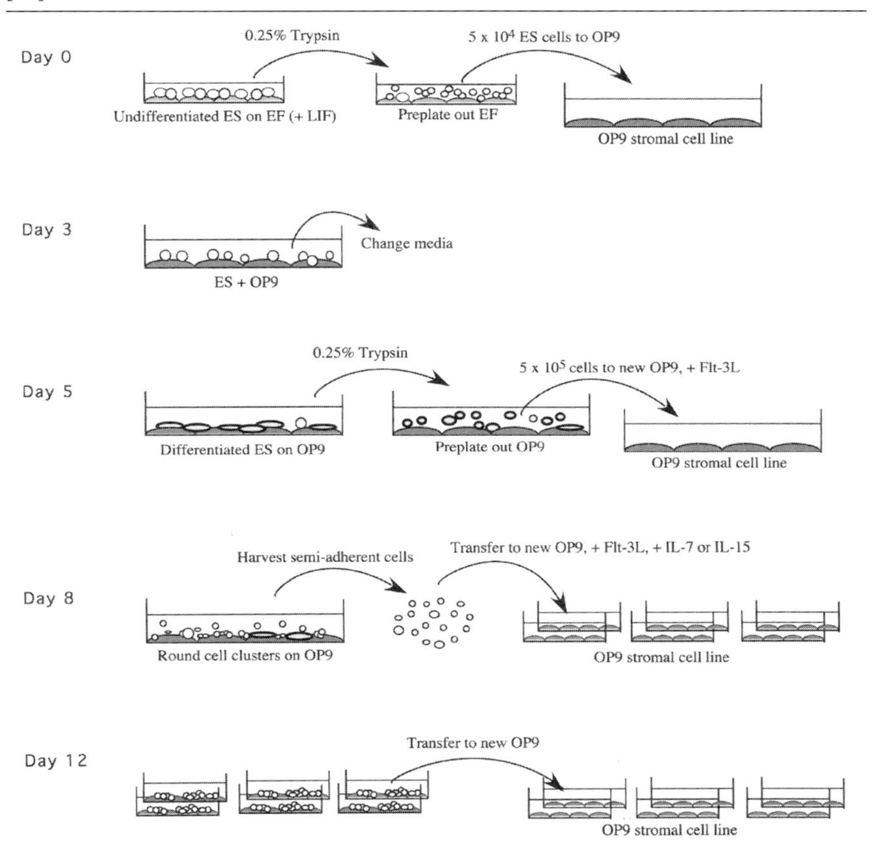

FIG. 1. Method for generating lymphocytes in ES/OP9 cocultures. ES cells are induced to differentiate on the bone marrow-derived stromal cell line, OP9. Initial seeding of ES cells onto OP9 cells is designated as day 0. After 5 days of coculture, mesoderm-like colonies are disaggregated by treatment with 0.25% trypsin, and replated onto new OP9 monolayers with the addition of exogenous Flt-3L. On day 8 of coculture, semi-adherent cells are harvested by gently washing the surface of the coculture plates, and the cells are transferred to new OP9 monolayers with the addition of exogenous Flt-3L plus IL-7 for B lymphocytes, and IL-15 for NK lymphocytes. Cytokines are re-added with media changes. Lymphopoiesis can be observed from as early as day 12 of cocultures, and predominates after day 16.

Make OP9 media with the test lots and follow the method described below for ES/OP9 coculture. Select the lot of FBS from the coculture that generates the most B cells between days 14 and 16. When a serum lot for OP9 media has expired, new lots are tested that have been cross-matched to the previous one. FBS lots that have been used previously but are no longer available are Hyclone AHM9368 and Hyclone FLC13226.

5. *High glucose DMEM* (Sigma D-5671).
6. *αMEM* (Gibco 12000-022). αMEM is reconstituted as needed. To reconstitute, add 1 package of powder and 2.2 g of Sodium Bicarbonate to 1 liter of media-grade dH_2O in a glass cylinder with stir bar (do not use the glass cylinder or stir bar for any other purpose), stir for 5 min. Filter using a Millipore 0.22 μm Stericup apparatus (Millipore SCGPU05RE). Rinse the glass cylinder and stir bar with water (no detergents) and ethanol after each use, and autoclave. Alternatively, αMEM in liquid form is also available from Gibco (Cat No. 12561 or 32561).
7. *1X Phosphate Buffered Saline (PBS)* (Gibco 14190-144) without Ca^{2+}, without Mg^{2+}.
8. *HSG solution.* Aliquot 5 ml HEPES (Gibco 15630-080; 100X or 1 *M*), 5 ml sodium pyruvate (Gibco 11360-070; 100X or 100 m*M*), and 0.5 ml Gentamicin (Gibco 15750-060; 1000X or 50 mg/ml) each into 14 ml tubes. Store at 4°C.
9. *PG2 solution.* Aliquot 5 ml penicillin/streptomycin (Gibco 15140-122; 100X or 10,000 U/ml penicillin and 10,000 μg/ml streptomycin), 5 ml Glutamax (Gibco 35050-061; 100X or 200 m*M*), and 0.5 ml beta-mercaptoethanol (Gibco 21985-023; 1000X or 55 m*M*) each into 14 ml tubes. Store at −20°C.
10. *2.5% Trypsin* (Gibco 15090-046). Diluted with PBS to 0.25% solution as needed.
11. *Leukemia Inhibitory Factor (LIF)* (R&D Systems 449-L, StemCell Technologies 02740, or Chemicon LIF2010) reconstituted in media at 5 μg/ml (1000X). Store at −80°C in small aliquots.
12. *Flt-3 Ligand* (R&D 308-FK) reconstituted in media at 5 μg/ml (1000X). Store at −80°C in small aliquots.
13. *IL-7* (R&D Systems 407-ML) reconstituted in media at 1 μg/ml (1000X). Store at −80°C in small aliquots.
14. *IL-15* (R&D Systems 247-IL) reconstituted in media at 10 μg/ml (1000X). Store at −80°C in small aliquots.
15. *Tissue culture.* All procedures are carried out with standard aseptic tissue culture practices using sterile plasticware (6-well plates, 6 cm dishes, 10 cm dishes, 14 ml centrifuge tubes and 50 ml centrifuge tubes). All incubations are carried out at 37°C in a humidified incubator containing 5% CO_2 in air.

16. *Centrifuge*. Cells are pelleted at approximately 500*g* (1500 rpm in a Beckman Allegra 6R) for 5 min.

Methods

1. *ES media*. High glucose DMEM is supplemented with 15% FBS, 10.5 ml HSG solution, and 10.5 ml PG2 solution.
2. *OP9 media*. Supplement freshly reconstituted αMEM with 20% FBS and 1X penicillin/streptomycin solution. This media is used to culture OP9 cells and is also used throughout the ES/OP9 coculture.
3. *Thawing and maintaining ES cells*. Thaw a vial of ES cells in a 37°C water bath, transfer cells to a 14 ml centrifuge tube containing 10 ml ES media, pellet the cells, resuspend cells in 3 ml ES media and add to a 6 cm dish of irradiated EF cells. Add 3 μl of LIF (1000X), gently shake the dish to mix. Change the media the next day by removing all the media and adding fresh media and LIF. Two days after thawing ES cells, trypsinize (see below, trypsin passage) the cells and replate onto new EF cells. To maintain ES cells in culture, trypsin passage ES cells every other day, and change the media on alternate days. Maintain the ES cells at less than 80–90% confluence (i.e., 80–90% of the surface area is covered with ES cells).
4. *Trypsin passage of ES cells*. Remove media from the dish of ES cells. Gently add 4 ml of PBS, gently swirl the plate to wash cells, and remove PBS. Add 2 ml of 0.25% trypsin to the dish and return to the incubator for 5 min. Disaggregate the cells by vigorous pipetting and transfer to a 14 ml centrifuge tube. Wash the dish with 6 ml of ES media and transfer to the same 14 ml tube. Pellet the cells (500*g* for 5 min) and resuspend in 3 ml of ES media. Remove media from a 6 cm dish of confluent irradiated EF cells, add the resuspended ES cells, and add 3 μl of LIF, gently shake the plate to evenly disperse the cells and LIF.
5. *Freezing ES cells*. Trypsinize, wash, and pellet the ES cells. Resuspend the ES cells in cold (4°C) freezing medium (90% FBS + 10% DMSO, tissue culture grade), transfer 1 ml to cryovials (2–4 vials per 6 cm dish of 80% confluent ES cells), and store vials in −80°C freezer overnight. The next day, transfer vials to a liquid nitrogen container.
6. *Thawing and maintaining OP9 cells*. Thaw a vial of OP9 cells in a 37°C water bath, transfer cells to a 14 ml centrifuge tube containing 10 ml OP9 media, pellet the cells, resuspend cells in 10 ml

OP9 media and plate onto a 10 cm dish. Change the media the next day. When the 10 cm dish of OP9 cells is 85–95% confluent, trypsin passage the cells and split the 10 cm dish into 4 dishes—any combination of four 10 cm dishes or 6-well plates—(OP9 cells in 6-well plates are used on day 8 of coculture, and subsequently). Trypsin passage the OP9 cells every 2 to 3 days. It is important to not let the OP9 cells become overconfluent. If they are overconfluent, they can be used for cocultures immediately, but should not be used to further expand OP9 stocks. OP9 cells will lose their ability to induce lymphohematopoiesis from ES cells after prolonged culture and/or if they are grown overconfluent. Keep reserve stocks of OP9 cells that have been expanded from an early thawing. For example, once the thawed OP9 cells become 85–95% confluent, split (trypsin passage) a 10 cm dish into four 10 cm dishes. When these four 10 cm dishes are 85–95% confluent, split them into sixteen 10 cm dishes, and freeze the cells into 16 vials when the 16 dishes become 85–95% confluent. From each of these 16 expansion stock vials, repeat the procedure to generate 16 working stock vials and use these for ES/OP9 cocultures. Older stocks of OP9 cells that may not be sufficient for initiating cocultures can still be frozen and thawed for use at day 8 and 12 of ES/OP9 cocultures.

7. *Trypsin passage of OP9 cells.* Remove media from the 10 cm dish of OP9 cells. Add 10 ml of PBS, gently swirl the plate to wash cells, and remove PBS. Add 4 ml 0.25% trypsin to the dish and return to the incubator for 5 min. Disaggregate the cells by vigorous pipetting and transfer to a 14 ml centrifuge tube. Wash the dish with 10 ml of OP9 media and transfer to the same tube. Pellet the cells, resuspend in 12 ml OP9 media and aliquot 3 ml each to four 10 cm dishes. Add 7 ml OP9 media to bring the volume in each dish to 10 ml, gently shake the dish to evenly distribute the cells.
8. *Freezing OP9 cells.* Trypsin passage, wash, and pellet the OP9 cells. Resuspend the OP9 cells in cold (4°C) freezing medium (90% FBS + 10% DMSO), add to cryovials (1 vial per 10 cm dish of 85–95% confluent OP9 cells), and transfer vials to a −80°C freezer overnight. The next day, transfer vials to a liquid nitrogen container.
9. *ES/OP9 coculture.* See Fig. 1 for a schematic representation of the procedure.
   a. *Preparing cells.* Thaw ES cells onto irradiated EF cells 4–6 days prior to beginning of coculture. Passage the cells as for maintaining ES cells. Thaw OP9 cells 4 days prior to beginning of coculture. Two days prior to beginning the coculture, trypsin passage a confluent 10 cm dish of OP9 cells into 4×10 cm dishes.

b. *Day 0.* OP9 cells should be 85–95% confluent. Aspirate off old media from OP9 cells in a 10 cm dish, and add 10 ml of OP9 media. Aspirate off the media from the ES cell dish. Wash ES cells once in PBS. Trypsinize the ES cells. After disaggregating the cells, add 6 ml of ES media and transfer the cells to a new 10 cm dish (not an EF monolayer dish) in order to preplate out the EFs. Incubate at 37°C for 30 min. After the incubation, collect the nonadherent cells into a tube, pellet, resuspend in 3 ml ES media, count the cells, and seed 5×10^4 ES cells onto a monolayer of OP9 cells in a 10 cm dish containing newly added media.
c. *Day 3.* Change media of cocultures. Aspirate off old OP9 media without disturbing the monolayer. Add 10 ml of OP9 media.
d. *Day 5.* 50–100% of the ES cell colonies should have differentiated into mesoderm-like colonies[4] in the coculture. Aspirate off the old media and wash with 10 ml PBS without disturbing the cells. Aspirate off the PBS, add 4 ml 0.25% Trypsin and incubate for 5 min at 37°C. Disaggregate the cells by vigorous pipetting. Add 8 ml OP9 media to neutralize the trypsin, transfer to a 10 cm dish, and incubate for 30 min at 37°C to preplate out the OP9 cells. Collect the nonadherent cells and transfer into a 14 ml tube, pellet, resuspend in fresh OP9 media, count the cells, and transfer 5×10^5 cells to a new confluent 10 cm dish of OP9 cells. Add Flt-3L to a final concentration of 5 ng/ml.
e. *Day 8.* Small clusters (typically 4–10 cells) of round blast-like cells should be visible. Using a 10 ml pipette, gently wash the surface of the plate and transfer cells to a 50 ml tube. Repeat the gentle wash with 10 ml of PBS, taking care to leave the OP9 monolayer with other differentiated colonies intact. Check the dish by microscope to see that very few clusters of round cells remain. Pellet the loosely-adherent harvested cells, resuspend in 3 ml of OP9 media and transfer to new OP9 monolayers in 6-well plates. Transfer the cells from one 10 cm dish of day 8 cocultures into one well of a 6-well plate with OP9 cells. Add Flt-3L to the passaged cells. For B cell differentiation, add IL-7 to a final concentration of 1 ng/ml. For NK cells, add IL-15 to a final concentration of 10 ng/ml. All the appropriate cytokines should be re-added at each subsequent passage or media change.
f. *Day 10.* Change media by transferring supernatants from the cocultures into tubes (without disrupting the monolayers). Add fresh media to wells to prevent the cells from drying out. Pellet cells from the supernatant, resuspend in fresh media and gently pipette

back to the same well. Continue adding the appropriate cytokines, i.e., Flt-3L, and IL-7 or IL15.

g. *Day 12*. Passage the cells by treating with trypsin. Disaggregate the cells with vigorous pipetting, transfer cells to a new 10 cm dish to preplate out the OP9 cells, incubate for 30 min, and transfer nonadherent cells to new OP9 monolayers. Alternatively, passage cells without trypsin by disaggregating cultures with vigorous pipetting, pass through a 70 μm Nylon mesh filter into a tube, pellet, resuspend in new media, and transfer to new OP9 monolayers. We have not observed major differences with or without trypsin treatment. Trypsin treatment will yield more cells; however, trypsin passaging may not be desirable for some later applications.

h. *Beyond day 12*. To continue cocultures beyond day 12, transfer cells to new OP9 cell monolayers every 4 to 10 days with media changes every 2 to 3 days. Follow the passage procedure for day 12. For efficient hematopoietic differentiation it is advisable to leave the proliferating colonies undisturbed as long as possible. However, it is necessary to passage to new OP9 monolayers as the confluent OP9 cells will differentiate (see below) into cells that will no longer support further hematopoiesis effectively, and because cocultures will begin to "roll up" at the edges as overconfluent OP9 cells detach from the tissue culture plates.

Comments

1. *Flow cytometry*. Hematopoietic cells expressing the receptor tyrosine phosphatase CD45 (leukocyte common antigen, LCA) can be detected as early as day 5 of ES/OP9 coculture.[13] Progression of lymphocytes differentiating from ES cells can be readily assessed by flow cytometry (Fig. 2). Stages of B cell differentiation have been characterized by the expression of various cell surface molecules. Hematopoietic progenitors express the receptor tyrosine kinase CD117 (c-kit)[16] and those with restricted B cell lineage potential can be identified by expression of CD19 and CD45R (B220) on the cell surface[17] (Fig. 2). Early B cell progenitors, termed pro-B cells, are further characterized by the surface expression of CD43 in contrast to later progenitors, termed pre-B cells, which lack expression of this

[16] K. Ikuta, N. Uchida, J. Friedman, and I. L. Weissman, *Annu. Rev. Immunol.* **10**, 759–783 (1992).

[17] S. K. Cho, A. Bourdeau, M. Letarte, and J. C. Zúñiga-Pflücker, *Blood* **98**, 3635–3642 (2001).

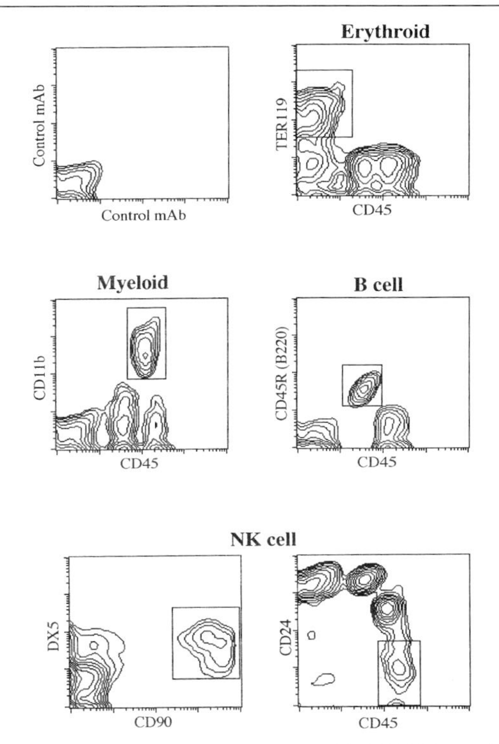

FIG. 2. Multiple hematopoietic lineages are generated in ES/OP9 cocultures. Hematopoietic lineages generated from day 14 ES/OP9 cocultures analyzed by flow cytometry for cell surface expression of the indicated lineage-associated markers. Cells expressing lineage markers for each lineage, as indicated, are boxed. This figure was adapted from Cho *et al.* (2001).[17]

marker.[18,19] Differentiation of pre-B lymphocytes into immature B cells can be detected by surface expression of the immunoglobulin B cell receptor IgM, and mature B cells can be detected by cell surface expression of IgM and IgD.[20] Natural killer (NK) lymphocytes can be detected by positive cell surface expression of the pan-NK cell marker, DX5, and the lack of CD24 (heat stable antigen, HSA) cell surface expression[17,21] (Fig. 2). Fluorochrome- or

[18] Y. S. Li, R. Wasserman, K. Hayakawa, and R. R. Hardy, *Immunity* **5**, 527–535 (1996).

[19] R. R. Hardy, C. E. Carmack, S. A. Shinton, J. D. Kemp, and K. Hayakawa, *J. Exp. Med.* **173**, 1213–1225 (1991).

[20] F. Melchers, A. Rolink, U. Grawunder, T. H. Winkler, H. Karasuyama, P. Ghia, and J. Andersson, *Curr. Opin. Immunol.* **7**, 214–227 (1995).

[21] J. R. Carlyle, A. M. Michie, S. K. Cho, and J. C. Zúñiga-Pflücker, *J. Immunol.* **160**, 744–753 (1998).

biotin-labeled antibodies for flow cytometric analysis can be purchased from Pharmingen or eBiosciences.

2. *OP9 differentiation.* OP9 cells that have been maintained in good condition have the morphology of veiled (large cytoplasm) adherent cells with short dendritic-like extensions, and have an overall elongated star-like shape. With prolonged culture the cells become larger with fewer cytoplasmic extensions, and take on a triangular (scalene) shape. The latter will not suffice for lymphohematopoietic induction from ES cells but should still support lymphopoiesis from tissue explants such as fetal liver or bone marrow. In addition, when OP9 cells are maintained in culture for prolonged periods and/or are grown overconfluent, they will differentiate into large cells containing many fat droplets. During the course of ES/OP9 cocultures, cells are seeded onto nearly confluent OP9 monolayers that become overconfluent within days. Thus, the appearance of some cells containing fat droplets during ES/OP9 coculture is normal. However, if these cells predominate, the OP9 monolayer will not efficiently support hematopoietic differentiation.
3. *Differentiation kinetics of hematopoiesis during ES/OP9 coculture.* Hematopoietic cells expressing CD45 can be observed as early as day 5 of coculture. A wave of erythropoiesis (TER119$^+$ cells) and myelopoiesis (CD11b$^+$ cells) is observed between days 10 and 14, followed by a wave of lymphopoiesis. When cocultures are maintained beyond day 20, only lymphopoiesis is typically observed, indicating that multilineage differentiation does not persist in ES/OP9 cocultures under the protocol conditions.

Final Remarks

ES cells can efficiently differentiate into B and NK lymphocytes *in vitro* by coculture on the stromal cell line, OP9. Evidence supports that these lymphocytes are functionally analogous to lymphocytes *in vivo*.[13] Thus, ES/OP9 cocultures can serve as a model system for the study of lymphocyte differentiation and for functional aspects of B cell and NK cell biology. However, the extent to which lymphopoiesis in ES/OP9 cocultures can fully recapitulate lymphocyte development *in vivo* remains to be determined. It is unclear, for example, whether lymphocytes generated in ES/OP9 cocultures are derived from embryonic-type hematopoietic progenitors and/or from definitive hematopoietic stem cells representing adult-type hematopoiesis. Further studies should yield insight to the molecular mechanisms that regulate differentiation steps during lymphopoiesis, and to the role that microenvironments play in establishing developmental programs during lymphopoiesis.

Acknowledgment

This work was supported by funds from the National Cancer Institute of Canada. SKC was supported by a fellowship from the Lady Tata Memorial Fund. J. C. Z. P. is supported by an Investigator Award from the Canadian Institutes of Health Research.

[12] Probing Dendritic Cell Function by Guiding the Differentiation of Embryonic Stem Cells

By PAUL J. FAIRCHILD, KATHLEEN F. NOLAN, and HERMAN WALDMANN

Introduction

Dendritic cells (DC) are unique among populations of antigen presenting cells by virtue of their capacity to direct the outcome of antigen recognition by naive T cells.[1] By serving as messengers direct from the site of infection, DC inspect the T cell repertoire, identifying those cells specific for the cargo of antigens they display on their surface as peptide-MHC complexes. Having encountered T cells with a complementary receptor for antigen, DC deliver instructions directing the expansion of relevant clones and their deployment to the front line of the immune response. Should these T cells pose a threat to the integrity of the host, however, DC may decommission them from active service, thereby minimizing any damage from friendly fire.

The role played by DC in fine-tuning these opposing forces of self-tolerance and immunity, makes them attractive candidates for immune intervention in a variety of disease states. If their tolerogenicity could be reliably exploited in the clinic, their administration to the recipients of organ allografts or individuals suffering from autoimmune disease, would help negotiate a lasting cease-fire, favoring the reemergence of peripheral tolerance.[2] Alternatively, by specifically enhancing their immunogenicity, DC might be coerced into breaking the natural state of self-tolerance to tumor-associated antigens (TAA) and

[1] J. Banchereau and R. M. Steinman, *Nature* **392**, 245 (1998).
[2] P. J. Fairchild and H. Waldmann, *Curr. Opin. Immunol.* **12**, 528 (2000).

0076-6879/2003 $35.00

directing the full force of the immune response towards transformed cells and established tumors.[3]

Despite their clinical potential, surprisingly little is known of the molecular basis of DC function which permits them to fulfill these conflicting roles. As a trace population of leukocytes in both lymphoid organs and interstitial tissues, DC have traditionally proven difficult to isolate in sufficient numbers and purity to facilitate their study. Consequently, the publication of protocols for the generation of large numbers of immature DC from precursors present in mouse bone marrow[4] served as a welcome breakthrough, permitting their analysis on a scale that was previously unattainable. Indeed, it is primarily from the study of bone marrow-derived DC (bmDC) that a far greater understanding of the DC life cycle has emerged, together with the changes they undergo at successive stages of maturation. Nevertheless, the intrinsic resistance of terminally differentiated DC to genetic modification continues to limit the field, confounding attempts to investigate the function of novel genes that have begun to emerge from global gene expression profiling of mouse and human DC.[5,6]

In order to address this issue, we have developed an approach to the study of DC which draws on the unique features of embryonic stem (ES) cells: their self-renewal, pluripotency and tractability for genetic modification.[7,8] By deciphering the pathway of differentiation of DC from their earliest possible progenitors, we have combined the benefits offered by large yields of primary, untransformed DC, with prospects for their genetic modification at source. Indeed, by exploiting the propensity for transfection, selection and cloning of ES cells, we have demonstrated the feasibility of producing a permanent resource which may be differentiated on demand into stable lines of DC, uniformly expressing a defined, mutant phenotype.[9] Here, we describe in detail the protocols involved and discuss the advantages such an experimental system provides over the current art.

[3] L. Fong and E. G. Engleman, *Annu. Rev. Immunol.* **18**, 245 (2000).

[4] K. Inaba, M. Inaba, N. Romani, H. Aya, M. Deguchi, S. Ikehara, S. Muramatsu, and R. M. Steinman, *J. Exp. Med.* **176**, 1693 (1992).

[5] S. Hashimoto, T. Suzuki, H.-Y. Dong, S. Nagai, N. Yamazaki, and K. Matsushima, *Blood* **94**, 845 (1999).

[6] J. H. Ahn, Y. Lee, C. Jeon, S.-J. Lee, B.-H. Lee, K. D. Choi, and Y.-S. Bae, *Blood* **100**, 1742 (2002).

[7] H. Waldmann, P. J. Fairchild, R. Gardner, and F. Brook, International Patent Application No: PCT/GB99/03653 (1999).

[8] P. J. Fairchild, F. A. Brook, R. L. Gardner, L. Graça, V. Strong, Y. Tone, M. Tone, K. F. Nolan, and H. Waldmann, *Curr. Biol.* **10**, 1515 (2000).

[9] P. J. Fairchild, K. F. Nolan, S. Cartland, L. Graça, and H. Waldmann, *Transplantation*, in press.

Directed Differentiation of ES Cells: Protocols and Pitfalls

Maintenance of the Parent ES Cell Line

The appeal of directed differentiation as an approach to probing DC function, lies in the distinctive properties of ES cells and, as such, is wholly reliant on maintaining their integrity during protracted periods in tissue culture. The pluripotency of ES cells and their capacity for self-renewal are maintained primarily through the action of leukemia inhibitory factor (LIF) which must, therefore, be supplied at levels that are not in danger of becoming limiting. Although recombinant LIF or media conditioned by Buffalo rat liver cells, known to secrete the cytokine, may be used for short-term culture, their prolonged use risks compromising the capacity of ES cells to differentiate and should, therefore, be avoided. Consequently, we prefer to culture the parent ES cell line on monolayers of primary embryonic fibroblasts that provide not only an abundant source of LIF, but a variety of additional, poorly defined growth factors, implicated in maintenance of the pluripotent state.

We routinely derive fibroblast feeder cells from C57Bl/6 embryos at day 12 or 13 of gestation using well established protocols.[10] A single pregnancy normally yields 10–12 vials of fibroblasts which may be stored under liquid nitrogen for several months at a time. Upon thawing, each vial will expand to fill a 75 cm^2 flask within 1 or 2 days and may be passaged at least five times before succumbing to replicative senescence. The fibroblast stock may be cultured in parallel with the ES cell line in DMEM (Gibco) supplemented with 10% fetal calf serum (FCS), 2 mM L-glutamine, 50 U/ml of penicillin, 50 μg/ml of streptomycin and 5×10^{-5} M 2-mercaptoethanol. In order to passage the ES cell line, a confluent flask of fibroblasts must first be mitotically inactivated, either by exposure to 300 rads of γ irradiation, or by culture for 2 hr at 37°C in 10 μg/ml of mitomycin C (MMC; Sigma). If MMC is preferred, fibroblasts should be washed thoroughly in PBS to prevent any carry-over which might inhibit proliferation of the ES cell line. Once mitotically inactivated, fibroblasts should be released by incubating for 3 min with PBS containing 0.05% trypsin (Gibco) and 0.02% EDTA. Vigorous agitation of the flask will produce a single cell suspension which may be pelleted and divided equally among two 25 cm^2 flasks. After 2 hr incubation, the fibroblasts should have produced a confluent monolayer, competent to receive the ES cell line.

Although each line of ES cells has distinctive characteristics, including its doubling time in culture, we find that the majority may be passaged

[10] W. Wurst and A. L. Joyner, *in* "Gene Targeting: A Practical Approach" (A. L. Joyner,, ed.), p. 33. IRL Press, Oxford, 1993.

routinely every $2\frac{1}{2}$ to 3 days. In order to prepare a single cell suspension of ES cells, cultures should be incubated for 4 min at 37°C in trypsin–EDTA and shaken vigorously to resuspend the cells. After washing to remove any traces of trypsin, these cells may be used to seed the awaiting flasks of fibroblasts at two distinct dilutions, determined empirically from the density of colonies in the original stock. For 24 hr after plating, ES cells remain camouflaged among the monolayer of feeder cells into which they readily integrate. After 2 days, however, nascent colonies begin to appear which rapidly increase in size, forming characteristic amorphous structures by day 3 in which the boundaries of individual cells are difficult to distinguish (Fig. 1A).

Compared to fibroblasts, the nutrient requirements of most ES cell lines demand a richer medium for optimal growth composed of DMEM supplemented with 15% FCS, 2 m*M* L-glutamine, 1 m*M* sodium pyruvate and 5×10^{-5} *M* 2-mercaptoethanol. Although some laboratories routinely add antibiotics to the culture medium, the sensitivity of ES cells to their presence makes it advisable to avoid their use if at all possible. Unlike

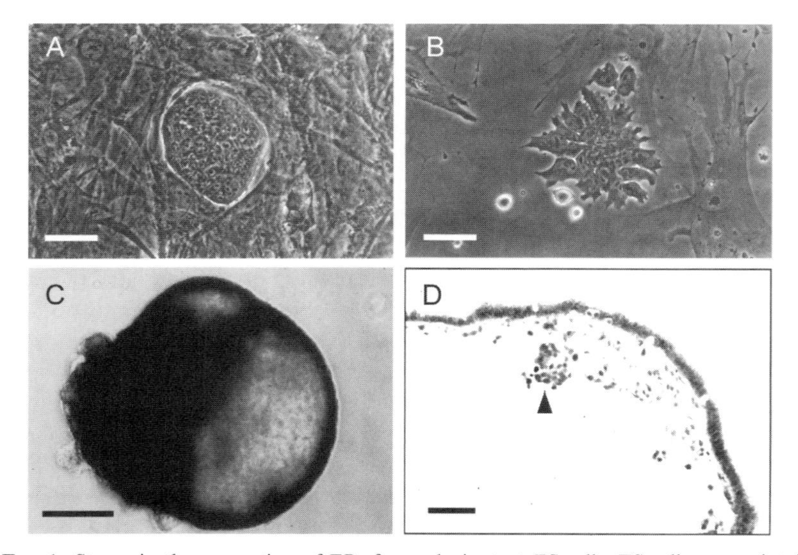

FIG. 1. Stages in the generation of EBs from pluripotent ES cells. ES cells are maintained long-term on a monolayer of primary embryonic fibroblasts on which they appear as discrete, amorphous colonies (A) (bar represents 50 μm). Once passaged in gelatinized flasks containing a source of rLIF, ES cell colonies adopt a less compact structure, permitting the identification of individual cells (B) (bar represents 50 μm). (C) Cystic EB after 14 days in suspension culture showing polarity and a characteristic fluid-filled structure (bar represents 300 μm). (D) Section through an EB showing a nascent blood island closely associated with the mesothelium (arrowhead) (bar represents 50 μm).

primary embryonic fibroblasts which adapt to almost any source of FCS, ES cells are rather more discerning, necessitating the periodic testing of batches for toxicity and plating efficiency[11] and the purchase of the most appropriate serum in bulk. Although pandering to the individual whims of each ES cell line greatly increases the chances of their subsequent differentiation, even the most rigorous regime for their maintenance is unlikely to prevent a proportion of cells becoming karyotypically abnormal. Occasionally, such random events confer on ES cells a growth advantage, causing them to progressively dominate the culture. For this reason, many laboratories do not propagate ES cells beyond passage 20, although this arbitrary boundary has been set based on the cumulative experience of obtaining germline transmission, for which requirements are rather more stringent than for their differentiation *in vitro*.

Generation of Embryoid Bodies

Although various approaches have been described for investigating the spontaneous and directed differentiation of ES cells, our own studies have focussed on the formation of embryoid bodies (EB), derived from the sustained proliferation and concomitant differentiation of ES cells, maintained in suspension culture. These macroscopic structures recapitulate many of the early events of ontogeny, including formation of the visceral yolk sac,[12] the most primitive of all hematopoietic tissues and a ready source of the hematopoietic stem cells from which DC derive.

In order to encourage the formation of EBs from the parent ES cell line, we first remove most of the embryonic fibroblasts, which, as a potent source of LIF, would otherwise impede the differentiation process. This is best achieved by passaging the ES cells twice in a source of exogenous LIF to circumvent the need for feeder cells: since fibroblasts, carried over from the parent stock, have been mitotically inactivated, the majority are lost from the culture by serial dilution. To sustain ES cells in the absence of a monolayer of fibroblasts, it is essential to provide a matrix to which they can adhere, gelatin being the reagent of choice for most laboratories. We routinely gelatinize flasks by incubating for 30 min at 37°C with 0.1% gelatin, after which they are washed thoroughly with PBS before use. In order to ensure the pluripotency of the ES cells during this transition period, they should be cultured in complete medium further supplemented with

[11] E. J. Robertson, *in* "Teratocarcinomas and Embryonic Stem Cells: A Practical Approach" (E. J. Robertson,, ed.), p. 71. IRL Press, Oxford, 1987.

[12] T. Doetschman, H. Eistetter, M. Katz, W. Schmidt, and R. Kemler, *J. Embryol. Exp. Morphol.* **87**, 27 (1985).

1000 U/ml of recombinant LIF (Chemicon), or an equivalent level of activity supplied in the form of a conditioned medium. To conserve stocks of rLIF, we passage the ES cells in vented, 12.5 cm^2 flasks (Falcon), each of which requires as little as 4 ml of supplemented medium. When maintained under these conditions, ES cells continue to form discrete colonies, although their morphology is quite distinct from those that develop on a monolayer of fibroblasts, the individual cells being easily distinguished by their prominent nuclei and nucleoli (Fig. 1B).

Having substantially reduced the proportion of fibroblasts, ES cells may be harvested using trypsin–EDTA and introduced into suspension cultures. This may be achieved by plating them at very low density in petri dishes composed of bacteriological plastic (Sarstedt) which have not been treated with extracellular matrix proteins to which the ES cells might otherwise adhere. We normally seed 4×10^5 ES cells per plate in 20 ml of complete medium. The removal of exogenous LIF at this stage encourages the spontaneous differentiation of ES cells, while their low density helps prevent their aggregation. Under these conditions, proliferation of individual ES cells gives rise to structures that become macroscopic by day 4–5 of culture and continue to increase in size over the ensuing 10 days, often reaching several millimeters in diameter. During this period, it is essential to prevent the depletion of nutrients: should the medium become acidic, the EBs should be transferred to a 50 ml Falcon tube and allowed to settle under unity gravity, before removing the spent supernatant and resuspending them in fresh medium.

The chaotic manner in which differentiation proceeds in developing EBs results in considerable heterogeneity in their size and morphology. While the majority remain simple EBs, lacking any polarity, a proportion becomes cystic, forming a large fluid-filled structure (Fig. 1C) in which discrete blood islands can occasionally be observed as foci of proliferating cells, actively synthesizing fetal hemoglobin (Fig. 1D). Although evidence for ongoing hematopoiesis might be expected to favor cystic EBs as a source of DC, our experiments have revealed no apparent correlation between morphology of the EB and its capacity to support DC differentiation, both cystic and simple EBs serving as equally potent sources of this cell type.

Differentiation of DC from EBs

Although the EB provides an optimal microenvironment for the spontaneous differentiation of a range of tissues derived from each of the three embryonic germ layers, the generation of many cell types requires additional signals in the form of growth factors and cytokines that must be supplied exogenously. Our own studies have revealed the essential role played by interleukin 3 (IL-3) and granulocyte-macrophage colony stimulating factor (GM-CSF) in guiding the early commitment of progenitors towards the DC lineage.[8] Skewing of the differentiation program towards this population may be facilitated by plating EBs onto tissue culture plastic in the same medium used for the culture of the parent ES cell line but further supplemented with 25 ng/ml of rGM-CSF and 200 U/ml of rIL-3 (World Health Organization Units; R&D Systems). Although there is considerable latitude in the age of EBs that support DC growth, we have found 14 days in suspension culture to be optimal: EBs maintained for 21 days may prove permissive but cell yields begin to decrease if they are cultured much beyond this time frame. Furthermore, EBs plated after only 7 days in culture may likewise support the differentiation of DC, but the proportion capable of doing so is relatively low.

In order to encourage the differentiation of DC, EBs should be harvested by transfer to a conical 50 ml tube (Falcon) and allowed to sediment. These may then be resuspended in complete medium and used to seed 90 mm tissue culture dishes (Corning) containing medium supplemented with GM-CSF and IL-3. After 24–48 hr incubation, the majority of EBs will have adhered to the extracellular matrix proteins, with which the plastic is treated, and will have supported the outgrowth of colonies of terminally differentiated cell types in a radial fashion. Although many morphologically distinct cell types may be observed in these nascent colonies, one which is readily identifiable is the cardiomyocyte which appears only a few days after plating.[13] The unique properties of these cells cause the parenchyma of many EBs to contract rhythmically; indeed, it is not unusual to observe discrete areas of cardiomyocytes within colonies, which contract independently and out of synchrony with one another. Since cardiomyocytes develop spontaneously in such cultures, without the need for exogenous growth factors, their appearance is an early and reliable indication that culture conditions are optimal for eliciting the normal differentiation program.

The first cells belonging to the DC lineage appear as early as day 5 after plating EBs in an appropriate cytokine milieu, although there is considerable variability in the timing of their emergence, even between individual EBs cultured from the same passage of the ES cell line. The DC may be easily identified by virtue of their distinctive location around the perimeter of most colonies (Fig. 2A), forming a "halo" that is highly refringent under darkfield illumination. This population expands rapidly with time, initially forming clusters of irregular-shaped cells with

[13] V. A. Maltsev, J. Rohwedel, J. Hescheler, and A. M. Wobus, *Mech. Dev.* **44**, 41 (1993).

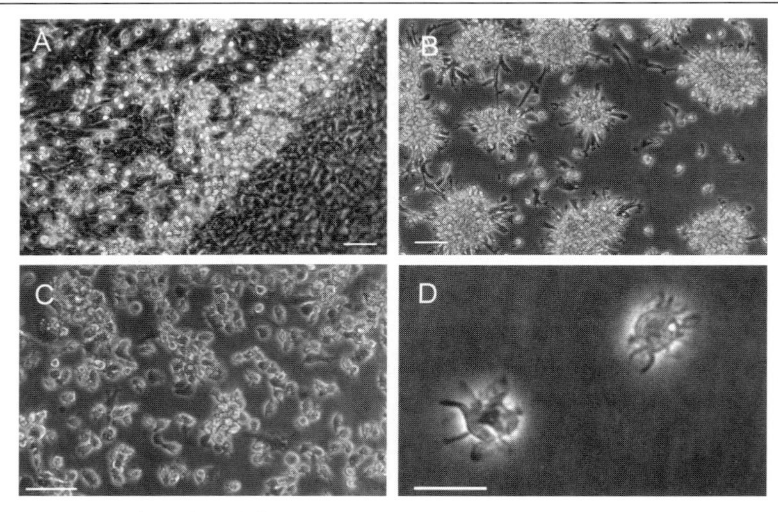

FIG. 2. Generation of esDC from EBs. (A) The edge of a colony of differentiated cell types derived from a single EB showing the early appearance of immature esDC as a "halo" of phase-bright cells (bar represents 50 μm). (B) Clusters of esDC typically observed early during the culture period, close to the perimeter of most EBs (bar represents 50 μm). (C) Accumulation of esDC at high densities in areas of the tissue culture vessel unoccupied by EBs (bar represents 50 μm). (D) ES cell-derived DC induced to mature in response to LPS, showing a pronounced dendritic morphology with characteristic veils of cytoplasm (bar represents 10 μm).

characteristic veils of cytoplasm (Fig. 2B), which are highly reminiscent of immature DC cultured from bone marrow progenitors. Although ES cell-derived DC (esDC) are closely associated with EBs during the first week or 10 days of culture, their migratory properties cause them to dissociate from clusters and colonize areas of the culture dish free of underlying stromal cells, where they continue to expand in number, eventually attaining very high densities (Fig. 2C). Since it is these areas between EBs which contribute most esDC to the final yield, the number of EBs introduced into each plate is critical. We find that 20–30 EBs provides sufficient progenitors to seed the esDC population while ensuring optimal space between colonies into which esDC may migrate and proliferate.

The capacity for expansion of esDC greatly surpasses that of their bone marrow-derived counterparts. In experiments aimed at quantifying the yield of esDC that might be obtained from a single ES cell, we have cultured the parent line at very low densities to ensure the clonality of the resulting EBs. When single EBs from these cultures were micromanipulated into 150 mm tissue culture plates, they frequently filled the dish with esDC within 3 or 4 weeks of culture, with yields in excess of 10^7 cells per plate. Moreover, replacement of the medium with a fresh source of nutrients and growth factors supported the emergence of a second cohort of esDC; indeed, these cultures could be harvested 4 or 5 times in succession without compromising the properties of the esDC, although yields were found to progressively decrease with age.[14]

Although many factors may subtly affect the yield of esDC, the source of FCS is one variable we have identified as having a major impact. Consequently, it is highly advisable to screen batches from a variety of sources for their efficacy at supporting the differentiation of esDC and their subsequent colonization of the tissue culture plastic. Importantly, batches of FCS that are optimal for ES cell growth need not be effective for the generation and expansion of esDC, although it is certainly worth testing the ability of batches to fulfill both roles.

Maturation of esDC

While esDC share many of the phenotypic and functional properties of immature bmDC[8,14] (Table I) they are nevertheless distinct in a number of respects. Arguably one of the most intriguing differences is the lack of spontaneous maturation that characterizes cultures of bmDC. Indeed, esDC appear arrested at an immature stage of the DC life cycle unless specifically induced to mature through exposure to bacterial products, such as lipoteichoic acid and lipopolysaccharide (LPS).

In order to promote maturation in a coordinated manner, we first harvest esDC by replacing the spent medium with 10 ml of fresh medium supplemented with GM-CSF and IL-3. The weakly adherent esDC may be released by gently pipetting over the surface of the dish using a P1000 Gilson pipette, the use of light-to-moderate force leaving most stromal cells attached. Since, like the esDC, the original EBs are only loosely adherent, a proportion may be inadvertently released, but may be subsequently removed by filtration of the suspension over a 70 μm cell strainer (Falcon). The suspension of esDC should be transferred to a fresh 60 mm tissue culture plate, preferably without centrifugation which may cause them to aggregate into clumps that are difficult to dissociate. After overnight incubation, we pulse cultures with 1 μg/ml of LPS purified from *E. coli* Serotype 0127 : B8 (Sigma). Within a few hours of exposure to LPS, esDC become firmly adherent with commensurate changes in their morphology. After overnight incubation, however, a proportion is released from the tissue culture plastic, forming a population of free-floating cells with

[14] P. J. Fairchild, F. A. Brook, R. L. Gardner, L. Graça, V. Strong, Y. Tone, M. Tone, K. F. Nolan, and H. Waldmann, *in* "Pluripotent Stem Cells: Therapeutic Perspectives and Ethical Issues" (B. Dodet and M. Vicari, eds.), p. 25. John Libbey Eurotext, Paris, 2001.

TABLE I
PHENOTYPIC AND FUNCTIONAL PROPERTIES OF esDC

	Immature esDC	Mature esDC	Immature bmDC	Mature bmDC
Surface Phenotype				
CD11b	+	+	+	+
CD11c	−	+	+	+
CD40	−	+	−	+
CD44	+ +	+ +	+ +	+ +
CD45	+ +	+ +	+ +	+ +
CD54	+	+ +	+	+ +
CD80	−	+ +	+	+ +
CD86	−	+ +	+	+ +
F4/80	+	+	+	±
MHC Class I	+	+ +	+	+ +
MHC Class II	−	+ +	+	+ +
Functional Properties				
Endocytosis	+ +	+	+	±
Antigen processing	+ +	+	+ +	±
Presentation of Antigen	+ +	+	+ +	±
Stimulation of naive T cells	±	+ +	+	+ +
Secretion of NO	+ +	N.D.	+	−

dramatic veils of cytoplasm and extensive dendrites (Fig. 2D). These may be harvested by transferring the supernatant to a fresh tissue culture dish and replacing the medium. Interestingly, even without the addition of further LPS, the original cultures will release two further waves of mature esDC on days 2 and 3 following exposure to the maturation stimulus, all three cohorts being routinely pooled in our laboratory for use in experiments. We have shown these cells to express a surface phenotype largely indistinguishable from populations of mature DC *ex vivo* and have demonstrated their capacity to stimulate potent primary T cell responses in mixed leukocyte cultures.[8] Whereas a proportion of esDC adopt the classic properties of mature DC following exposure to LPS, a significant population remains resistant to maturation. Although the genetic basis of this maturation arrest remains uncertain, its future elucidation may provide important insights into the mechanisms governing the acquisition of immunogenicity.

Validation of the Differentiation Pathway by Global Gene Expression Profiling

In an attempt to validate the protocols we have established, we have made use of gene expression profiling to chart global changes in gene transcription occurring during directed differentiation of ES cells. Serial analysis of gene expression (SAGE) is a powerful tool for defining and comparing the molecular signatures of distinct cell types[15,16] and is based on the analysis of short sequence tags, 14 bp in length, derived from a defined position within the mRNA transcripts from the cell types of interest. Encoded within these tags is sufficient information to uniquely identify most known genes and to facilitate the cloning and characterization of genes that have not previously been described. As such, SAGE libraries provide a comprehensive and unbiased blueprint of a given cell type, the information they provide being both qualitative and quantitative in nature. Indeed, the frequency with which a particular tag appears in the library reflects the relative abundance of the corresponding transcript, permitting meaningful comparisons to be made between libraries derived from functionally distinct cell types.

Using this approach, we have generated SAGE libraries from the ES cell line, ESF116, and purified esDC, differentiated from these cells using the protocols described above. When compared with a range of other libraries using algorithms similar to those employed to define the phylogenetic distance between species of plants and animals, esDC may be seen to fall into the same clade as both immature and mature bmDC (Fig. 3A). This suggests that, at the level of global gene expression, these two populations are more similar to one another than they are to any other cell type represented, including a variety of other leukocyte lineages. In contrast, the parent ES cell line is more closely allied, at the level of gene transcription, to cell types of nonimmunological origin, such as whole brain tissue and 3T3 fibroblasts (Fig. 3B).

When the expression of individual gene tags is compared between the parent ES cells and their DC derivatives, clear evidence can be found for a progressive constriction in the differentiation potential from a population of pluripotent cells to a terminally differentiated cell type, devoted primarily to antigen processing and presentation. Figure 3B represents a scatter plot of gene tags, demonstrating their relative abundance in either SAGE library: those tags confined to the central region are considered, on statistical grounds, to be expressed at comparable levels by either cell type (open circles) and include many known or suspected housekeeping genes (grey circles). Conversely, tags lying outside this area are differentially expressed at a 95% level of confidence (black circles), those that fall along the axes being unique to one or other cell type. Among those genes unique to the parent ES cell line is Oct-3/4, a POU family transcription factor, normally

[15] V. E. Velculescu, L. Zhang, B. Vogelstein, and K. W. Kinzler, *Science* **270**, 484 (1995).

[16] V. E. Velculescu, B. Vogelstein, and K. W. Kinzler, *Trends Genet.* **16**, 423 (2000).

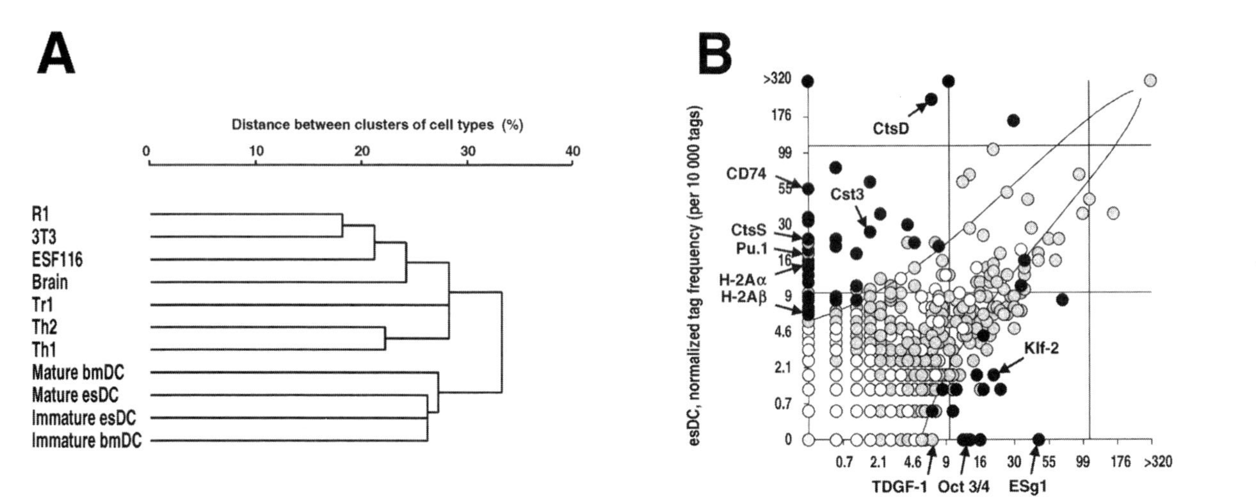

FIG. 3. Comparative analysis of gene expression by the ES cell line ESF116 and esDC derived from them, using custom-written SAGEclus software.[17] (A) Dendrogram showing the degree of similarity of the ESF116 cell line and both immature and matured esDC to a number of other cell types, as determined by clustering on the basis of similarity of gene expression. (B) Direct comparison of ESF116 and immature esDC gene expression profiles presented in the form of a scatter plot. Tags lying outside the coned area are differentially expressed (95% confidence of > 1.2-fold difference) and are represented as black spots, while tags which do not vary statistically across the set of SAGE libraries displayed in (A), are shown in grey. Annotated gene transcripts were identified using the following SAGE tags: CtsD CCTCAGCCTG; CtsS ATAGCCCCAA; Cst3 CCTTGCTCAA; CD74 GTTCAAGTGA; Pu.1 CCCGGCCTGG; H-2Aα GAAGAAGTGG; H-2Aβ GCACTATTGT; Klf-2 CGCGACTGTG; TDGF-1 AATATGCACA; Oct-3/4 CATTCAAACT; ESg-1 AAGACCCTGG.

restricted in its expression to early embryonic tissues, germ line cells and undifferentiated cell types. In keeping with its downregulation during differentiation, the sustained function of Oct-3/4 has been shown to be necessary for the maintenance of pluripotency and the capacity for self-renewal.[18] Kruppel-like factor-2 (Klf2) is a zinc finger transcription factor whose function is less well defined but which is differentially expressed by ESF116, consistent with the high level of expression previously reported from SAGE analysis of the R1 ES cell line.[19] In addition to transcription factors, our SAGE libraries also reveal a number of genes expressed by ESF116 which are associated with the early stages of embryogenesis. The gene encoding teratocarcinoma derived growth factor-1 (TDGF-1) is, for instance, critically involved in mesoderm development and, when disrupted by gene targeting, results in postgastrulation lethality.[20] Importantly, this EGF-like extracellular protein has also been reported to be expressed at high levels in pluripotent cells, as is embryonal stem cell specific gene-1 (ESg-1), whose expression is rapidly downregulated upon the onset of differentiation.

In sharp contrast, the end-point of terminal differentiation is characterized by the expression of transcription factors, such as Pu.1, implicated in myeloid development (Fig. 3B). Furthermore, many differentially expressed genes are intimately involved in the up-take of foreign antigen and its presentation to the immune system, the constitutive expression of MHC class II genes (H-2A) being one of the classic hallmarks of the DC. In concert with class II genes, the invariant chain (CD74) is also upregulated, together with cystatin C (Cst 3) and cathepsin S (Cts S), responsible for controlling its proteolytic degradation during antigen processing.[21] Lysosomal enzymes, such as cathepsin D (Cts D), are likewise highly expressed, consistent with the preoccupation with antigen processing, anticipated of a population of cells arrested at an immature stage of the DC life cycle.

Genetic Modification of esDC

While SAGE provides evidence for the validity of the differentiation program induced by the protocols we have established, its use also

[17] D. Zelenika, E. Adams, S. Humm, L. Graça, S. Thompson, S. P. Cobbold, and H. Waldmann, *J. Immunol.* **168**, 1069 (2002).

[18] H. Niwa, J. Miyazaki, and A. G. Smith, *Nat. Genet.* **24**, 372 (2000).

[19] S. V. Anisimov, K. V. Tarasov, D. Tweedie, M. D. Stern, A. M. Wobus, and K. R. Boheler, *Genomics* **79**, 169 (2002).

[20] M. M. Shen and A. F. Schier, *Trends Genet.* **16**, 303 (2000).

[21] P. Pierre and I. Mellman, *Cell* **93**, 1135 (1998).

highlights the large number of genes differentially expressed by esDC that have yet to be cloned or whose function remains unknown. Arguably the most important feature of esDC is the prospect they raise for elucidating the function of such genes by genetic modification of the parent ES cell line. Unlike ES cells, terminally differentiated populations of DC are peculiarly resistant to genetic manipulation, making electroporation and lipid-based strategies wholly ineffective. While the use of viral vectors for the introduction of heterologous genes has proven far more effective, their use risks compromising DC function, as evidenced by their impact on the process of maturation.[22] Vectors based on the Herpesviridae, for example, inhibit maturation of DC, arresting their differentiation at an immature stage of the life cycle. In contrast, many adenoviral vectors are intrinsically stimulatory, prematurely inducing the maturation of the DC they infect and precluding a comparison of gene function at distinct stages of the maturation pathway. By exploiting the tractability of ES cells for genetic modification, we have, however, overcome many of these limitations.

As a proof of principle, we have exploited enhanced green fluorescent protein (EGFP) as a convenient reporter gene. Using a standard lipid-based approach, normally inadequate for introducing genes into *bona fide* DC, we have been able to stably transfect the parent ES cells and clone the resulting population. Each of the clones obtained was able to support the development of EBs under selection conditions and generated large numbers of esDC, of which up to 95% expressed the desired, mutant phenotype[9] (Fig. 4). These results compare favorably with transduction efficiencies of 35%, reported for alternative approaches, likewise intended to avoid the use of viral vectors by exploiting the properties of cationic peptides.[23] Most importantly, however, by using ES cells as a vehicle for the introduction of genes into DC, we have demonstrated how the functional integrity of the resulting population is preserved: esDC expressing the EGFP gene remained immature for prolonged periods of time, yet responded readily to bacterial products by acquiring a mature, immunocompetent phenotype and the capacity to stimulate primary T cell responses.[9]

In order to produce lines of "designer" DC, uniformly expressing a gene of interest, the quality of the parent ES cells is of paramount importance. Since the process of transfection, selection and cloning may require many

[22] L. Jenne, G. Schuler, and A. Steinkasserer, *Trends Immunol.* **22**, 102 (2001).

[23] A. S. Irvine, P. K. E. Trinder, D. L. Laughton, H. Ketteringham, R. H. McDermott, S. C. H. Reid, A. M. R. Haines, A. Amir, R. Husain, R. Doshi, L. S. Young, and A. Mountain, *Nat. Biotechnol.* **18**, 1273 (2000).

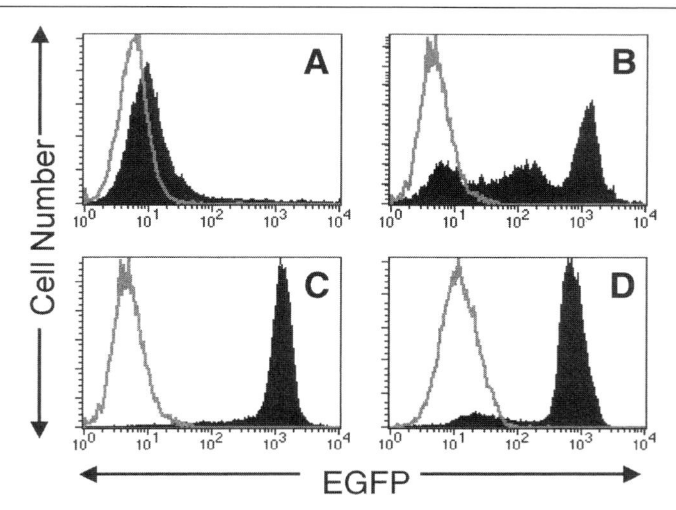

FIG. 4. Derivation of "designer" DC by genetic modification of the parent ES cell line. Flow cytometric analysis of the expression of a reporter gene, EGFP (filled histograms). (A–C) Levels of transgene expression by transfected ES cells before (A) and after (B) selection in G418 and of a representative ES cell line, cloned from this heterogeneous population (C). Open histograms represent the level of background fluorescence emitted by control mock-transfected ES cells. (D) Faithful expression of EGFP by esDC differentiated from the ES cell clone shown in (C) (filled histogram), compared with endogenous levels of fluorescence emitted by esDC derived from mock-transfected cells (open histogram).

passages, it is advisable to ensure that the starting population has as low a passage number as possible. The design of eukaryotic expression vectors should take account both of the nature of the promoter and the preferred means of selecting transfectants. For the high level expression of a transgene, we have found the promoter driving expression of elongation factor-1α (EF-1α)[24] to be preferable to the use of the CMV promoter, which appears to be silenced in terminally differentiated esDC (unpublished observation). The selection strategy adopted is limited by the requirement for embryonic fibroblast feeder cells that share with the transfectant, resistance to the selecting agent. Being transgenic for β-galactosidase, Rosa 26 mice also express the neomycin resistance gene, rendering embryonic fibroblasts derived from them intrinsically resistant to this antibiotic. The widespread availability of this mouse strain therefore acts as the underlying rationale for the use of G418 selection systems by many laboratories.

For the purpose of lipofection, the parent ES cells should be weaned from fibroblast feeder cells, as described above, and plated at 10^5 cells per

[24] S. Mizushima and S. Nagata, *Nucl. Acids Res.* **18**, 5322 (1990).

well of a gelatinized 6-well plate (Falcon) in 4 ml of complete medium supplemented with 1000 U/ml of rLIF. After 48 hr incubation, the ES cells will have adhered to the substrate and begun to form discrete colonies. We find that transfection is most successful when these colonies reach 40% confluency, since any greater density risks their overgrowth and extensive cell death before selection can be applied. Although many commercial reagents are available for lipofection, we obtain reproducibly good results with LipofectAMINE Plus (Life Technologies), when used according to the manufacturer's instructions. Two days after lipofection, stable transfectants may be selected by addition of the neomycin analogue G418 to the culture medium: although the sensitivity of the ES cells should be investigated in advance, 600 μg/ml of G418 has proven cytotoxic for all lines tested in our laboratory. As transfectants begin to emerge from the cell debris, they may be cloned in gelatinized 96-well flat bottomed plates at 0.5 and 5.0 cells per well, the clonality of the resulting colonies being readily verified by inverted, phase-contrast microscopy. Individual colonies may be progressively expanded in 24-well plates, before being reintroduced onto Rosa 26 embryonic fibroblasts. Once clones have been screened for expression of the transgene and frozen stocks prepared, they can be differentiated into esDC using the protocols outlined above, modified only by the addition to the culture medium of a "holding dose" of 300 μg/ml of G418 to maintain transgene expression.

Discussion

The properties of DC which distinguish them from other populations of antigen presenting cells have been appreciated for some years, yet the contribution made by specific genes to their functional phenotype remains only poorly understood. The intransigence of terminally differentiated DC to genetic manipulation, has limited the application of molecular biological approaches to their study. Although viral vectors have enjoyed some level of success, their use risks corrupting the very properties that make DC unique, as evidenced by their adverse impact on the normal process of maturation. By contrast, the use of ES cells as a conduit for the introduction of heterologous genes, offers significant advantages over the current art.

Since ES cells are far more amenable to genetic modification than their DC progeny, a desired mutant phenotype may be readily obtained by selection, cloning and screening of transfected cells. Once an appropriate clone has been identified, its capacity for self-renewal permits a permanent resource to be established, which may be differentiated, on demand, into lines of untransformed DC, expressing an identical, mutant phenotype. By consistently drawing on the same stock of engineered ES cells, the need for successive transduction of DC is avoided, significantly improving the reproducibility of experiments. Furthermore, the efficiency with which the transgene is expressed by esDC renders the need for purification of the modified population largely redundant. The demonstrable lack of effect that the process of genetic modification has on maturation of esDC, enables the systematic investigation of gene function at successive stages of the DC life cycle, a goal that has so far proven elusive.

Although the over-expression of heterologous genes may reveal important insights into the genetic basis of DC function, the application of protocols for targeting both alleles of a gene in ES cells by homologous recombination,[25] raises prospects for the construction of esDC, functionally deficient in a gene of interest. Alternatively, the use of RNA interference[26] may provide a rapid and effective means of abrogating the expression of candidate genes and assessing their involvement in aspects of DC function as diverse as their capture of foreign antigen, immunogenicity and patterns of migration *in vivo*. Such a high throughput system may ultimately prove attractive for the identification of novel targets for immune intervention in a variety of disease states.

While the protocols we have described are capable of generating impressive yields of DC that are unequivocally of myeloid origin by morphological, phenotypic and functional criteria, recent years have witnessed a renaissance of the DC field, fueled by the identification of distinct subsets whose lineage allegiances are less well defined.[27] Lymphoid-derived DC in the mouse are thought to share a common ancestry with T cell progenitors whereas the derivation of plasmacytoid DC remains strangely enigmatic. Given the pluripotency of ES cells, it might be anticipated that subtle changes in the culture conditions and provision of growth factors may skew differentiation along distinct pathways, allowing a comparison of gene function in distinct subsets while providing a powerful system for clarifying their lineage relationships. Beyond these advantages, however, it is the advent of human ES cell lines,[28] amenable to genetic modification,[29] that may ultimately pave the way for the application of esDC within the clinic. By overcoming the species barrier, it is our hope that

[25] R. M. Mortensen, D. A. Conner, S. Chao, A. A. Geisterfer-Lowrance, and J. G. Seidman, *Mol. Cell Biol.* **12**, 2391 (1992).

[26] B. R. Cullen, *Nat. Immunol.* **3**, 597 (2002).

[27] Y.-J. Liu, *Cell* **106**, 259 (2001).

[28] J. A. Thomson, J. Itskovitz-Eldor, S. S. Shapiro, M. A. Waknitz, J. J. Swiergiel, V. S. Marshall, and J. M. Jones, *Science* **282**, 1145 (1998).

[29] R. Eiges, M. M. Schuldiner, M. Drukker, O. Yanuka, J. Itskovitz-Eldor, and N. Benvenisty, *Curr. Biol.* **11**, 514 (2001).

the use of esDC may one day progress beyond the molecular dissection of their function, to the rational design of DC for therapeutic intervention.

Acknowledgment

We are grateful to Steve Cobbold for help with bioinformatics, Siân Cartland for technical support and to Richard Gardner and Frances Brook for the provision of ES cell lines. Work in the authors' laboratory was funded by a program grant from the Medical Research Council (UK).

[13] Gene Targeting Strategies for the Isolation of Hematopoietic and Endothelial Precursors from Differentiated ES Cells

By WEN JIE ZHANG, YUN SHIN CHUNG, BILL EADES, and KYUNGHEE CHOI

Introduction

In 1981, investigators successfully derived pluripotent embryonic stem (ES) cells from the preimplantation stage of mouse embryos; the blastocyst.[1,2] Subsequently, ES lines from many different species including human have been derived.[3–5] The derivation of ES cells is quite straightforward in that blastocysts are plated onto a feeder layer of fibroblasts. Under this condition, the inner cell mass of blastocysts will give rise to colonies of undifferentiated cells (ES colonies), which can be isolated and further expanded. Once established, ES cells can be maintained as pluripotent stem cells on a feeder layer of fibroblasts. One of the factors that is responsible for maintaining ES cells as stem cells is the leukemia inhibitory factor (LIF).[6,7] In fact, ES cells can be maintained as stem cells with LIF alone without

[1] M. J. Evans and M. H. Kaufman, *Nature* **292**, 154 (1981).

[2] G. R. Martin, *Proc. Natl. Acad. Sci. USA* **78**, 7634 (1981).

[3] J. A. Thomson, J. Kalishman, T. G. Golos, M. Durning, C. P. Harris, R. A. Becker, and J. P. Hearn, *Proc. Natl. Acad. Sci. USA* **92**, 7844 (1995).

[4] J. A. Thomson, J. Kalishman, T. G. Golos, M. Durning, C. P. Harris, and J. P. Hearn, *Biol. Reprod.* **55**, 254 (1996).

[5] J. A. Thomson, J. Itskovitz-Eldor, S. S. Shapiro, M. A. Waknitz, J. J. Swiergiel, V. S. Marshall, and J. M. Jones, *Science* **282**, 1145 (1998).

[6] A. G. Smith, J. K. Heath, D. D. Donaldson, G. G. Wong, J. Moreau, M. Stahl, and D. Rogers, *Nature* **336**, 688 (1988).

[7] R. L. Williams, D. J. Hilton, S. Pease, T. A. Willson, C. L. Stewart, D. P. Gearing, E. F. Wagner, D. Metcalf, N. A. Nicola, and N. M. Gough, *Nature* **336**, 684 (1988).

0076-6879/2003 $35.00

the feeder cells. When reintroduced into a blastocyst, ES cells can contribute to all tissues with the exception of the extra embryonic endoderm and trophoblast of the developing embryo.[8,9] It is this particular trait that makes ES cells a valuable tool for genetic engineering (Fig. 1). In addition, ES cells can be differentiated *in vitro* into many different somatic cell types,[10–30] allowing for the possibility that ES-derived cells can be used as a potential source for cell transplantation or cell based therapy (Fig. 1). Therefore, ES cells have gained much scientific and general public attention.

[8] A. Bradley, M. Evans, M. H. Kaufman, and E. Robertson, *Nature* **309**, 255 (1984).
[9] R. S. Beddington and E. J. Robertson, *Development* **105**, 733 (1989).
[10] M. V. Wiles and G. Keller, *Development* **111**, 259 (1991).
[11] G. Keller, M. Kennedy, T. Papayannopoulou, and M. V. Wiles, *Mol. Cell Biol.* **13**, 473 (1993).
[12] T. Nakano, H. Kodama, and T. Honjo, *Science* **265**, 1098 (1994).
[13] A. J. Potocnik, P. J. Nielsen, and K. Eichmann, *EMBO J.* **13**, 5274 (1994).
[14] W. Risau, H. Sariola, H. G. Zerwes, J. Sasse, P. Ekblom, R. Kemler, and T. Doetschman, *Development* **102**, 471 (1988).
[15] R. Wang, R. Clark, and V. L. Bautch, *Development* **114**, 303 (1992).
[16] D. Vittet, D. M. H. Prandini, R. Berthier, A. Schweitzer, H. Martin-Sisteron, G. Uzan, and E. Dejana, *Blood* **88**, 3424 (1996).
[17] J. Yamashita, H. Itoh, M. Hirashima, M. Ogawa, S. Nishikawa, T. Yurugi, M. Naito, K. Nakao, and S. Nishikawa, *Nature* **408**, 92 (2000).
[18] V. A. Maltsev, J. Rohwedel, J. Hescheler, and A. M. Wobus, *Mech. Dev.* **44**, 41 (1993).
[19] V. A. Maltsev, A. M. Wobus, J. Rohwedel, M. Bader, and J. Hescheler, *Circ. Res.* **75**, 233 (1994).
[20] T. C. Doetschman, H. Eistetter, M. Katz, W. Schmidt, and R. Kemler, *J. Embryol. Exp. Morphol.* **87**, 27 (1985).
[21] M. Drab, H. Haller, R. Bychkov, B. Erdmann, C. Lindschau, H. Haase, I. Morano, F. C. Luft, and A. M. Wobus, *FASEB J.* **11**, 905 (1997).
[22] J. Rohwedel, V. Maltsev, E. Bober, H. H. Arnold, J. Hescheler, and A. M. Wobus, *Dev. Biol.* **164**, 87 (1994).
[23] G. Bain, D. Kitchens, M. Yao, J. E. Huettner, and D. I. Gottlieb, *Dev. Biol.* **168**, 342 (1995).
[24] C. Strubing, G. Ahnert-Hilger, J. Shan, B. Wiedenmann, J. Hescheler, and A. M. Wobus, *Mech. Dev.* **53**, 275 (1995).
[25] A. Fraichard, O. Chassande, G. Bilbaut, C. Dehay, P. Savatier, and J. Samarut, *J. Cell Sci.* **108**, 3181 (1995).
[26] S. Liu, Y. Qu, T. J. Stewart, M. J. Howard, S. Chakrabortty, T. F. Holekamp, and J. W. McDonald, *Proc. Natl. Acad. Sci. USA* **97**, 6126 (2000).
[27] C. Bagutti, A. M. Wobus, R. Fassler, and F. M. Watt, *Dev. Biol.* **179**, 184 (1996).
[28] C. Dani, A. G. Smith, S. Dessolin, P. Leroy, L. Staccini, P. Villageois, C. Darimont, and G. Ailhaud, *J. Cell Sci.* **110**, 1279 (1997).
[29] L. D. Buttery, S. Bourne, J. D. Xynos, H. Wood, F. J. Hughes, S. P. Hughes, V. Episkopou, and J. M. Polak, *Tissue Eng.* **7**, 89 (2001).
[30] J. Kramer, C. Hegert, K. Guan, A. M. Wobus, P. K. Muller, and J. Rohwedel, *Mech. Dev.* **92**, 193 (2000).

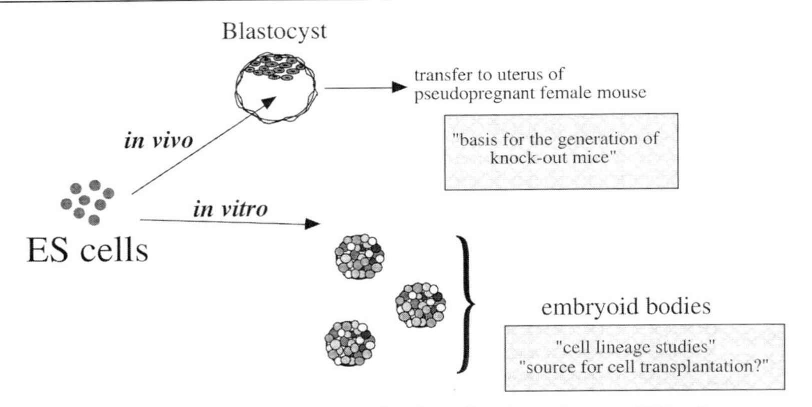

FIG. 1. *In vivo* and *in vitro* application of embryonic stem (ES) cells.

ES Maintenance

Mouse ES cells grow rapidly with an average division time of about 8 hr. Therefore, ES cells require frequent splitting. We normally split ES cells every 2 days and do not keep ES cells in culture for a long time after the cells are thawed. Typically, a new vial of cells is thawed after 5–6 passages. We found that ES cells maintained on STO feeder cells give consistent *in vitro* differentiation behavior. Feeder cells are prepared as follows:

1. Day 1, split STO cells at a density of 50,000 cells/cm^2 (i.e., add 1.25×10^6 cells per T25 flask).
2. Day 2, treat STO cells with mitomycin C for 2–3 hr, wash, and add fresh new media. Typically, 4 ml of media containing mitomycin C is added per T25 flask.
3. Day 3, remove media and add ES cells (8×10^5 to 1×10^6 cells per T25 flask) in ES-DMEM (Dulbecco Modified Eagle's Medium) media.

Reagents and Materials

1. Tissue culture flasks/dishes should be pretreated with 0.1% gelatin in PBS for a minimum of 20 min at room temperature. Simply add the appropriate amount (i.e., 3 ml per T25 or 7 ml per T75, just enough to cover the bottom of the flask/dish) of 0.1% gelatin in PBS and keep it at room temperature. Aspirate off gelatin right before adding cells. Flasks/dishes with 0.1% gelatin in PBS can be stored at 4°C as well.
2. 0.1% gelatin. Dissolve gelatin (Sigma G-1890) at 0.1% in PBS and autoclave.

3. 100X mitomycin C solution. Dissolve mitomycin C (Sigma M-0503) at 1 mg/ml in PBS. Keep at 4°C in dark.
4. DMEM. Dissolve 1 package of DMEM (Gibco 12100-046) powder in distilled water. We normally use distilled water from Millipore (Bedford, CA) Milli-Q purification system (QTUM000EX). Add 3.024 g $NaHCO_3$ (Sigma S5761), 10 ml Penicillin/Streptomycin (10,000 U Gibco/BRL 6005140PG), and 25 ml 1 *M* HEPES buffer (Gibco/BRL 380-5630 PG). Bring up to 1 liter with distilled water, filter through 0.22-micron filter, and keep at 4°C.
5. ES-DMEM media. 15% FCS (preselected), 1.5% LIF (leukemia inhibitory factor), and 1.5×10^{-4} *M* MTG in DMEM.
6. FCS for ES culture. We normally prescreen FCS for ES culture. Typically, ES cells adapted to grow without feeder cells are used for serum test. ES cells are maintained in test serum for 5–6 passages and scored for morphology, differentiated or undifferentiated. A good lot of serum should keep the ES cells in undifferentiated state.
7. Conditioned media from Chinese Hamster Ovary (CHO) cells transfected with LIF gene (Genetics Institute) is used as a source for LIF. Typically LIF protein is secreted from the cells into the conditioned media at ~5 μg/ml.
8. 1.5×10^{-4} *M* MTG solution is prepared by diluting MTG (Sigma, M-6145) 1:10 in DMEM and adding 12.4 μl per 100 ml of ES-media. β-Mercaptoethanol (BME, Sigma M-7522) is alternatively used at 1×10^{-4} *M* (make 100X stock solution by adding 72 μl of 14 *M* BME to 100 ml of 1X PBS, and add 1 ml per 100 ml of ES-media). Make sure that MTG or BME is made fresh.
9. Feeder/STO media. STO cells are grown in 10% FCS (the same serum used for ES culture) in DMEM. MTG (add 6.2 μl per 100 ml of media of MTG that is prediluted 1:10 in DMEM) or BME (at 5×10^{-5} *M*, add 0.5 ml of 100X stock per 100 ml media) is also added.
10. When splitting ES cells, it is important to count ES cells only. STO cells are easily recognized, as they are granular and somewhat bigger in cell size.
11. Trypsin/EDTA. Dissolve 2.5 g of trypsin (Sigma T-4799) in 1X PBS. Add 2.16 ml of 0.5 *M* EDTA and bring up to 1 liter with 1X PBS. Filter sterilize through 0.22-micron filter. Store aliquots at −20°C.
12. 10X PBS. Add 160 g NaCl, 4 g KCl, 28.8 g Na_2HPO_4, and 4.8 g KH_2PO_4 per 2 liters of distilled water. Heat dissolve and adjust the pH to 7.4. Make 1X by diluting 1:10 in distilled water and autoclave.

13. ES cell freezing media: 90% FCS with 10% DMSO (Sigma D2650). ES cells are frozen at a density of 2–3 $\times 10^6$ cells/ml of freezing media. Add 1 ml of cells to each freezing vial.

In Vitro Differentiation of ES Cells

Mouse ES cells have been instrumental in determining the *in vitro* potential of generating numerous differentiated cell types from an ES cell. Somatic cells that have been successfully generated from *in vitro* differentiated ES cells include hematopoietic,[10–13] endothelial,[14–17] vascular smooth muscle, cardiac, myogenic,[17–22] neuronal, astrocytes, oligodendrocytes,[23–26] epithelial,[27] adipocytes,[28] osteoblasts,[29] and chondrocytes.[30] Many different methods of *in vitro* differentiation of ES cells can efficiently be used to generate the progeny of all three primary germ layers—the endoderm, ectoderm, and mesoderm. The most typical method is to differentiate ES cells in a stromal cell independent manner to give rise to a three-dimensional, differentiated cell mass called embryoid bodies (EBs, Fig. 1).[10,11] ES cells can also be differentiated two-dimensionally on stromal cells (typically OP9 cells) or on type IV collagen coated tissue culture plates without intermediate formation of the EB structure.[12,31] The developmental kinetics of various cell lineages including hematopoietic, endothelial, and neuronal precursors within EBs and molecular and cellular studies of these cells have demonstrated that the development of various cell lineages within EBs recapitulates the normal mouse embryo development.[32–34] In addition, EBs provide a large number of cells representing an early/primitive stage of development which are otherwise difficult to access in an embryo.[35,36] Therefore, the *in vitro* differentiation model of ES cells is an ideal system for obtaining and studying primitive progenitors of all cell lineages.

ES Differentiation to EBs

The following protocol is an example of differentiating ES cells via three-dimensional EBs. We found that liquid differentiation is good for

[31] S. I. Nishikawa, S. Nishikawa, M. Hirashima, N. Matsuyoshi, and H. Kodama, *Development* **125**, 1747 (1998).

[32] G. M. Keller, *Curr. Opin. Cell Biol.* **7**, 862 (1995).

[33] S. I. Nishikawa, *Curr. Opin. Cell Biol.* **13**, 673 (2001).

[34] K. Guan, H. Chang, A. Rolletschek, and A. M. Wobus, *Cell Tissue Res.* **305**, 171 (2001).

[35] M. Kennedy, M. Firpo, K. Choi, C. Wall, S. Robertson, N. Kabrun, and G. Keller, *Nature* **386**, 488 (1997).

[36] K. Choi, M. Kennedy, A. Kazarov, J. C. Papadimitriou, and G. Keller, *Development* **125**, 725 (1998).

obtaining early EBs (up to day 5–6) and methylcellulose differentiation suitable for obtaining late EBs (day 6–14). The methylcellulose (Fluka, 64630) media contains the same reagents as liquid differentiation media, except that methylcellulose is added to 1% of the final volume.

1. Two days prior to set up differentiation, split ES cells (4×10^5 ES cells per T25 flask) into ES-IMDM medium without feeder cells in T-25 flasks. All flasks should be gelatinized.
2. Change media the next day.
3. Set up differentiation as follows:
   A. Aspirate the medium from the flask.
   B. Add 1 ml of trypsin, swirl, and remove quickly.
   C. Add 1 ml trypsin and wait until cells start to come off. It usually takes about 1–2 min. Do not over-trypsinize cells.
   D. Stop the reaction by adding 1 ml FCS (which is to be used for differentiation) and 4 ml IMDM and pipette up and down to make single cell suspensions. It is important not to have cell clumps. Transfer to a 14 ml snap cap tube (Falcon 352059).
   E. Centrifuge for 5–10 min at 1000 rpm.
   F. Wash the cell pellet in 10 ml IMDM (without FCS). Spin at 1000 rpm for 5–10 min.
   G. Resuspend the cell pellet in 5 ml IMDM (with 10% FCS) and count viable ES cells using 2% eosin solution in PBS. Make sure to count ES cells only. Add 6000–10,000* ES cells per ml of differentiation media to obtain day 2.75–3 EBs. Add 4000–5000* cells per ml to obtain day 4–5 EBs. Add 500–2000* cells per ml to obtain day 6–10 EBs.
      *Add higher cell number for ES lines that differentiate poorly.
   H. Primary differentiation is set up based on the cell number required for subsequent experiments. Total EB cell numbers typically obtained are as follows: Day 2.75, $\sim 5\text{–}1 \times 10^6$ EB cells/10 ml of differentiation; day 4, $\sim 2\text{–}3 \times 10^6$ EB cells/10 ml of differentiation; day 6, $\sim 3\text{–}5 \times 10^6$ EB cells/10 ml of differentiation.

Differentiation Media

	Liquid	*Methylcellulose*
2X methylcellulose	–	1%
FCS (preselected)	15%	15%
Ascorbic acid (5 mg/ml)	50 μg/ml	50 μg/ml
L-Glutamine (200 m*M*)	2 m*M*	2 m*M*

	Liquid	*Methylcellulose*
MTG	4.5×10^{-4} *M*	4.5×10^{-4} *M*
IMDM	up to 100%	up to 100%

Reagents and Materials

1. IMDM. Dissolve 1 package of IMDM powder (Gibco/BRL, 12200-036) in distilled water. Add 3.024 g of $NaHCO_3$ and 10 ml of Penicillin/Streptomycin. Bring up to 1 liter and filter through 0.22-micron filter.
2. FCS for ES differentiation. We normally prescreen FCS for ES differentiation. Typically, ES cells are differentiated in test serum and analyzed by FACS for Flk-1 staining or hematopoietic replating (see below). A good lot of serum should give rise to 30–50% of Flk- 1^+ cells by day 3–4 EB stage and good hematopoietic replating.
3. ES-IMDM: 15% FCS (preselected), 1.5% LIF (leukemia inhibitory factor), and 1.5×10^{-4} *M* MTG in IMDM.
4. 4.5×10^{-4} *M* of MTG is prepared by diluting 26 μl of MTG into 2 ml of IMDM and adding 3 μl of diluted MTG per ml of differentiation media.
5. Add Kit-Ligand (KL, final 1%) and IL-3 (final 1%) for day >6 EBs. Feed for day >9 EBs on day 6 (4–5 ml per 100 mm petri dish). Make the same methylcellulose cocktail with 0.5% methylcellulose instead of 1%. KL is from media conditioned by CHO cells transfected with KL expression vector (Genetics Institute, Cambridge, MA). IL-3 is from medium conditioned by X63 AG8-653 myeloma cells transfected with a vector expressing IL-3.[37]
6. Make ascorbic acid (Sigma A-4544) solution fresh each time you set up a differentiation. Dissolve ascorbic acid at 5 mg/ml in H_2O and filter sterilize (0.22 μm).
7. Differentiation is done in bacterial petri-dishes (Valmark #900). Do not use tissue culture dishes.
8. 2X methylcellulose is prepared as follows (1 liter preparation):
   A. Weigh a sterile Erlenmeyer flask.
   B. Add ~450 ml of sterile water.
   C. Bring to boil on a hot plate and keep boiling for 3–4 min.
   D. Add 20 g of methylcellulose, swirl quickly, and return flask to the hot plate.
   E. Remove from hot plate and swirl again when it starts to boil. Return flask to the hot plate. Repeat 3–4 times.

[37] H. Karasuyama and F. Melchers, *Eur. J. Immunol.* **18**, 97 (1988).

F. Weigh and add sterile water (room temperature) up to 500 ml of methylcellulose.
G. Let it sit on bench to cool down to room temperature.
H. In a separate flask, make 500 ml of 2X IMDM and filter sterilize.
I. Slowly add 2X IMDM to methylcellulose and mix vigorously.
J. Put the mixture on ice until the right consistency is achieved.
K. Make ~100 ml aliquots and keep frozen at −20°C. When ready to use, thaw and use a syringe to disperse methylcellulose (do not use pipettes).

ES Differentiation to Hematopoietic and Endothelial Cells

As discussed earlier, the *in vitro* differentiation model of ES cells has proven to be valuable for studies of cell lineage development. Hematopoietic cells develop within EBs faithfully following *in vivo* developmental kinetics.[35,36,38] For example, the development of hematopoietic and endothelial cells within EBs mimics *in vivo* events so strongly that yolk sac blood island-like structures with vascular channels containing hematopoietic cells can be found within cystic EBs.[20] As in the developing embryo, primitive erythroid cells develop prior to definitive hematopoietic populations.[11,37] Endothelial cells within EBs also follow similar kinetics, in that they develop from Flk-1$^+$ mesodermal cells.[16,17,39] Hematopoietic and endothelial progenitors present within EBs can be successfully analyzed by flow cytometry and by direct replating EB cells into methylcellulose cultures to measure the frequency of hematopoietic progenitors. Additionally, EB cells can be sorted for early hematopoietic and endothelial cell markers to further analyze their hematopoietic and/or endothelial cell potential.

FACS Analysis

A powerful tool for identifying, analyzing and isolating specific EB cell populations is the flow cytometer, which detects individual cells passing directly through a laser beam. Typically, a flow cytometer, or fluorescence activated cell sorter (FACS), utilizes monoclonal antibodies against cell surface proteins or intracellular markers to study the cell subsets. Flow cytometers use one or more lasers and the concept of hydrodynamic focusing to align single cell suspensions to the laser light.

[38] J. Palis, S. Robertson, M. Kennedy, C. Wall, and G. Keller, *Development* **126**, 5073 (1999).
[39] S. I. Nishikawa, S. Nishikawa, M. Hirashima, N. Matsuyoshi, and H. Kodama, *Development* **125**, 1747 (1998).

Sample suspensions, a mixture of labeled cells with specific antibodies tagged with fluorescent dyes or cells with intracellular dye markers, are injected by a small jet nozzle into the center of a chamber of isotonic saline known as sheath fluid. The flow cytometer determines sample pressure against sheath pressure to form laminar flow. The intended result is a single file march of cells consistently aligned to the laser beam. As each cell passes through a laser beam, it scatters the laser light. In addition, any dyes bound to the cell surface or intracellular dyes will be excited and will fluoresce. A detector placed directly in line with the laser beam opposite the cell intercept point measures forward scatter and the level of this low angle scatter is directly relative to the size of the cell. Side scatter is ideally measured 90° from the laser beam at the point where it intersects the cell. Collection optics focus the side scatter into a light–tight chamber where this very low intensity emission is measured to determine the internal complexity of the cell, also known as its granularity. Fluorescent emissions from the cell are collected from the same direction into the same chamber, the only difference being the optical filtering to create restrictive color bandwidths optimized for specific fluorescent dyes. This gives the flow cytometer the ability to gather information on the binding of several different labeled monoclonal antibodies simultaneously. The actual devices used to collect the emission are vacuum tube, grid-array photomultipliers. The analog electrical pulses generated by these devices are extremely low amplitude. They must be amplified and converted to digital channels before they can be graphed by computer acquisition software. If two or more adjacent fluorescent parameters are being analyzed at the same time, the spectral overlap is compensated, so as not to allow artifact fluorescent expression. The flow cytometer's computer software can present graphs that can be gated to include or exclude scatter and fluorescent characteristics of the cell populations, as well as deriving a wide range of useful statistical data. The digital bitmapping used to gate and analyze regions can also be used to make sort decisions that can allow a sorting flow cytometer to physically separate a target cell population from the rest. The ultimate goal is to collect information on cellular characteristics and the expression of cell surface proteins by each cell.

Flk-1 and SCL as Tools for Studying Hematopoietic and Endothelial Cell Differentiation

Gene targeting studies have shown that Flk-1, a receptor tyrosine kinase, and SCL, a basic helix-loop-helix transcription factor, are critical for both hematopoietic and endothelial cell development. Flk-1 deficient mice are embryonic lethal due to defects in the formation of blood islands and vasculatures.[40] SCL deficient mice display an embryonic lethality due to hematopoietic defects.[41,42] Subsequent studies have shown that *Scl*$^{-/-}$ ES cells fail to contribute to remodeling of the primary vascular plexus in the yolk sac indicating that SCL is also required for endothelial cell development.[43] As Flk-1 and SCL deficient mice show defects in both primitive and definitive hematopoietic programs as well as endothelial cell lineage development, studies of Flk-1 and SCL expressing cells should be voluable for further delineating cellular and molecular pathways leading to hematopoietic and endothelial cell lineage development.

Cells expressing Flk-1 can be easily identified and isolated by utilizing antibodies against the extracellular domain of Flk-1. However, it is rather difficult to isolate cells expressing cytoplasmic proteins or transcription factors. One way of isolating SCL expressing cells is to knock in a reporter gene, such as a green fluorescent protein (GFP) or a nonfunctional surface protein, into the *Scl* locus. GFP or antibodies against a nonfunctional surface protein can then be used to identify and isolate SCL expressing cells. In an attempt to study SCL expressing cells, we have knocked in a nonfunctional human CD4 into the *Scl* locus (Fig. 2).[44] Briefly, the 5′ homology arm containing ~7.4 kb BamHI to NotI DNA fragment was isolated from the *Scl* 2A genomic clone (kindly provided by Dr. S. Orkin at Harvard Medical School). The extra cellular and transmembrane domain of the human CD4 (hCD4) gene[45] was knocked in in-frame into the NotI site of the exon IV. This NotI site is 10 nucleotides downstream of the initiation codon. The 3′ homology arm containing ~4 kb NotI to SalI DNA fragment was isolated from the *Scl* 2A genomic clone. The 3′ arm was first blunt ligated into the XhoI site located downstream of the PGK-neomycin cassette of the pLNTK. The 5′ homology-hCD4 fragment was blunt ligated into the SalI site of the pLNTK/3′ homology arm. The targeting construct also contains a thymidine kinase gene. R1 ES cells were electroporated with a linearized SCL targeting vector construct and selected with

[40] F. Shalaby, J. Rossant, T. P. Yamaguchi, M. Gertsenstein, X. F. Wu, M. L. Breitman, and A. C. Schuh, *Nature* **376**, 62 (1995).

[41] R. A. Shivdasani, E. L. Mayer, and S. H. Orkin, *Nature* **373**, 432 (1995).

[42] L. Robb, I. Lyons, R. Li, L. Hartley, F. Kontgen, R. P. Harvey, D. Metcalf, and C. G. Begley, *Proc. Natl. Acad. Sci. USA* **92**, 7075 (1995).

[43] J. E. Visvader, Y. Fujiwara, and S. H. Orkin, *Genes Dev.* **12**, 473 (1998).

[44] Y. S. Chung, W. J. Zhang, E. Arentson, P. D. Kingsley, J. Palis, and K. Choi, *Development* **129**, 5511 (2002).

[45] P. Bedinger, A. Moriarty, R. C., 2nd, Donovan, N. J. von Borstel, K. S. Steimer, and D. R. Littman, *Nature* **334**, 162 (1988).

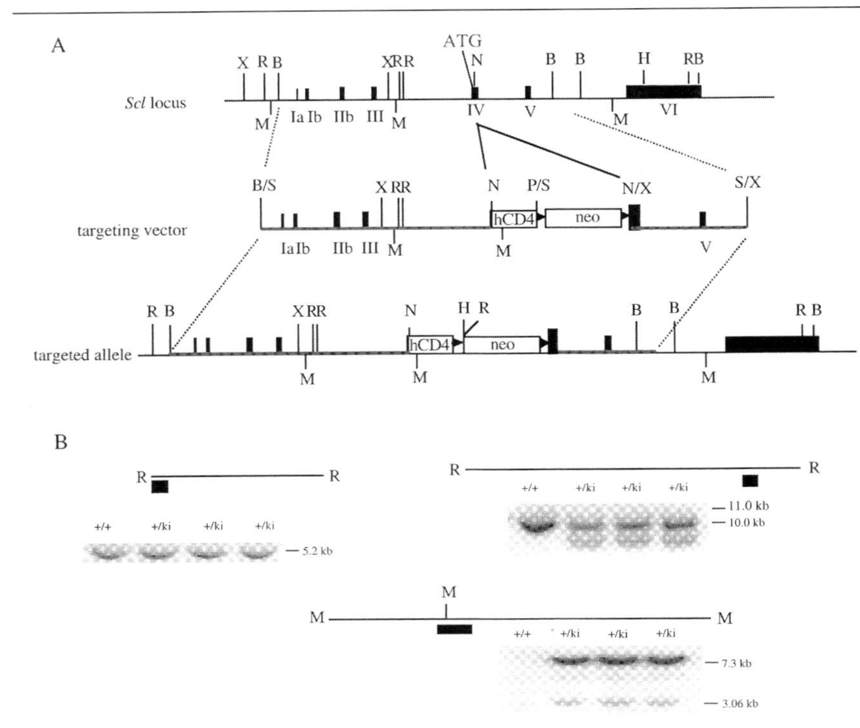

FIG. 2. Generation of human CD4 (hCD4) knock in ES cells. (A) Targeting strategy used to insert hCD4 gene into the *Scl* locus is shown. The endogenous mouse *Scl* locus, targeting construct, and the targeted allele are shown. The black boxes and numbers below indicate the exons. The ATG codon starts within exon IV. (B) Southern blot analysis of the targeted allele. DNA was digested with enzymes indicated and run on an agarose gel. Left, genomic DNA was digested with EcoRI and probed with genomic DNA as indicated below. Both targeted and wild type alleles generated a 5.2 kb DNA band. Middle, genomic DNA was digested with MunI and probed with human CD4 gene. Only the targeted allele gave 7.3 and 3 kb DNA bands. Right, genomic DNA was digested with EcoRI and probed with exon 6 probe as indicated below. The wild type allele (upper band) and the targeted allele (lower band) are shown. The enzymes used are as follows: X, XhoI; R, EcoRI; N, NotI; B, BamHI; H, HindIII; and M, MunI.

500 μg/ml G418 and 2×10^{-6} *M* gancyclovior (FIAU). G418 and FIAU resistant clones were picked after ~10 days of selection and expanded for further analysis (read a chapter by Ramirez-Solis *et al.* for detailed protocol for gene targeting procedures in ES cells).[46]

The following is the protocol for direct FACS analysis of EB cells for Flk-1 and hCD4 (SCL) expression.

[46] R. Ramirez-Solis, A. C. Davis, and A. Bradley, *Methods Enzymol.* **225**, 855 (1993).

1. Collect EBs in a 50 ml tube (Fisher, #14-432-22) by centrifugation at 1000 rpm for 1 min or by settling down at room temperature for 10–20 min.
2. Remove the supernatant and treat EBs with 3–5 ml of 7.5 m*M* EDTA (BioRad, #161-0729)/PBS (pH 7.4) for 2 min. Vortex quickly and spin down cells by centrifugation at 1000 rpm for 5 min, re-suspend the cell pellet in 5 ml of staining buffer (4% FCS in PBS), pass through a 20-gauge needle (Fisher, #14826-5C) 4–5 times to generate a single cell suspension, and count the cell number.
3. After centrifugation, resuspend the cells to a density of 5×10^6 cells/ml in 2.4G2 supernatant to block antibodies from binding to Fc receptors II and III (CD 16 and CD32).[47] Afterwards, place cells into each well of a V-shaped 96-well plate (Fisher, #07-200-96) at 5×10^5 cells/well followed by incubation on ice for 30 min.
4. Subsequently, add 100 μl of biotinylated mouse anti-human CD4 monoclonal antibody (CALTAG, #MHCD0415-4), freshly diluted (1 : 100) in staining buffer (4% FCS in PBS), into each well and incubate on ice for 15 min. After incubation, spin cells down at 1000 rpm for 5 min, remove the supernatant, wash cells in 150 μl of staining buffer, and spin down again.
5. After three washes, add 100 μl of freshly diluted streptavidin–allophycocyanin (1 : 1000, Pharmingen, #554067) and phycoerythrin (PE)-conjugated anti-Flk-1 monoclonal antibody (1 : 200, Pharmingen, #555308) in staining buffer and incubate on ice for 15 min in the dark.
6. After incubation, wash cells three times, re-suspend in 300 μl of staining buffer, and transfer to a 5 ml polypropylene tube (VWR, #60818-500) for analysis. Cells are analyzed on a FACS Caliber (Becton-Dickinson). FACS data are analyzed with CellQuest software (Becton-Dickinson).

*A three-color FACS analysis of Flk-1, human CD4, and endothelial or hematopoietic markers is carried out by staining cells first with endothelial/hematopoietic markers. Add directly FITC-conjugated anti-mouse CD31 (1 : 200, Pharmingen, #553372), FITC-conjugated anti-mouse CD34 (1 : 200, Pharmingen, #09434D), or FITC-conjugated anti-mouse CD45 (1 : 200, Pharmingen, #01114D). When cells are stained

*Use syringes for dispersing methylcellulose. 16 gauge needles are used when adding cells to petri dishes.

[47] J. C. Unkeless, *J. Exp. Med.* **150**, 580 (1979).

with nonlabeled anti-mouse VE-cadherin (1 : 200, Pharmingen, #28091D) and anti-mouse Ter-119 antibodies (1 : 200, Pharmingen, #553671), use FITC-conjugated goat anti-rat IgG (1 : 200, CALTAG, #R40001) and FITC-conjugated goat anti-rat IgG_{2b} (1 : 200, Pharmingen, #553884), respectively, to amplify the signals. Cells are subsequently stained with anti-human CD4 and anti-Flk-1 antibodies as described above.

Hematopoietic Replating

Hematopoietic progenitors present within EBs can be assayed by directly replating EB cells or by replating sorted cell populations. When EB cells are directly replated, day 2.75–3 EBs are typically used for blast colony assay,[35,36] day 4 EBs for a primitive erythroid colony, and day 6–10 EBs for definitive erythroid and myeloid progenitor analysis. When sorted cells are used, EBs from the desired stage are sorted for hCD4 (SCL) or Flk-1 expressing cells and then replated for hematopoietic colony formation. The following is the protocol for direct EB replating.

1. Harvest EBs.
   A. *EBs in liquid:* Transfer media containing EBs into 50 ml tubes. Wash the plate with IMDM. Let it sit at room temperature for about 10–20 min. EBs will settle down to the bottom of the tube.
   B. *EBs in methylcellulose:* Add equal volume of cellulase (2 units/ml, final 1 unit/ml) and incubate 20 min at 37°C. Collect EBs in 50 ml tubes. Wash the plate with IMDM. Let it sit at room temperature for about 10–20 min. EBs will settle down at the bottom of the tube.
2. Aspirate off the media, add 3 ml trypsin, and incubate for 3 min at 37°C (use water bath). Vortex quickly and add 1 ml FCS (use the same serum for differentiation). Dissociate through a 20 gauge needle by passaging 4–5 times. Transfer to a 14 ml snap cap tube and spin for 5–10 min at 1000 rpm. Use collagenase for older EBs (day >9, for example). When collagenase is used, incubate EBs for 1 hr at 37°C.
3. Resuspend the cell pellet in 0.3–1 ml of IMDM (with 10% FCS). Count the viable cells. At this point, there should be no cell clumps.
4. Add $3–6 \times 10^4$ EB cells per 1 ml of methylcellulose replating media. Add 1.25 ml of methylcellulose mix to each 35 mm bacterial dish for blast colony assays. Otherwise, add 1 ml of methylcellulose mixture to each dish. Prepare three replica dishes for each sample, i.e., make 4.5 ml of methylcellulose replating media for blast colony assay and make 4 ml methylcellulose replating media for erythroid and myeloid colony assays.

Methylcellulose mix

	Blast	*Primitive erythroid*	*Definitive erythroid and myeloid*
methylcellulose	1%	1%	1%
FCS	10%	–	–
PDS (plasma derived serum)	–	10%	10%
Ascorbic acid	12.5 μg/ml	12.5 μg/ml	12.5 μg/ml
L-glutamine	2 m*M*	2 m*M*	2 m*M*
transferrin	200 μg/ml	200 μg/ml	200 μg/ml
MTG	4.5×10^{-4} *M*	4.5×10^{-4} *M*	4.5×10^{-4} *M*
D4T C.M.*	20%	–	–
VEGF	5 ng/ml	–	–
KL	1%	–	1%
Epo	–	2 units/ml	2 units/ml
PFHM-II	–	5%	5%
IL-1	–	–	5 ng/ml
IL-3	–	–	1%
IL-6	–	–	5 ng/ml
IL-11	–	–	5–25 ng/ml
G-CSF	–	–	2–30 ng/ml
GM-CSF	–	–	3–5 ng/ml or 20 u/ml
M-CSF	–	–	2–5 ng/ml
IMDM	to 100%	to 100%	to 100%

Reagents and Materials

1. 2X Cellulase. Dissolve cellulase (Sigma C-1794) in PBS at 2 units/ml. Filter sterilize through 0.45 micron filter.
2. Collagenase. Dissolve 1 g of collagenase (Sigma C0310) in 320 ml 1X PBS. After filter sterilization, add 80 ml of FCS. Make an aliquot, and keep at −20°C.
3. D4T conditioned medium (C.M.). D4T endothelial cells[35,36] are cultured in 10% FCS in IMDM. Remove media and change to 4% FCS in IMDM when it becomes ~80% confluent. Culture an additional 72 hr, and collect the supernatant. Spin down for 5 min at 1000 rpm to remove cell debris, and filter sterilize the supernatant by utilizing a 0.45 mm filter unit. Make 5–10 ml aliquots, and keep at −80°C. Once thawed, D4T C.M. is kept at 4°C for about one week.
4. Erythropoietin, Amgen Epogen NDC 55513-126-10.
5. VEGF, R&D Systems 293-VE.

6. KL (kit-ligand), R&D Systems 455-MC.
7. IL-1, R&D Systems 401-ML.
8. IL-3, R&D Systems 403-ML.
9. IL-6, R&D Systems 406-ML.
10. IL-11, R&D Systems 418-ML.
11. G-CSF, R&D Systems 414-CS.
12. GM-CSF, R&D Systems 415-ML.
13. M-CSF, R&D Systems 416-ML.
14. PFHM-II (protein free hybridoma medium), Gibco 12040-077.

Cell Sorting and in vitro *Culture of Sorted Cell Populations*

Double-color staining and sorting for Flk-1^+ and human $CD4^+$ (SCL^+) cells are performed the same way as for FACS analysis. Prior to sorting, filter stained cells through 40 μm nylon-mesh. Cells are sorted using FACS MoFlo (Becton-Dickinson) into 14 ml tubes (Fisher, #14-959-49B) containing 2 ml of FCS. Reanalyze the sorted cells on a FACS Caliber (Becton-Dickinson) to determine the sorting efficiency.

Endothelial Cell Culture and Immunohistochemical Staining

Day 6 EBs are normally used for cell sorting for endothelial cell assays. Endothelial cells from sorted Flk-1^+ or $hCD4^+$ (SCL^+) cells are typically assayed as follows:

1. Plate the sorted cells onto type IV collagen coated, 24-well plates in IMDM media containing 15% preselected FCS, ascorbic acid (50 μg/ml), L-glutamine (2 m*M*), MTG (4.5×10^{-4} *M*), and VEGF (50 ng/ml) at a cell density of 2×10^4/well.
2. Culture cells in humidified 37°C incubator with 5% CO_2 for 3–4 days.
3. For immunohistochemical staining, wash adherent cells with 1X PBS, fix for 10 min in 1X PBS containing 4% para-formaldehyde (PFA) at 4°C, and wash twice (10 min each) in PBS.
4. Following the wash, quench the endogenous peroxidase in methanol/30% hydrogen peroxide/10% sodium azide (50 : 10 : 1) for 1 hr at 4°C. Wash cells twice, and block with PBS containing 1% goat serum, 0.2% bovine serum albumin, and 2% skim milk for 1 hr.
5. Incubate cells with biotinylated anti-mouse CD31 overnight at 4°C. After three washes, incubate cells with streptavidin–horseradish peroxidase for 1 hr at room temperature.
6. Wash cells three times, and incubate with a DAB kit to develop the color.

Reagents and Materials

1. Type IV collagen: Dissolve type IV collagen (Sigma C-5533) at 25 μg/ml in PBS.
2. 4% para-formaldehyde solution: Dissolve para-formaldehyde (Sigma P-6148) at 4% in PBS.
3. Methanol, Fisher A412-4.
4. 30% hydrogen peroxide, Sigma H-1009.
5. Goat serum, DAKO X553371.
6. Bovine serum albumin, Gibco 15260-037.
7. Streptavidin–horseradish peroxidase, Zymed 43-4323.
8. CD31, Pharmingen 553371.
9. DAB kit, Vector SK-4100.

General Considerations When Sorting EB Cells

EB cells are sticky, and every precaution must be made to get them into a single cell suspension. For pure sorts with good yields, the sample must be as close to an absolute single cell suspension as entirely possible. Adequate isolation from cell clumps and lysed membranes by washing and filtering through a 40–50 micron cell strainer (for example, B-D Falcon 352340, or Partec 04-004-2327, which is our favorite) is pretty much a standard requirement. Keeping the sample on ice is recommended unless the cell line prevents such. The samples must be aspirated from B–D Falcon (part #352063) 5 ml polypropylene tubes. VWR poly tubes do not seal well on the aspirate stage, and any polystyrene or glass tube cannot be used at the pressures employed.

As a suspension buffer, the use of Hank's Balanced Salt Solution (HBSS) with no calcium or magnesium and no phenyl red is recommended. For cell nutrition, the use of fetal calf serum (FCS) or bovine serum albumin (BSA) is recommended depending on the cell line, which should be between 1 and 2%. The use of EDTA is optional based on the cell line and its aggregation characteristics, and normally 0.1 m*M* is all that is necessary.

Cell density depends on three factors: (1) yield requirement, or the number of cells needed from the sort; (2) the percentage of the sample to be sorted; and (3) the expression or amount of separation between negative and positive staining. If several million cells are to be sorted and/or have a low percentage of target cells, the density must be as high as possible and still not clump. For example, if 20 million sorted cells are needed and the target population is 5%, cells should be suspended around 10–25 million per ml. This should be adequate to sort at about 10–25 thousand cells per second, and to collect 20 million cells in 3.5–4.5 hr. If "sticky, friendly" cells like EB cells are to be sorted, cell density is recommended at 10–15 million cells/ml.

The expression level of a given molecule has as much to do with the sorting success. If the expression level is low and cannot be separated well from the negative cell population, the electronics of the sorter works a great deal harder to bitmap the target and differentiate it from nontarget populations. In this case, cells must be sorted more slowly and at more dilute suspensions.

Sorted sample receiving containers should contain culture media or fetal calf serum equal to one fifth of its volume. Receiving containers vary and can be of any substance, glass, polystyrene, or polypropylene. Sorting can be done into 5, 14, or 50 ml tubes or into well plates varying from 24 to 1536 wells/plate. It is actually better to use glass tubes on very long duration sorts so as not to build up static electricity, known to be an issue with plastic tubes. Eventually, a static "wall" forms, and sorted cells get "repelled" from the receiving tubes.

Acknowledgment

We thank the lab members for help and support. This work was supported by grants from NIH and American Cancer Society.

[14] Establishment of Multipotent Hematopoietic Progenitor Cell Lines from ES Cells Differentiated *In Vitro*

By Leif Carlsson, Ewa Wandzioch, Perpétua Pinto do Ó, and Åsa Kolterud

Introduction

The mechanisms controlling the self-renewal and differentiation process of hematopoietic stem cells (HSCs) and the mechanisms governing the formation and expansion of the hematopoietic system during embryonic development are largely unknown. Increased knowledge in these closely related areas is important to fully understand the regulation of the hematopoietic system. However, the low levels of HSCs in hematopoietic tissue hamper direct studies of HSCs, and molecular and cellular analyses of the hematopoietic system during embryonic development are complicated by the limited access to tissue at this stage. We have developed an experimental system that makes it possible to address both of these aspects of the hematopoietic system. This system is based on the generation of immortalized hematopoietic progenitor cell (HPC) lines from *in vitro* differentiation of mouse embryonic stem (ES) cells genetically modified to over-express the LIM-homeobox gene *Lhx2*. The HPC lines share many basic characteristics with normal HSCs, such as responsiveness to specific

0076-6879/2003 $35.00

growth factors/growth factor combinations, interaction with stromal cells and expression of specific transcription factors.[1,2] The unique properties of the HPC lines offers a valuable tool for directly addressing basic properties of HSCs, such as signal transduction, cytokine/growth factor response, lineage commitment and adhesion to support cells. Moreover, *Lhx2* appears to play an important role in the establishment of the hematopoietic system during embryonic development, since *Lhx2*$^{-/-}$ mice die *in utero* due to severe anemia.[3] Thus, characterization of these cell lines might also be useful for elucidating mechanisms involved in embryonic development of the hematopoietic system.

The procedure to generate hematopoietic cells from ES cells differentiated *in vitro* is carried out essentially according to the two-step differentiation procedure described by Keller *et al.*,[4] with minor modifications. In the first differentiation step the ES cells are differentiated into three-dimensional structures termed embryoid bodies (EBs). This step mimics the gastrulation process leading to differentiation into cell types representing the three germ layers including mesodermal cell types such as hematopoietic progenitor cells.[5] In the second differentiation step the hematopoietic progenitor cells formed within the EBs are stimulated to differentiate/proliferate in clonal (colony) assays using specific cytokines/growth factors.

Here we describe a detailed experimental approach for generating HPC lines from genetically modified ES cells that has been allowed to differentiate *in vitro*. The estimated time to generate HPC lines, including genetic modification of ES cells, is approximately 2 months. The general outline of the experimental approach and the estimated time for each step of the procedure is shown in Fig. 1. In our original protocol,[1] we used the feeder-independent CCE ES cell line derived from the 129/Sv mouse strain.[6] Using a similar approach we have also generated multipotent HPC lines from the bone marrow of a mouse with a different genetic background (C57BL/6-cast).[7] Thus, expression of *Lhx2* specifically immortalize immature hematopoietic progenitor/stem cells, suggesting that any ES cell line capable of differentiating into hematopoietic progenitor cells *in vitro* could be used for generating HPC lines.

[1] P. Pinto do Ó, Å. Kolterud, and L. Carlsson, *EMBO J.* **17**, 5744 (1998).
[2] P. Pinto do Ó, E. Wandzioch, Å. Kolterud, and L. Carlsson, *Exp. Hematol.* **29**, 1019 (2001).
[3] F. Porter, J. Drago, Y. Xu, S. Cheema, C. Wassif, S.-P. Huang, E. Lee, A. Grinberg, J. Massalas, D. Bodine, F. Alt, and H. Westphal, *Develop.* **124**, 2935 (1997).
[4] G. Keller, M. Kennedy, T. Papayannopoulou, and M. Wiles, *Mol. Cell. Biol.* **13**, 473 (1993).
[5] G. Keller, *Curr. Opin. Cell Biol.* **7**, 862 (1995).
[6] E. Robertson, A. Bradley, M. Kuehn, and M. Evans, *Nature* **323**, 445 (1986).
[7] P. Pinto do Ó, K. Richter, and L. Carlsson, *Blood* **99**, 3939 (2002).

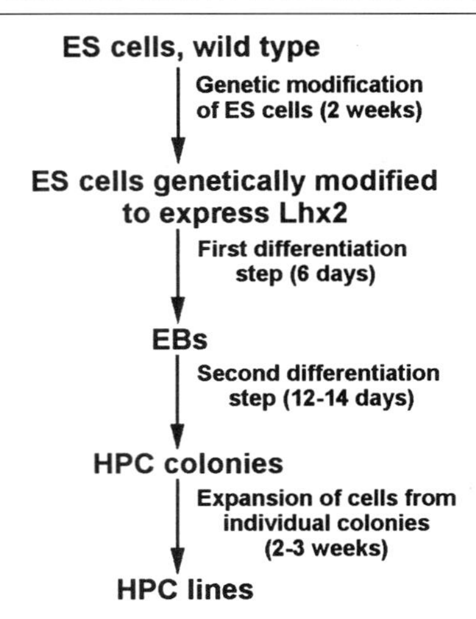

FIG. 1. Schematic representation of the approach to generate HPC lines from genetically modified ES cells differentiated *in vitro*. Approximate time to carry out the respective step is indicated.

Material and Methods

Procedures

Maintenance of ES Cells

All cell culturing described herein was carried out in a humidified environment with 5% CO_2 in air at 37°C. CCE ES cells are maintained in an undifferentiated state on gelatinized 25 cm^2 tissue culture flasks (Falcon 353014) in DMEM ES medium. Passage ES cells when they are approximately 50–70% confluent (usually every 2–3 days) by trypsinizing the ES cells and transfer one-tenth of the cells to a new flask. The ES cells should not be allowed to reach too high (>70%) confluence or to be at too low (<20%) confluence at the time of passage since it negatively affects their ability to differentiate *in vitro*.

Production of Retrovirus

Retroviral Vectors and Virus-Producing Cell Line

Onco-retroviral vectors containing the selectable marker gene Neo are used to genetically modify the ES cells. We have used the murine stem cell virus (MSCV) vector developed by Hawley *et al.*[8] successfully.[1] However,

[8] R. Hawley, F. Lieu, A. Fong, and T. Hawley, *Gene Therapy* **1**, 136 (1994).

we find that another retroviral vector denoted MND-X-SN developed by Robbins *et al.*,[9] is more effective than the MSCV vector in generating HPC lines upon differentiation of transduced ES cells. Although both vectors can be used for the protocol outlined herein, the time consuming subcloning procedures of ES cells prior to differentiation as described in Pinto do Ó *et al.*,[1] is not required if the MND-X-SN vector is used.

cDNA containing the coding region of mouse *Lhx2* was inserted into the EcoRI site upstream of the pgk-neo cassette in the MND-X-SN vector and the resulting vector is referred to as MND-*Lhx2*. The virus-producing cell line BOSC-23 is routinely used for virus production and the protocol for producing virus is essentially as described by Pear *et al.*,[10] with minor modifications. The BOSC-23 cells are maintained on gelatinized 25 cm^2 tissue culture flasks in DMEM medium. For optimal virus production the BOSC-23 cells should be cultured in gpt selective medium prior to transfection and subsequent collection of virus particles.

Transfection of BOSC-23 Cells with Retroviral Vectors and Collection of Virus

Day 1: Plate 3×10^6 cells on a gelatinized 60 mm tissue culture dish (Falcon 353003) or 2.7×10^6 cells on a gelatinized 25 cm^2 tissue culture flask in 4 ml of DMEM medium.

Day 2: The BOSC-23 cells should be approximately 70–80% confluent for optimal virus production. If less than 30% confluent it is recommendable to increase the number of cells plated at day 1. Just prior to transfection, change the medium to 4 ml of DMEM medium supplemented with 25 μM chloroquine (SIGMA C6628) since it increases the virus titre significantly. For transfection, resuspend 6–10 μg of ethanol precipitated retroviral vector plasmid DNA (MND-*Lhx2*) in 438 μl H_2O, add 62 μl 2 *M* $CaCl_2$ and mix. Add 500 μl of 2xHBS, mix and add to the BOSC-23 cells. Replace this medium with fresh DMEM medium after 6 hr.

Day 3: Replace the DMEM medium with 3 ml DMEM ES medium.

Day 4: Collect the virus-containing medium and filter it through a 0.45 μm filter. If the BOSC-23 cells are not confluent at this point collection of virus can be postponed for an additional day. The viral supernatant can be used immediately for transducing ES cells or frozen and maintained at $-80°C$ in 100–200 μl aliquots.

[9] P. Robbins, X.-J. Yu, D. Skelton, K. Pepper, R. Wasserman, L. Zhu, and D. Kohn, *J. Virol.* **71**, 9466 (1997).

[10] W. Pear, G. Nolan, M. Scott, and D. Baltimore, *Proc. Natl. Acad. Sci. USA* **90**, 8392 (1993).

Genetic Modification of ES Cells

Day 1 (afternoon): Plate 5×10^5 ES cells on gelatinized 85 mm tissue culture dishes in 10 ml DMEM ES medium. Plate an additional dish with ES cells that will not be transduced. This will serve as a negative control for the selection procedure.

Day 2 (morning): Replace the medium on the ES cells with the 3 ml of DMEM ES medium containing 100–200 μl of viral supernatant and 4 μg/ml hexadimethrine bromide (SIGMA H9268), incubate for 5 hr and subsequently replace this medium with 10 ml of DMEM ES medium.

Day 4: Start the selection for transduced ES cells by replacing the medium with DMEM ES medium supplemented with 500 μg/ml G418 (Geneticin®, GIBCO 11811-064). Add the same medium to the dish with nontransduced ES cells.

Depending on the virus titre the transduced ES cells should be passaged 1–4 days after initiation of selection to prevent the ES cells from becoming too confluent. Maintain the ES cells in selective medium until no viable cells are present in the control dish with nontransduced cells. The selection procedure usually takes less than 7 days. G418-resistant ($G418^R$) ES cells can be used directly for differentiation into EBs, or frozen in 1 ml aliquots in Freezing medium. A 50–70% confluent 85 mm tissue culture dish of ES cells can be frozen in 3 aliquots. Transduced ES cells should not be kept in culture for more than 1–2 weeks prior to inducing differentiation for optimal results.

Differentiation of ES Cells

First Differentiation Step, Formation of EBs

Two days prior to initiation of differentiation the ES cells should be passaged into IMDM ES medium on gelatinized flasks. Make two dilutions when passaging the cells into IMDM ES medium aiming for approximately 50% confluence after 2 days of culture (a 1 : 10 and a 1 : 20 dilution are usually sufficient). Trypsinize the ES cells for 5 min at room temperature, wash cells twice in Standard medium before resuspending the cells in 0.5 ml Standard medium. Determine cell concentration using a hemocytometer, viability should be above 95% and the resuspended cells should be present as single cells. The differentiation is carried out in suspension culture grade dishes (not tissue culture grade) with 5 ml medium per 60 mm dish. Perform the differentiation at cell densities of 500–2000 cells/ml in IMDM ES differentiation medium in five 60 mm dishes (Corning, 25060-60), which usually yield sufficient number of EB cells for subsequent experiments.

Suspension culture grade dishes are used to avoid adherence of the EBs to the plastic since it interferes with hematopoietic commitment during ES cell differentiation. Many different suppliers of cell culture dishes have Suspension culture grade dishes, but it should be noted that there is a large variability between different brands with respect to adherence of EBs to the plastic. The degree of adherence can also vary within the same brand of petri dishes. It is therefore recommendable to test every new batch of petri dishes even though they have worked in the past. Another important aspect of the differentiation process is to determine optimal cell concentration. The exact cell concentration for optimal differentiation has to be determined empirically since it depends on many different parameters such as ES cell line, length of differentiation and FCS batch. Since too low cell concentration have a negative effect on proper EB formation we recommend to use as high cell concentration as possible without allowing the culture medium to become too acidic at the time when EBs are harvested.

Harvest of EBs

Collect the medium containing EBs at day 6 of differentiation and transfer to a 50 ml tube, let the EBs sediment for 10–15 min. When the EBs have settled at the bottom of the tube, carefully remove the supernatant. Add 3 ml of Trypsin-EDTA solution and incubate 3 min at 37°C. Add 2 ml of FCS and disperse the EBs into single cells by passaging up and down through a 20-gauge needle thrice using a 5 ml syringe. Add 5 ml of Standard medium and centrifuge 10 min at 1000 rpm. Resuspend the cells in 1–2 ml of Standard medium and determine the cell concentration using a hemocytometer. The cell concentration should be such that less than 100 μl of EB cell suspension is required per triplicate in the clonal assay described below since the volume of the cell suspension is neglected in the Methylcellulose medium.

Second Differentiation Step, Clonal Assays of EB Cells

The clonal assays are performed in 35 mm petri dishes (suspension culture grade, Falcon 1008) in Methylcellulose medium. EB cells are plated at 10^5 and 2×10^5 cells/dish in a final volume of 1.25 ml Methylcellulose medium/dish in triplicates. The total volume (V_{tot}) of Methylcellulose medium needed for an experiment can be calculated as follows (in ml): $V_{tot} = [(3.5\times1.25)\times A] + 2$, where A is the total number of triplicates.

Add the volume of EB cells that corresponds to the number of cells in 3.5 dishes to the bottom of a 14 ml snap cap tube (Falcon 352059). For example, add 3.5×10^5 cells if 10^5 cells/dish are plated. Add approximately 4.4 ml (1.25 ml$\times$3.5) of Methylcellulose medium to the tube using a 5 ml syringe with an 18-gauge needle and mix well using a Vortex. Let the

Methylcellulose medium settle for 10–15 min (do not centrifuge), and dispense 1.25 ml per dish using a 5 ml syringe with an 18-gauge needle. Make sure that the Methylcellulose medium covers the whole dish by tilting it.

Generation and Maintenance of Multipotent Hematopoietic Progenitor (HPC) Lines

An important step when generating HPC lines is to be able to identify the correct colony type in the clonal assays of the EB cells. Clonal assays of EB cells obtained from control ES cells (nontransduced or transduced with empty vector) using the Methylcellulose medium described herein gives two distinct types of colonies containing different types of erythroid cells.[1] One type of colony contains 50–100 relatively large hemoglobinized (i.e., red) cells that are of the nucleated primitive erythroid type (Fig. 2A). This type of colony appears after 4–6 days in culture and the frequency of the colony forming cell (CFC) generating this colony should be more than 1000 per 10^5 EB cells. The other type of colony is considerably larger and contains mainly small, hemoglobinized and anucleated cells that are of the definitive (or adult) erythroid type (Fig. 2B). This type of colony appears after 8–11 days in culture and the frequency of the CFC generating this colony should be more than 5 per 10^5 cells. EB cells obtained from ES cells transduced with *Lhx2* generate an additional type of colony containing few or no hemoglobinized cells, and consequently do not appear as a "red" colony (Fig. 2C). These "non-red" colonies usually appear after 8 days in culture but cannot be distinguished from a colony containing definitive erythroid cells until 12–14 days in culture. The "non-red" colonies mainly contain cells with an immature blast-like morphology and a few megakaryocytes, and is referred to as HPC colonies.[1] The frequency of the CFC generating the HPC colony should be more than 1 per 10^5 EB cells. The cells within the HPC colonies express *Lhx2* from the retroviral vector and HPC lines are established from cells present in this colony.[1]

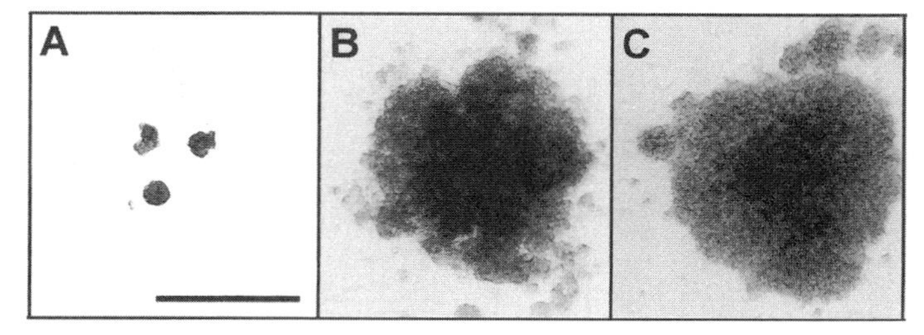

FIG. 2. Morphology of different colony types appearing in the clonal assays of day 6 EB cells. Colonies shown are those containing primitive erythroid cells (A), definitive (adult) erythroid cells (B) and an HPC colony (C). Scale bar indicates 0.25 mm. (See Color Insert.)

Collect individual HPC colonies (see Fig. 2C) from the Methylcellulose cultures using a pipette set at approximately 50 μl. Transfer each colony into one well in a 24-well plate (Falcon 35047 or equivalent) containing 1 ml of HPC medium. Since these cells will be kept in culture for a long period of time, it is important to transfer the HPC colonies under aseptic condition as far as possible. Clonal cell lines are obtained if each colony is expanded separately. The culture period in the 24-well plate is critical to the establishment of HPC lines. The cells should therefore be left in this well until obvious proliferation has occurred and a dense culture has been obtained. The time required to obtain a dense culture can vary but the usual time period is 7–9 days. The cells should be in suspension and very few, if any, cells adhering to the plastic should be present. Since the HPC lines secrete mediators promoting their own proliferation,[2] the cells should be kept at a cell density above 10^6 cells/ml during the whole establishment phase. Collect the cells in the well when they have reached this density and distribute them into three wells in a 24-well plate (1 ml of HPC medium per well). When the cell density in these three wells is above 10^6 cells/ml, the cells in all three wells can be transferred to one 25 cm^2 culture flask in a total volume of 5 ml. Aggregates of 5–10 cells are frequently observed in these cultures. At this stage the expanded cells can be tested if they represent an HPC line by staining cytospun cells with May-Grünwald Giemsa stains. If the majority of cells ($>90\%$) are blast-like cells with a large nuclei and a small rim of cytoplasm (see Fig. 3A) it is an HPC line and can be continuously expanded in culture, provided that the cell density is kept above 5×10^5 cells/ml. Expansion of mast cells can occur under these conditions and these cultures are difficult to distinguish from HPC lines, unless a May-Grünwald-Giemsa staining is performed on the cells. If a majority of the expanded cells are mast cells (see Fig. 3B), HPC lines cannot be established from such cultures. The HPC lines can be frozen and stored in Freezing medium at $2\text{–}3 \times 10^6$ cells per vial. The HPC lines are thawed as normal cells but immediately upon thawing they should be cultured in 2–3

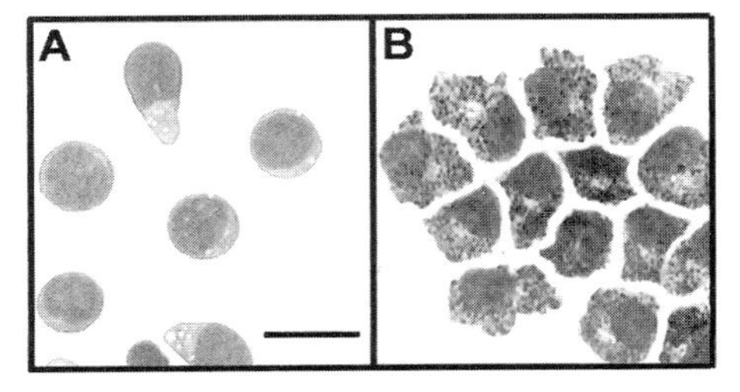

FIG. 3. May-Grünwald-Giemsa staining of cytospun cells showing the cellular morphology of cells derived from an expanded HPC line (A) and expanded mast cells (B). Scale bar indicates 10 μm.

wells in a 24-well plate in 1 ml HPC medium per well. When the cells have recovered they can be expanded as described above. Once the HPC lines have been established the FCS content in the medium can be lowered to 5%, e.g., the Standard medium supplemented with Stem cell factor.

Characteristics of the HPC Lines

Approximately 10 different HPC lines have been established and 6 of these have been characterized to some extent. The characteristics of these 6 HPC lines are summarized below:

1. Each HPC line has a unique retroviral integration site in their genome.[2]
2. Each HPC line express *Lhx2* from the retroviral vector (Refs. 1, 2 and unpublished observation).
3. Each HPC line self-renew by a cell nonautonomous mechanism, e.g., they proliferate poorly or not at all at cell densities below 10^5 cells/ml (Ref. 2 and unpublished observation).
4. Each HPC line is strictly dependent on Steel factor/Stem cell factor for maintenance in an immature state.[1,2]
5. A relatively broad mixture of early acting growth factors (Stem cell factor, thrombopoietin, IL-3, Flt3L, IL-6) is required to induce efficient proliferation and differentiation of the HPC lines in clonal assays.[1,2]
6. Each HPC line is multipotent, e.g., upon stimulation with appropriate growth factors in clonal assays, individual cells generate at least, erythrocytes, megakaryocytes, macrophages and neutrophilic granulocytes.[1,2]
7. Each HPC line can interact with support cells since they generate cobblestone areas when cocultured with stromal cells (Ref. 2 and unpublished observation).
8. The cell surface marker expression of individual HPC lines are relatively homogeneous, although slight differences have been observed between different HPC lines. Most HPC lines are c-kit^{+}, CD29^{+}, CD44^{+}, CD41^{+}, Sca-1^{-}, AA4.1$^{-/low}$, Thy1.2^{-} and Lin^{-}.[1] The expression of CD34 and CD45 can vary between individual HPC lines. The HPC lines can either be CD34^{-}, CD34^{low} or CD34^{+} (unpublished observation). Although all HPC lines are clearly hematopoietic, we have obtained both CD45^{+} and CD45^{-} HPC lines (unpublished observation). This is somewhat surprising since CD45 is considered to be a pan-leukocyte marker,[11] indicating that

[11] P. Johnson, A. Maiti, and D. Ng, *in* "Weir's Handbook of Experimental Immunology" (L. Herzenberg, D. Weir, L. Herzenberg, and C. Blackwell, eds.), Vol. 2, p. 62.1. Blackwell Science, Cambridge, MA, 1997.

$CD45^-$ hematopoietic progenitor cells might be present during early embryonic development.

Cell Culture Media, Buffers, Reagents

10x phosphate buffered saline (PBS) stock solution: 80 g NaCl, 2 g KCl, 14.4 g Na_2HPO_4 2.4 g KH_2PO_4. Heat to dissolve, set pH to 7.4, bring to 1 liter and filter through 0.22 μm filter. Make working solutions (1x) by diluting ten times in (cell culture tested) water and autoclave.

Gelatin solution: 0.1% (w/v) of Gelatin (SIGMA, G1890) is added to 1x PBS, which will dissolve during autoclaving. Gelatinize the cell culture dish/flask by covering the culture surface area with the Gelatin solution for at least 20 min prior to culturing the cells. Cell culture dishes/flasks with Gelatin solution can be stored at 4°C for several weeks. Remove excess Gelatin solution before cells and medium is added.

Trypsin–EDTA solution: For 1 liter working solution add 2.5 g Trypsin (SIGMA, T4799) and 2.16 ml 0.5 *M* EDTA to 1x PBS and filtrate through a 0.22 μm filter. This Trypsin-EDTA solution is used both for trypsinizing/passaging adherent cells and for trypsinizing EBs. For trypsinizing adherent cells remove the medium and add sufficient Trypsin–EDTA solution to cover the cells (1.5–2 ml depending on cell culture area). Incubate at 37°C for 5 min and make sure that the cells are in a single cell suspension. Add medium (3.5–8 ml) and transfer the desired volume to another tissue culture flask/dish containing the appropriate cell culture medium.

DMEM ES medium: DMEM (Dulbecco's modified Eagle's medium, GIBCO 31965-023), 1.5×10^{-4} *M* monothioglycerol (MTG, SIGMA M1753), 1 m*M* Na-pyruvate (GIBCO 11360-039), 25 m*M* HEPES (GIBCO 15630-056), 100 units/ml penicillin, 100 μg/ml streptomycin (PEST, GIBCO 15140-122), 15% fetal calf serum (pretested for its ability to maintain ES cells in an undifferentiated state), Leukemia Inhibitory Factor (LIF). The source of LIF can either be conditioned media from a cell line transfected with LIF cDNA or commercially available recombinant LIF. Optimal concentration of LIF for maintaining ES cells in an undifferentiated state needs to be tested empirically for each ES cell line and individual batch of LIF.

DMEM medium: DMEM, 1.5×10^{-4} *M* MTG, 1 m*M* Na-pyruvate, PEST, 10% FCS.

gpt selective medium: DMEM medium, 0.25 mg/ml xanthine (SIGMA X3627), HAT (hypoxanthine, aminopterin, thymidine, GIBCO 21060-017, 50x stock solution), 25 μg/ml mycophenolic acid (SIGMA M3536).

2x HEPES buffered saline (HBS): 50 m*M* HEPES (pH 7.05), 10 m*M* KCl, 12 m*M* D-glucose, 280 m*M* NaCl, 1.5 m*M* Na_2HPO_4.

Freezing medium: 10% (v/v) Dimethylsulfoxide (DMSO, SIGMA D5879) 90% (v/v) FCS.

IMDM ES medium: IMDM (Iscove's modified Dulbecco's medium, GIBCO 42200-014 prepared according to supplier and supplemented with PEST, this is used in all media containing IMDM described below), 1.5×10^{-4} *M* MTG, 15% FCS (same as for DMEM ES medium), LIF (same as for DMEM ES medium).

Standard medium: IMDM, 1.5×10^{-4} *M* MTG, 5% FCS.

IMDM ES differentiation medium: IMDM, 10 μl/ml m*M* L-glutamine (200 m*M*, GIBCO 25030-024), 4.5×10^{-4} *M* MTG, 25 μg/ml ascorbic acid (SIGMA A4403), 5% PFHM II (GIBCO 12040-051), 15% FCS (pretested for optimal hematopoietic commitment during ES cell differentiation *in vitro*).

Methylcellulose stock solution: Weigh a sterile 2 liter Erlenmeyer flask. Add 450 ml of autoclaved water and bring to a boil on a hot plate. Add 20 g of methylcellulose (Fluka 64630) and let it come to boil 3–4 times (prevent it from over boiling). Swirl vigorously between boils until no clumps of methylcellulose are apparent. Allow this mixture to cool to room temperature. Meanwhile, prepare 500 ml of 2x IMDM supplemented with PEST and 124 μl of MTG diluted 1 : 10. Thus, add ingredients required for 1 liter of IMDM but make only 500 ml. Filter the 2x IMDM through a 0.22 μm filter and subsequently add this to the methylcellulose and swirl vigorously. Weigh the Erlenemeyer flask again and add sterile water until the weight of the Methylcellulose stock solution equals 1000 g. Incubate the Methylcellulose stock solution at 4°C over night, aliquot in sterile 100 ml glass bottles (Schott, Duran®) and store at −20°C. Use syringes (without needle) for measuring and dispensing Methylcellulose stock solution and syringes with an 18-gauge needle for measuring and dispensing Methylcellulose medium.

Methylcellulose medium: 50% 2x Methylcellulose stock solution, 10% plasma-derived serum (PDS, Animal Technologies, Tyler, Tx), 10 μl/ml L-glutamine (200 m*M*, GIBCO 25030-024), 300 μg/ml Transferrin (Boehringer, 30 mg/ml, 652202), 100 ng/ml Steel factor/Stem cell factor (R&D Systems, 455-MC), 4 IU/ml Erythropoietin (Eprex®, Janssen-Cilag), 5% PFHM II (GIBCO 12040-051). Bring to final volume by adding IMDM. Prepare the Methylcellulose medium in 50 ml tubes (Falcon) and start by adding the Methylcellulose stock solution using a syringe. Mix thoroughly and quickly centrifuge down the Methylcellulose medium.

HPC medium: IMDM, 1.5×10^{-4} *M* MTG, 100 ng/ml Steel factor/Stem cell factor, 10% FCS.

May-Grünwald-Giemsa staining: Cytospin cells onto glass slides and let air-dry. Incubate the glass slides for 5 min in May-Grünwald solution (MERCK, 1.01424.0500) followed by a 15 min incubation in Giemsa solution (MERCK, 1.09204.0500) diluted 1 : 20 in water. Rinse the glass slides by incubating them in water for 10–15 min. After the glass slides have been air-dried the stained cells can be monitored in a microscope.

Fetal calf sera, types and batches: Different batches and types of fetal bovine sera have considerable impact on the outcome of the experiments described herein. We therefore routinely use two different types of fetal calf sera, normal fetal calf serum (FCS) and plasma-derived serum (PDS). We use two different batches of FCS that have been pretested for different applications, whereas the application where PDS is used (see below) is batch independent. One of the FCS batches has been pretested for its ability to maintain ES cells in an undifferentiated state and this FCS is used in the DMEM ES and IMDM ES media. Batches of FCS pretested for this application can be purchased from different serum suppliers (for example, SIGMA, Boehringer and HyClone). The second batch of FCS used has been empirically tested for its ability to support hematopoietic commitment during ES cell differentiation *in vitro* and is therefore used in the IMDM ES differentiation medium. We have noted that addition of PFHM II to the IMDM ES differentiation medium makes commitment to hematopoiesis during ES cell differentiation less dependent on the FCS batch. However, it is still recommended to test at least five different FCS batches in the IMDM ES differentiation medium in order to select for a batch that support maximum commitment to hematopoiesis during ES cell differentiation. Commitment to hematopoiesis is determined by measuring the frequency of CFCs generating primitive erythroid colonies in day 6 EBs, which should be at least 1000 CFCs per 10^5 EB cells. If the frequency is much less than 1000 CFCs per 10^5 EB cells, commitment to hematopoiesis has been too inefficient during the first differentiation step. It is therefore highly unlikely that an HPC colony is obtained since the frequency of the progenitor cell generating this colony is below detection in the EBs. This batch of FCS is also used in DMEM medium, Standard medium and HPC medium. The PDS (also referred to as platelet-poor serum) is exclusively used in the clonal assays of the EB cells, because it is the only type of serum that supports development of colonies containing mature primitive erythroid cells. Proper assessment of this type of colony is important since the frequency of this precursor is used for determining the efficiency of hematopoietic commitment during the first step of ES cell differentiation.

Concluding Remarks

We have described an approach to generate multipotent hematopoietic progenitor cell lines from genetically modified ES cells differentiated *in vitro* in a reproducible manner. These HSC-like cell lines will facilitate the identification of molecular, cellular and biochemical properties of hematopoietic progenitor/stem cells since they represent an unlimited source of such cells. Furthermore, the *Lhx2* gene appears to play an important role *in vivo* during the development of the hematopoietic system since the *Lhx2*$^{-/-}$ mice die of a severe anemia during embryonic development.[3] Thus, these cell lines have the potential to uncover molecular mechanism responsible for the embryonic development of the hematopoietic system. Moreover, the generation of HPC lines from ES cells with different targeted mutations would allow for direct studies on hematopoietic progenitor cells with defined mutations, provided that the mutation does not interfere with the self-renewal process of the HPC line.

Acknowledgment

We wish to thank Drs. Staffan Bohm and Anna Berghard for critical reading of this manuscript. This work was supported by the Swedish Cancer Society, Swedish National Board for Laboratory Animals, Umeå University Biotechnology Fund, and EU Regional Fund, Objective 1.

[15] Vasculogenesis and Angiogenesis from *In Vitro* Differentiation of Mouse Embryonic Stem Cells

By Olivier Feraud, Marie-Hélène Prandini, and Daniel Vittet

Introduction

Embryonic stem (ES) cell *in vitro* differentiation provides a powerful model system for studies of cellular and molecular mechanisms of vascular development. This *in vitro* system recapitulates most of the endothelial differentiation program as observed *in vivo* during vasculogenesis in the embryo. Several studies have mentioned the formation of vascular structures evoking blood islands and the presence of tubular channels in embryoid bodies (EBs).[1–3] Furthermore, ES cell differentiation into the endothelial lineage was characterized to occur through successive

[1] T. C. Doetschman, H. Eistetter, M. Katz, W. Schmidt, and R. Kemler, *J. Embryol. Exp. Morph.* **87**, 27 (1985).

[2] W. Risau, H. Sariola, H.-G. Zerwes, J. Sasse, P. Ekblom, R. Kemler, and T. Doetschman, *Development* **102**, 471 (1988).

[3] R. Wang, R. Clark, and V. L. Bautch, *Development* **114**, 303 (1992).

0076-6879/2003 $35.00

maturation steps that recapitulates murine vasculogenesis *in vivo* and leads to the formation of vascular cords evoking a primitive vascular network.[4] Further maturation of ES-derived EBs in three dimensional collagen gels also displays many features of the sprouting angiogenic process.[5] Endothelial sprouts exhibiting an angiogenic phenotype can arise from the preexisting EB primitive vascular network when plated into type I collagen in the presence of angiogenic growth factors. These endothelial outgrowths, in which formation can be inhibited by known angiostatic agents, were observed to give rise to tube-like structures in some circumstances with concomitant differentiation of α-smooth muscle actin positive cells. The ES/EB system thus exhibits the unique property to allow the study of both vasculogenesis and angiogenesis. Potentialities of the *in vitro* ES cell differentiation system include the analysis of gene function and vascular morphogenesis during endothelial differentiation of genetically manipulated ES cells. Indeed, the analysis of the consequences of genetic modifications, that can be easily introduced in these cells, offers excellent and complementary alternatives to *in vivo* studies on transgenic animals,[6–9] especially when these mutations are lethal for the embryos, preventing analysis of further distal developmental events such as vascular system maturation and remodeling. The ES/EB model also appears useful for the evaluation of the effects of external stimuli on both vasculogenesis and angiogenesis, the analysis of the expression patterns of endothelial genes, the generation and the phenotypic and functional characterization of early endothelial precursors, as well as arterio-venous cell fate specification. Finally, with the isolation of their human counterparts, ES cells may provide a source of endothelial cells for transplantation and gene therapy.

ES Cell Differentiation into the Endothelial Lineage

Differentiation Methods

Various protocols have been developed to promote endothelial differentiation from ES cells. They include culture in semi-solid

[4] D. Vittet, M.-H. Prandini, R. Berthier, A. Schweitzer, H. Martin-Sisteron, G. Uzan, and E. Dejana, *Blood* **88**, 3424 (1996).

[5] O. Féraud, Y. Cao, and D. Vittet, *Lab. Invest.* **81**, 1669 (2001).

[6] D. Vittet, T. Buchou, A. Schweitzer, E. Dejana, and P. Huber, *Proc. Natl. Acad. Sci. USA* **94**, 6273 (1997).

[7] A. C. Schuh, P. Faloon, Q.-L. Hu, M. Bhimani, and K. Choi, *Proc. Natl. Acad. Sci. USA* **96**, 2159 (1999).

[8] V. L. Bautch, S. D. Redick, A. Scalia, M. Harmaty, P. Carmeliet, and R. Rapoport, *Blood* **95**, 1979 (2000).

[9] J. B. Kearney, C. A. Ambler, K.-A. Monaco, N. Johnson, R. G. Rapoport, and V. L. Bautch, *Blood* **99**, 2397 (2002).

methylcellulose-based media, suspension cultures, attached culture in monolayers, and the hanging drop method. These ES cell culture systems have been described in detail in previous papers.[10,11] With the exception of attached cultures, in all other assays, ES cells are unable to adhere to the dishes and generate three-dimensional cell aggregates known as embryoid bodies (EBs) that contain differentiated cell populations. These EBs can be used in different ways (gene expression analysis, cell content analysis after enzymatic dissociation,... etc...) at various stages of their maturation. Endothelial differentiation appeared to occur in reproducible and predictable pattern regardless of the procedure used. In fact, the choice of the protocol to be used will depend on the purpose of the study since some procedures are better suited to particular situations. For example, when isolation of EBs is to be performed at early stages of differentiation (before day 4), differentiation in suspension culture will facilitate the harvest of small EBs. However, a major limitation in suspension culture is EBs aggregation whereby two or more EBs fuse to form a larger aggregates of cells, thus preventing analysis at developmental stages later than day 6. The recent introduction of the spinner flask technique in suspension culture has partly solved this problem since it has been shown to allow the obtainment of large amounts of EBs until day 8,[12] that may be suitable for some routine screening. On the other hand, the differentiation in monolayer culture is recommended for the isolation and the replating of vascular progenitor cells since dissociation of EBs by enzymatic collagenase treatment seriously affect the antigenicity and the viability of endothelial cells. Nevertheless, when performed in two-dimensional attached cultures, the nature of the support appears essential to promote efficient endothelial differentiation. Indeed, comparable results of endothelial differentiation as those obtained in EB culture can only be achieved when ES cells were allowed to attach and to differentiate on type IV collagen.[13]

Focusing on the EB system, we will concentrate on methylcellulose differentiation cultures that favor single cell derived EB formation.[11] In addition, a better synchrony in the differentiation process is achieved towards suspension culture since EBs exhibit a narrower size distribution.[12]

[10] G. M. Keller, *Curr. Opin. Cell Biol.* **7**, 862 (1995).

[11] S. M. Dang, M. Kyba, R. Perlingeiro, G. Q. Daley, and P. W. Zandstra, *Biotechnol. Bioeng.* **78**, 442 (2002).

[12] M. Wartenberg, J. Günther, J. Hescheler, and H. Sauer, *Lab. Invest.* **78**, 1301 (1998).

[13] M. Hirashima, H. Kataoka, S. Nishikawa, N. Matsuyoshi, and S.-I. Nishikawa, *Blood* **93**, 1253 (1999).

Procedure for Endothelial ES Cell Differentiation in Semi-Solid Methylcellulose Medium

This procedure has been adapted from the *in vitro* differentiation protocols used for *in vitro* hematopoietic differentiation of ES cells.[14] We recommend the use of ES-screened products. Products for growth of undifferentiated ES cells and for *in vitro* differentiation of ES cells are now all commercially available. However, methylcellulose stock solutions can easily be prepared in the laboratory at a lower cost according to the procedure described by Wiles.[14] In addition, we recommend to test and to select fetal calf serum (FCS) for the optimal yield of EB formation and generation of endothelial cells.

To get reproducible results, a new vial of ES cells at a similar low passage number is thawed in each new experiment and we rigorously followed the same protocol. For the expansion and the routine culture of ES cells, refer to the basic considerations by Wiles.[14] After the thawing of a new vial, ES cells are cultured for two passages before initiation of differentiation. ES cells are cultured without feeder layer on gelatin-coated tissue culture dishes in Dulbecco's modified Eagle's medium (DMEM with high glucose) supplemented with 15% FCS, 1% nonessential aminoacids, 1% antibiotic solution (ATAM, Invitrogen, Paisley, Great Britain) and 150 μM monothioglycerol, in the presence of approximately 1000 Units/ml recombinant Leukemia inhibitory factor (LIF) (ESGRO; Chemicon, Temecula, CA) to keep them undifferentiated. To induce differentiation, subconfluent ES cells are harvested using trypsin/EDTA, resuspended in Iscove's modified Dulbecco's medium (IMDM) and then cultured in the absence of LIF in 1% methylcellulose-based semi-solid medium. Final 1x differentiation medium composition is IMDM containing 15% FCS, 10 μg/ml insulin, 50 U/ml penicillin, 50 μg/ml streptomycin and 450 μM monothioglycerol (freshly prepared from a 11.6 M stock solution, since diluted solutions rapidly deteriorate). Routinely, we prepare 2% methylcellulose in IMDM and 2.2x concentrated differentiation medium. Then, final cell suspension in semi-solid medium is obtained by mixing (proportions per 1 ml): 0.5 ml 2% methylcellulose in IMDM, 0.45 ml 2.2x concentrated differentiation medium and 0.05 ml ES cells in IMDM. All ingredients are first homogenized by vigorous vortexing before the addition of cells, then mixed with cells by multiple and gentle tube inversion. Take care not to overfill tubes to facilitate easy and adequate mixing of cells. After allowing air bubbles to dissipate, 2 ml of methylcellulose medium containing cells are distributed with a 1 ml syringe equipped with an

[14] M. V. Wiles, *Meth. Enzymol.* **225**, 900 (1993).

18-gauge needle and the plates are swirled to ensure the uniform distribution of the viscous medium. Routinely, we use 35 mm bacterial dishes or 6-well bacterial plate and cultures are maintained without further feeding at 37°C in a humidified 5% CO_2 atmosphere. As it will be described later, this medium can be supplemented with "angiogenic" growth factors to optimize endothelial cell differentiation. When differentiation is performed with the addition of exogenous growth factors, 20 μl of 100x concentrated factors are previously distributed per dish before the addition of complete semi-solid medium with cells.

The procedure given here for endothelial differentiation is for cell lines adapted to grow on gelatin-coated culture ware. It has worked extremely well on 129/Sv-derived ES cells: R1 and CJ7. Some variability may exist among ES cell lines in their ability to form EBs *in vitro*. The cell plating density at the initiation of the differentiation might then be adjusted to obtain consistent and reproducible EB formation (between 150 and 300 EBs per 2 ml of medium is recommended). We have found that plating 1500–2500 cells per 35 mm plate of 2 ml medium to be the appropriate range when using R1 or CJ7 ES cell line adapted for growth on gelatin without embryonic fibroblast feeder layer. However, these densities should only be considered as approximate starting points and each cell line should be evaluated at various concentrations to monitor EB formation efficiency. Once this problem is solved, no major difference in endothelial differentiation have been seen regardless of the ES cell line used. For a full description of reagents required and major troubleshooting for ES cell culture in methylcellulose medium, see review by Wiles.[14]

Growth Factors

Spontaneous endothelial differentiation which occurs after LIF removal can be seriously improved by selected cytokines. However, these growth factors were unable to exclusively direct the differentiation into the endothelial lineage but rather improved it. They can be applied either directly on ES cells but also at different steps of the differentiation process. Consistent with its pivotal role in the first steps of endothelial development *in vivo*,[15,16] VEGF was found to favor endothelial commitment and differentiation, as assessed by the formation of PECAM positive vascular

[15] P. Carmeliet, V. Ferreira, G. Breier, S. Pollefeyt, L. Kieckens, M. Gertsenstein, M. Fahrig, A. Vandenhoeck, K. Harpal, C. Eberhardt, C. Declercq, J. Pawling, L. Moons, D. Collen, W. Risau, and A. Nagy, *Nature* **380**, 435 (1996).

[16] N. Ferrara, K. Carver-Moore, H. Chen, M. Dowd, L. Lu, K. S. O'Shea, L. Powell-Braxton, K. J. Hillan, and M. W. Moore, *Nature* **380**, 439 (1996).

cords in ES-derived EBs.[4,8,17] Other factors seem to help since endothelial differentiation and vascular structure formation were found best supported by a growth factor cocktail including VEGF-A, FGF2, IL6 and Erythropoietin.[4] Although additional experiments are necessary to characterize the precise contribution of each of these factors, we systematically include 50 ng/ml human VEGF, 100 ng/ml human FGF2, 2 U/ml mouse erythropoietin and 10 ng/ml human interleukin 6 (final concentrations) to improve endothelial differentiation.

Endothelial Gene Expression Analysis

Harvest and dissociation of EBs from methylcellulose for scoring endothelial differentiation by flow cytometry analysis:

The development inside the endothelial lineage is accurately monitored by the identification and the counting of endothelial cells. This can be achieved by cell immunostaining with PECAM antibodies, PECAM being used as a membrane marker of the endothelium,[18] and flow cytometry quantitation. For this purpose, disruption of the EBs to a single cell suspension is required. The procedure for EB dissociation is based on enzymatic digestion in a solution of 0.2% collagenase B (Roche Diagnostics, Cat. No. 1-088-815). It can be equally performed after varying periods of time in culture. Briefly, after counting the mean EB number per dish, EBs are harvested by dilution of the methylcellulose medium with 3 ml PBS without Mg^{2+} or Ca^{2+}. A further rinse with 2 ml PBS is performed to ensure the recovery of all EBs. After centrifugation (220*g*, 5 min) and two washes in PBS, the EB pellet is resuspended in a dissociation solution containing 0.2% collagenase B, 10% Fetal calf serum and 200 U/ml DNase I in PBS. 0.5 ml collagenase solution is added per dish EB content. The suspension is incubated for 1 hr at 37°C while flushing through a 1 ml tip every 15 min. At the end of the incubation, the suspension is passed through a 21-gauge needle three times (up and down) in order to complete the dissociation of remaining aggregates. After further dilution of the cell suspension with PBS, cells are pelleted by centrifugation (180*g*, 5 min), washed twice with PBS and counted. On an average, the obtaining of 750,000 total cells can be expected at day 11 from the EBs content of a 35 mm dish resulting from the differentiation of 2500 ES cells initially plated. After counting, cell concentration is adjusted to a density of 5×10^5 to 1×10^6 cells/ml before processing for immunostaining with PECAM antibodies

[17] S. Marchetti, C. Gimond, K. Iljin, C. Bourcier, K. Alitalo, J. Pouysségur, and G. Pagès, *J. Cell Sci.* **115**, 2075 (2002).

[18] H. S. Baldwin, H. M. Shen, H.-C. Yan, H. M. DeLisser, A. Chung, C. Mickanin, T. Trask, N. E. Kirschbaum, P. J. Newman, S. M. Albelda, and C. A. Buck, *Development* **120**, 2539 (1994).

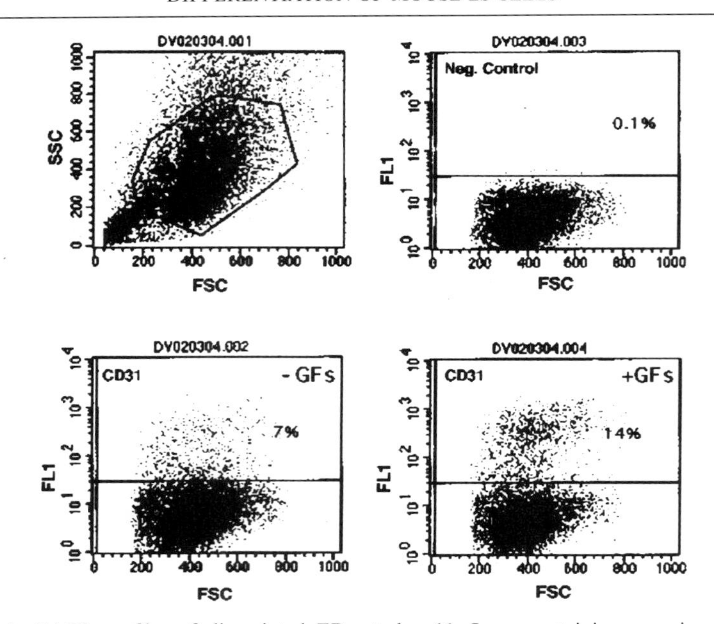

FIG. 1. FACS profiles of dissociated EBs at day 11. Immunostaining experiments were performed after collagenase dissociation of 11-day-old EBs resulting from ES cell differentiation in semi-solid methylcellulose culture medium in the absence (−GFs) or in the presence (+GFs) of the angiogenic growth factor cocktail as described in the text. Representative profiles of PECAM (CD31) immunoreactivity in the gate cell content are illustrated. PECAM positive cell percentage is indicated on each profile. Neg Control, corresponding negative control.

(1 ml per test). FACS profiles of the PECAM staining of dissociated day 11 EBs resulting from ES cell differentiation in the absence or in the presence of endothelial growth factors are illustrated in Fig. 1. Trypsin/EDTA solutions should not be used for EB disruption since they severely affect PECAM antigen reactivity for the immunodetection.[19] Up to now, we have not obtained good results with other antibodies but we do not exclude that some other pairs of antigen/antibody may work well. However, it appears necessary to check the immunoreactivity of the antibodies on control endothelial cells in the same treatment conditions to ensure that they can be used for endothelial cell quantitation after EB disruption by collagenase. On the other hand, the purification of intermediate endothelial differentiation stages appear more advisable after cell recovery by nonenzymatic cell dissociation solution.[20] Indeed, the collagenase procedure may not be

[19] A. Vecchi, C. Garlanda, M. G. Lampugnani, M. Resnati, C. Matteuci, A. Stoppaciaro, H. Schnurch, W. Risau, L. Ruco, A. Mantovani, and E. Dejana, *Eur. J. Cell. Biol.* **63**, 247 (1994).

[20] J. Yamashita, H. Itoh, M. Hirashima, M. Ogawa, S. Nishikawa, T. Yurugi, M. Naito, K. Nakao, and S.-I. Nishikawa, *Nature* **408**, 92 (2000).

recommended for further replating of dissociated cells as the survival is greatly reduced.

RT-PCR Analysis of Endothelial Gene Expression

Alternatively, endothelial gene expression can be measured by RT-PCR analysis after harvest of the EBs at different periods of time after the differentiation initiation. Most commercially available RNA extraction kits that has been tested have given good results for yield and purity of total RNA. As a rule, one can expect the obtaining at day 11 of 10 μg total RNA from the ES-derived EBs content of a 35 mm dish resulting from the differentiation of 2500 ES cells initially plated. Standard protocols are used for DNase treatment of the RNA samples and cDNA synthesis. PCR experiments are performed with the equivalent of 200 ng reverse-transcribed RNA using exponential amplification conditions. Sequences of primers and PCR conditions for each primer pair are indicated in Table I. The reader can refer to previously published methods for additional details concerning the procedures used.[4,5,14]

Vascular Morphogenesis in ES-derived EBs

The extent of cord-like vascular structure formation in EBs can be assessed by immunostaining experiments with specific endothelial markers on sections of EBs. Various commercially available antibodies can be used to monitor ES cell endothelial differentiation on cryostat sections of EBs.[4,6] They include rat monoclonal antibodies against PECAM (clone MEC13.3; Pharmingen, San Diego, CA), Flk-1 (clone Avas 12α1, Pharmingen), VE-cadherin (clone 11D4.1, Pharmingen), CD34 (clone RAM34, Pharmingen), Panendothelial cell antigen (clone MECA-32, Pharmingen) or polyclonal antibodies against von Willebrand Factor (A082, Dako, Glostrup, Denmark) and VE-cadherin (BMS158, Bender Medsystems, Vienna, Austria). In addition, whole mount immunostaining experiments with PECAM antibody allow the obtaining of a three-dimensional picture of endothelial cell organization in the EB.

For embedding and cryostat sections, EBs are fixed in fresh 4% paraformaldehyde in PBS at 4°C for 1 hr. After two washes in PBS, EBs are cryoprotected by transfer in 10%, then 20% sucrose in 1x PBS at 4°C, 1 hr each. After centrifugation (50*g* to 100*g*, 5 min), supernatant is discarded and the remaining EBs are transferred into a plastic mold. After further centrifugation of the mold, the excess liquid is aspirated using a blunt-end 24 gauge needle, before adding a drop of Tissue-Tek OCT compound (Sakura Finetek, Zoeterwoude, The Netherlands). After further homogenization of EBs, the mold is transferred to a mixture of dry ice/ EtOH at

TABLE I

Oligonucleotides Primers and Probes for Endothelial Marker Gene Expression Analysis by RT-PCR

Gene	Sense 5′ to 3′	Antisense 5′ to 3′	Annealing temperature (°C)	Cycle number	Size (bp)	Probe 5′ to 3′
PECAM	GTCATGGCCATGGT CGAGTA	CTCCTCGGCATCT TGCTGAA	55	25	260	GGCACACCTGTAG CCAACTT
Flk-1	TCTGTGGTTCTGCGT GGAGA	GTATCATTTCCAA CCACCCT	55	27	269	GATTAACTTGCAGGG GACAGC
Flt-1	CTCTGATGGTGA TCGTGG	CATGCCTCTGG CCACTTG	60	32	316	AGAAGCCCCGCCT AGACA
tie-1	TCTCTATGTGCAC AACAGCC	ACACACACATTCG CCATCAT	55	30	391	TGACGGGCGTTTT CAACTGC
tie-2	CCTTCCTACC TGCTA	CCACTACACCTTT CTTTACA	57	32	441	CTGTTCACCTCA GCCTTCAC
VE-cadherin	GGATGCAGAGGCTCA CAGAG	CTGGCGGTTCACG TTGGACT	55	25	226	ACATGCTACC TGCCCA
HPRT	GCTGGTGAAAAGGA CCTCT	CACAGGACTAGAA CACCTGC	55	30	248	TCACGTTTGTGTC ATTAGTG

−70°C and hold steady until it starts to freeze. OCT compound is then added to fill the mold and freezing is completed. Frozen blocks can then be stored at −20°C for several months before being sectioned 10 μm thick. Indirect immunofluorescence experiments are performed according to standard protocols.[4]

For PECAM whole mount immunostainings, the protocol has been adapted from the procedures described for immunohistochemistry of whole mount embryos.[21,22] EBs are fixed in freshly prepared MeOH/DMSO (4 : 1) by overnight incubation at 4°C with gentle stirring, then transferred in 100% MeOH and stored at −20°C. For immunostainings, EBs are rehydrated in the following series: 50% MeOH, 20% MeOH then PBS, 15 min incubation each. After 30 min saturation in PBS containing 2% BSA and 0.1% Tween 20, incubation with the PECAM antibody (MEC13.3) is proceeded for 3 hr at ambient temperature. PECAM immunoreactivity is then revealed using the APAAP procedure and Naphtol-AS-MX-phosphate/fast red TR salt as a substrate.[23] Briefly, EBs are first incubated with rabbit anti-rat immunoglobulins (Dako, 1/100 diluted) then with APAAP rat monoclonal (Dako; 1/100 diluted) for 1 hr each at ambient temperature before incubation with the chromogenic substrate to visualize the staining. Vascular morphogenesis as it can be observed on sections of 11-day-old EBs or after whole mount immunostaining are shown in Fig. 2.

Angiogenesis during EB Maturation in Three Dimensional Collagen Matrix

Some reports have mentioned that the ES/EB system can recapitulate many aspects of the angiogenic process. Experiments performed by plating EBs primarily grown in suspension cultures onto gelatinized dishes or onto matrigel showed the formation of branching vascular structures indicative of vascular morphogenesis.[24,25] Recently, we have demonstrated that the ES/EB system provide a useful tool to investigate angiogenesis molecular mechanisms when ES-derived EBs are embedded in a three dimensional

[21] C. A. Davis, *Meth. Enzymol.* **225**, 502 (1993).

[22] B. Hogan, R. Beddington, F. Costantini, and E. Lacy, "Manipulating the Mouse Embryo. A Laboratory Manual," 2nd ed. Cold Spring Harbor Laboratory Press, Cold Spring Harbor, New York, 1994.

[23] J. L. Cordell, B. Falini, W. N. Erber, A. K. Ghosh, Z. Abdulaziz, S. MacDonald, K. A. F. Palford, H. Stein, and D. Y. Mason, *J. Histochem. Cytochem.* **32**, 219 (1984).

[24] M. Bielinska, N. Narita, M. Heikinheimo, S. B. Porter, and D. B. Wilson, *Blood* **88**, 3720 (1996).

[25] W. Bloch, E. Forsberg, S. Lentini, C. Brakebusch, K. Martin, H. W. Krell, U. H. Weidle, K. Addicks, and R. Fässler, *J. Cell Biol.* **139**, 265 (1997).

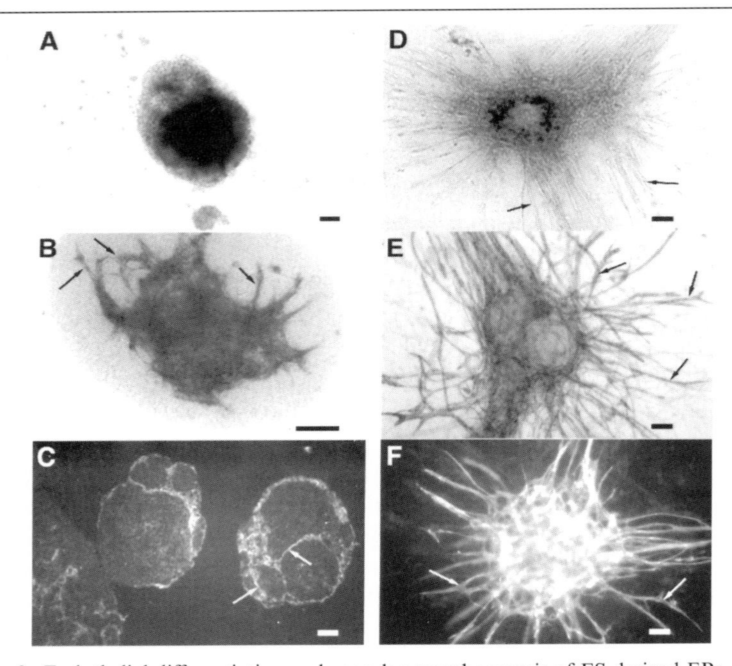

FIG. 2. Endothelial differentiation and vascular morphogenesis of ES-derived EBs. 11-day-old EBs (A–C) or 11-day-old EBs embedded into type I collagen gel after three days in the presence of the angiogenic growth factor cocktail (D–F) are illustrated. A and D, phase constrast microscopy illustration of the EBs morphology; arrows indicate examples of budding outgrowths invading the collagen matrix. B and C, endothelial cord-like structures in 11-day-old EBs revealed after either PECAM whole mount immunostaining (B) or immunofluorescence experiments on EB cryostat sections (C). In both cases, arrows point to PECAM positive vascular structures. E and F, endothelial sprouts visualization by PECAM immunostainings revealed either by the APAAP staining procedure (E) or by immunofluorescence (F). Arrows point to examples of expanding PECAM positive endothelial outgrowths which connected into a network. Scale bar: 100 μm for all panels.

type I collagen matrix.[5] The differentiation procedure involves a two-step protocol that allows the analysis of both vasculogenesis and sprouting angiogenesis. In the first step, ES cells are induced to differentiate in semi-solid methylcellulose medium that leads to endothelial development and capillary-like structure formation within ES-derived EBs. In the second step, 11-day-old EBs are cultured into a type I collagen gel containing angiogenic factors which promote endothelial outgrowth formation. Endothelial sprouts displaying an angiogenic phenotype were found to rapidly develop from the initial EB cord-like vascular structures. In addition, they were observed to be specifically induced or inhibited by known angiogenesis activators and inhibitors, respectively. Such an

ES-based experimental system may then be suitable to study molecular events involved during sprouting angiogenesis.

Procedure for EB Secondary Culture into Type I Collagen Gels

The start of the protocol is a continuation of the ES primary differentiation protocol in methylcellulose semi-solid medium. To induce ES differentiation, proceed as usual (see previously described procedure). At day 11 of differentiation, the number of EBs per 35 mm petri dish is counted. EBs are harvested from methylcellulose culture by dilution of the medium with IMDM. The culture is flooded with the medium and mixed gently to avoid the EBs disruption, then transferred into a 15 ml tube. The dish is washed with another 1 ml IMDM to ensure that all EBs are collected. EBs are recovered by low speed centrifugation (180*g*, 5 min), washed twice with IMDM, then resuspended into fresh IMDM at a concentration of 500 EBs/ml (refer to the initial counting).

Then, the EBs are gently mixed with the collagen medium at a final concentration of 50 EBs/ml and 1.2 ml is poured into a 35-mm bacterial grade petri dish. The plate is swirled to allow the suspension to recover all the EBs on surface of the dish. It is important to ensure that EBs are evenly distributed in the dish. The flasks are transferred into a 37°C incubator with an atmosphere of 5% CO_2, 95% air. Complete gel formation is achieved within 30–60 min. It is important not to disturb the cultures during this time. Further analysis of EBs sprouting can be performed between one to four days later. First signs of sprouting were evident 24 hr later and optimal outgrowth development occurred 72–96 hr after secondary plating.

All the ingredients of the collagen medium with the exception of collagen are mixed and stored on ice before harvest of the EBs. Rat tail type I collagen, at a final concentration of 1.25 mg/ml, is added and mixed just before the addition of the EBs to avoid prior gelation of the medium. Indeed, the collagen starts to gel within several minutes following addition of the complete medium and the EB suspension. Thus, if more than 6 dishes are being set up, prepare only the corresponding volume to be poured and dispensed into the dishes before proceeding with another series. Final EBs suspension in the collagen medium are obtained by mixing (proportions per 1 ml): 0.18 ml IMDM 5x concentrated, 0.33 ml Medium A 3x concentrated (Medium A 1x: 15% fetal calf serum, 450 μM monothioglycerol, 10 μg/ml insulin, 50 U/ml penicillin and 50 μg/ml streptomycin), 0.01 ml angiogenic growth factors 100x concentrated, 0.1 ml of an EB suspension at 500 EBs/ml in IMDM, and 0.38 ml of a collagen solution at 3.3 mg/ml in H_2O. Note that optimal final collagen concentrations may vary depending on the batch used and have been observed in

TABLE II
TROUBLESHOOTINGS for EBs CULTURE IN COLLAGEN GELS

Problem	Possible cause	Recommendation
Gelation starts rapidly after collagen addition, and before the end of the distribution.	Temperature too high for the ingredients. Too much time spent between medium preparation and distribution into the dishes.	Work on ice. Prepare complete medium containing cells only for a limited number of plates.
EBs concentrates at the periphery of the dish.	Uneven distribution of EBs.	Take care to uniformly distribute EBs. Do not move the plate before gelation is completed.
Poor signs of EBs sprouting.	The gel display too compact collagen fibril network.	Check the collagen concentration and perform the test at lower doses to decrease the rigidity of the gel.
Important gel retraction at day 2 of secondary culture.	Plating too many EBs in the gel leading to excessive traction on collagen fibrils and protease secretion.	Do not plate more than 50 EBs per 35 mm dish.
Few endothelial sprouts are revealed after immunostainings.	Insufficient or inactive growth factors used.	Check growth factor concentrations and use low binding tips to avoid loss of material.

our hands to range from 0.75 to 1.5 mg/ml. Listed in Table II are the most common problems associated with secondary culture in collagen gels and some of the possible solutions.

Visualization and Quantitation of Endothelial Sprouting

In addition to qualitative studies of sprouting angiogenesis, the ES/EB system also offers the possibility of quantitation of the angiogenic process by morphometric analysis after prior revelation of endothelial cells by immunostaining experiments. This can be achieved after whole mount PECAM immunostaining in the petri dish following the same procedure as described above for EBs. As an alternative, we prefer to transfer the gel on a glass slide and dehydrate it before processing the slide for immunostainings experiments. This latter procedure bring facilities for both storage and manipulations. Briefly, the 35 mm gel containing dish is turned over a 50 mm×75 mm glass slide. The collagen gel is gently laid out on the slide and the excess of liquid left around the gel is removed by pipetting with a dispenser. The gel is then dehydrated using a nylon linen and absorbent filter cards. Take care not to press down onto the gel since it may damage the EBs. Slides are air-dried overnight then wrapped in an aluminum foil and

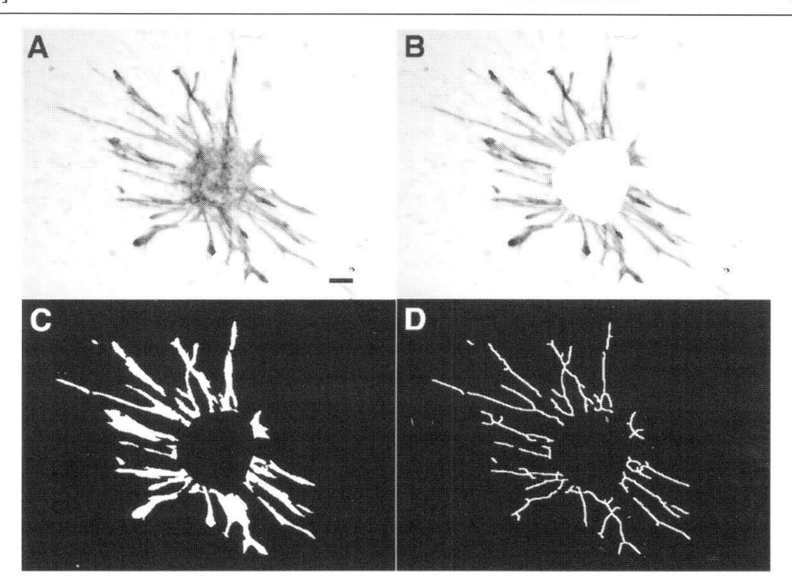

FIG. 3. Image processing for endothelial sprouting quantitation. (A) Initial image of PECAM staining captured at a magnification of 10×. (B) Image A from which the area corresponding to the body mass is masked. (C) Image B binarized to black and white. (D) Final skeletonized image of the PECAM staining.

stored at −20°C for use, days later or processed immediately for immunostainings using standard immunocytochemical or immunofluorescence techniques after fixation. Representatives sprouting EBs are illustrated in Fig. 2. The length of the endothelial sprouts is quantified by morphometric analysis using image analysis software visiolab@2000 (BIOCOM, Les Ullis, France). After PECAM whole mount immunostainings of the EBs, digital images are captured with a computer-supported digital camera at a magnification of 10× and processed for morphometric analysis (Fig. 3). After masking manually the EB area, the image of the staining is binarized to black and white according to visual inspection of the initial staining. The threshold for binarization was checked for each image, since there may be variability in the contrast level of the staining above background. The binarized images are subsequently skeletonized. To check the quality of the entire procedure, the skeletonized image is systemically superposed with the initial image of the staining. For calibration, the image of a grid micrometer is acquired at the same magnification. Mean values for the total length of endothelial sprouts in the presence of the GFs cocktail were in the range 3000–7000 μm for CJ7 and R1 ES-derived EBs.

Further Considerations

This system appears useful to help to characterize modulators of the earliest stages of differentiation and of mitogenesis inside the endothelial lineage. In addition, it may also be useful to study regulators of angiogenesis since the ES/EB system recapitulate many features of the angiogenic process. Thus, some future applications may include the optimization of a screening pharmacological angiogenesis assay with green fluorescent protein-expressing ES-derived endothelial cells. *In vitro* endothelial differentiation has recently been reported from human ES cells.[26] The development of such an assay with human ES cells may then constitute a new approach for the development of angioactive drugs. In addition to displaying complementary data to *in vivo* gene knock-out on the consequences of specific mutations, the use of inducible expression systems under control of endothelial specific promoters would certainly bring important new information on the consequences of a specific gene overexpression during the normal ES cell differentiation program.

Acknowledgment

The authors are indebted to Mrs Annie Schweitzer for help in establishing these protocols. The authors also thank the financial support provided by INSERM (contract APEX no. 99-01) and from the Association pour la Recherche sur le Cancer (subvention no. 9586) during the assembly of this work.

[26] S. Levenberg, J. S. Golub, M. Amit, J. Itskovitz-Eldor, and R. Langer, *Proc. Natl. Acad. Sci. USA* **99**, 4391 (2002).

[16] ES Cell Differentiation to the Cardiac Lineage

By KENNETH R. BOHELER

Introduction

Advances in our understanding of cardiomyocyte cell biology and development have until recently depended upon our ability to generate primary cultures from enzymatically dispersed fetal, neonatal or adult hearts. Primary cardiomyocyte cultures recapitulate many of the physiologic and molecular characteristics of intact hearts, but are not ideally suited for *in vitro* developmental studies.

0076-6879/2003 $35.00

In 1981, Evans and Kaufman were the first to isolate embryonic stem (ES) cells.[1] ES cells are derived from the preimplantation embryo (inner cell mass or epiblast), which distinguishes these cells from embryonic carcinoma (EC) and embryonic germ (EG) cells. ES cells are characterized by the capacity to proliferate *in vitro* (self-renew) and have an ability to generate large numbers of differentiated cell progeny from all three primary germ lineages, including mesoderm-derived cardiomyocytes. The history of ES cell differentiation *in vitro* began with Drs. Wobus and Doetschman.[2,3] These investigators independently observed that ES cells grown in suspension form three-dimensional cell aggregates termed embryoid bodies (EBs) that readily differentiate into many recognizable cell types. We now know that selected cell lines and specific cultivation conditions must be employed to maximize differentiation of ES cells to cardiomyocytes.

In vitro differentiation of embryonic stem (ES) cells therefore provides an alternative source to study cardiomyocytes in culture. Numerous studies have demonstrated that cardiogenic induction in ES cells faithfully recapitulates many of the physical and molecular properties of developing myocardium *in vivo*. *In vitro* derived populations of cardiomyocytes are, however, not uniform but consist of multiple cell types typical of primary myocardium, atrium, ventricle, sinus node or the His-Purkinje system,[3–5] some of which can show enhanced selection based on culture conditions.[6]

ES cell differentiation to the cardiac lineage facilitates developmental analyses of heart cells. In addition, these techniques can be employed *in vitro*: (1) to test the effects of differentiation factors or xenobiotics on embryogenesis, (2) to investigate pharmacological effects on differentiating cardiomyocytes that otherwise would not be readily available from cultivated cells, (3) to analyze electrophysiological characteristics of immature and developing cardiomyocytes of different lineages or genetic makeup, and (4) to establish strategies for cell and tissue therapy. For the latter two, specific isolation protocols have been developed to facilitate these analyses and involve either simple dissection, FACS

[1] M. J. Evans and M. H. Kaufman, *Nature* **292**, 154 (1981).
[2] A. M. Wobus, H. Holzhausen, P. Jakel, and J. Schoneich, *Exp. Cell Res.* **152**, 212 (1984).
[3] T. C. Doetschman, H. Eistetter, M. Katz, W. Schmidt, and R. Kemler, *J. Embryol. Exp. Morphol.* **87**, 27 (1985).
[4] A. M. Wobus, G. Wallukat, and J. Hescheler, *Differentiation* **48**, 173 (1991).
[5] V. A. Maltsev, J. Rohwedel, J. Hescheler, and A. M. Wobus, *Mech. Dev.* **44**, 41 (1993).
[6] A. M. Wobus, K. Guan, S. Jin, M. C. Wellner, J. Rohwedel, G. Ji, B. Fleischmann, H. A. Katus, J. Hescheler, and W. M. Franz, *J. Mol. Cell. Cardiol.* **29**, 1525 (1997).

analysis[7] or antibiotic selection.[8,9] Because ES cells can be readily cultivated, genetically modified and tested *in vitro* for effects on cardiomyocyte differentation, this model has proven extremely useful in gaining insights into all of these processes. In this article, I present methods to cultivate and differentiate mouse ES cells into cardiomyocytes *in vitro*. The principles of cardiomyocyte *in vitro* differentiation and the characterization of these cells have been covered in recent reviews, which serve as excellent companions to the methods presented here.[10,11]

General Considerations

The maintenance of high quality pluripotent, undifferentiated ES cells is critical to *in vitro* differentiation protocols. Many methods for the cultivation of ES cells have been reported, but a number of parameters consistently influence the differentiation potential of mouse ES cells to cardiomyocytes. These include: (1) the ES cell lines employed, (2) the media and its components (fetal bovine serum (FBS), growth factors and additives), (3) the culture conditions, including the number of cells present in forming EBs, and (4) the timing of EB plating.

Most ES cell lines require inactivated fibroblast feeder layers for good growth and maintenance in the undifferentiated state.[1,12] Some ES cell lines require both feeder cells and leukemia inhibitory factor (LIF), and both should be employed with cell lines that have not previously been tested for differentiation to cardiomyocytes; however, it is advisable to culture ES cells in the manner in which they were derived. Feeder cell-independent growth of ES cells is also possible by cultivation in the presence of LIF or conditioned media, but long-term, these conditions often prove inadequate. Most EC cell lines do not require feeder cells, but EG cell, like ES cell lines, generally require feeder cells for continued growth

[7] M. Muller, B. K. Fleischmann, S. Selbert, G. J. Ji, E. Endl, G. Middeler, O. J. Muller, P. Schlenke, S. Frese, A. M. Wobus, J. Hescheler, H. A. Katus, and W. M. Franz, *FASEB J.* **14**, 2540 (2000).

[8] M. G. Klug, M. H. Soonpaa, G. Y. Koh, and L. J. Field, *J. Clin. Invest.* **98**, 216 (1996).

[9] K. R. Boheler and A. M. Wobus, *in* "Stem Cells: A Cellular Fountain of Youth?" (M. P. Mattson and G. van Zant, eds.), p. 141. Elsevier Science, 2002.

[10] A. M. Wobus, K. Guan, H.-T. Yang, and K. R. Boheler, In K. Turksen, ed.), "Methods Mol. Biol," Vol. 185, pp. 127. Humana Press, Totowa, New Jersey, 2002.

[11] K. R. Boheler, J. Czyz, D. Tweedie, H.-T. Yang, S. V. Anisimov, and A. M. Wobus, *Circ. Res.* **91**, 189 (2002).

[12] G. R. Martin, *Proc. Natl. Acad. Sci. USA* **78**, 7634 (1981).

in the undifferentiated state. Some EG cell lines (EG-1 cells[13]) require both feeder cells and LIF to maintain the undifferentiated state.

Several types of media have proven useful for cultivation and differentiation of ES cells to cardiomyocytes. Either Dulbecco's modified Eagle's medium (DMEM) or Iscove's modified Dulbecco's medium (IMDM) can be employed for cultivation of undifferentiated ES cells. Differentiation with IMDM may require fewer starting cells than differentiation with DMEM, but this tendency is sometimes subtle. In general and when starting with the same number of starting cells, IMDM promotes cardiomyocyte formation and differentiation more rapidly than DMEM. Good quality FBS is also critical for long-term culture and differentiation of ES cells, and failure to adequately test serum represents one reason why ES cells may not differentiate as effectively to cardiomyocytes. The most convenient tests for sera include: (i) comparative plating efficiencies at 10, 15, and 30% serum concentrations, (ii) alkaline phosphatase activity in undifferentiated ES cells, and (iii) *in vitro* differentiation capacity after five passages in selected batches of serum. Ideally, ES cell lines are cultivated without antibiotics, but the addition of penicillin and streptomycin mixtures generally proves useful.

Several mouse ES cell lines and the derivatives of these lines have proven highly effective in the formation of cardiomyocytes *in vitro*: D3, R1, Bl17, AB1, AB2.1, CCE, HM1, and E14.1.[10] Other ES cell lines are likely to prove equally effective, but the efficiency must be tested, and not assumed. The number of ES cells employed for efficient differentiation to cardiomyocytes varies, and depending upon the cultivation media, EBs formed in suspension (e.g., hanging drops) with 400 to 1000 starting cells usually work well. EBs not formed in suspension generally produce low yields of beating cardiomyocytes. The timing of EB body plating is another critical factor in cardiomyocyte formation. EBs must progress through a certain degree of differentiation prior to plating. Within the developing EB, cardiomyocytes develop between an epithelial layer and a basal layer of mesenchymal cells.[14] Premature plating (insufficient development) or delayed plating (necrosis and other cell death) of EBs may result in cell populations containing very few to no cardiomyocytes.

Finally, it should be kept in mind that good aseptic/sterile tissue culture techniques should be routinely practiced. Unless noted otherwise, all

[13] C. L. Stewart, I. Gadi, and H. Bhatt, *Dev. Biol.* **161**, 626 (1994).

[14] J. Hescheler, B. K. Fleischmann, S. Lentini, V. A. Maltsev, J. Rohwedel, A. M. Wobus, and K. Addicks, *Cardiovasc. Res.* **36**, 149 (1997).

manipulations are performed in a laminar flow cabinet, and cells are incubated in a humidified incubator (37°C, 5% CO_2). Both cultivation of ES cells and differentiation of ES cells to cardiomyocytes can be adversely affected by mycoplasm, yeast or bacterial contaminations, changes in CO_2 levels and cellular pH, inadequate feeding or passaging of ES cells, etc. With these considerations in mind, the following methods result in high yields of EBs with beating cardiomyocytes that are readily identifiable within 1–4 days after plating.

Solutions and Media

(All solutions should be autoclaved or filter sterilized.)

Phosphate-buffered saline (PBS), pH 7.2: Without Ca and Mg.

Trypsin solution: 0.2% trypsin (1 : 250, Gibco BRL) in PBS.

Trypsin:EDTA solution: 0.25% trypsin, 1 m*M* EDTA–4 Na.

β-mercaptoethanol stock solution: 10 m*M* β-ME in PBS. Prepare fresh and store at 4°C for 2–3 days.

Gelatin solution: 0.1% gelatin in PBS.

Mitomycin C (MC) stock: 0.2 mg/ml Mitomycin C in PBS, filter sterilize. Prepared weekly (light sensitive).

Monothioglycerol (MTG) solution: Prepare a fresh 150 m*M* stock solution of MTG in IMDM.

BRL Culture Media: Dulbecco's modification of Eagle's medium (DMEM, 4.5 g/liter glucose) (or IMDM) supplemented with 10% FBS, 50 U penicillin, 50 μg/ml streptomycin.

Cultivation medium I: DMEM (4.5 g/liter glucose) supplemented with 15% fetal bovine serum (FBS) for feeder layer cells, 50 U penicillin, 50 μg/ml streptomycin.

Cultivation medium II: DMEM (4.5 g/liter glucose) supplemented with 15% FBS, 2m*M* L-glutamine, 10 μM β-mercaptoethanol, 0.1 m*M* nonessential amino acids (100X, Gibco BRL, Cat. No. 11140-035), 50 U/ml penicillin, 50 μg/ml streptomycin.

Differentiation medium: DMEM (4.5 g/liter glucose) supplemented with 20% FBS and 2 m*M* L-glutamine, 10 μM β-mercaptoethanol, 0.1 m*M* nonessential amino acids, 50 U/ml penicillin, 50 μg/ml streptomycin.

BRL-conditioned media (BRL-CM): DMEM (or IMDM) supplemented with BRL-stock (see text for details) (60 ml stock per 100 ml total), 15% FBS (9 ml/100 ml. The BRL stock contains 10% FBS), 0.01 mg/ml transferrin, 2 m*M* L-glutamine, 10 μM β-mercaptoethanol, 0.1 m*M* nonessential amino acids, 50 U/ml penicillin, 50 μg/ml streptomycin. 10 ng/ml LIF is optional. [If IMDM is used, replace β-mercaptoethanol with 450 μM MTG.]

Retinoic acid: Prepare a 10^{-3} *M* stock solution of all-*trans* retinoic acid in 96% ethanol. Store aliquots at −20°C in the dark and use once after thawed.

Collagenase-supplemented low-Ca^{2+} solution: 120 m*M* NaCl, 5.4 m*M* KCl, 5 m*M* $MgSO_4$, 5 m*M* Na pyruvate, 20 m*M* glucose, 20 m*M* taurine, 10 m*M* HEPES-NaOH, pH 6.9 at 24°C, supplemented with 1 mg/ml collagenase B (Roche) and 30 μM $CaCl_2$. Make this solution with HPLC-grade H_2O, because the collagenase activity is highly sensitive to the calcium concentration.

KB solution: 85 m*M* KCl, 30 m*M* K_2HPO_4, 5 m*M* $MgSO_4$, 1 m*M* EGTA , 5 m*M* Na pyruvate, 5 m*M* creatine, 20 m*M* taurine, 20 m*M* glucose, freshly added 2 m*M* Na_2ATP; pH 7.2 at 24°C.

Methods

Primary Embryonic Fibroblast Feeder Cells

Pluripotent ES cells must be cultivated on monolayers of mitotically-inactivated fibroblast feeder cells,[3,12] on gelatinized tissue culture dishes in Buffalo rat liver (BRL) cell conditioned medium,[15] or in the presence of LIF.[16,17] For long-term culture and maintenance, mitotically inactivated feeder layers are optimal. Both primary mouse embryonic fibroblasts and STO fibroblast cell lines are commonly employed as feeder layers, but primary feeder layers are generally preferable to established cell lines. Previously frozen primary fibroblast feeder cells are also inferior to freshly cultivated feeder layers, and the reliability of STO cells to maintain optimal growth conditions of ES cells varies considerably. The technique presented here relies wholly on freshly prepared fibroblast feeder layers from CD-1 outbred mice. We find that feeder cells produced with this outbred strain last longer in culture than those produced with inbred lines of mice (e.g., C57Bl/6). The major disadvantage of primary fibroblasts is their relative short life span, since after about 15–20 divisions (2 weeks in culture with passaging), these cells go into "crisis" and are no longer suitable for passage or

[15] A. G. Smith and M. L. Hooper, *Dev. Biol.* **121**, 1 (1987).

[16] A. G. Smith, J. K. Heath, D. D. Donaldson, G. G. Wong, J. Moreau, M. Stahl, and D. Rogers, *Nature* **336**, 688 (1988).

[17] R. L. Williams, D. J. Hilton, S. Pease, T. A. Willson, C. L. Stewart, D. P. Gearing, E. G. Wagner, D. Metcalf, N. A. Nicola, and N. M. Gough, *Nature* **336**, 684 (1988).

maintenance of ES cell cultures. Although this technique may be more laborious than using frozen cells or stable cell lines, the results are consistent and maintain ES cells in an undifferentiated, pluripotent state.

Procedure

1. Select a pregnant mouse (CD-1 outbred strain) with embryos at day $\sim$14.5 to 17.5 of pregnancy. Kill the mouse by cervical dislocation or CO_2 asphyxiation and place in a beaker containing absolute ethanol.
2. Transport mouse to a laminar flow hood. Aseptically remove the uterus with the fetuses and place in a sterile petri dish containing PBS (10 ml). Remove the placenta, fetal membranes and decapitate the fetuses. Remove all viscera (heart, liver, intestine, etc.) and as much blood as possible.
3. Rinse the carcasses in PBS and place in a petri dish containing trypsin solution (10 ml) to remove any remaining blood or viscera.
4. Transfer the carcasses to another petri dish with trypsin (10 ml) and mince the tissue. Using a large bore pipette (e.g., disposable 25 ml), transfer the finely minced tissues to an Erlenmeyer flask containing a stir bar, glass beads (5 mm diameter), and 2–3 ml of trypsin solution per fetus.
5. Stir on magnetic stir plate at room temperature (RT) for 25–30 min (increase the digestion time when the fetuses are older i.e., >16.5 dpc).
6. Filter the cell suspension through a sieve, screen, or cheesecloth. Add an equal volume of cultivation medium I to the filtered suspension, mix and centrifuge 5 min at 500*g*.
7. Gently remove the supernatant and resuspend the pellet in about 3 ml of cultivation medium I. Plate cells on 10 cm tissue culture plates (about 2×10^6 cells per 10 cm dish) containing 10 ml cultivation medium I.
8. Change the medium the following morning to remove blood cells, dead cells, and cellular debris. Cultivate for an additional 1–2 days.
9. To passage the primary culture of mouse embryonic fibroblasts, remove media and rinse once with PBS (10 ml) at room temperature or prewarmed (37°C).
10. Add 3–6 ml of prewarmed (37°C) trypsin solution. Leave the trypsin on the cells for 45–60 sec (the time decreases as passage number increases) and remove.

11. As the cells begin to loosen from the plate surface, add 3 ml cultivation medium I. Dissociate cells by trituration and split 1 : 2 to 1 : 4 on 10 cm tissue culture plates, containing 10 ml of cultivation medium I.
12. Passage the primary mouse embryonic fibroblasts every 2–3 days. Embryonic fibroblasts at passages 2–4 are most suitable as feeder layers for undifferentiated ES cells. After about 2 weeks, the cells will go into crisis and are not suitable for continued cultivation.

Inactivation of Primary Embryonic Fibroblast Feeder Cells

The following protocol utilizes mitomycin C to inactivate primary mouse embryonic fibroblast feeder. Mitomycin C is a DNA cross-linker and is a suspected carcinogen. It is light sensitive and should be made fresh on a weekly basis. As an alternative, γ-irradiation at 6000–10,000 rads can be used to inhibit cell growth, but special equipment is required that may not be readily available. Both treatments cause the fibroblasts to accumulate in the G2 phase of the cell cycle, preventing further division.

Procedure

1. For mitomycin C treatment, treat a confluent 10 cm plate of embryonic fibroblast cells with 0.01 mg/ml MC for 2.5 hr (2–3 hr range).
2. Coat 6 cm tissue culture dishes with gelatin solution for 0.5–1 hr. Plates can also be prepared 24 hr in advance and stored at 4°C before use.
3. Remove the MC-containing solution and wash the cells three times with 10 ml PBS.
4. Flood the feeder cells with 3–5 ml trypsin solution and remove almost immediately. Once the cells begin to dislodge, add 1 ml of Cultivation medium I and triturate. Passage the cells 1 : 4 to 1 : 6 onto 6 cm gelatin-coated plates containing Cultivation medium I. (Passages 2–4 primary fibroblasts can be passaged 1 : 5 to 1 : 6, but later passages of untreated feeder cells can only be passaged 1 : 3 to 1 : 4.)
5. Swirl plates in a "figure 8" motion several times on the bench to obtain a uniform distribution of cells and transfer to incubator.
6. Allow cells to attach a minimum of 3 hr before coculture with ES cells. Feeder cell layers prepared one day prior to ES cell subculture are optimal.

Preparation of BRL-Conditioned Medium

In the absence of feeder layers, ES cells can be maintained in an undifferentiated state by cultivation in the presence of Buffalo Rat Liver conditioned medium (BRL-CM)[15] or in the presence of LIF. I do not recommend long-term cultivation of ES cells in the absence of feeder layers. These additives alone are insufficient to maintain pluripotentiality of ES cell lines. ES cells accumulate genetic and epigenetic changes during *in vitro* culture, and this process may be exacerbated when grown in the absence of feeder layers. The active factor secreted by BRL cells is identical to LIF, but we find that differentiation to cardiomyocytes is improved when ES cell cultures are grown in the presence of both BRL-CM and LIF. When cultivated on gelatinized plates in the absence of feeder cells, ES cells have a more flattened morphology, but after 1 to 2 passages back on feeder layers, the ES colony morphology returns to its original form.[18] Buffalo Rat Liver cells can be purchased from ATCC.

Procedure

1. Cultivate BRL (Buffalo Rat Liver) cells in 75 cm^2 tissue culture flasks with 20 ml BRL culture media. Grow until confluent with media changes as needed.
2. Once confluent, change the media and incubate each flask with 20 ml BRL culture media for 3 days.
3. Collect the media of several plates every three days and filter sterilize (if necessary, spin down before filtration to remove floating cells and debris). Store the filtered medium at 4°C for one week, or at −80°C until required. The BRL monolayer cultures can be used to condition medium for up to 2 weeks. This media is referred to as BRL-stock.

Cultivation of Undifferentiated ES Cells

The culture conditions described in this section are employed routinely in the lab for mouse R1 and D3 embryonic stem cells. These protocols are designed to maintain ES cells in an undifferentiated, pluripotent state suitable for differentiation to cardiomyocytes. The method of choice is cultivation on feeder cells, but short-term feeder layer independent methods are also described. It is important to passage ES cells cultivated on feeder layers every 24–48 hr. Do not cultivate longer than 48 hr without passaging, or the cells may begin to differentiate and be unsuitable for differentiation studies.

[18] S. V. Anisimov, K. V. Tarasov, D. Tweedie, M. D. Stern, A. M. Wobus, and K. R. Boheler, *Genomics* **79**, 169 (2002).

Procedure

1. Thaw 1 ml (1×10^6 cells) of frozen ES cells (in Cultivation Medium II, but containing 20% FBS and 8% dimethylsulfoxide) rapidly at 37°C and transfer the cells to 10 ml of Cultivation Medium II.
2. Centrifuge the cells at 300*g* for 3–5 min and remove supernatant.
3. Resuspend the pellet in 1 ml of Cultivation Medium II and add to 6 cm tissue culture plates previously coated with 0.1% gelatin and containing either
   a. A monolayer of MC-treated feeder cells containing 3–4 ml of Cultivation Medium II. (Ideally, ES cells are passaged at least one time on feeder layers after thawing before preparation of EBs.) or
   b. 4 ml of Cultivation Medium II + 10 ng/ml LIF or BRL-CM or BRL-CM + LIF.
4. Change the medium after the first overnight incubation and passage as needed every 24–48 hr.
5. Prior to passaging, change the medium of ES cell cultures (cultivated either in the presence or absence of feeder cells) 1 to 2 hr before trypsinization.
6. Remove the medium, and wash 1X with PBS. Add 0.75 ml of prewarmed (37°C) trypsin : EDTA, incubate at room temperature for 30–60 sec. Add 0.75 ml cultivation medium II.
7. Disperse the cell suspension by vigorous trituration through the narrow bore of a 2 ml pasteur pipette. It is important to get a good cell suspension. Centrifuge (300*g*) and either passage or proceed to *in vitro* differentiation.
8. If passaging, resuspend the cells in Cultivation medium II and passage 1 : 5 to 1 : 10 onto MC-treated (6 cm) feeder cells (as 3a above) or passage 1 : 3 to 1 : 6 onto gelatin-coated plates (6 cm) containing either BRL-CM + 10 ng/ml LIF or cultivation medium II + 10 ng/ml LIF.

Differentiation Protocols

For ES cell differentiation to most phenotypes (except neuronal), ES cells must be cultivated in three-dimensional aggregates or "embryoid bodies" (EB), by the "hanging drop" method, by "mass culture," or by differentiation in methylcellulose (see Ref. 10 for references). The hanging drop method generates EBs of a defined cell number (and size!). This technique is particularly useful for developmental studies. Mass cultures of EBs may be employed for generation of a large number of differentiated cells, and can be performed through various techniques (suspension in dishes, spinner flasks, etc.). This technique is less efficient than the hanging drop technique, but many more cell aggregates can be produced in less time. Methylcellulose has been employed primarily for clonal selection or ES cell differentiation to hematopoietic cell lineages and not to cardiac lineages. Here I describe differentiation protocols utilizing the hanging drop method and mass cultures in petri dishes. Both methods can be employed with ES cells grown on feeder layers, in BRL-conditioned media or in medium containing LIF. During the actual differentiation process, LIF must be omitted from the cultures, because it will inhibit the formation of cardiomyocytes.

Hanging Drop Method

Procedure

1. From step 7 of cultivation of undifferentiated ES cells, dilute the suspended cells in Differentiation medium. To a small aliquot of these cells, add one-fifth volume Trypan Blue to a small number of cells and count with the aid of a hemocytometer. (ES cells are small and round; whereas, any contaminating feeder cells are generally much larger in size.)
2. Prepare a cell suspension containing a defined ES cell number of 400 to 1000 cells per 20 μl of Differentiation medium. Use of 400–600 cells of ES cell lines D3 or R1 for preparation of EBs is optimal for cardiac differentiation. (ES cells grown on feeder layers should contain >100-fold more ES cells than fibroblasts, because the presence of fibroblasts to the developing EBs can affect the differentiation to cardiomyocytes.)
3. Place 20 μl drops ($n = 50$ to 60) of the ES cell suspension on the lids of 100 mm bacteriological petri dishes containing 5–10 ml PBS.
4. Cultivate the ES cells in hanging drops for 2 days and transfer to the incubator. The cells will aggregate and form one EB per drop. Do not disturb the developing EB during this period.
5. Rinse the aggregates carefully from the lids with 2 ml of Cultivation medium II, transfer into a 60 mm suspension culture dishes with 5 ml of Cultivation medium II. Continue cultivation in suspension for 2–5 days until the time of plating. "Good" suspension culture dishes for differentiating EBs can be purchased from Kord Products Inc. (Ontario, Canada. Cat. no. 901).
6. To plate, coat tissue culture dishes with 0.1 % gelatin solution for 1–24 hr at 4°C before use. Aspirate gelatin solution from pretreated 24-well microwell plates or 6 cm tissue culture dishes and add Cultivation medium II.

7. Transfer a single EB into each well of gelatin-coated 24 well plates for morphological analysis, or transfer 20–40 EBs per dish onto 6 cm tissue culture dishes containing 4 cover slips (10×10 mm) for immunofluorescence, or 15–20 EBs onto 6 cm tissue culture dishes for RT-PCR analysis or dissections of EB outgrowths. For the investigation of early cardiac stages, plate EBs at day 5, but otherwise, plate on days 6 to 7. The first beating clusters in EBs can be seen in 7-day-old EBs, but maximal cardiac differentiation is achieved after EB plating.
8. Change the medium every second or third day. To characterize the EB outgrowths morphologically, calculate the percentage of EBs (from at least 24 wells) with beating areas to calculate the percentage of EBs differentiating to cardiomyocytes. We routinely obtain $>90\%$ with this technique.

Variants

1. To enhance the development of ventricular-cardiomyocytes,[6] add retinoic acid (10^{-9} *M*) between days 5 and 15 of culture.
2. If P19 embryonic carcinoma (EC) cells are employed, induce cardiac differentiation by treatment with 1% dimethylsulfoxide (DMSO).[19] EC cells are cultivated without feeder cells and result after DMSO induction in a high number of cardiac cells.
3. EG cells are cultivated and differentiated similarly to that described for the ES cells.

Mass Culture Procedure

For mass culture, 5×10^5 to 2×10^6 cells (depending on ES cell lines used) are plated into bacteriological petri dishes containing initially either cultivation medium or differentiation medium, depending on the cells. Spinner flasks can also be employed for the generation of very large numbers of cardiomyocytes.[20] Once EBs have formed, the cells should be cultured in Cultivation medium II until the time of plating. Under these conditions and in suspension, cell aggregates of various sizes will form, but ensuring single cell suspensions can minimize these variations. This method is a variation of that described by Miller-Hance *et al.*[21]

[19] M. W. McBurney, E. M. Jones-Villeneuve, M. K. Edwards, and P. J. Anderson, *Nature* **299**, 165 (1982).

[20] J. Hescheler, M. Wartenberg, B. K. Fleischmann, K. Banach, H. Acker, and H. Sauer, *Methods Mol. Biol.* **185**, 169 (2002).

[21] W. C. Miller-Hance, M. LaCorbiere, S. J. Fuller, S. M. Evans, G. Lyons, C. Schmidt, J. Robbins, and K. R. Chien, *J. Biol. Chem.* **268**, 25244 (1993).

Procedure

1. Follow step 1 of Hanging drop protocol.
2. For differentiation, add 500,000 ES cells to a 10 cm bacteriological petri dish containing 10 ml of Differentiation medium.
3. After 48 hours of culture, remove Differentiation medium and add 10 ml of Cultivation medium II.
4. From this point on, change the medium every second day but add $\frac{1}{2}$ volume of Cultivation medium II (5 ml) on the intervening alternate days (ectoderm-like layers usually appear between 4 and 6 days of culture).
5. On day 7 or 8, plate cells, and maintain in Cultivation medium II. These conditions give rise to cell aggregates that display spontaneous contractions by 8 day of suspension culture or within one day after plating.

Single-cell Isolation of ES Cell-derived Cardiac Cells

For electrophysiologic studies or analysis of sarcomeric or other intracellular proteins, ES cell-derived cardiomyocytes should be isolated as single cells. A number of isolation protocols have been developed to facilitate analyses of ES cell derived cardiomyocytes, and involve principally either simple dissection, or in the case of genetically modified ES cells, FACS analysis[7] or antibiotic selection.[8,9] The following dissection procedure is a variation of that described by Maltsev *et al.*[22]

Procedure

1. Isolate the beating areas of EBs ($n = 10$–20) mechanically using a pulled 2 ml Pasteur pipette with a beveled end or micro-scalpel under an inverted microscope. Collect the cells in a centrifuge tube containing PBS at room temperature. Centrifuge at 1000*g* for 1 min and remove the supernatant.
2. Incubate the pellet in collagenase B-supplemented low-Ca^{2+} solution at 37°C for 25–45 min depending on the collagenase activity. The collagenase solution can be prepared in advance, but should not be

[22] V. A. Maltsev, A. M. Wobus, J. Rohwedel, M. Bader, and J. Hescheler, *Circ. Res.* **75**, 233–244 (1994).

kept longer than 3–4 days. For the isolation of cardiac clusters, shorten the incubation time to 10–20 min.

3. Pellet the cells (1000*g*) and remove the enzyme solution. Resuspend the cell pellet in about 800 μl of KB solution and incubate at 37°C for 30–60 min.
4. Transfer the cell suspension to tissue culture plates containing gelatin-coated slides and incubate in Cultivation medium II at 37°C overnight. The KB medium is diluted at least 1 : 10 with the medium.
5. After an overnight incubation, change the medium and usually within 1–2 hr, cardiomyocytes begin rhythmic contractions and are ready for study.

Acknowledgment

I am grateful to David Crider for expert technical assistance and for reviewing the text to ensure the accuracy of the procedures. This work was supported by the NIH.

[17] Bone Nodule Formation via *In Vitro* Differentiation of Murine Embryonic Stem Cells

By SARAH K. BRONSON

Introduction

For more than a decade numerous laboratories have differentiated osteoprogenitors from neonatal calvaria or marrow stromal cells/mesenchymal stem cells (MSCs) from the bone marrow into mature osteoblasts in tissue culture.[1,2] During this process a single-cell or colony-forming unit (CFU) is able to proliferate, secrete and mineralize bone matrix yielding a discrete raised nodule that protrudes into the medium and readily absorbs the silver nitrate in a von Kossa stain after fixation. Observing the similarities between the differentiation potential

[1] C. G. Bellows, J. E. Aubin, J. N. M. Heersche, and M. E. Antosz, *Calcified Tissue Int.* **38**, 143–154 (1986).

[2] J. N. Beresford, S. E. Graves, and C. A. Smoothy, *Am. J. Med. Genet.* **45**, 163–178 (1993).

0076-6879/2003 $35.00

of rodent and human MSCs and murine embryonic stem (ES) cells, we began to explore the ability of ES cells to form mineralized nodules *in vitro* under similar conditions. The large majority of *in vitro* differentiation protocols for murine ES cells first allow the ES cells to differentiate in the form of an embryoid body (EB). We utilize a fairly simple method for allowing EB formation and culture the EBs for only two days, after which individual cells are plated under conditions typically used for the *in vitro* differentiation of bone nodules from stem and progenitor cells from other sources.

Principle of Method

An ES cell is a nearly totipotential cell.[3] It is likely that the cell immediately preceding the osteoprogenitor is a cell with more restricted potential and increased responsiveness to specific differentiation stimuli. Several of the described methods for allowing EB formation start with single-cell suspensions of an ES cell culture. We experimented with such a method but became frustrated by the extremely variable quality of EBs, i.e., EBs formed large aggregates and adhered to the untreated plastic dish. We have settled on a method that takes advantage of the ease with which ES cell colonies can be flushed from the accompanying feeder layer. These undispersed colonies are then plated on bacterial grade plastic in the absence of LIF and feeders. This yields spherical EBs of relatively uniform size and results in minimal adherence to each other or the plate.

Materials and Reagents

We have used five independently derived stock murine ES cell lines as well as several genetically altered ES cell lines for these experiments. BK4 was subcloned by Beverly Koller, Ph.D from E14TG2a[4] (itself a 6-TG resistant sublcone of E14). HM1 is a rederivation of E14TG2a.[5] R1 is a hybrid 129 line.[6] W3-1A is an MRL/mpJ line,[7] and DBA-252, a DBA-1LacJ line.[8] All cell lines are maintained on irradiated murine

[3] A. G. Smith, *Annu. Rev. Cell. Dev. Biol.* **17**, 435–462 (2001).
[4] M. Hooper, K. Hardy, A. Handyside, S. Hunter, and M. Monk, *Nature* **326**, 292–295 (1987).
[5] T. M. Magin, J. McWhir, and D. W. Melton, *Nucl. Acids Res.* **20**, 3795 (1992).
[6] A. Nagy, J. Rossant, R. Nagy, W. Abramow Newerly, and J. C. Roder, *Proc. Natl. Acad. Sci. USA* **90**, 8424–8428 (1993).
[7] J. L. Goulet, C.-U. Wang, and B. H. Koller, *J. Immunol.* **159**, 4376–4381 (1997).
[8] M. L. Roach, J. L. Stock, R. Bynum, B. H. Koller, and J. D. McNeish, *Exp. Cell Res.* **221**, 520–525 (1995).

embryonic fibroblast feeder layers. Earlier passages of all of the lines have been demonstrated to be capable of germline transmission.

ES Cell Medium:	500 ml DMEM (Invitrogen 11995-065) 75 ml ES Cell Qualified FBS (Invitrogen 10439-024) 6 ml L-Glutamine (Invitrogen 25030-081) 4 μl 2-Mercaptoethanol (Sigma M-7522) ~1000 u/ml Leukemia Inhibitor Factor (LIF)
Trypsin Solution:	80 ml 0.25% Trypsin (Invitrogen 15050-065) 420 ml Solution B; single-use aliquots stored at −20°C
Solution B:	4 g NaCl, 0.2 g KCl, 0.175 g $NaHCO_3$, 0.5 g glucose 0.1 g EDTA (free acid) Add above to 475 ml ES cell quality water and stir overnight to fully dissolve; add NaOH to pH 7.0 Bring to 500 ml, filter through 0.22 μm filter, store at 4°C
EB Medium:	500 ml DMEM (Invitrogen 11995-065) 75 ml ES Cell Qualified FBS (Invitrogen 10439-024) 6 ml L-Glutamine (Invitrogen 25030-081) 4 μl 2-Mercaptoethanol (Sigma M-7522)
Osteoprogenitor Medium:	500 ml IMDM (Invitrogen 12440-053) 50 ml FBS (Invitrogen 10439-024) 6 ml L-Glutamine (Invitrogen 25030-081) 4 μl 2-Mercaptoethanol (Sigma M-7522) 0.6 ml 1000x Ascorbic Acid stock
1000x Ascorbic Acid	0.5 g Ascorbic Acid (Invitrogen 13080-023) in 10 ml ES cell quality water Filter through 0.22 μm filter, make fresh as needed.
100x β-Glycerophosphate	29.7 g β-Glycerophosphate Sigma (G-9891) 100 ml ES quality H_2O
Collagenase Solution	75 mg Collagenase Type II (Invitrogen 17101-015) per ml HBSS (Invitrogen 14175-087)
Collagenase Buffer	10 mM Sorbitol, 111.2 mM KCl, 1.3 mM $MgCl_2$ 13 mM Glucose, 21.3 mM Tris HCl pH 7.4 Filter sterilized and stored at 4°C

10x PBS (Invitrogen 70011-044) and 70 μm Cell Strainer (Falcon 2350).

The ES cell lines were all cultured similarly[9] except for the DBA-252 line that was cultured as described.[8] Typically cells are split 1 : 10–1 : 3 after trypsinization of a sub-confluent plate to a nearly single-cell suspension, every 2–3 days. The ES cell medium contains supernatants from Cos1 cells transiently transfected with a LIF expressing plasmid that were titrated using ESGRO (Chemicon International). Lots of fetal bovine serum are tested for performance in ES cell cultures and heated at 56°C for 30 min prior to freezing aliquots. Cells are used within two passages from the thaw. No antibiotics are used in any of the cultures. Care is taken that medium with ascorbic acid is used within 5 days.

Generation of Embryoid Bodies

Wash an ES cell-containing plate twice with 3 ml 1x PBS. Trypsinize cells with 2 ml trypsin solution and incubate 3–5 min at 37°C. Knock plate to release colonies and add 2 ml ES cell medium. Pipet to create a nearly single-cell suspension, bring volume to 10 ml and mix well. Plate cells densely (1 : 1–1 : 3) back onto the original plate with no additional feeders and incubate overnight at 37°C. After 24 hr remove medium and pipet off colonies with 12 ml of EB Medium and transfer to 2–4 100 mm petri dishes (untreated/bacterial grade). Plates can be washed again with EB medium and the additional wash divided among the same plates. Bring the volume in each 100 mm dish to 8 ml. Incubate 2 days at 37°C. (If the medium changes pH after the first 24 hr you should plate them more sparsely, they can be split after the first day but the EBs are fragile and will do better without this manipulation.)

Preparation of EB derived Cell Populations for Plating in the Osteoprogenitor Culture System

Pipet the EBs off the dish and then pipet the suspension into a 70 μm cell strainer over the original petri dish; discard flow through. Turn the cell strainer upside-down and pipet 10 ml Osteoprogenitor Medium over it onto a new petri dish. (You could pipet the EBs directly into a 50 ml conical here but putting them into a dish allows you to look at them under the microscope and sort them by mouth pipet if you desire.) Transfer EBs into a 50 ml conical tube and allow EBs to settle in the conical tube about 3–5 min and then carefully aspirate the medium off. Wash the EBs twice with 1x PBS. Add 100 μl trypsin solution (for approximately 200 EBs) and pipet until

[9] A. Mohn and B. H. Koller, *in* "DNA Cloning 4 Mammalian Systems" (D. M. Glover and B. D. Hames, eds.) Chapter 5. IRL Press, New York, 1995.

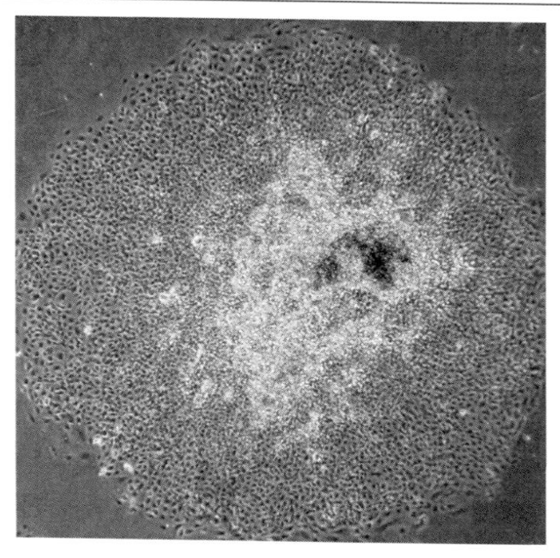

FIG. 1. 40X phase contrast image of osteoblast colony 7 days postplating. Outer portion of colony is still proliferating while center of colony is more differentiated and shows increased matrix secretion and increased height.

solution is cloudy and there are no visible EBs. Add an equal volume of Osteoprogenitor Medium and pipet to mix. Pull cells gently through an 18 gauge needle three times. Pipet cells through a 70 μm Cell Strainer into the 50 ml conical tube and discard material remaining in strainer. Bring volume to 5–20 ml with Osteoprogenitor Medium and pipet to mix. Count an aliquot and centrifuge the rest of the cells for 10 min at 1000 rpm. Calculate amount of Osteoprogenitor Medium in which to resuspend the cell pellet and plate at the desired concentration(s) in tissue culture plates. We typically plate 2×10^3 cells per 100 mm dish to allow accurate counting of CFUs at 21–28 days postplating. Cultures are maintained in a humidified 5% CO_2 tissue culture chamber at 37°C for 3–4 weeks.

Feed the plates with osteoprogenitor medium every 2–3 days. Add β-Glycerophosphate (βGP) to a 1x concentration after one week to aid mineralization (adding phosphate at time of initial plating seems to inhibit proliferation). Within 1 week there are colonies that have characteristic osteoblast morphology, see Fig. 1. Colonies are typically symmetrical with an outer proliferative zone and a central area showing increased matrix deposition.

Fixation and Staining of 3–4 Week Cultures

Medium is aspirated from the plate and the plate is washed with 3 ml cold 1x PBS two times. Add 5–10 ml 10% neutral buffered formalin (Fisher

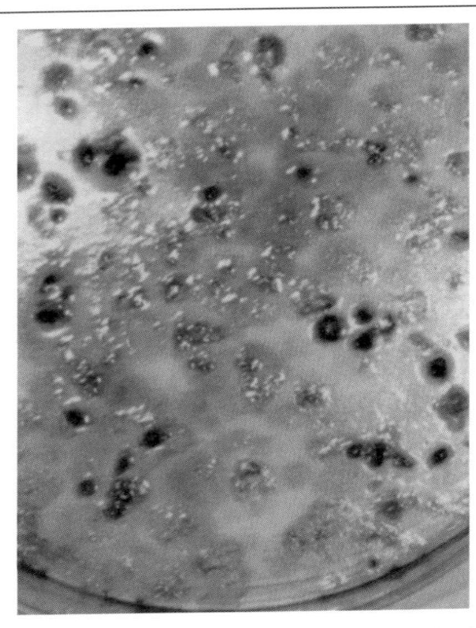

FIG. 2. Densely plated culture to demonstrate extent of mineralized nodule formation visible after 21 days. Stained using von Kossa method with toluidine blue counterstain. (See Color Insert.)

SF-100-4), and fix at room temperature for 2 hr. (Alternatively plates can be placed at 4°C indefinitely.)

For von Kossa stain and toluidine blue counterstain:

Rinse with deionized H_2O (dH_2O). Add 5–10 ml of 2.5% silver nitrate Sigma S-0139 in dH_2O and let plates sit 30 min at room temperature. (Make fresh each day and beware, it stains black!) Wash twice with dH_2O. Add 5–10 ml 0.1% toluidine blue O (Sigma T-0394) in dH_2O. Swirl for 10–15 sec then aspirate and rinse three times with dH_2O. Invert plates and air dry.

Figure 2 shows a relatively densely plated culture stained after 21 days. Typically cells are plated more sparsely for quantifying CFUs. With this method of EB preparation we observe 40–60 total CFUs on a plate originally seeded with 2×10^3 cells with approximately 40–60% of these colonies showing significant mineralization, for a CFU-O frequency in the range of 1 in 80 cells plated. It is of interest to note that the addition of exogenous dexamethasone from 10^{-12} to 10^{-8} M has no impact on the total number of CFUs or the percent of the CFUs that show mineralization. This is in contrast to what has been observed in bone marrow-derived osteoprogenitor cultures.[10] We have looked at two different parameters with respect to the EB population used to seed the osteoprogenitor cultures.

[10] A. Herbertson and J. E. Aubin, *J. Bone Miner. Res.* **10**, 285–294 (1995).

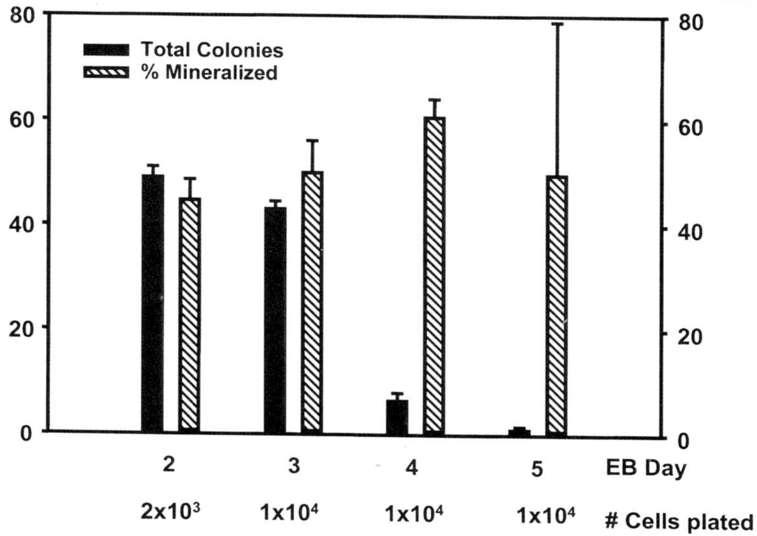

FIG. 3. Total colony formation and percent of colonies mineralized for 21 day HM1 EB-derived cultures. Note the five-fold difference in the number of cells plated between Day 2 and Days 3–5.

One parameter is the number of cells per EB. We have tested populations of EBs that range in size from 150 to 2500 cells and find that the CFU-O frequency is not significantly different. Given that our CFU-O frequency is on the order of 1 in 80 cells, an individual EB could contain anywhere from 1 to 25 CFU-Os. Another parameter is the number of days the EBs are cultured prior to disruption and plating of cells in the osteoprogenitor culture. Day 2 yields the highest number of colonies with osteoblast morphology at one week, and as shown in Fig. 3, Day 2 also yields the greatest number of CFUs. In one experiment where we included ES cells and Day 1 EBs we saw very few colonies with osteoblast morphology on these plates but were able to observe mineralization of confluent monolayers. We conclude that the time spent in the EB form has a greater impact on the CFU-O frequency than the actual number of cells per EB.

Figure 4 is representative of several experiments performed using various stock ES cell lines as starting material for EB cultures. The percentage of CFUs forming mineralized nodules is typically the same for all the lines, i.e., 40–60%. While not included in the experiment in Figs. 4 and 5, we observe significantly reduced overall colony numbers, as much as two-fold fewer CFUs, using the DBA and MRL lines.

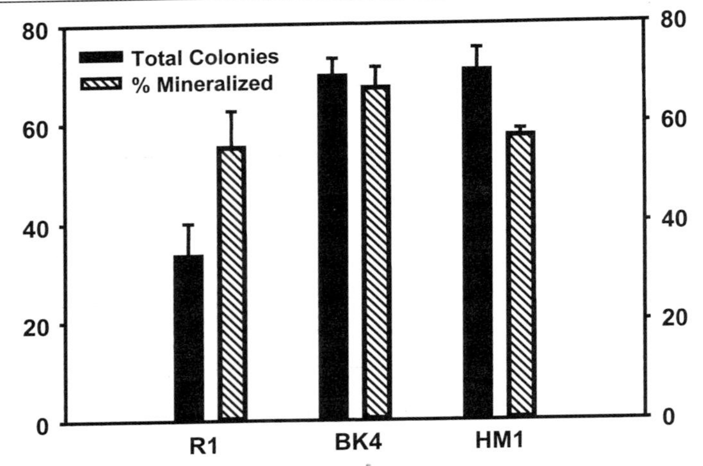

FIG. 4. Total colony formation and percent of colonies mineralized for 28 day EB-derived cultures from three different stock ES cell lines.

Characterization of Gene Expression in Osteoprogenitor Cultures

In addition to colony morphology and mineralization, we have also assessed gene expression by real time PCR analysis of RNA harvested from whole plates. Despite the fact that the starting population is a mix of colony types with only approximately 50% of the colonies typically progressing to mineralized nodules, we are able to see gene expression characteristic of the osteoblast lineage. The analyzed material in Fig. 5 was taken from plates identical to those fixed and stained in Fig. 4, and is representative of several similar experiments. In similar but independent experiments comparing gene expression by real time PCR in day 8 versus day 28 cultures we observed increases at 28 days of 10-fold for Cbfa1, 17-fold for Type I collagen, and 35-fold for osteocalcin.

Concluding Remarks

It is worth noting that there appear to be differences between the nodule formation from our ES cell-derived method and our marrow-derived cultures. Marrow-derived cultures require exogenous glucocorticoid for nodule formation and are delayed approximately a week in comparison to ES cell-derived cultures with regard to characteristic osteoblast differentiation and nodule formation and mineralization. This is supported by unpublished experiments with murine marrow-derived cultures in our laboratory utilizing the same Osteoprogenitor Medium, and thus is unlikely to be due to high levels of glucocorticoid in the serum. Earlier observations comparing the performance of calvaria-derived

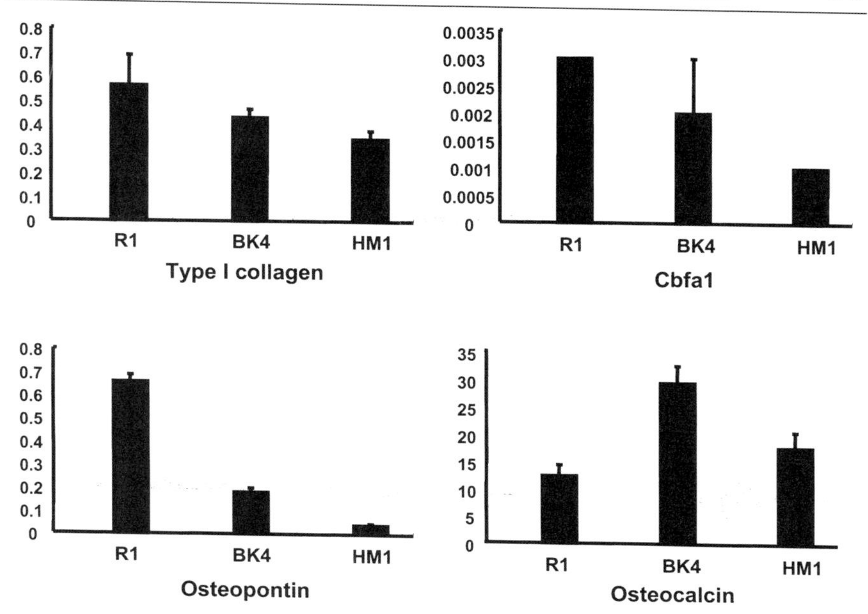

FIG. 5. Gene expression quantified by real time PCR for 28 day EB-derived cultures from the same experiment shown in Fig. 4. Cultures were harvested with collagenase [To PBS washed 100 mm dish add 2 ml Collagenase Buffer and 40 μl Collagenase Solution and incubate at 37°C for 15 min. Pipet to remove material, add 10 ml Osteoprogenitor Medium and wash once with PBS] prior to preparation of RNA using TriReagent (Molecular Research Center TR-118) according to the manufacturer's directions. The following primer sequences were used for the real time PCR amplification: Cbfa1Primer 1 TGC TTC ATT CGC CTC ACA AA, Primer 2 TGC TGT CCT CCT GGA GAA AGT T, Probe AAC CAC AGA ACC ACA AGT GCG GTG C; Type I Collagen Primer 1 CTG CCG ATG TCG CTA TCC A, Primer 2 AGG TGA TGT TCT GGG AGG CC, Probe TCC TGC GCC TGA TGT CCA CCG, Osteocalcin primer 1 CGG CCC TGA GTC TGA CAA A, Primer 2 GCC GGA GTC TGT TCA CTA CTA CCT T, Probe CCT TCA TGT CCA AGC AGG AGG GCA; Osteopontin Primer 1 GGC ATT GCC TCC TCC CTC, Primer 2 TGC AGG CTG TAA AGC TTC TCC, Probe CGG TGA AAG TGA CTG ATT CTG GCA GCT C. All assays were normalized to 18S rRNA.

cultures and marrow-derived cultures in rodents suggested that differences between these two culture systems reflected the relatively earlier and later progenitors found in marrow and calvaria, respectively.[11] Clearly the ES cell-derived progenitors are an example of yet a third type of progenitor. Differences in the presence or absence of competing lineages from the three progenitor sources may alter the CFU-O frequency in ways we do not yet understand. Two other groups have published related

[11] J. E. Aubin and J. N. M. Heershe, *Drug Dev. Res.* **49**, 206–215 (2000).

work on the ability of murine ES cells to differentiate along this pathway. Buttery *et al.*[12] used cells from day 5 EBs as a starting progenitor population, but they plate these cells over 1000-fold more densely than we describe here. They also report a significant increase in mineralized nodules in the presence of an equal number of fetal osteoblasts or exogenous dexamethasone or retinoic acid. In contrast, we are able to plate progenitors sparsely enough to count isolated colonies, and see no effect of exogenous dexamethasone. Phillips *et al.*[13] used a hanging drop method for EB generation and then further cultured EBs in suspension in the presence of retinoic acid for 3 days, for a total of 5 days EB culture. At this point, EBs were plated without disruption in the presence of the statin, compactin, or BMP2, one or the other of which was necessary for mineralized nodule formation. It is not clear why we are able to see a high frequency of mineralized nodules in the absence of the exogenous factors. One possibility, as mentioned above, is that Buttery *et al.*,[12] by plating at an increased density, and Phillips *et al.*,[13] by allowing the EBs to remain intact, are actually preserving competing lineages. For example, EBs are known to contain progenitors capable of differentiating into numerous other lineages. These other lineages may either consume medium components necessary for osteoblast development or may produce factors that are inhibitory to osteoblast development. Despite the differences in the methodology, it is clear that there is a progenitor population in the EB that is capable of becoming an osteoprogenitor and forming a mineralized nodule *in vitro*. This is extremely important because ES cell-derived osteoblasts and their progenitors show excellent therapeutic promise for osteoporosis, rheumatic diseases, developmental bone defects, and severe fractures.[14–16] Also, *in vitro* differentiation of ES cells along the osteoblast lineage is a valuable tool for addressing pertinent questions about proliferation, differentiation, survival, and intercellular communication between cells of the bone lineage *in vitro*. And as with any ES cell *in vitro* differentiation system, it provides a mechanism to screen various genetic alterations prior to transferring them to the germline and determining the impact of the alteration *in vivo*.

[12] L. D. K. Buttery *et al.*, *Tissue Eng.* **7**, 89–99 (2001).
[13] B. W. Phillips, N. Belmonte, C. Vernochet, G. Ailhaud, and C. Dani, *Biochem. Biophys. Res. Commun.* **284**, 478–484 (2001).
[14] D. J. Prockop, *Science* **276**, 71–74 (1997).
[15] S. P. Bruder, D. J. Fink, and A. I. Caplan, *J. Cell. Biochem.* **56**, 283–294 (1994).
[16] R. F. Pereira *et al.*, *Proc. Natl. Acad. Sci. USA* **95**, 1142–1147 (1998).

Acknowledgment

The author thanks Dr. David Melton at the University of Edinburgh for the HM1 ES cell line, Drs. Janet Rossant and Andras Nagy at the University of Toronto for the R1 ES cell line, Dr. John McNeish and Pfizer for the DBA-252 ES cell line, and Dr. Beverly Koller at the University of North Carolina at Chapel Hill for the W3-1A ES cell line. The author thanks Drs. H. Donahue and Z. Li for helpful discussions and assistance in studying gene expression; current and previous members of her laboratory: Miguel Barthelery, Kimberly Dunham, Jason Heaney, Nicole Renteria, and Karyn Sheaffer for their contributions; and NIH/NIAMS AR47365 for support.

[18] *In Vitro* Differentiation of Mouse ES Cells: Bone and Cartilage

By JAN KRAMER, CLAUDIA HEGERT, and JÜRGEN ROHWEDEL

Introduction

The pluripotency of mouse embryonic stem (ES) cells is defined by their origin, the inner cell mass (ICM) of the embryonic blastocyst.[1,2] Although the presence of these pluripotent cells is transient *in vivo*, permanent ES cell lines with these properties can be obtained *in vitro* by cocultivation of the ICM cells with embryonic fibroblasts or in the presence of leukemia inhibitory factor (LIF).[3,4] Generation of mouse germline chimeras by injection of ES cells into blastocysts demonstrates that these cells are pluripotent. ES cells also spontaneously differentiate *in vitro* into cell types of all three germ layers. *In vitro* differentiation of ES cells which closely recapitulates embryonic cell differentiation processes[5–8] has been used as a model system to analyse cell differentiation. This is particularly an attractive alternative way to investigate the consequences of loss of function mutations after gene targeting if knock-out mice cannot be generated. Because of their broad differentiation capacity ES cells are also discussed as an experimental approach to generate cells for transplantation. Cardiomyocytes, neuronal

[1] G. R. Martin, *Proc. Natl. Acad. Sci. USA* **78**, 7634 (1981).

[2] M. J. Evans and M. H. Kaufman, *Nature* **292**, 154 (1981).

[3] A. G. Smith, J. K. Heath, D. D. Donaldson, G. G. Wong, J. Moreau, M. Stahl, and D. Rogers, *Nature* **336**, 688 (1988).

[4] R. L. Williams, D. J. Hilton, S. Pease, T. A. Willson, C. L. Stewart, D. P. Gearing, E. F. Wagner, D. Metcalf, N. A. Nicola, and N. M. Gough, *Nature* **336**, 684 (1988).

[5] G. M. Keller, *Curr. Opin. Cell Biol.* **7**, 862 (1995).

[6] P. D. Rathjen, J. Lake, L. M. Whyatt, M. D. Bettess, and J. Rathjen, *Reprod. Fertil. Dev.* **10**, 31 (1998).

[7] K. Guan, J. Rohwedel, and A. M. Wobus, *Cytotechnology* **30**, 211 (1999).

[8] J. Rohwedel, K. Guan, C. Hegert, and A. M. Wobus, *Toxicol. In Vitro* **15**, 741 (2001).

0076-6879/2003 $35.00

cells and insulin-secreting cells derived from ES cells *in vitro* have been transplanted into animals and shown to integrate into host tissues.[9–14] A common method to differentiate ES cells *in vitro* is their cultivation as cell aggregates, so-called embryoid bodies (EBs). We studied differentiation of ES cells via EBs into chondrogenic and osteogenic cell types *in vitro* and found this to be an appropriate model to analyse the process of chondrogenesis and endochondral ossification.[15,16] In contrast to other mesodermal cell types derived from ES cells such as cardiac or skeletal muscle cells, the efficiency to generate chondrogenic and osteogenic cells was comparatively low. However it could be stimulated by growth factors of the TGF-β-family.[15] The efficiency of ES cell-derived chondrocyte differentiation was found to be influenced by several parameters such as the batch of fetal calf serum used for cultivation or the size of EBs. For a reproducible pattern of chondrogenic and osteogenic differentiation it was an important prerequisite to generate EBs of the same size. This was achieved by differentiation of ES cells via hanging drop-cultivation. Here we describe some basic techniques of ES cell cultivation, the methods used for chondrogenic and osteogenic differentiation of ES cells, for characterization of the specific cell types and for isolation of chondrocytes from EBs. Furthermore, we briefly summarize approaches to enhance the differentiation efficiency.

Cultivation and Differentiation of ES Cells *In Vitro*

Material, Media, Additiva, and Buffer Solutions

Material

- 24-well micro-well plates, 2-chamber Lab Tek chamber slides (21.3×20 mm/well), 60 and 100 mm tissue culture dishes

[9] M. G. Klug, M. H. Soonpaa, G. Y. Koh, and L. J. Field, *J. Clin. Invest* **98**, 216 (1996).

[10] O. Brüstle, K. N. Jones, R. D. Learish, K. Karram, K. Choudhary, O. D. Wiestler, I. D. Duncan, and R. D. McKay, *Science* **285**, 754 (1999).

[11] J. W. McDonald, X. Z. Liu, Y. Qu, S. Liu, S. K. Mickey, D. Turetsky, D. I. Gottlieb, and D. W. Choi, *Nat. Med.* **5**, 1410 (1999).

[12] B. Soria, E. Roche, G. Berna, T. Leon-Quinto, J. A. Reig, and F. Martin, *Diabetes* **49**, 157 (2000).

[13] S. Liu, Y. Qu, T. J. Stewart, M. J. Howard, S. Chakrabortty, T. F. Holekamp, and J. W. McDonald, *Proc. Natl. Acad. Sci. USA* **97**, 6126 (2000).

[14] S. Arnhold, D. Lenartz, K. Kruttwig, F. J. Klinz, E. Kolossov, J. Hescheler, V. Sturm, C. Andressen, and K. Addicks, *J. Neurosurg.* **93**, 1026 (2000).

[15] J. Kramer, C. Hegert, K. Guan, A. M. Wobus, P. K. Müller, and J. Rohwedel, *Mech. Dev.* **92**, 193 (2000).

[16] C. Hegert, J. Kramer, G. Hargus, J. Müller, K. Guan, A. M. Wobus, P. K. Müller, and J. Rohwedel, *J. Cell Sci.* **115**, 4617 (2002).

(Nunc, Wiesbaden, FRG) were used after coating with gelatine (see below).
- 100 mm bacteriological petri dishes (Greiner, Frickenhausen, FRG), uncoated.
- 15 and 50 ml centrifugation tubes and 2 ml cryoconservation vials (Nunc).
- Sterile dissection instruments (scissors, forceps and a sieve with a pore diameter of about 0.5 mm).
- A sterile 100 ml Erlenmeyer flask with a stirring bar and approximately 50 glass pearls with a diameter of about 4 mm.
- Sterile 0.2 μm micro-filters (Sartorius, Göttingen, FRG).
- 2, 5, 10, and 20 ml pipettes and a micropipettor for sterile 100 μl filter tips (Eppendorf, Cologne, FRG).

Media and additiva

Selected batches of fetal calf serum (FCS; Sigma, Taufenkirchen, FRG) were used to supplement the medium. FCS was inactivated by heating to 54°C for 30 min. To select an appropriate FCS batch, ES cells were differentiated via EBs (see below) using media supplemented with different FCS batches and analyzed for their chondrogenic differentiation efficiency by Alcian blue-staining (see below). The basic medium was Dulbecco's modified Eagle's medium (DMEM; Life Technologies, Eggenstein, FRG).

- Additiva:
  - 2 mM L-glutamine (Life Technologies; 1 ml per 100 ml medium)
  - 5×10^{-5} M β-mercaptoethanol (Serva, Heidelberg, FRG; 1 ml β-ME-stock solution (see below) per 100 ml medium)
  - Non-essential amino acids (Life Technologies; 1 ml per 100 ml medium)
  - Penicillin and streptomycin (Life Technologies; 1 ml per 100 ml medium)

- Media:
  - Medium 1 for cultivation of embryonic fibroblasts: DMEM supplemented with 15% FCS and additiva
  - Medium 2 for ES cell cultivation: DMEM supplemented with 15% FCS, additiva and LIF (5 ng/ml; Life Technologies)
  - Medium 3 for cryoconservation: DMEM supplemented with 20% FCS and 8% dimethyl sulfoxide (DMSO; Sigma)
  - Medium 4 for ES cell differentiation: DMEM supplemented with 20% FCS and additiva

Buffer solutions

- Phosphate buffered saline (PBS) — g/l *Aqua tridest*

	g/l *Aqua tridest*
NaCl (Merck, Darmstadt, FRG)	10.00
KCl (Merck)	0.25
$Na_2HPO_4 \times H_2O$ (Fluka, Deisenhofen, FRG)	1.44
KH_2PO_4 (Merck)	0.25

  Adjust to pH 7.2 and sterilize by filtration.
- 0.2% trypsin (Biochrom, Berlin, FRG) in PBS (trypsin solution). 0.02% EDTA (Merck) in PBS (EDTA-solution). A 1 : 1-mixture of trypsin and EDTA-solutions was used to trypsinize cells.
- For gelatine-coating of tissue culture plastic a 1% gelatine (Fluka) solution in PBS was prepared as stock solution and autoclaved. Tissue culture dishes, micro-well plates and chamber slides were treated with a 0.1% gelatine solution (stock solution diluted 1 : 10 with PBS) one day before cell or EB-plating. The solution was left on the plates overnight at 4°C and sucked off immediately before use.
- 2 mg mitomycin C (MMC; Serva, Heidelberg, FRG) in 10 ml PBS (MMC-stock solution) were sterilized by micro-filtration. MMC-stock (300 μl) solution was diluted in 6 ml of medium 1 to obtain a MMC-working solution.
- 7 μl of β-mercaptoethanol (β-ME; Serva) in 10 ml PBS (β-ME-stock solution) were sterilized by micro-filtration.

ES cell lines

ES cells of lines BLC6,[17] D3,[18] E14[19] or R1[20] were used.

Procedures

All cultivations were done in a CO_2-incubator at 37°C in a humidified 5% CO_2 atmosphere.

Preparation and cultivation of embryonic fibroblasts

Embryonic fibroblasts used as feeder layer cells for ES cell cultivation were prepared and cultivated according to an established method.[21]

[17] A. M. Wobus, R. Grosse, and J. Schöneich, *Biomed. Biochim. Acta* **47**, 965 (1988).

[18] T. C. Doetschman, H. Eistetter, M. Katz, W. Schmidt, and R. Kemler, *J. Embryol. Exp. Morphol.* **87**, 27 (1985).

[19] M. Hooper, K. Hardy, A. Handyside, S. Hunter, and M. Monk, *Nature* **326**, 292 (1987).

[20] A. Nagy, J. Rossant, R. Nagy, W. Abramow-Newerly, and J. C. Roder, *Proc. Natl. Acad. Sci. USA* **90**, 8424 (1993).

[21] A. M. Wobus, H. Holzhausen, P. Jäkel, and J. Schöneich, *Exp. Cell Res.* **152**, 212 (1984).

Embryos at day 14–16 p.c. were removed from pregnant mice and rinsed under sterile conditions in PBS. The placenta and fetal membranes were cut off, head and liver were removed and the remaining carcass was rinsed first in PBS and second in trypsin solution. Finally, the embryonic tissue was minced in 5 ml fresh trypsin solution. These 5 ml were transferred to a sterile Erlenmeyer flask with glass pearls and stirred for 25 to 45 min. The resulting cell suspension was filtered through a sterile sieve and 10 ml medium 1 were added. The cells were spinned down, the pellet resuspended in 3 ml medium 1 and plated onto gelatine-coated 100-mm tissue culture dishes filled with 10 ml medium 1 (approximately 2×10^6 cells, isolated from about 2 embryos were plated per dish). The feeder layer cells were cultivated for 1–2 days and either trypsinized and frozen in medium 3 (1 dish per freezing vial) or split 1 : 3 onto gelatine-coated 100 mm-tissue culture dishes and cultivated for one more day in medium 1 for further use. Mitomycin C was used for growth-inactivation of the subcultured or frozen feeder cells. Frozen cells were thawed rapidly, plated onto two gelatine-coated 100-mm tissue culture dishes and cultivated overnight. For inactivation, cells were incubated with MMC-working solution for about $2\frac{1}{2}$ hr at 37°C in a CO_2-incubator. After aspirating the MMC-working solution the cells were washed thrice with PBS followed by trypsinization. Finally, the cells were plated onto gelatine-coated 60-mm culture dishes with 4 ml medium 1. A confluent monolayer was obtained by using about 1.5×10^5 cells per 60 mm-culture dish. The growth-inactivated feeder layer can be used up to 4 days after preparation. Optimal for cocultivation with ES cells are 1 day old feeder layer cells.

Cocultivation of ES Cells with Growth-Inactivated Feeder Layer Cells

ES cells are maintained in the undifferentiated state by cultivation on feeder layer cells (Fig. 1A) and under the influence of LIF. Two hours before the ES cells were thawed, medium 1 on the feeder layer was replaced by medium 2. The vial with the ES cells was thawed rapidly at 37°C. The cold cell suspension was diluted 1 : 10 with medium 2, spinned down, resuspended in medium 2 and transferred onto the prepared feeder layer. It is critical to subculture the ES cells carefully for maintaining the cells in the pluripotent state. Therefore ES cells should be subcultivated every 24 (max. 48) hr. One to two hours before subcultivation medium 1 on the feeder layer was replaced by medium 2. For subcultivation medium 2 was removed from the ES cells and cells were washed twice with PBS. Two milliliter trypsin/EDTA-solution were added to the ES cells and incubated for 30–60 sec at room temperature. The cells were suspended in trypsin/EDTA-solution, transferred to a centrifugation tube with 10 ml medium 2 and spinned

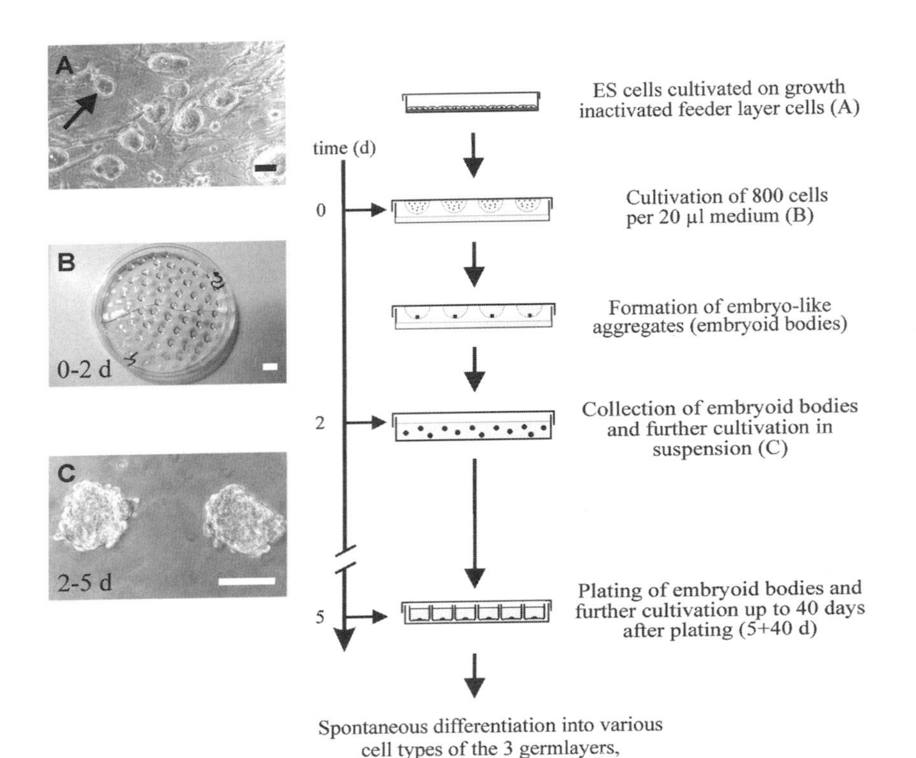

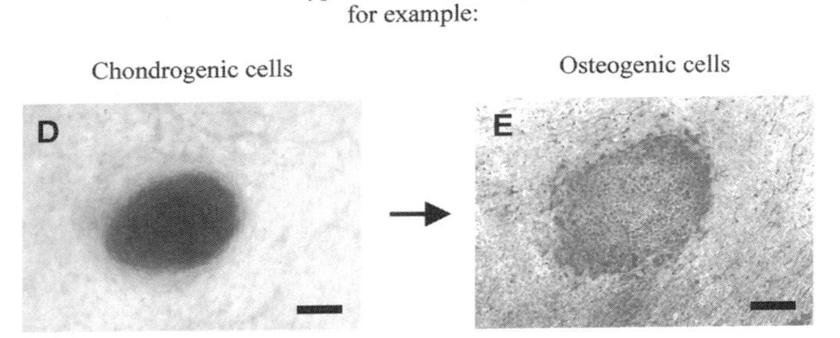

FIG. 1. Cultivation and differentiation of ES cells *in vitro*. To keep ES cells undifferentiated and pluripotent they were cultivated in the presence of leukemia inhibitory factor (LIF) and cocultivated with growth-inactivated embryonic fibroblasts used as feeder layer (A; arrow = ES cell-colony). After trypsinization ES cells were cultivated as hanging drops (B) in differentiation medium without LIF and formed cell aggregates. These aggregates, called embryoid bodies (EBs), were collected after 2 days of hanging drop-cultivation (0–2 d) and transferred into suspension culture (C). Cultivation in suspension was performed for 3 days (2–5 d). On the fifth day of differentiation (5 d) EBs were plated onto gelatine-coated dishes. During further differentiation up to 40 days (5 + 40 d) various cell types of all the three germlayers could be detected in the EB outgrowths. For example, cartilage nodules were detected by Alcian blue-staining around 10 days after plating of EBs derived from the ES cell line BLC6 (D). At a very late stage of ES cell differentiation around 30 days after plating these nodules disintegrate and the cells express alkaline phosphatase (E) and other marker for osteogenic cells (see also Fig. 2). Bars = 50 μm (A), 1000 μm (B), 100 μm (C, D, E). (See Color Insert.)

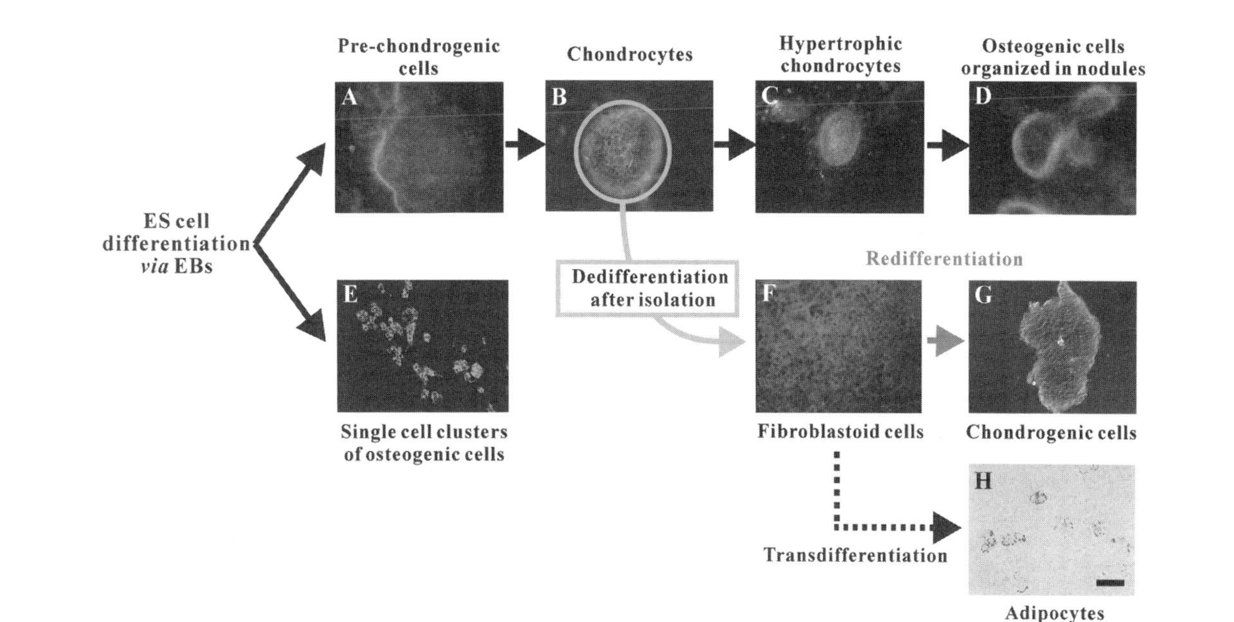

FIG. 2. ES cell-derived chondrogenic and osteogenic differentiation *in vitro*. Differentiation of ES cells *in vitro* via EBs into chondrogenic and osteogenic cells recapitulates chondrogenic and osteogenic differentiation processes. Prechondrocytic cells expressing the cartilage-associated transcription factor scleraxis were detected in EB outgrowths (A) as demonstrated by mRNA *in situ* hybridization. These cells later formed areas of high cell density, called nodules, and expressed the cartilage matrix protein collagen II (B) as demonstrated by mRNA *in situ* hybridization for scleraxis (green) coupled with immunostaining for collagen II (red). These chondrocytes progressively developed into hypertrophic, collagen X-expressing cells as revealed by mRNA *in situ* hybridization (C). Finally, immunostaining showed that chondrocytes localized in nodules expressed osteogenic proteins, such as osteopontin (D). Furthermore, direct differentiation of ES cells into single cell clusters of osteopontin-positive cells, bypassing the chondrocytic stage, was observed (E). Chondrocytes isolated from EBs showed a distinct differentiation plasticity. The isolated cells initially dedifferentiated into collagen I-positive fibroblastoid cells (F). These cells redifferentiated in culture into chondrocytes reexpressing collagen II (green) and a reduced amount of collagen I (red) (G). However, occasionally other mesenchymal cell types such as adipocytes appeared in these cultures as demonstrated by Sudan III-staining for adipogenic cells (H). Bar = 100 μm. (See Color Insert.)

down. The cell pellet was resuspended in medium 2 using a 2 ml glass pipette. Finally, the single cell suspension was transferred 1 : 3 onto 60-mm tissue culture dishes with MMC-inactivated feeder layer.

For freezing the undifferentiated ES cells, cells were trypsinized as described and after resuspension in medium 2, centrifuged and resuspended in 1.5 ml medium 3. The single cell suspension was transferred to a freezing vial and slowly frozen in a styropor box at −80°C over night. During the next days the vials were transferred into liquid nitrogen.

In Vitro *Differentiation of ES Cells via Embryoid Bodies (EBs)*

Pluripotent ES cells are able to differentiate spontaneously into cells of all three primary germ layers. Several protocols for differentiation of ES cells into various cell types were established. For chondrogenic and osteogenic differentiation we performed ES cell cultivation via EBs generated by hanging drops. The major steps (Fig. 1) are: (a) generation of EBs by hanging drop-cultivation of ES cells for 2 days (0–2 d), (b) cultivation of the EBs in suspension for 3 days (2–5 d) and (c) plating of the EBs at the fifth day of differentiation (5 d) followed by further cultivation up to 40 days (5 + 40 d).

Hanging drop-cultivation of ES cells (0–2 d; Fig. 1B). Undifferentiated ES cells cocultivated with feeder layer in medium 2 were trypsinized and the cell suspension was centrifuged in medium 4. After resuspension in 2 ml medium 4 the cell number was determined. For differentiation into chondrogenic and osteogenic cells we used 800 ES cells per EB. Therefore, cells were diluted in medium 4 to obtain a cell suspension with 4×10^4 cells/ml. Twenty microliter aliquots of the cell suspension were placed on the inner side of a 100-mm bacteriological petri dish-lid using a micropipette with 100 μl filter tips. The dish was filled with 10 ml PBS to avoid evaporation of the medium. The hanging drops were cultivated for 2 days (0–2 d), initiating spontaneous differentiation.

Cultivation of the EBs in suspension (2–5 d; Fig. 1C). On the second day of hanging drop-culture the drops with the EBs were collected with 2 ml medium 4 using a 5-ml glass pipette and transferred into a 100-mm bacteriological petri dish containing 8 ml medium 4. Two lids with about 50 drops per lid were collected in one dish. EBs were further cultivated in suspension for 3 days (2–5 d). It is important to select a batch of bacteriological petri dishes which enables suspension culture. In some batches the EBs attach. If this happens it is sometimes possible to resuspend the EBs again by carefully rinsing the dishes with medium 4.

Plating of the EBs (5 d). On the fifth day of differentiation (5 d) EBs were plated onto gelatine-coated 60-mm tissue culture dishes (for Alcian blue-staining and RT-PCR), chamber slides (for immunostaining, *in situ* hybridization and alkaline phosphatase-staining) and micro-well plates (for morphological analysis). After adding medium 4 (4 ml per 60-mm dish, 1.5 ml per chamber and 1 ml per well) EBs were transferred using a micropipette with sterile 100-μl filter tips. 10–15 EBs per 60 mm-dish, 5 EBs per chamber and 1 EB per well were plated. The medium was initially changed after 4 and 8 days. During further cultivation up to 40 days after plating (5 + 40 d) it was renewed every second day. To study the differentiation processes samples were analyzed every second day. Spontaneous differentiation of chondrogenic and osteogenic cell types can be monitored by Alcian blue-staining (see below) during early (Fig. 1D) and Alkaline phosphatase-staining (see below) during later stages (Fig. 1E) of EB cultivation.

Characterization of ES Cell-derived Chondrogenic and Osteogenic Differentiation

The model system of ES cell differentiation recapitulates the complete process of chondrogenic differentiation from undifferentiated, pluripotent cells up to terminally differentiated chondrogenic and osteogenic cells[15,16] (Fig. 2). In EB outgrowths we found prechondrogenic cells expressing the transcription factor scleraxis (Fig. 2A), which later formed a distinct condensed group of cells so-called nodules. These nodules stained intensively with Alcian blue (see Fig. 1D) and contained collagenous matrix as confirmed by immunostaining for collagen II (Fig. 2B). Cells in these nodules also produced cartilage oligomeric matrix protein (COMP) as demonstrated by immunostaining.[15] Furthermore, semiquantitative RT-PCR analysis demonstrated that genes encoding transcription factors involved in mesenchymal differentiation, such as Pax-1, scleraxis and Sox-9 as well as extracellular matrix proteins of cartilage tissue, such as aggrecan and collagen II were expressed during ES cell differentiation.[15] ES cell-derived chondrogenic cells progressively developed into hypertrophic and calcifying cells.[16] The composition of the extracellular matrix in the nodules changed during differentiation, indicated by a loss of Alcian blue-staining. This was confirmed by immunostaining. At late differentiation stages the cells localized in nodules expressed proteins such as collagen X (Fig. 2C), osteopontin (Fig. 2D) and bone sialoprotein[16] characteristic for hypertrophic and calcifying chondrocytes. Finally, at a terminal differentiation stage the nodules stained positive for AP (see Fig. 1E). In addition, expression of genes such as collagen X, osteocalcin and cbfa-1 was detected by RT-PCR. The expression pattern of collagen marker genes during the process of endochondral ossification *in vivo* was nicely recapitulated during EB cultivation.[16] The osteogenic cells appeared either, as described above, by transdifferentiation from hypertrophic chondrocytes organized in nodules or directly from osteoblast precursor cells.[16] This direct differentiation of ES cells into osteoblastic and osteogenic cells, bypassing the chondrocytic stage, was indicated by a population of osteogenic cells differentiating as single cell clusters (Fig. 2E).

Detection of Gene Expression by Semiquantitative RT-PCR Analysis

To analyze the expression of cartilage-associated genes during EB cultivation samples of 10–15 EBs plated per 60 mm-tissue culture dish were collected at different stages of differentiation and washed twice with PBS. Total RNA was isolated using the RNeasy Mini Kit (Qiagen, Hilden, FRG). The RNA concentrations were determined by measuring the absorbance at 260 nm. RNA samples were adjusted to 50 ng/μl and reverse transcribed using oligo-dT primer and Superscript II reverse transcriptase (Life Technologies) as follows:

10 μl	RNA (500 ng)
1 μl	oligo-dT primer
1 μl	water

Incubate for 10 min at 70°C, add the following:

4 μl	first strand buffer
2 μl	0.1 *M* DTT
1 μl	10 m*M* dNTPs (dATP, dCTP, dGTP, dTTP each at a concentration of 10 m*M*)

Incubate for 2 min at 42°C, add 1 ml reverse transcriptase, incubate for 50 min at 42°C. Terminate the reaction by incubation for 5 min at 95°C. Aliquots of 1 μl from RT-reactions were used for amplification of transcripts using primer-specific for the analyzed genes and Taq DNA polymerase (Roche, Mannheim, FRG) according to the manufacturer's instructions. RT-reactions were denatured for 2 min at 95°C, followed 34–45 cycles of 40 sec denaturation at 95°C, 40 sec annealing at the primer specific temperature (Table I) and 50 sec elongation at 72°C. Expression of several marker genes was studied using sequence-specific primer (Table I).

RNA from limb buds or limbs of 10 and 16 day p.c. old mouse embryos were used as positive controls. Distilled water was always included as a negative control. Electrophoretic separation of PCR products was carried out in 2% agarose gels. The fragments were analyzed by computer-assisted densitometry in relation to HPRT or β-tubulin gene expression. To obtain

TABLE I

OLIGONUCLEOTIDE SEQUENCES OF PRIMERS USED TO STUDY CHONDROGENIC AND OSTEOGENIC GENE EXPRESSION DURING ES CELL DIFFERENTIATION *IN VITRO* BY RT-PCR

Gene	Antisense-primer	Sense-primer	Ann. temp.	Fragment length
Scleraxis[22]	5′-GTGGACCCTCCTCCTTCTAATTCG-3′	5′-GACCGCACCAACAGCGTGAA-3′	63°C	375 bp
Pax-1[23]	5′-TTCTCGGTGTTTGAAGGTCATTGCCG-3′	5′-GATGGAAGACTGGGCGGGTGTGAA-3′	60°C	318 bp
Sox-9[24]	5′-TCTTTCTTGTGCTGCACGCGC-3′	5′-TGGCAGACCAGTACCCGCATCT-3′	57°C	135 bp
Cbfa-1[25]	5′-ATCCATCCACTCCACCACGC-3′	5′-AAGGGTCCACTCTGGCTTTGG-3′	63°C	371 bp
Aggrecan[26]	5′-TCCTCTCCGGTGGCAAAGAAGTTG-3′	5′-CCAAGTTCCAGGGTCACTGTTACCG-3′	60°C	270 bp
Collagen II[27]	5′-AGGGGTACCAGGTTCTCCATC-3′	5′-CTGCTCATCGCCGCGGTCCTA-3′	60°C	432 bp (A) 225 bp (B)
Collagen X[28]	5′-ATGCCTTGTTCTCCTCTTACTGGA-3′	5′-CTTTCTGCTGCTAATGTTCTTGACC-3′	61°C	164 bp
Osteocalcin[29]	5′-ATGCTACTGGACGCTGGAGGGT-3′	5′-GCGGTCTTCAAGCCATACTGGTC-3	64°C	330 bp
β-tubulin[30]	5′-GGAACATAGCCGTAAACTGC-3′	5′-TCACTGTGCCTGAACTTACC-3′	54°C	317 bp
HPRT[31]	5′-GCCTGTATCCAACACTTCG-3′	5′-AGCGTCGTGATTAGCGATG-3′	63°C	507 bp

The expression of cartilage- and bone-associated marker genes, listed together with a reference for the gene sequence, was studied by RT-PCR using sequence-specific antisense- and sense-primer at different annealing temperatures (ann. temp.). The lengths of the amplified fragments are given. Primer used to detect collagen II expression amplified two fragments representing a juvenile (A) and adult (B) splice variant.

TABLE II

ANTIBODIES USED FOR IMMUNOSTAINING TO DETECT CHONDROGENIC AND OSTEOGENIC CELL TYPES DURING ES CELL DIFFERENTIATION *IN VITRO*

Protein	Antibody designation	Dilution	Antibody type
Collagen II[32]	II-II6B3	1:20	mAb
Osteopontin[33]	MPIIIB10$_1$	1:10	mAb
Collagen X[34]	X-AC9	1:20	mAb
Bone sialoprotein I and II[33]	WVID1(9C5)	1:10	mAb
Cartilage oligomeric matrix protein (COMP)[35]	–	1:20	pAs
Collagen I	–	1:100	pAs

Monoclonal antibodies (mAbs) or polyclonal antisera (pAs) against several cartilage and bone marker proteins were used in different dilutions in PBS for indirect immunostaining of EBs. Monoclonal antibodies were obtained from the Developmental Studies Hybridoma Bank, University of Iowa, USA; antiserum against COMP was a gift from M. Paulsson (University of Cologne); an antiserum against collagen I was commercially available (Chemicon, Temecula, CA, USA).

reliable expression data we would recommend to collect data from at least two independent differentiation experiments, to repeat the RT-PCR reactions at least three times per experiment and to evaluate the data using the Student's t-test. Even then the standard deviations are often relatively large. For some genes the deviations could be reduced using a

[22] P. Cserjesi, D. Brown, K. L. Ligon, G. E. Lyons, N. G. Copeland, D. J. Gilbert, N. A. Jenkins, and E. N. Olson, *Development* **121**, 1099 (1995).

[23] U. Deutsch, G. R. Dressler, and P. Gruss, *Cell* **53**, 617 (1988).

[24] V. Lefebvre, P. Li, and B. de Crombrugghe, *EMBO J.* **17**, 5718 (1998).

[25] P. Ducy, R. Zhang, V. Geoffroy, A. L. Ridall, and G. Karsenty, *Cell* **89**, 747 (1997).

[26] E. Walcz, F. Deak, P. Erhardt, S. N. Coulter, C. Fulop, P. Horvath, K. J. Doege, and T. T. Glant, *Genomics* **22**, 364 (1994).

[27] M. Metsäranta, D. Toman, B. de Crombrugghe, and E. Vuorio, *J. Biol. Chem.* **266**, 16862 (1991).

[28] K. Elima, I. Eerola, R. Rosati, M. Metsäranta, S. Garofalo, M. Perälä, B. de Crombrugghe, and E. Vuorio, *Biochem. J.* **289**, 247 (1993).

[29] C. Desbois, D. A. Hogue, and G. Karsenty, *J. Biol. Chem.* **269**, 1183 (1994).

[30] D. Wang, A. Villasante, S. A. Lewis, and N. J. Cowan, *J. Cell Biol.* **103**, 1903 (1986).

[31] D. S. Konecki, J. Brennand, J. C. Fuscoe, C. T. Caskey, and A. C. Chinault, *Nucl. Acids Res.* **10**, 6763 (1982).

[32] T. F. Linsenmayer and M. J. Hendrix, *Biochem. Biophys. Res. Commun.* **92**, 440 (1980).

[33] M. A. Dorheim, M. Sullivan, V. Dandapani, X. Wu, J. Hudson, P. R. Segarini, D. M. Rosen, A. L. Aulthouse, and J. M. Gimble, *J. Cell Physiol.* **154**, 317 (1993).

[34] T. M. Schmid and T. F. Linsenmayer, *J. Cell Biol.* **100**, 598 (1985).

[35] E. Hedbom, P. Antonsson, A. Hjerpe, D. Aeschlimann, M. Paulsson, E. Rosa-Pimentel, Y. Sommarin, M. Wendel, A. Oldberg, and D. Heinegard, *J. Biol. Chem.* **267**, 6132 (1992).

RT-PCR method,[36,37] including the oligonucelotide primer for the internal standard and the analyzed gene in the same reaction. But this was not always the case.

Indirect Immunostaining

EBs were analyzed for differentiated cells by immunostaining. Samples of 5 EBs cultivated on chamber slides were rinsed twice in PBS, fixed for 5 min with methanol : acetone (7 : 3) at −20°C, rinsed thrice in PBS again and incubated for 15 min in 10% goat serum at room temperature. Specimens were then incubated with the primary antibodies (Table II) for 1 hr at 37°C in a humidified chamber. Slides were rinsed thrice with PBS and incubated for 45 min at 37°C with either, FITC (1 : 200)- or Cy3 (1 : 400)-labeled anti-mouse IgG (Dianova, Hamburg, FRG), respectively. Finally slides were washed thrice in PBS and once in distilled water and embedded in Vectashield mounting medium (Vector, Burlingame, USA). For microscopical analysis the fluorescence microscope AXIOSKOP (Zeiss, Oberkochen, FRG) was used.

Whole-mount Fluorescence In Situ *Hybridization for Scleraxis or for Collagen X-mRNA Coupled with Immunostaining for Collagen II*

To localize transcripts of cartilage-associated genes in EBs a modified whole-mount procedure for fluorescence mRNA *in situ* hybridization[38] was used. Samples of 5 EBs cultivated on chamber slides were analyzed at different developmental stages. Slides were rinsed twice with PBS and cells were fixed with 4% (w/v) paraformaldehyde, 4% (w/v) sucrose in PBS for 20 min at room temperature. Specimens were washed twice with PBS for 5 min, incubated at 70°C in 2x SSC (20x SSC stock solution: 175.3 g NaCl and 88.2 g sodium citrate in 1000 ml distilled water) for 15 min, washed once again with PBS and 2x SSC and fixed again for 5 min as described. Washing with PBS and 2x SSC was repeated and the EBs were subsequently dehydrated at room temperature for 2 min each in 50%, 70%, 95%, and twice in 100% ethanol. For prehybridization a buffer containing 5x SSC, 5x Denhardt's (50x Denhardt's stock solution: 5 g Ficoll, 5 g polyvinylpyrrolidone, 5 g bovine serum albumin in 500 ml distilled water), 50% formamide, 250 μg/ml

[36] H. Wong, W. D. Anderson, T. Cheng, and K. T. Riabowol, *Anal. Biochem.* **223**, 251 (1994).

[37] A. M. Wobus, G. Kaomei, J. Shan, M. C. Wellner, J. Rohwedel, G. Ji, B. Fleischmann, H. A. Katus, J. Hescheler, and W. M. Franz, *J. Mol. Cell Cardiol.* **29**, 1525 (1997).

[38] G. Yamada, C. Kioussi, F. R. Schubert, Y. Eto, K. Chowdhury, F. Pituello, and P. Gruss, *Biochem. Biophys. Res. Commun.* **199**, 552 (1994).

yeast-t-RNA, 250 μg/ml denatured salmon sperm DNA and 4 mM EDTA was used. EBs were prehybridized for 3 hr in a humid chamber at 45°C. Hybridization was performed in prehybridization buffer without salmon sperm DNA at 45°C in a humid chamber overnight with digoxigenin-labeled sense- or antisense-probes at a concentration of 1 ng/μl. After hybridization, specimens were washed twice with 2x SSC for 15 min, once with 0.2x SSC for 15 min and twice with 0.1x SSC for 15 min at 45°C and rinsed in PBS. For immunostaining on the same slide the monoclonal antibody II-II6B3 against collagen II (see Table I) was then applied followed by incubation in a humidified chamber for 1 hr at 37°C. EBs were washed three times with PBS at room temperature and FITC-conjugated sheep F(ab) fragments against digoxigenin (Roche) and Cy3-conjugated goat anti-mouse secondary antibodies (Dianova) for indirect detection of collagen II, both diluted 1 : 800 in PBS were added and the slides incubated for 1 hr at 37°C. Specimens were washed three times in PBS and once in distilled water and embedded in Vectashield mounting medium (Vector). For microscopical analysis the fluorescence microscope AXIOSKOP (Zeiss) was used.

Hybridization Probes

For mRNA *in situ* hybridization, we amplified a fragment of scleraxis- or collagen X-cDNA from RNA isolated from 16 day p.c. mouse limb buds according to the RT-PCR protocol described above using sequence-specific primer (Table I). The PCR fragments were cloned into the plasmid vector pCR-BluntII-TOPO using the TOPO II cloning kit (Invitrogen, Groningen, NL) according to the manufacturer's protocol and the sequences were verified by sequencing. From linearized plasmids of the cloned cDNA fragments the digoxigenin-labeled RNA probes of either sense- or antisense-orientation of scleraxis or collagen X were synthesized by *in vitro* transcription using the T7- or SP6-RNA polymerase (Roche) following the protocol supplied by the manufacturer.

Isolation of Chondrogenic Cells from EBs

To analyze their differentiation behavior in culture chondrogenic cells were isolated from EBs by micro-dissection. EBs were cultivated as described above and chondrogenic nodules were cut from the EB outgrowths with a micro-scalpel under sterile conditions and collected in PBS. To characterize the differentiation stage of these cells, samples of these nodules were investigated for expression of collagen marker proteins by immunostaining.[16] To this end, isolated nodules were embedded in

Tissue-Tek O.C.T. (Sakura Finetechnical, Tokyo, Japan) and frozen at −20°C. Cryosections (10 μm) were prepared with a cryostat (Leica, Bensheim, FRG) and plated onto Vectabond (Vector)-coated slides. After air-drying, the sections were fixed in acetone for 10 min at −20°C and finally washed in PBS. Indirect immunostaining was performed after fixation as described above.

To isolate the cells, nodules were incubated in 0.1% collagenase solution for 50 min at 37°C under moderate shaking to obtain a single cell suspension. Cells were centrifuged for 5 min, resuspended in medium 4 and plated at high density onto gelatine- or collagen II- (Sigma) coated 60-mm tissue culture dishes (at a density of $1–2 \times 10^5$ cells) or onto chamber slides (at a density of 2.1×10^4 cells/well) for total RNA isolation or immunostaining and histochemical staining, respectively.

Initially these cells showed dedifferentiation in culture (Fig. 2F) but later redifferentiated into mature chondrocytes (Fig. 2G) as demonstrated by RT-PCR and immunostaining for collagen markers[16] as described above. However, we observed additional cell types such as adipogenic cells, epithelial cell and skeletal muscle cells in the cultures.[16] This was demonstrated by Sudan III-staining (Fig. 2H; see the following subsection) or by immunostaining for cytokeratins using anti-pan-cytokeratin (Sigma) diluted 1 : 100 and the monoclonal antibody EA-53 (Sigma) against sarcomeric α-actinin diluted 1 : 20 following the method described above. Obviously, the isolated chondrogenic cells showed a distinct differentiation plasticity and it became evident that transdifferentiation of dedifferentiated chondrocytes at least into adipogenic cells occurred.[16]

Histochemical Stainings

a) *Alcian blue-staining*

As a fast screen for chondrogenic nodules EBs can be stained with Alcian blue.

a1) *Solution*

Dissolve 4.5 g NaCl and 6.4 g $MgCl_2$ in 400 ml 3% acetic acid, add 0.25 g Alcian blue 8GX (Sigma), adjust to pH 1.5 and adjust the volume to 500 ml. Stir for 2–3 hr and clear by filtration.

a2) *Procedure*

Alcian blue is a water soluble phthalocyanin-stain described to be selective for muco-substances, particularly for cartilage proteoglycans. EBs cultivated on 60 mm-tissue culture dishes were fixed for 30 min in a 3.7% formaldehyde (Merck) solution, followed by washing in PBS for three times and rinsing in distilled water before incubation was carried out in Alcian blue-solution overnight. Finally, specimens were rinsed briefly with PBS, 5% acetic acid and PBS again. Leave PBS on dishes for analysis using an invert miroscope.

b) *Alkaline Phosphatase (AP)-Staining*
AP-staining was used to detect osteogenic cells. A staining-kit (Sigma) including all solutions was used. EBs cultivated on chamber slides were washed with PBS and fixed for 30 sec with a citrate-formaldehyde-solution. Before and after incubation in the dark with the staining-solution for 15 min by room temperature the specimens were rinsed with distilled water. Finally, a staining-step with hematoxylin-solution, followed by a further rinsing-step, was performed.

c) *Sudan III-staining*
In cultures of chondrocytes isolated from EBs we found many cells carrying large droplets. To demonstrate that these are adipogenic cells the lipid-stain Sudan III (Sigma) was used. Cells grown on chamber slides were washed with PBS followed by staining for 3 min with a 0.2% solution of Sudan III in 70% ethanol.

Enhancing the Efficiency of ES Cell-derived Chondrogenic Differentiation

Spontaneous differentiation of ES cells into a distinct cell type is rarely optimal. The number of chondrogenic nodules spontaneously appearing in the EB outgrowths was found to be influenced by different parameters. Remarkable differences regarding the efficiency of chondrogenic differentiation were observed in EBs derived from various ES cell lines.[39] For example, in EB outgrowths derived from ES cell line BLC6 the number of chondrogenic nodules was approximately five-fold higher compared to those derived from ES cell line D3. Furthermore, the amount of ES cells used for EB formation was decisive for the development of cartilage nodules. Among the EBs prepared from 200, 500, and 800 cells the highest number of chondrogenic nodules was detected in outgrowths of EBs derived from 800 cells. Other parameters influencing the differentiation potential were the basic cultivation medium, the batch of serum, the time of EB cultivation in suspension and the day of EB-plating. For example more nodules were detected in outgrowths of EBs cultivated in DMEM compared to EBs cultivated in Iscove's modified Dulbecco's medium.[40] Previous studies already showed that

[39] J. Rohwedel, unpublished results (2002).
[40] J. Kramer, unpublished results (2001).

TABLE III
EFFECT OF GROWTH AND DIFFERENTIATION FACTORS ON THE EFFICIENCY OF CHONDROGENIC DIFFERENTIATION OF MOUSE ES CELLS *IN VITRO*

Factor	ES cell line	Concentration	Application period	Effect
Retinoic acid	D3	10^{-8} *M*	0–2 d	±
			2–5 d	±
			5–5+4 d	–
TGF-β1	D3	2 ng/ml	1–5+30 d	–
BMP-2	D3	2 ng/ml	1–5+30 d	+
		10 ng/ml	0–2 d	±
		10 ng/ml	2–5 d	+
		10 ng/ml	5–5+4 d	±
BMP-4	D3	10 ng/ml	1–5+30 d	+
TGF-β1+FGF-2	D3	2 ng/ml, each	2–5 d	–
TGF-β3	D3, BLC6	10 ng/ml	0–2 d	+
			2–5 d	+
			5–5+4 d	–

Several factors were tested for their potential to activate chondrogenesis at different concentrations and during different stages of mouse ES cell differentiation *in vitro* using different mouse ES cell lines. The factors either, showed no effect (±), or repression (–), or induction (+) of chondrogenic nodule formation as judged by Alcian blue-staining of EBs.

optimal differentiation into the mesenchymal direction was managed by cultivation in suspension for 3 days and plating at the fifth day of differentiation.[41,42]

For optimizing the differentiation efficiency growth factors and signaling molecules might be useful. Specific growth factors can be added to the differentiation medium in various concentrations and during different stages of the cultivation process. We studied the effect of several factors on chondrogenic differentiation (Table III) and found that ES cell-derived chondrogenesis could be activated by growth factors of the TGF-β-family.[15] Studies about the influence of growth factors on differentiation should be performed under serum-free conditions, whereas we used selected batches of serum. This was necessary because mesodermal differentiation is almost completely inhibited under serum-free conditions. Treatment of EBs with BMP-2 at a concentration of 2 ng/ml or BMP-4 at a concentration of 10 ng/ml applied to the EBs during the entire cultivation time (from 1 d

[41] J. Rohwedel, V. Maltsev, E. Bober, H. H. Arnold, J. Hescheler, and A. M. Wobus, *Dev. Biol.* **164**, 87 (1994).

[42] J. Rohwedel, T. Kleppisch, U. Pich, K. Guan, S. Jin, W. Zuschratter, C. Hopf, W. Hoch, J. Hescheler, V. Witzemann, and A. M. Wobus, *Exp. Cell Res.* **239**, 214 (1998).

up to 5 + 35 d) increased the development of chondrogenic cells organized in nodules approximately 2 to 2.5-fold. Furthermore, BMP-2 enhanced stage-dependently the chondrogenic differentiation of ES cells. The time-dependent effect of BMP-2 to increase chondrogenic differentiation was limited to the cultivation-step of EBs in suspension (2–5 d). These results are in line with studies demonstrating that this period was sensitive for the influence of differentiation factors on mesodermal cell differentiation.[44,45] In fact, it has been shown that this early stage of ES cell differentiation is a decisive period of early mesodermal development.[38,46] Finally, TGF-β3 which inhibited redifferentiation of chondrogenic cells isolated from EBs in culture[16] showed a slightly inductive effect on the development of chondrogenic cells when applied at a concentration of 2 ng/ml to EB cultures from 0–2 d or 2–5 d.

Acknowledgment

The skillful technical assistance of S. Vandersee and B. Lembrich is gratefully acknowledged. The work was supported by a grant from the Deutsche Forschungsgemeinschaft to J. R. (Ro 2108/1-1 and 1-2).

[44] A. M. Wobus, J. Rohwedel, V. Maltsev, and J. Hescheler, *Roux's Archive of Developmental Biology* **204**, 36 (1994).
[45] C. Dani, A. G. Smith, S. Dessolin, P. Leroy, L. Staccini, P. Villageois, C. Darimont, and G. Ailhaud, *J. Cell Sci.* **110**, 1279 (1997).
[46] J. Rohwedel, K. Guan, W. Zuschratter, S. Jin, G. Ahnert-Hilger, D. Fürst, R. Fässler, and A. M. Wobus, *Dev. Biol.* **201**, 167 (1998).

[19] Development of Adipocytes from Differentiated ES Cells

By BRIGITTE WDZIEKONSKI, PHI VILLAGEOIS, and CHRISTIAN DANI

Introduction

In this chapter we describe an *in vitro* system that allows access to development of adipocytes from pluripotent mouse ES cells. Because there are no specific markers that can be used to identify adipoblasts and preadipocytes *in vivo*, the white adipose tissue can be identified only when it contains adipocytes (i.e., terminally differentiated cells) and cannot be detected macroscopically during embryogenesis. Therefore, very little is known about the ontogeny of adipose tissue. Mainly, the *in vitro* system used to study adipogenesis is immortal preadipocyte cell lines which

0076-6879/2003 $35.00

represent already determined cells.[1] In the recent years, transcriptional factors, such as the nuclear receptor Peroxisome Proliferator-Activated Receptor (PPAR) γ, and cell surface receptors, such as the receptor for the Leukemia Inhibitory Factor (LIF-R), have been shown to play a regulatory role in the terminal differentiation process. Both PPARγ and LIF-R knockout mice are not viable precluding the study of their role in the formation of fat. The generation of LIF-R$^{-/-}$ ES cells or of PPAR$\gamma^{-/-}$ ES cells combined with conditions of culture to commit stem cells into adipogenic lineage allowed us and others to circumvent the lethality problem and to provide evidence that LIF-R and PPARγ play a key role in the conversion of preadipocytes into adipocytes.[2–4] The use of *in vitro* differentiation of ES cells to analyse the effects of mutations on adipogenesis is just beginning and master genes that commit progression from pluripotent stem cells to the adipogenic lineage have not yet been identified.

ES cells have been previously shown to differentiate spontaneously into various lineages in culture.[5] The first morphological observation of adipocyte-like cells derived from ES cells was reported by Field *et al.*[6] However, the number of ES cell-derived adipocytes was low because spontaneous commitment of ES cells into adipogenic lineage is rare. In order to use ES cells to study adipocyte development it was essential to determine conditions of culture that commit ES cells into the adipogenic lineage at a high rate. This step has been achieved by showing that a prerequisite is to treat ES cell-derived embryoid bodies (EBs) at an early stage of their differentiation with all-*trans*-retinoic acid (RA) for a short period of time.[7] Two phases can be distinguished in the development of adipogenesis from ES cells (Fig. 1). The first phase, between day 2 and 5 after EB formation, corresponds to the permissive

[1] E. D. Rosen and B. M. Spiegelman, *Annu. Rev. Cell Dev. Biol.* **16**, 145–171 (2000).

[2] E. D. Rosen, P. Sarraf, A. E. Troy, G. Bradwin, K. Moore, D. S. Milstone, B. M. Spiegelman, and R. M. Mortensen, *Mol. Cell.* **4**, 611–617 (1999).

[3] J. Aubert, S. Dessolin, N. Belmonte, M. Li, F. R. McKenzie, L. Staccini, P. Villageois, B. Barhanin, A. Vernallis, A. G. Smith, G. Ailhaud, and C. Dani, *J. Biol. Chem.* **274**, 24965–24972 (1999).

[4] C. Vernochet, D. S. Milstone, C. Iehle, N. Belmonte, B. Phillips, B. Wdziekonski, P. Villageois, E. Z. Amri, P. E. O'Donnell, R. M. Mortensen, G. Ailhaud, and C. Dani, *FEBS Lett.* **510**, 94–98 (2002).

[5] T. C. Doetschman, H. Eistetter, M. Katz, W. Schmidt, and R. Kemler, *J. Embryol. Exp. Morphol.* **87**, 27–45 (1985).

[6] S. J. Field, R. S. Johnson, R. M. Mortensen, V. E. Papaioannou, B. M. Spiegelman, and M. E. Greenberg, *Proc. Natl. Acad. Sci. USA* **89**, 9306–9310 (1992).

[7] C. Dani, A. Smith, S. Dessolin, P. Leroy, L. Staccini, P. Villageois, C. Darimont, and G. Ailhaud, *J. Cell. Sci.* **110**, 1279–1285 (1997).

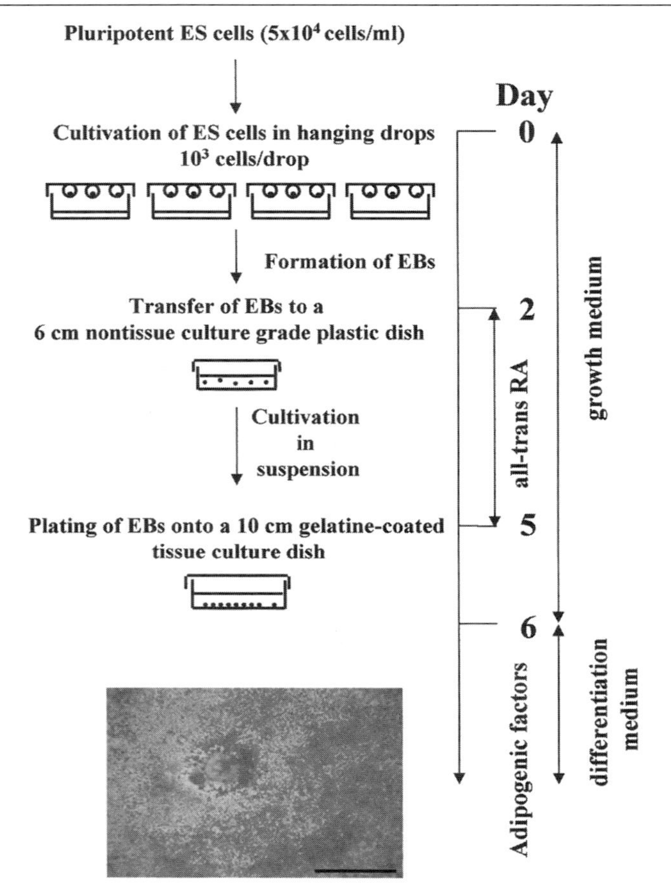

FIG. 1. Overview of the protocol used for *in vitro* differentiation of ES cells into adipocytes and photomicrographic record of a 20 day-old outgrowth derived from RA-treated EBs. Adipocytes stained with Oil-Red O for fat droplets are shown. Bar: 1000 μM. (See Color Insert.)

period for the commitment of ES cells which requires RA. The second phase corresponds to the permissive period for terminal differentiation and is influenced by adipogenic factors as previously shown for the differentiation of cells from preadipose clonal lines. The treatment leads to 60–80% of outgrowths containing adipose cells compared to 2–5% in the absence of RA treatment. A 20-day-old outgrowth derived from RA-treated EBs is shown (Fig. 1): large adipocyte colonies, stained with Oil-Red O, which developed out of the center of the body can be distinguished. Adipocytes derived from ES cells display both lipogenic and lipolytic activities in response to insulin and to β-adrenergic agonists respectively, indicating that mature and functional adipocytes are formed.

We describe a protocol here for culture conditions in which ES cells undergo adipocyte differentiation at a high rate and in a highly reproducible fashion. This culture system should provide a powerful means for studying the first step of adipose cell development and a route to identify regulatory genes involved in adipogenesis.

Methods and Materials

Maintenance of ES Cells on Gelatine-coated Tissue Culture Plastic

We describe conditions to maintain CGR8 ES cells on gelatine-coated tissue culture flasks in the absence of feeder cells. Coating is performed by covering the surface of a 25 cm^2 flask with 5 ml of 0.1% gelatine for 15 min at room temperature and cells are maintained in a pluripotent undifferentiated state by adding LIF. To subculture cells from a 25 cm^2 flask, aspirate medium off and wash twice with 5 ml of PBS. Aspirate off the PBS and add 1 ml of trypsin solution. Ensure the trypsin covers the cell monolayer and incubate at 37°C, 5% CO_2, for 2–3 min. Check under an inverted microscope that cells are correctly dissociated. It is critical to produce a single cell suspension for subcultures. This is achieved by knocking the flask several times to ensure complete dissociation during the trypsin treatment. Once the dissociation is completed, add 5 ml of growth medium to stop trypsinization and suspend the cells by vigorous pipetting. Transfer the cells to a sterile tube and centrifuge at 250*g* for 5 min at room temperature. Aspirate the medium off and resuspend the cell pellet with 5 ml of growth medium by pipetting up and down 2–3 times. Count cells. Add 10^6 ES cells into 10 ml of prewarmed growth medium containing LIF then transfer to a freshly gelatinized 25 cm^2 flask. Attachment of ES cells to the substratum is susceptible to change according to the tissue culture material. Change medium every day. Cultures should be split before cells have reached confluence, usually every two or three days. Pluripotent ES are identified microscopically by the appropriate cellular morphology. Pluripotent stem cells are small, have large nucleus containing prominent nucleoli structures, have minimum cytoplasm and grow as tight colonies. Cells that have lost their pluripotency are flat and grow as individual cells. The maintenance of pluripotent ES cell cultures with a minimum proportion of cells that have lost their pluripotency is crucial for subsequent differentiation. This is achieved by addition of an appropriate quantity of LIF in a high quality culture medium with an adequate batch of serum. A 25 cm^2 flask seeded with 10^6 cells should yield $6–10\times10^6$ cells at each passage.

Materials required:

(1) LIF is required to maintain pluripotent ES cells. LIF is commercially available (named ESGRO by Chemicon) or can be prepared in the laboratory. To produce LIF, Cos7 cells are transiently transfected with a LIF-expressing construct (using standard techniques) and after 4 days of growth the medium is collected. A titration of LIF activity in the medium conditioned by transfected Cos7 cells is then performed by testing several dilutions of the medium on ES cells plated at 5×10^3 cells/well of 24-well plates. Five days after, with no change of the medium, cell morphology is inspected. At clonal density, a 10^4 dilution of the conditioned medium should maintain pluripotent ES cells and we use a 10^3 dilution to routinely maintain ES cells in 25 cm^2 as described above. (2) growth medium: to 450 ml 1X Glasgow MEM (Biomedia GMEMNAO2052) add 5 ml of 100X nonessential amino acids (Gibco BRL 11140035), 10 ml of 200 m*M* glutamine, 100 m*M* sodium pyruvate (Gibco BRL 25030024 and 11360039), 50 ml of selected fetal calf serum (FCS, Bio Media SA France or Dutscher SA France), 0.5 ml of 0.1 *M* 2-mercaptoethanol. To prepare 0.1 *M* 2-mercaptoethanol add 100 μl 2-mercaptoethanol (Sigma M-7522) to 14 ml of sterile H_2O. Store up to 3 weeks at +4°C. (3) Phosphate-buffered saline (PBS) calcium and magnesium free (Gibco BRL 14190-094). (4) Trypsin solution: Trypsin 1X is prepared by adding 1 ml of 2.5% trypsin (Gibco BRL 2509008) plus 1 ml of EDTA 100 m*M* and 1 ml of chicken serum to 100 ml PBS. Aliquot (10 ml) and store at −20°C. Thawed aliquots are stored at +4°C. (5) Gelatine 0.1%: purchase gelatine 2% (Sigma G-1393) and dilute to 0.1% with PBS. Store at +4°C. (6) Tissue culture 25 cm^2 flasks (Greiner 690175 or Corning 430168).

Differentiation of ES Cell-derived Embryoid Bodies into Adipocytes

Adipocyte differentiation is initiated by aggregation of ES cells to form embryoid bodies (EBs). The formation of EBs in mass culture by maintaining ES cells in suspension at a high density in nontissue culture grade plastic is rapid and give rise to a high number of EBs formed. However, in our hands, this method leads subsequently to a low number of outgrowths containing adipocyte colonies. We routinely used the hanging drop method for the formation of EBs. A schematic representation of the experimental protocol is shown in Fig. 1. Dilute ES cells to 5×10^4 cells/ml of growth medium without LIF and supplemented with antibiotics. Place aliquot of 20 μl of the suspension onto the lid of a 10 cm nontissue culture grade dish. This is defined as day 0 of EB formation.

We use a multipipettor with a sterile combitip (Eppendorf) to form drops of 20 μl. Approximately 80 drops can be fitted on a lid. Invert the lid and place it over the bottom of the plate filled with 8 ml PBS containing few drops of FCS. It is essential to cover the bottom of the plate with liquid to prevent the evaporation of the hanging drops. When the lid is inverted, each drop hangs, and the cells fall to the bottom of the drop where they aggregate into a single clump. Two days after EB formation, remove the lid, invert it and collect drops in a conical sterile tube. Let it stand for 5 min at room temperature to allow the cell aggregates to sediment. Aspirate the supernatant and resuspend the pellet in 4 ml of growth medium supplemented with retinoic acid (RA). Owing to the high instability of RA, the concentration of RA able to commit ES cells into the adipogenic lineage at a high rate should be determined for each new preparation of RA (try 10^{-8} *M* to 10^{-6} *M*). Transfer the suspension into 6-cm nontissue culture dishes. Usually we pool two day-old EBs from 4 lids of 10-cm petri dishes into one 6 cm bacteriological grade petri dish. Cultivate EBs in suspension for 3 days in the presence of RA, changing the medium every day. Bacterial grade petri dishes are used to prevent cell attachment to the substrate. However, 4–5 day-old EBs have a tendency to attach to the bottom of the plastic dish. EBs that are firmly attached to the dish should be eliminated as these kind of EBs seem to have no adipogenic capacity. Attachment of EBs on bacteriological petri dishes is eliminated by coating the dish with poly(2-hydroxyethylmethylacrylate).[8] At day 5 after EB formation, change the medium without the addition of RA and plate 2–4 EBs per cm^2 in gelatinized-tissue culture dishes to allow EBs to attach. EBs contained in one 6-cm bacterial grade petri dish are plated into one 10-cm tissue culture dish coated with gelatine. A higher density of EBs can lead to a decrease in the number of EB containing adipocyte colonies. A day after plating, change the growth medium to differentiation medium. Change medium every other day. After 10–20 days in the differentiation medium, at least 60–80% of EB outgrowths should contain adipocyte colonies. A wide variety of differentiated derivatives such as neurone-like cells, fibroblast-like cells and unidentified cell types appear over this period. Spontaneously beating cardiomyocytes should appear 1–5 days after plating the control culture (untreated with RA). At least 40% of EBs should contain beating cardiomyocytes. In contrast, few EBs should contain beating cells from RA-treated cultures. It is important to note that the treatment of EBs with RA is a prerequisite for adipogenesis. In serum-containing medium (growth medium) and in serum-free medium (20% Knockout SR from

[8] J. Folkman and A. Moscona, *Nature* **273**, 345–349 (1978).

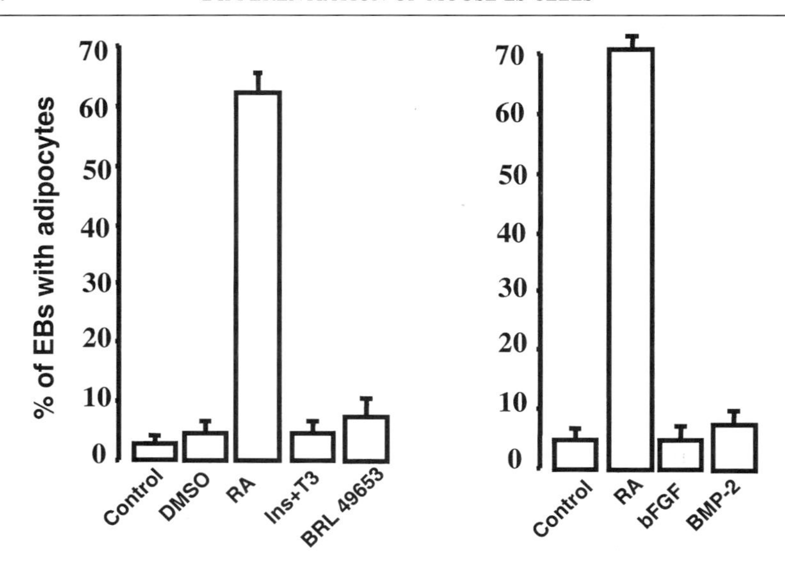

FIG. 2. Critical role of RA on the adipocyte differentiation of ES cells. EBs were treated from day 2–5 with indicated compounds in growth medium (left panel) or in serum-free medium (right panel). The percentage of EBs containing adipocyte colonies after 25 days in differentiation medium is shown. DMSO: 0.1%, RA: 10^{-7} *M*; Insulin: 85 n*M*, T3: 2 n*M*, BRL496543: 0.5 μM, bFGF: 10 ng/ml, BMP-2: 20 ng/ml.

Life Technologies, cat. no. 10828), RA cannot be substituted by insulin, triiodothyronine, dexamethasone, bFGF, BMP-2 or potent activators of PPARs known to be important for terminal differentiation (Fig. 2). The addition of a PPARγ activator, such as thiazolidinedione BRL49653 or ciglitazone, in the differentiation medium dramatically stimulates the terminal differentiation of RA-treated EBs into adipocytes.

Materials required:

(1) Poly(2-hydroxyethylmethylacrylate) is supplied as Cellform Polymer (ICN Biomedicals, 150207). Coating is performed by covering the surface of the dish with 1.2% of the polymer in ethanol. Leave 1–2 hr in a tissue culture hood to evaporate ethanol. Rinse twice with PBS. Treated dishes can be stored for extended periods. (2) Retinoic acid: *All trans* retinoic acid (RA, Sigma R-2625) is diluted in the dark into dimethyl sulfoxide to prepare a 10 m*M* stock solution. Aliquot and store at –20°C. Subsequent dilutions of RA are performed in ethanol and used for one experiment only. After dilution into the culture media, the concentration of ethanol never exceeds 0.1%. RA is light-sensitive. (3) Differentiation medium: This medium

consists of growth medium supplemented with 0.5 μg/ml insulin (Sigma I-6634), 2 n*M* triiodothyronine (Sigma T-6397), 0.5 μ*M* thiazolidinedione BRL49653 (SmithKline Beecham Pharmaceuticals, this compound is not commercially available) or 2 μ*M* ciglitazone (Biomol, 034GR205) and 5 μg/ml antibiotics (Penicillin/Streptomycin, Bio Media PEST0502012). (4) Insulin is prepared at 1 mg/ml in cold HCl 0.01 N. Mix gently and sterilize by filtration. Aliquot (1 ml) and store at −20°C. Thawed aliquots are stored at +4°C. (5) Triiodothyronine is prepared at 2 m*M* in ethanol (stock solution). Store at −20°C. (6) Bacteriological grade 10 cm and 6 cm petri dishes (Greiner 633185 and 628102). (7) Tissue culture 10 cm dishes (Corning, 664160).

Assessment of ES Cell Differentiation into Adipocytes

Adipose cells with lipid droplets are easily visualized microscopically, especially under bright field illumination. As nonadipose cells containing structures resembling droplets are often detectable in untreated and RA-treated cultures, it is essential to identify droplet-like structure as triglyceride droplets. Staining of cultures with Oil-Red O, a specific stain for triglycerides, and counting the percentage of EB outgrowths containing adipocyte colonies give a good indication of adipocyte differentiation. Aspirate medium and wash cells once with PBS. Fix cells for 15 min at room temperature. Wash twice with PBS and stain with Oil-Red O solution for 15 min at room temperature. Wash twice with H_2O and cover cells with a film of storage solution. ES cell differentiation into adipocytes can be monitored by the detection of a specific mRNA marker such as a-FABP (adipocyte-Fatty Acid Binding Protein, also named ALBP or aP2).[9] It is interesting to note that as the concentration of RA increased, a switch from skeletal myogenesis to adipogenesis occurs. In our hands, a-FABP is the best candidate to monitor adipogenesis and Myogenin to monitor skeletal myogenesis. Expression of a-FABP gene is specifically induced in terminally differentiated cells whereas Myogenin is expressed in skeletal myocytes only.[10] Both genes can be detected in 20 day-old EB outgrowths by Northern-blotting using 20 μg of total RNA (see Fig. 3). Detection in early differentiated outgrowths requires a more sensitive method such as RT-PCR. Total RNA is prepared with TRI Reagent according to manufacturer's instructions. We routinely get 100–200 μg RNA from one 10 cm tissue culture dish containing 20 day-old EB outgrowths.

[9] C. R. Hunt, J. H. Ro, D. E. Dobson, H. Y. Min, and B. M. Spiegelman, *Proc. Natl. Acad. Sci. USA* **83**, 3786–3790 (1986).
[10] W. E. Wright, D. A. Sassoon, and V. K. Lin, *Cell* **56**, 607–617 (1989).

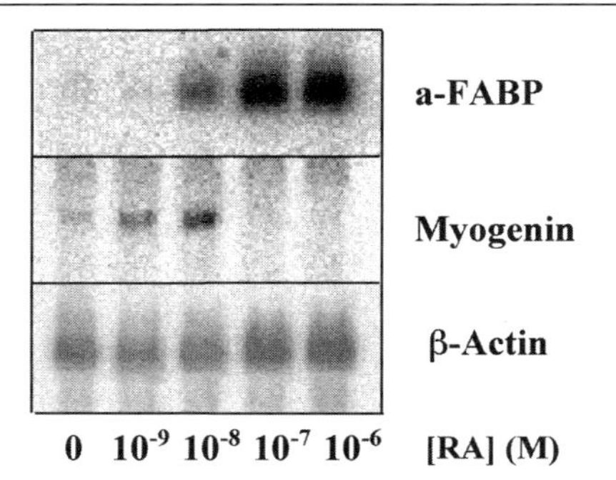

FIG. 3. Northern-blot analysis of expression of myogenin and a-FABP in 20 day-old outgrowths derived from EBs treated from day 2–5 with indicated concentration of RA.

Materials:

(1) Fix buffer: 0.20% glutaraldehyde (Sigma, G-5882) in PBS. Store at +4°C. (2) Oil-Red O solution. We prepare at stock solution at 0.5% Oil-Red O (Sigma, O-0625) in isopropanol. Rotate overnight. To prepare the working solution, mix 6 volumes of stock solution with 4 volumes H_2O. Mix and filter. Store at room temperature. (3) Store solution: 70% glycerol in H_2O (v/v). (4) TRI Reagent (Euromedex, TR-118). (5) For Northern-blot analysis, cDNA of a-FABP was obtained from B. Spiegelman (Dana-Farber Cancer Institute, Boston, USA) and cDNA of myogenin was obtained from E.N. Olson (Southwestern Medical Center, University of Texas, Dallas, USA). (6) Pairs of primers used for PCR reactions and temperatures of annealing are the following:

adipocyte-Fatty Acid Binding Protein:

5′GATGCCTTTGTGGGAACCTGG3′ and 5′TTCATCGAATTCCAC GCCCAG3′, annealing temperature: 56°C, size of the expected cDNA: 213 bp, size of genomic DNA-derived contaminated band: 2400 bp.

Myogenin:

5’AGCTCCCTCAACCAGGAGGA3’ and 5’GGGCTCTCTGGACTCCA TCT3’, annealing temperature: 60°C, size of the cDNA: 500 bp, size of genomic DNA band: 1500 bp

Hypoxanthine Phosphoribosyltransferase (HPRT, as a standard to balance the amount of RNA and cDNA used):

5′GCTGGTGAAAAGGACCTCT3′ and 5′CACAGGACTAGAACACCT GC3′, annealing temperature: 60°C, size of the cDNA: 249 bp, size of genomic DNA band: 1100 bp

Acknowledgment

This work was supported by the Association pour la Recherche Contre le Cancer (grant no. 9982 and 4625 to C. D.).

[20] *In Vitro* Differentiation of Embryonic Stem Cells into Hepatocytes

By TAKASHI HAMAZAKI and NAOHIRO TERADA

Introduction

The ability to differentiate into a wide range of cell types *in vitro* has drawn much attention to embryonic stem cells (ES cells) as a novel source for potential cell therapy,[1] especially after cultures of human ES cells and human embryonic germ cells were established.[2,3] Since cell replacement therapy using primary hepatocytes has been shown to be effective in animal models of hepatic failure and liver-based metabolic diseases,[4–7] establishing a method to produce functional hepatocytes from ES cells has been greatly anticipated.

Generation of certain cell types from ES cells has been achieved by various approaches. The most classical but still ideal methods are by addition of various growth factors or chemicals to the cell culture medium. For example, pure populations of mast cell precursors can be obtained from mouse ES cells using interleukin-3 and Stem Cell Factor (c-kit ligand).[8] Alternatively, tissue-specific precursors can be sorted using fluorescence-activated cell sorter (FACS) based on expression of specific markers on the

[1] G. Keller and H. R. Snodgrass, *Nat. Med.* (*London*) **5**, 151 (1999).

[2] J. A. Thomson, J. Itskovitz Eldor, S. S. Shapiro, M. A. Waknitz, J. J. Swiergiel, V. S. Marshall, and J. M. Jones, *Science* **282**, 1145 (1998).

[3] M. J. Shamblott, J. Axelman, S. Wang, E. M. Bugg, J. W. Littlefield, P. J. Donovan, P. D. Blumenthal, G. R. Huggins, and J. D. Gearhart, *Proc. Natl. Acad. Sci. USA* **95**, 13726 (1998).

[4] A. A. Demetriou, A. Reisner, J. Sanchez, S. M. Levenson, A. D. Moscioni, and J. R. Chowdhury, *Hepatology* **8**, 1006 (1988).

[5] V. Roger, P. Balladur, J. Honiger, M. Baudrimont, R. Delelo, A. Robert, Y. Calmus, J. Capeau, and B. Nordlinger, *Ann. Surg.* **228**, 1 (1998).

[6] S. Gupta, G. R. Gorla, and A. N. Irani, *J. Hepatol.* **30**, 162 (1999).

[7] Y. Yoshida, Y. Tokusashi, G. H. Lee, and K. Ogawa, *Gastroenterology* **111**, 1654 (1996).

[8] T. P. Garrington, T. Ishizuka, P. J. Papst, K. Chayama, S. Webb, T. Yujiri, W. Sun, S. Sather, D. M. Russell, S. B. Gibson, G. Keller, E. W. Gelfand, and G. L. Johnson, *EMBO J.* **19**, 5387 (2000).

0076-6879/2003 $35.00

cell surface. For instance, flk1-positive cells from mouse ES cells were demonstrated to serve as vascular progenitors.[9] When there is no antibody available to sort, tissue-specific promoter-derived drug selection or green fluorescence protein (GFP) expression has been used to purify certain cell types. ES cell-derived cardiac myocytes and insulin-secreting cells were isolated through this method.[10–12] In other cases, forced expression of transcription factors in ES cells has been proven to promote specific lineage differentiation, such as GATA 4 and 6 for primitive endoderm,[13] HoxB4 for hematopoietic precursors,[14] and Nurr 1 for dopaminergic neurons.[15] ES cells also can be differentiated into specific lineages by coculture with other cells. Differentiation into hematopoietic cells and dopaminergic neurons, for instance, was induced when mouse ES cells were cultured on feeder layers of OP9 and PA6 cells respectively.[16,17] Despite intensive studies of *in vitro* differentiation of ES cells into other lineages as described above, hepatic differentiation of ES cells has not been well characterized until very recently.

Visceral Endoderm and Hepatic Differentiation in Embryoid Bodies

Mouse ES cells can be maintained *in vitro* indefinitely in an undifferentiated state in a medium containing leukemia inhibitory factor (LIF).[18,19] The *in vitro* differentiation of mouse ES cells is simply induced by removing LIF from the medium. ES cells aggregate into structures termed embryoid bodies, in which all three germ layers develop and

[9] J. Yamashita, H. Itoh, M. Hirashima, M. Ogawa, S. Nishikawa, T. Yurugi, M. Naito, K. Nakao, and S. Nishikawa, *Nature (London)* **408**, 92 (2000).

[10] M. G. Klug, M. H. Soonpaa, G. Y. Koh, and L. J. Field, *J. Clin. Invest.* **98**, 216 (1996).

[11] M. Muller, B. K. Fleischmann, S. Selbert, G. J. Ji, E. Endl, G. Middeler, O. J. Muller, P. Schlenke, S. Frese, A. M. Wobus, J. Hescheler, H. A. Katus, and W. M. Franz, *FASEB J.* **14**, 2540 (2000).

[12] B. Sonia, E. Roche, G. Berna, T. Leon-Quinto, J. A. Reig, and F. Martin, *Diabetes* **49**, 157 (2000).

[13] J. Fujikura, E. Yamato, S. Yonemura, K. Hosoda, S. Masui, K. Nakao, J. Miyazaki, and H. Niwa, *Genes Dev.* **16**, 784 (2002).

[14] C. D. Helgason, G. Sauvageau, H. J. Lawrence, C. Largman, and R. K. Humphries, *Blood* **87**, 2740 (1996).

[15] J. H. Kim, J. M. Auerbach, J. A. Rodriguez-Gomez, I. Velasco, D. Gavin, N. Lumelsky, S. H. Lee, J. Nguyen, R. Sanchez-Pernaute, K. Bankiewicz, and R. McKay, *Nature (London)* **418**, 50 (2002).

[16] T. Nakano, H. Kodama, and T. Honjo, *Science* **272**, 722 (1996).

[17] H. Kawasaki, K. Mizuseki, S. Nishikawa, S. Kaneko, Y. Kuwana, S. Nakanishi, S. I. Nishikawa, and Y. Sasai, *Neuron* **28**, 31 (2000).

[18] M. J. Evans and M. H. Kaufman, *Nature (London)* **292**, 154 (1981).

[19] G. R. Martin, *Proc. Natl. Acad. Sci. USA* **78**, 7634 (1981).

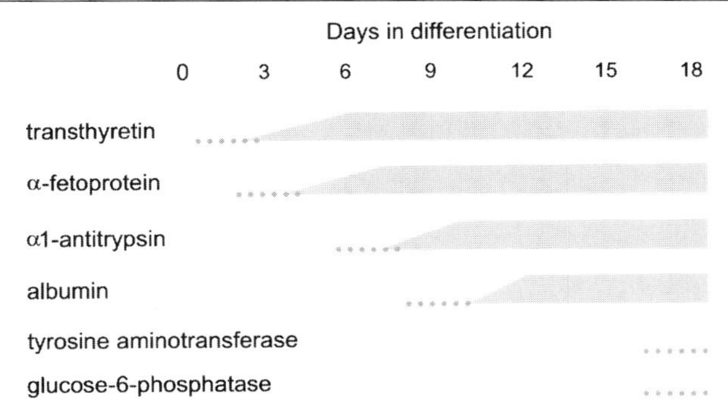

FIG. 1. Expression pattern of visceral endoderm and hepatic differentiation marker genes in embryoid bodies.

interact with each other. Outer layer cells of embryoid bodies tend to express α-fetoprotein and transthyretin showing visceral endoderm phenotype.[20,21] Bone morphogenic protein (BMP) 4, which is secreted from inner core of embryoid body, promotes maturation of visceral endoderm.[22] These processes are compatible with the transition from epiblast stage to egg-cylinder stage of mouse embryo, supporting the idea that differentiating ES cells *in vitro* recapitulate the process of early embryonic development.

When visceral endoderm and hepatic differentiation are examined by gene expression during *in vitro* ES cells differentiation, an early marker such as transthyretin is detected within 3 days after removal of the LIF and gradually increases until Day 18. α-Fetoprotein and α1-antitrypsin are first detected around Days 6 to 9. Albumin mRNA expression first appears within Day 12. Expression of these genes, however, does not distinguish hepatic differentiation from visceral endoderm differentiation. Late differential markers of hepatocyte, tyrosine aminotransferase (TAT) and glucose-6-phosphatase (G6P) are often undetectable throughout the time course (up to Day 18) (Fig. 1). These data indicate that ES cells spontaneously differentiate toward visceral endodermal cells or early hepatic lineage cells, but they hardly differentiate into mature hepatocytes.

[20] S. Becker, J. Casanova, and L. Grabel, *Mech. Dev.* **37**, 3 (1992).
[21] K. Abe, H. Niwa, K. Iwase, M. Takiguchi, M. Mori, S. I. Abe, K. Abe, and I. Yamamura, *Exp. Cell Res.* **229**, 27 (1996).
[22] E. Coucouvanis and G. R. Martin, *Development* **126**, 535 (1999).

During embryonic development, an initial step of hepatic lineage specification occurs on embryonic day 9 (E9) in mice. In this early stage, fibroblast growth factors (FGFs), derived from adjacent cardiac mesoderm, commit the foregut endoderm to form the liver primordium.[23] BMPs, which are expressed from septum transversum, are also involved in hepatic lineage specification.[24] Over the next two days, the liver bud proliferates and migrates into surrounding septum transversum, which consists of loose connective tissue containing collagen.[25] Vasculogenic endothelial cells promote the growths of the liver bud.[26] During and after the mid stage of hepatogenesis, hepatocyte growth factor (HGF), which is secreted from surrounding mesenchymal cells, is essential for fetal hepatocytes.[27,28] From E12 through E16, the fetal liver becomes the major site for hematopoiesis.[29] During this late stage, hematopoietic cells produce oncostatin M that induces maturation of murine fetal hepatocytes.[30] In addition to the studies above, transforming growth factor-β (TGF-β) was shown to augment albumin production in prenatal hepatocytes, implicating TGF-β in the regulation of hepatic differentiation[31] (Table I).

Further differentiation of ES cells toward mature hepatocytes can be induced by sequential addition of growth factors to embryoid bodies, which are critical for hepatic lineage differentiation during embryonic development as discussed above. These include acidic FGF (Day 9 to Day 12), HGF (Day 12 to Day 18), and oncostatin M, dexamethasone, and mixture of insulin, transferrin and selenious acid (Day 15 to Day 18). This combination of the growth factors indeed enhanced the expression of albumin mRNA up to 9.5-fold (real-time PCR) or 7-fold (gene chip analysis) at Day 18 when compared to control embryoid bodies without addition of the factors. Moreover, G6P and TAT genes, indicators of hepatocyte maturation, are now clearly detectable in Day 18 embryoid

[23] J. Jung, M. Zheng, M. Goldfarb, and K. S. Zaret, *Science* **284**, 1998 (1999).

[24] J. M. Rossi, N. R. Dunn, B. L. Hogan, and K. S. Zaret, *Genes Dev.* **15**, 1998 (2001).

[25] S. Cascio and K. S. Zaret, *Development* **113**, 217 (1991).

[26] K. Matsumoto, H. Yoshitomi, J. Rossant, and K. S. Zaret, *Science* **294**, 559 (2001).

[27] C. Schmidt, F. Bladt, S. Goedecke, V. Brinkmann, W. Zschiesche, M. Sharpe, E. Gherardi, and C. Birchmeier, *Nature (London)* **373**, 699 (1995).

[28] Y. Uehara, O. Minowa, C. Mori, K. Shiota, J. Kuno, T. Noda, and N. Kitamura, *Nature (London)* **373**, 702 (1995).

[29] T. Kinoshita, T. Sekiguchi, M. J. Xu, Y. Ito, A. Kamiya, K. Tsuji, T. Nakahata, and A. Miyajima, *Proc. Natl. Acad. Sci. USA* **96**, 7265 (1999).

[30] A. Kamiya, T. Kinoshita, Y. Ito, T. Matsui, Y. Morikawa, E. Senba, K. Nakashima, T. Taga, K. Yoshida, T. Kishimoto, and A. Miyajima, *EMBO J.* **18**, 2127 (1999).

[31] A. Sanchez, A. M. Alvarez, M. Benito, and I. Fabregat, *J. Cell Physiol.* **165**, 398 (1995).

TABLE I
CELLS, EXTRACELLULAR FACTORS, INTRACELLULAR MOLECULES CONSIDERED TO BE CRITICAL FOR EMBRYONIC LIVER DEVELOPMENT

Embryonic development	Cells	Extracellular factors	Intracellular molecules
Early	Cardiac mesoderm Septum transversum Vasculogenic endothelial	FGF1, FGF2, FGF8 BMP4	GATA4
Mid	Mesenchym	HGF	SEK1 c-jun
Late	Hematopoietic cells	Oncostatin M glucocorticoid TGF-β	C/EBPα HNF4

bodies in the presence of the growth factors.[32] It should be noted that TAT was expressed in embryoid bodies without addition of growth factors in another study. The discrepancy could be explained by subtle differences among ES cell lines, variation of culture conditions for ES cell maintenance, or variation of fetal calf serum.[33]

Visceral endodermal/hepatic lineage differentiation can also be visualized in real-time by GFP expression under the control of the α-fetoprotein gene promoter. Within a few days after removal of LIF, GFP positive cells first appear at the outer layer of aggregated embryoid bodies, which is known to represent visceral endodermal cells[20,21] (Fig. 2a). When embryoid bodies are attached to a culture dish at day 5, these GFP positive cells gradually diminish GFP expression after attaching the culture dish. Independent of the initial GFP positive cells, a morphologically different GFP positive population appears and lines up as mono- to oligo-layer cells at the outer circumference of embryoid bodies (Fig. 2a, Day 7). Later, clusters of GFP positive cells then appear inside the embryoid bodies at around day 12 and gradually grow in cell population until Day 18 (Fig. 2a). The percentage of GFP positive cells was $1.2 \pm 0.9\%$ within Day

[32] T. Hamazaki, Y. Iiboshi, M. Oka, P. J. Papst, A. M. Meacham, L. I. Zon, and N. Terada, *FEBS Lett.* **497**, 15 (2001).

[33] R. Chinzei, Y. Tanaka, K. Shimizu-Saito, Y. Hara, S. Kakinuma, M. Watanabe, K. Teramoto, S. Arii, K. Takase, C. Sato, N. Terada, and H. Teraoka, *Hepatology* **36**, 22 (2002).

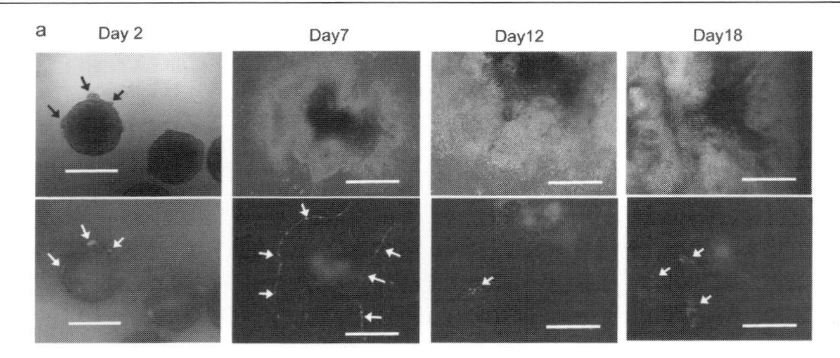

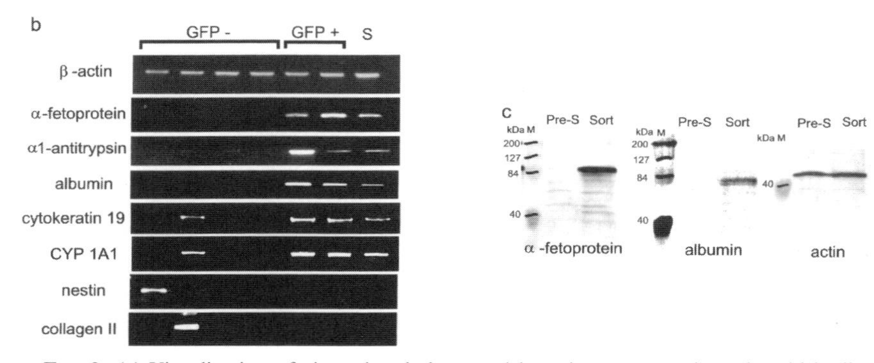

FIG. 2. (a) Visualization of visceral endoderm and hepatic precursors in embryoid bodies. GFP expression was driven by the α-fetoprotein gene promoter (-259 to $+24$). GFP positive cells first appeared at the outer layer of embryoid bodies around Day 2. After embryoid bodies were attached to the culture plate at Day 5, GFP positive cells lined up as mono- to oligo-layer cells at the outer circumference of embryoid bodies (Day 7). Clusters of GFP positive cells then appeared inside the embryoid bodies around Day 12 and gradually grew until Day 18. Arrows in the lower panels indicate GFP positive cells under a fluorescence microscope, and upper panels illustrate the corresponding pictures under a light microscope. Bars: 1.0 mm. (b) Visceral endoderm and hepatic marker expression in GFP positive and negative populations. Clusters of GFP positive cells in embryoid bodies (Day 18) were directly picked under a fluorescence microscope (GFP+). As controls, GFP negative cells were similarly harvested under a microscope from a GFP negative area (GFP−). GFP positive cells were also sorted using FACS (S). The mRNAs were extracted from these cells, and expression of hepatic lineage markers was examined by RT-PCR. Nestin and collagen II genes were also examined as markers for neural and mesodermal differentiation, respectively. (c) Western blotting analysis for sorted GFP positive cells. The same amount of protein (5 μg) from the sorted GFP positive cells (Sort) and presorted Day 18 embryoid bodies (Pre-S) were loaded on 10% polyacrylamide SDS gel. Protein expression of α-fetoprotein, albumin, and actin was examined using specific antibodies. (See Color Insert.)

18 embryoid bodies when determined by flow cytometry. Addition of the growth factors (acidic FGF, HGF, and oncostatin M) during differentiation increased population of the GFP positive cells up to $2.4 \pm 0.9\%$ ($p < 0.05$). These data indicate that visceral endoderm/hepatic lineage population is relatively small in embryoid bodies regardless of the additional growth factors. When the GFP positive cells were isolated, they expressed α-fetoprotein, α1-antitrypsin, and albumin exclusively (Fig. 2b). GFP negative cells did not express these genes, while some of them expressed other lineage markers such as nestin or collagen type II, which implicate neural precursor[34] or chondrocyte[35] differentiation, respectively. The GFP positive cells also expressed cytokeratin19, which is a marker of hepatoblasts[36] and bile ductal cells,[37] and CYP1A1, which is one of the isotype of P450 enzymes. Furthermore, α-fetoprotein and albumin protein were highly expressed in the sorted GFP positive cells when compared to presorted cells (Fig. 2c).

Hepatocytes can be distinguished from visceral endoderm in embryoid bodies by using other markers. Forrester *et al.* generated an ES cell line from transgenic mice carrying a gene trap vector insertion into an ankyrin repeat-containing gene.[38] The β-gal reporter gene in the transgenic embryos was first expressed in cells of embryonic foregut that were committed to hepatic fate and then exclusively in liver during early embryogenesis.[39] Importantly, no β-gal expression was observed in the extraembryonic endoderm of embryos carrying this gene trap integration. During *in vitro* differentiation, β-gal expression was observed in embryoid bodies at discrete regions by Day 9, often within the central core of the embryoid body. When the embryoid bodies were cultured on the plate, the β-gal-positive cells distributed like an arch shaped pattern, similar to the data shown above using α-fetoprotein promoter-derived GFP expression (Fig. 2a). The β-gal-positive cells were stained with albumin, α-fetoprotein and transferrin. The β-gal-positive cells at Day 12

[34] C. Andressen, E. Stocker, F. J. Klinz, N. Lenka, J. Hescheler, B. Fleischmann, S. Arnhold, and K. Addicks, *Stem Cells* **19**, 419 (2001).

[35] J. Kramer, C. Hegert, K. Guan, A. M. Wobus, P. K. Muller, and J. Rohwedel, *Mech. Dev.* **92**, 193 (2000).

[36] V. J. Desmet, P. Van Eyken, and R. Sciot, *Hepatology* **12**, 1249 (1990).

[37] J. Cocjin, P. Rosenthal, V. Buslon, L. Luk Jr, L. Barajas, S. A. Geller, B. Ruebner, and S. French, *Hepatology* **24**, 568 (1996).

[38] L. M. Forrester, A. Nagy, M. Sam, A. Watt, L. Stevenson, A. Bernstein, A. L. Joyner, and W. Wurst, *Proc. Natl. Acad. Sci. USA* **93**, 1677 (1996).

[39] A. J. Watt, E. A. Jones, J. M. Ure, D. Peddie, D. I. Wilson, and L. M. Forrester, *Mech. Dev.* **100**, 205 (2001).

also showed ultrastructual characteristics resembling primitive hepatic precursors (E10.5).[40]

To examine the function of hepatic precursors within embryoid bodies, Chinzei *et al.* measured urea synthesis activity in whole embryoid bodies. Although embryoid bodies contain various cell types other than hepatic lineage cells, Day 18 embryoid bodies showed low urea synthesis activity (about 40-fold lower compared to mouse primary hepatocytes). Additionally, dissociated embryoid bodies were injected into spleen of mice that were previously treated with 2-acetylaminofluorene and partial hepatectomy. Four weeks after injection of Day 9 embryoid bodies, integrations of ES-derived cells into the recipient liver were observed based on the presence of the Y-chromosome. The integrated cells demonstrated hepatocyte morphology and albumin expression determined by immunostaining.[33]

Yamada *et al.* demonstrated a cellular uptake of indocyanine green (ICG) within embryoid bodies, which is a functional characteristic of hepatocytes. Within 14 days after embryoid body formation, ICG-positive cells became apparent as clusters forming trabecula-like structure. These clusters were, however, lacking sinusoids, blood vessels and bile ducts. Electron microscopy revealed that the ICG-positive cells had numerous mitochondria, lysosomes, rough and smooth endoplasmic reticulum, and well-developed Golgi apparatus. Bile canaliculi were found between adjacent cells and sealed with tight junctions. Most of the ICG-positive cells were adjacent to the cardiac beating muscles, suggesting that embryoid bodies themselves provide an appropriate microenvironment that supports hepatogeneis *in vitro*.[41]

Induction of Hepatic Differentiation Without Forming Classical Embryoid Bodies

In the methods described above, hepatic differentiation of ES cells is dependent on embryoid body formation, which likely provides an appropriate microenvironment to mimic embryonic development. Although this system may serve as a useful *in vitro* model to study hepatic development, it may not be an ideal method to enrich and purify hepatic precursors because the yield of hepatic differentiation is minimum in embryoid bodies as described earlier. The following methods, in contrast, utilized embryoid

[40] E. A. Jones, D. Tosh, D. I. Wilson, S. Lindsay, and L. M. Forrester, *Exp. Cell Res.* **272**, 15 (2002).

[41] T. Yamada, M. Yoshikawa, S. Kanda, Y. Kato, Y. Nakajima, S. Ishizaka, and Y. Tsunoda, *Stem Cells* **20**, 146 (2002).

body formation either minimally or not at all to induce hepatic differentiation of ES cells.

Yin *et al.* also utilized the α-fetoprotein promoter, to identify liver progenitor cells from differentiated ES cells. They inserted the GFP gene downstream of the endogenous promoter of the α-fetoprotein gene using homologous recombination. When the targeted ES clones were induced to form embryoid bodies, green fluorescent positive cells were detected in a similar manner as described above with the truncated α-fetoprotein-promoter (Fig. 2a). Interestingly, they successfully enriched GFP positive cells using a methylcellulose culture method, a protocol often used for hematopoietic cell differentiation. In the protocol, ES cells were first grown in methylcellulose-containing medium (MethoCult M3134, StemCell Technologies, Inc.) for six days. At this point, less than 10% of the cells in these embryoid bodies were GFP positive. To further increase the yield, embryoid bodies were then dissociated into single-cell suspensions by collagenase and replated in the same methylcellulose-based medium. After another six day-culture, GFP positive cells constituted 30–50% of the cells by FACS analysis. Isolated GFP positive cells were transplanted into a liver of apolipoprotein-E- (ApoE) or haptoglobin-deficient mice. An engraftment of ES cell-derived cells was confirmed by β-gal staining. β-gal positive cells in the liver were also stained by albumin immnohistochemistry. Furthermore, ApoE protein was detected in hepatocytes and serum of transplanted ApoE-deficient mice. Similarly, haptoglobin protein was detected in serum of transplanted haptoglobin-deficient mice. The levels of haptoglobin and ApoE detected in transplanted animals were low due to limited engraftment efficiency (0.01%) of transplanted GFP positive cells in the host liver.[42]

Ishizaka *et al.* demonstrated an alternative method to induce functional hepatocytes *in vitro* independent of classical embryoid body formation. They forced expression of hepatocyte nuclear factor 3β (HNF-3β) in ES cells. The ES clones overexpressing HNF-3β were cultivated in α-MEM medium containing 10% FBS, FGF-2, dexamethasone, L-ascorbic-2-phosphate, and nicotinamide in 96-well round-bottomed Sumilon cell-tight spheroid plates for 1 month. The spheroid cells were maintained in the same medium without FGF2 but containing chitin fibers. In their protocol, albumin immunostaining was found in the surface cells of spheroids. The spheroids achieved three-fold higher urea synthesis rates and secreted 70-fold higher concentrations of triacylglycerol than primary cultured

[42] Y. Yin, Y. K. Lim, M. Salto-Tellez, S. C. Ng, C. S. Lin, and S. K. Lim, *Stem Cells* **20**, 338 (2002).

hepatocytes. When these cells were cultivated as spheroids, these functions were maintained up to 4 months.[43]

As another direct induction method of hepatic lineage differentiation from ES cells, Geron corporation invented the method to differentiate human ES cells into hepatocyte-like cells by using chemical compounds (International patent application WO 01/81549 A2). In their protocol, human ES cells were cultured in the medium containing 20% serum replacement and 1% DMSO for the first 4 days. For the next 6 days, 2.5% sodium butyrate was added into the medium. Sixty-three percent of these cells were expressing albumin, and more than 80% of them were α1-antitrypsin positive by immunohistochemistry. None of them expressed α-fetoprotein. Cytochrome P450 activity was demonstrated in these cells for another indicator for *in vitro* liver function.

Conclusion and Future Directions

The potential to generate functional hepatoctyes *in vitro* from ES cells has been demonstrated by several different methods as described above. Embryoid body formation provides an appropriate microenvironment for hepatic differentiation. Hence, it may be a useful model to study liver development *in vitro*. Especially when combined with gene modulation at ES cell level, these methods may be useful to rapidly determine the role of genes in hepatic lineage commitment and/or hepatic functions.[32,44] It may not be ideal, however, to enrich and purify hepatic precursors from embryoid bodies because relative population of hepatocyte precursors is likely limited in embryoid bodies regardless of additional growth factors.

Several methods have been introduced to generate hepatic precursors more efficiently from ES cells without forming classical embryoid bodies as described above. Such methods may be more practical when we consider clinical applications of ES-derived hepatic precursors in the future. ES cell differentiation using a chemical agent, without introducing exogenous gene, would be particularly promising. Multiple studies have demonstrated that ES cell-derived hepatocytes function *in vivo*. It should be noted, however, none of them have yet shown restoration of liver disease models after transplantation of ES cell-derived hepatocytes. Other remaining challenges will be reconstructing liver *in vitro* as organoid using tissue engineering

[43] S. Ishizaka, A. Shiroi, S. Kanda, M. Yoshikawa, H. Tsujinoue, S. Kuriyama, T. Hasuma, K. Nakatani, and K. Takahashi, *FASEB J.* **16**, 1444 (2002).

[44] T. Minamino, T. Yujiri, P. J. Papst, E. D. Chan, G. L. Johnson, and N. Terada, *Proc. Natl. Acad. Sci. USA* **96**, 15127 (1999).

technique.[45] Continued work is necessary for a complete understanding of the potential of this promising cell population.

Acknowledgment

The authors are indebted to J. M. Crawford and A. M. Meacham (University of Florida, Gainesville, FL) for critical reading of the manuscript. This work was supported partly by NIH-DK-59699 to N. T. and by American Liver Foundation Postdoctoral Research Fellowship to T. H.

[45] G. K. Michalopoulos, W. C. Bowen, K. Mule, J. C. Lopez-Talavera, and W. Mars, *Hepatology* **36**, 278 (2002).

[21] Differentiation of Mouse Embryonic Stem Cells into Pancreatic and Hepatic Cells

By Gabriela Kania, Przemyslaw Blyszczuk, Jaroslaw Czyz, Anne Navarrete-Santos, and Anna M. Wobus

Introduction

Pancreas and liver cells are derivatives of the definitive endoderm. During embryogenesis, the pancreas develops from dorsal and ventral regions of the foregut,[1,2] whereas the liver originates from the foregut adjacent to the ventral pancreas compartment.[3] The development of pancreas and liver is regulated by specific transcription factors and signaling molecules. Pancreas development is controlled by signals from the notochord including TGF-β family members, activinβB and FGF-2, factors repressing sonic hedgehog (shh), a negative regulator of the homeobox gene Pdx1. The transcription factor Pdx1 is expressed in pancreatic buds and becomes restricted to β cells in the adult animal.[4,5] Islet-1 (Isl-1) and neurogenin3 (ngn3) regulate the development of early endocrine

[1] B. S. Spooner, B. T. Walther, and W. J. Rutter, *J. Cell Biol.* **47**, 235–246 (1970).
[2] J. M. Slack, *Development* **121**, 1569–1580 (1995).
[3] K. S. Zaret, *Annu. Rev. Physiol.* **58**, 231–251 (1996).
[4] J. Jonsson, L. Carlsson, T. Edlund, and H. Edlund, *Nature* **371**, 606–609 (1994).
[5] M. Hebrok, S. K. Kim, B. St Jacques, A. P. McMahon, and D. A. Melton, *Development* **127**, 4905–4913 (2000).

0076-6879/2003 $35.00

progenitors.[6,7] Two members of the Pax gene family, Pax6 and Pax4, are essential for the specification of endocrine α[8]- and β[9]-cells, respectively.

The first evidence of liver development is the expression of albumin and α-fetoprotein in endodermal cells.[10] Growth factors of the FGF family released from cardiac mesoderm activate endoderm to express liver- instead of pancreas-specific genes.[11] BMP signaling from the septum transversum mesenchyme is required for the hepatogenic response to FGF and for secondary inductions resulting in the outgrowth of liver buds.[12] BMP signaling maintains the endodermal expression of the transcription factor GATA-4,[13] which together with HNF3β, is essential for the hepatic specification of endoderm.[14,15] Additional signaling molecules, such as hepatocyte-growth factor (HGF),[16] c-met,[17] c-jun[18] and β1-integrins[15] are involved in liver development. There is evidence that albumin-expressing cells from the endoderm (hepatoblasts) differentiate into hepatocytes and bile duct cells.[3]

Due to the high incidence of liver- and pancreas-related diseases in the human population, and the lack of suitable donor cells and tissues for transplantation, the need of generating a "surrogate" pancreatic and hepatic cell population has been strengthened. Several sources of cells for transplantation are considered including fetal and adult stem cells, or genetically modified cell lines. An alternative source for generating transplantable cells are embryonic stem (ES) cells which have an almost unlimited proliferation capacity, while retaining the potential to differentiate *in vitro* into cells of all

[6] U. Ahlgren, S. L. Pfaff, T. M. Jessell, T. Edlund, and H. Edlund, *Nature* **385**, 257–260 (1997).

[7] V. M. Schwitzgebel, D. W. Scheel, J. R. Conners, J. Kalamaras, J. E. Lee, D. J. Anderson, L. Sussel, J. D. Johnson, and M. S. German, *Development* **127**, 3533–3542 (2000).

[8] L. St Onge, B. Sosa-Pineda, K. Chowdhury, A. Mansouri, and P. Gruss, *Nature* **387**, 406–409 (1997).

[9] B. Sosa-Pineda, K. Chowdhury, M. Torres, G. Oliver, and P. Gruss, *Nature* **386**, 399–402 (1997).

[10] R. Gualdi, P. Bossard, M. Zheng, Y. Hamada, J. R. Coleman, and K. S. Zaret, *Genes Dev.* **10,** 1670–1682 (1996).

[11] J. Jung, M. Zheng, M. Goldfarb, and K. S. Zaret, *Science* **284**, 1998–2003 (1999).

[12] K. S. Zaret, *Curr. Opin. Genet. Dev.* **11**, 568–574 (2001).

[13] J. M. Rossi, N. R. Dunn, B. L. Hogan, and K. S. Zaret, *Genes Dev.* **15**, 1998–2009 (2001).

[14] N. Narita, M. Bielinska, and D. B. Wilson, *Development* **124**, 3755–3764 (1997).

[15] M. Grompe and M. J. Finegold, *in* "Stem Cell Biology" (D. R. Marshal, R. L. Garder, and A. Gottlieb, eds.), pp. 455–497. Cold Spring Harbor Laboratory Press, 2001.

[16] C. Schmidt, F. Bladt, S. Goedecke, V. Brinkmann, W. Zschiesche, M. Sharpe, E. Gherardi, and C. Birchmeier, *Nature* **373**, 699–702 (1995).

[17] F. Bladt, D. Riethmacher, S. Isenmann, A. Aguzzi, and C. Birchmeier, *Nature* **376**, 768–771 (1995).

[18] F. Hilberg, A. Aguzzi, N. Howells, and E. F. Wagner, *Nature* **365**, 179–181 (1993).

three primary germ layers.[19,20] It has been shown, that ES cells can be successfully differentiated into insulin-producing cells of mouse[21,22] and human[23] origin, as well as into functional hepatocyte-like cells *in vitro*.[24–27]

The close relationship between pancreas and liver development is substantiated by the findings that ductal epithelial pancreatic progenitor cells may be reprogrammed into liver cells[28] and, vice versa, liver stem cells may be directed into the pancreatic endocrine lineage.[29] Therefore, it was postulated that pancreatic and hepatic progenitor cells share a common stem/progenitor cell population.[30] Cells expressing the intermediate filament protein nestin may represent such a progenitor cell type, because *in vitro*, nestin-positive (nestin+) cells selected from ES cells have been differentiated into pancreatic endocrine[22,31–33] and hepatic[31] cells after addition of specific growth and extracellular matrix factors. *In vivo*,

[19] A. M. Wobus, *Mol. Aspects Med.* **22**, 149–164 (2001).

[20] A. M. Wobus, K. Guan, H.-T. Yang, and K. Boheler, *in* "Embryonic Stem Cells: Methods and Protocols" (K. Turksen, ed.), Methods in Molecular Biology, vol. 185: pp. 127–156. Humana Press Inc., Totowa, NJ, 2002.

[21] B. Soria, E. Roche, G. Berna, T. Leon-Quinto, J. A. Reig, and F. Martin, *Diabetes* **49**, 157–162 (2000).

[22] N. Lumelsky, O. Blondel, P. Laeng, I. Velasco, R. Ravin, and R. McKay, *Science* **292**, 1389–1394 (2001).

[23] S. Assady, G. Maor, M. Amit, J. Itskovitz-Eldor, K. L. Skorecki, and M. Tzukerman, *Diabetes* **50**, 1691–1697 (2001).

[24] T. Hamazaki, Y. Iiboshi, M. Oka, P. J. Papst, A. M. Meacham, L. I. Zon, and N. Terada, *FEBS Lett.* **497**, 15–19 (2001).

[25] E. A. Jones, D. Tosh, D. I. Wilson, S. Lindsay, and L. M. Forrester, *Exp. Cell Res.* **272**, 15–22 (2002).

[26] T. Yamada, M. Yoshikawa, S. Kanda, Y. Kato, Y. Nakajima, S. Ishizaka, and Y. Tsunoda, *Stem Cells* **20**, 146–154 (2002).

[27] R. Chinzei, Y. Tanaka, K. Shimizu-Saito, Y. Hara, S. Kakinuma, M. Watanabe, K. Teramoto, S. Arii, K. Takase, C. Sato, N. Terada, and H. Teraoka, *Hepatology* **36**, 22–29 (2002).

[28] M. D. Dabeva, S. G. Hwang, S. R. Vasa, E. Hurston, P. M. Novikoff, D. C. Hixson, S. Gupta, and D. A. Shafritz, *Proc. Natl. Acad. Sci. USA* **94**, 7356–7361 (1997).

[29] L. Yang, S. Li, H. Hatch, K. Ahrens, J. G. Cornelius, B. E. Petersen, and A. B. Peck, *Proc. Natl. Acad. Sci. USA* **99**, 8078–8083 (2002).

[30] G. Deutsch, J. Jung, M. Zheng, J. Lora, and K. S. Zaret, *Development* **128**, 871–881 (2001).

[31] H. Zulewski, E. J. Abraham, M. J. Gerlach, P. B. Daniel, W. Moritz, B. Muller, M. Vallejo, M. K. Thomas, and J. F. Habener, *Diabetes* **50**, 521–533 (2001).

[32] Y. Hori, I. C. Rulifson, B. C. Tsai, J. J. Heit, J. D. Cahoy, and S. K. Kim, *Proc. Natl. Acad. Sci. USA*. **99**, 16105–16110 (2002).

[33] P. Blyszczuk, J. Czyz, G. Kania, M. Wagner, U. Roll, L. St-Onge, and A. M. Wobus, *Proc. Natl. Acad. Sci. USA*. **100**, 998–1003 (2003).

nestin is transiently expressed in different cell types of embryonic and adult tissues, including the developing central nervous system,[34,35] skeletal muscle,[36] heart,[37] endothelial,[38] mesenchymal pancreatic,[39,40] and hepatic stellate[41] cells and is suggested to play a transient role in various proliferation and migration processes of progenitor cells.[42] Although, nestin+ cells may participate in the neogenesis of endocrine islet cells,[31,43] recent data showed that *in vivo*, nestin is expressed in mesenchymal and not in epithelial cells, where endocrine progenitors reside.[39] We therefore, speculate that nestin+ cells are characterized by a high developmental plasticity and represent a progenitor cell population which *in vitro* under the influence of genetic[33] or epigenetic factors[22] can be programmed into pancreatic endocrine[22,31] or into hepatic[29] cells. As known for many years, nestin+ cells also develop efficiently into neural[35] cell fates *in vitro*.

Here, we describe methods to direct ES cells into pancreatic and hepatic cells via the generation of nestin+ cells. In comparison to the normal ("basic") differentiation protocol,[20] selection of nestin+ cells resulted in qualitatively similar cell types, but the "nestin+ cell selection" protocol was much more efficient to generate insulin-[33] and albumin-producing cells. In addition, a histotypic differentiation model is presented. It is known for a long time that the three-dimensional organization of cells *in vitro* results in increased tissue-specific functions of various cell types.[44–48] The histotypic

[34] K. Frederiksen and R. D. McKay, *J. Neurosci.* **8**, 1144–1151 (1988).
[35] U. Lendahl, L. B. Zimmerman, and R. D. McKay, *Cell* **60**, 585–595 (1990).
[36] T. Sejersen and U. Lendahl, *J. Cell Sci.* **106**, 1291–1300 (1993).
[37] A. M. Kachinsky, J. A. Dominov, and J. B. Miller, *J. Histochem. Cytochem.* **43**, 843–847 (1995).
[38] J. Mokry and S. Nemecek, *Folia Biol. (Praha)* **44**, 155–161 (1998).
[39] L. Selander and H. Edlund, *Mech. Dev.* **113**, 189–192 (2002).
[40] J. Lardon, I. Rooman, and L. Bouwens, *Histochem. Cell Biol.* **117**, 535–540 (2002).
[41] T. Niki, M. Pekny, K. Hellemans, P. D. Bleser, K. V. Berg, F. Vaeyens, E. Quartier, F. Schuit, and A. Geerts, *Hepatology* **29**, 520–527 (1999).
[42] J. Mokry and S. Nemecek, *Acta Medica. (Hradec. Kralove)* **41**, 73–80 (1998).
[43] E. Hunziker and M. Stein, *Biochem. Biophys. Res. Commun.* **271**, 116–119 (2000).
[44] P. G. Layer, A. Robitzki, A. Rothermel, and E. Willbold, *Trends Neurosci.* **25**, 131–134 (2002).
[45] P. A. Halban, S. L. Powers, K. L. George, and S. Bonner-Weir, *Diabetes* **36**, 783–790 (1987).
[46] N. Koide, K. Sakaguchi, Y. Koide, K. Asano, M. Kawaguchi, H. Matsushima, T. Takenami, T. Shinji, M. Mori, and T. Tsuji, *Exp. Cell Res.* **186**, 227–235 (1990).
[47] J. Z. Tong, P. De Lagausie, V. Furlan, T. Cresteil, O. Bernard, and F. Alvarez, *Exp. Cell Res.* **200**, 326–332 (1992).
[48] J. R. Friend, F. J. Wu, L. K. Hansen, R. P. Remmel, and W.-S. Hu, *in* "Tissue Engineering Methods and Protocols" (J. R. Morgan and M. L. Yarmush, eds.), Methods in Molecular Medicine, vol. 18: pp. 245–252. Humana Press Inc., Totowa, NJ, 1999.

culture model enabled the generation of highly differentiated islet- and hepatocyte-like spheroids.

Material and Methods

Culture of Undifferentiated ES Cells

Embryonic stem (ES) cells of line R1[49] were cultivated on a feeder layer of primary mouse embryonic fibroblasts on gelatin (0.1%)-coated petri dishes (Falcon Becton Dickinson, Heidelberg, Germany) in Dulbecco's modified Eagle's medium (DMEM, Gibco BRL, Life Technologies, Eggenstein, Germany) supplemented with 15% heat-inactivated fetal calf serum (FCS, selected batches, Gibco), L-glutamine (Gibco, 2 m*M*), β-mercaptoethanol (Serva, Heidelberg, Germany, final concentration 5×10^{-5} *M*), nonessential amino acids (Gibco, stock solution diluted 1 : 100), penicillin–streptomycin (Gibco, stock solution diluted 1 : 100) and 10 ng/ml recombinant human leukemia inhibitory factor (LIF, for preparation, see Ref. 50) as described.[20]

Differentiation of ES cells into Pancreatic and Hepatic Cells

Two protocols were comparatively investigated, the "basic" protocol and the "nestin + cell selection" protocol:

1. According to the "basic" protocol, ES cells were differentiated as embryoid bodies (EB) via the "hanging drop" method[20] and after plating, the cells were cultured in the specific differentiation media. This protocol generates a population of different cell types including pancreatic[33] and hepatic cells.
2. By the "nestin + cell selection" protocol, ES cells were cultured as EB and after plating, nestin + progenitor cells were selected in growth factor-containing, but serum-free medium. Further, the nestin + cells were directed either into pancreatic or hepatic (or neural) cell types by specific differentiation factors. Whereas both protocols, in principle, gave rise to pancreatic and hepatic cell types, the "nestin + cell selection" protocol resulted in higher differentiation efficiency and an increased number of mature cells compared to the "basic" protocol.[33]

[49] A. Nagy, J. Rossant, R. Nagy, W. Abramow-Newerly, and J. C. Roder, *Proc. Natl. Acad. Sci. USA* **90**, 8424–8428 (1993).

[50] J. Rohwedel, U. Sehlmeyer, J. Shan, A. Meister, and A. M. Wobus, *Cell Biol. Int.* **20**, 579–587 (1996).

Selection of Nestin+ Cells

ES cells ($n = 200$) were cultivated as EBs in "hanging drops" for two days and after transfer into bacteriological plates (Greiner, Germany) for 2 days in suspension in Iscove's modification of DMEM (IMDM, Gibco) supplemented with 15% FCS, L-glutamine, nonessential amino acids (see above) and α-monothioglycerol (Sigma, Steinheim, Germany; final concentration 450 μM) instead of β-mercaptoethanol. Penicillin–streptomycin may be added to the cultures (see above). At day 4, EBs ($n = 30$) were plated onto gelatin-coated 6 cm tissue culture dishes (Nunc, Germany) and cultivated in IMDM supplemented by 10% FCS. The optimal density of EB-derived nestin+ cells is critical for the efficient generation of pancreatic and hepatic cell types. At day 4 + 1, the medium was exchanged by DMEM/F12 (Gibco) supplemented with 5 μg/ml insulin, 30 nM sodium selenite (both from Sigma), 50 μg/ml transferrin, 5 μg/ml fibronectin (both from Gibco) without FCS and cells were cultivated for further 7 days.[51]

ES Cell Differentiation into Pancreatic or Hepatic Cells

Differentiation into Pancreatic Cells

At day 4 + 8, EB outgrowths were dissociated by 0.1% trypsin (Gibco)/0.08% EDTA (Sigma) in PBS (1 : 1) for 1 min, collected by centrifugation and replated onto tissue culture plates (Nunc) in DMEM/F12 containing 20 nM progesterone, 100 μM putrescine, 1 μg/ml laminin, 10 mM nicotinamide (all from Sigma), B27 media supplement (Gibco), 25 μg/ml insulin, 50 μg/ml transferrin, 5 μg/ml fibronectin, and 30 nM sodium selenite (= "pancreatic differentiation medium") supplemented with 10% FCS and penicillin–streptomycin (see above). For immunofluorescence analysis, cells were plated onto poly-L-ornithine/laminin-coated cover slips, onto 3 cm culture dishes (ELISA), and onto 6 cm culture dishes for RT-PCR and further histotypic culture. At Day 4 + 9, FCS was removed and the cells were cultivated until day 4 + 8 + 20 or other stages (Fig. 1A).[33]

Differentiation into Hepatic Cell Types

At Day 4 + 8, EBs were dissociated by 0.1% trypsin (Gibco)/0.08% EDTA (Sigma) in PBS (1 : 1) for 1 min, collected by centrifugation and replated in Hepatocyte Culture Medium (HCM, "hepatic differentiation medium"). HCM composed of 500 ml Hepatocyte Basal Medium (Modified Williams' E), 0.5 ml ascorbic acid, 10 ml BSA-FAF

[51] A. Rolletschek, H. Chang, K. Guan, J. Czyz, M. Meyer, and A. M. Wobus, *Mech. Dev.* **105**, 93–104 (2001).

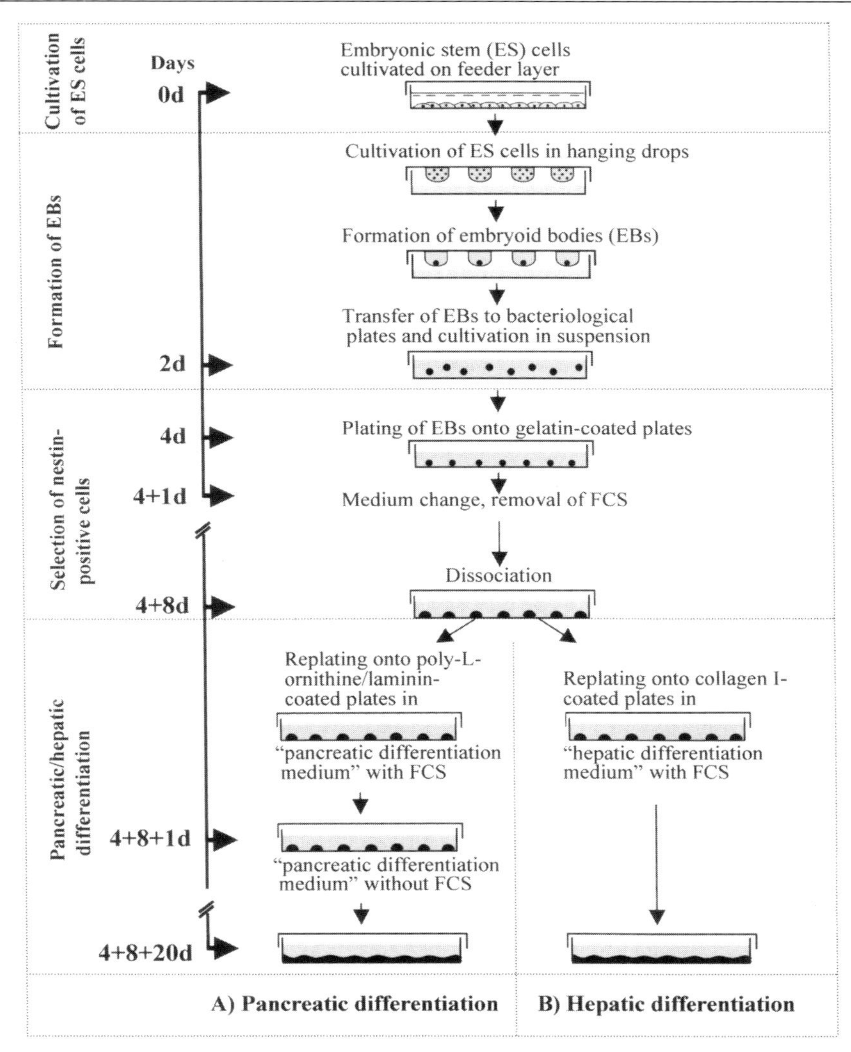

FIG. 1. Schematic presentation of the "nestin+ cell selection" protocol used to differentiate ES cells via embryoid bodies (EBs) and differentiation induction into pancreatic (A) and hepatic (B) cell types.

(bovine serum albumin-fatty acid free), 0.5 ml hydrocortisone, 0.5 ml transferrin, 0.5 ml insulin, 0.5 ml human epithelial growth factor (EGF) and 0.5 ml gentamycin-amphothericin (GA-1000; all from Clonetics, Bio-Whittaker; A Cambrex Company; Belgium) is supplemented by 20% FCS and penicillin–streptomycin (see above) and prepared immediately before use. For immunofluorescence, cells were transferred onto collagen type I-coated cover slips, onto 3 cm culture dishes (RT-PCR, ELISA) and onto 6 cm collagen type I-coated plates for further histotypic culture. The cells were cultivated until day 4+8+20 or other stages (Fig. 1B).

Histotypic Differentiation into Spheroids

The three-dimensional histotypic culture system was adapted for the differentiation and maturation of pancreatic and hepatic cells generated by the "nestin+ cell selection" protocol. During cultivation in "spinner culture" as spheroids, the cells aggregate, proliferate and establish specific cell-to-cell interactions. A comparison of spheroids with monolayer cultures showed that spheroids resulted in higher proliferation rates, and their differentiation properties closely resembled the *in vivo* situation.[44]

ES-derived cells cultivated according to the "nestin+ cell selection" protocol for 4+8+20 (pancreatic cells) or 4+8+14 days (hepatic cells), were dissociated for 1 min by 0.1% trypsin: 0.08% EDTA (1 : 1) in PBS (pancreatic cells) or 0.2% trypsin/0.02% EDTA (1 : 1) in PBS (hepatic cells), respectively.

Cells ($n = 1.5–2 \times 10^7$) were collected by centrifugation and transferred into 6 cm bacteriological petri dishes in "pancreatic" or "hepatic differentiation medium" (note that "hepatic differentiation medium" is supplemented by 10% FCS).

After 24 hr of static suspension culture, cell aggregates were transferred into 100 ml "spinner" culture flasks and cultivated in "pancreatic" or "hepatic differentiation medium" ("hepatic differentiation medium" supplemented by 10% FCS) by the CELLSPIN system (Cellspin, Integra Bioscience AG, Switzerland) at 30 rpm agitation at 37°C up to 10 days.

Analysis of Differentiated Phenotypes

The methods presented here allow the qualitative and quantitative determination of pancreas- and liver-specific markers at the mRNA (RT-PCR) and protein (immunofluorescence, immunohistochemistry) level. In addition, functional properties are detected by ELISA.

Semi-Quantitative RT-PCR Analysis

ES-derived cells were collected and suspended in lysis buffer composed of 4 *M* guanidinium thiocyanate, 25 m*M* sodium citrate (pH 7);

0.5% (w/v) sarcosyl and 0.1 *M* β-mercaptoethanol. Total RNA was isolated by the single step extraction method according to Chomczynski and Sacchi.[52] mRNA was reverse transcribed using PolyT tail primer Oligo d(T) and MuLV reverse transcriptase (Perkin Elmer, Überlingen, Germany).

cDNAs were amplified using oligonucleotide primers complementary to transcripts of the analyzed genes (see Table I) and AmpliTaq DNA polymerase (Perkin Elmer) as described.[20,53] The PCR reaction was electrophoretically separated on 2% (w/v) agarose gels containing 0.35 μg/ml of ethidium bromide and gels were illuminated with UV light, stored by E.A.S.Y. system (Herolab GmbH, Wiesloch, Germany) and analyzed by the TINA2.08e software (Raytest Isotopenmeßgeräte GmbH, Straubenhardt, Germany).

The intensity of the ethidium bromide fluorescence signals was determined from the area under the curve for each peak and the data of target genes were plotted as percentage changes in relation to the expression of the housekeeping gene β-tubulin. Expression of β-tubulin served to normalize the amounts of RT-products of all samples. The sample of the target gene with the highest ratio of the gene product in relation to β-tubulin was set as 100% and served as reference for the other samples of the target gene.[20]

Immunofluorescence Analysis

For immunofluorescence, EB outgrowths of ES cells growing on cover slips were either fixed with 4% paraformaldehyde (PFA) in PBS at room temperature (RT) for 20 min or in methanol:acetone (Met:Ac; 7:3) at −20°C for 10 min, depending on the antibody used (see Table II). After rinsing (3×) in PBS, bovine serum albumin (BSA, 1% in PBS) was used to inhibit unspecific labeling (30 min) at RT. Cells were incubated with the primary antibodies in specific dilutions (Table II) at 37°C for 60 min. Samples were washed (3×) in PBS and incubated with fluorescence-labeled secondary antibodies (diluted in 1% BSA in PBS) at 37°C for 60 min (Table III). To label the nuclei for a semi-quantitative estimation of immunofluorescence signals, cells were incubated in 5 μg/ml Hoechst 33342 in PBS at 37°C for 10 min. After washing (3×) in PBS and (1×) in Aqua dest., the specimens were embedded in mounting medium (Vectashield, Vector Laboratories Inc., Burlingame, CA, USA). Labeled cells were analyzed by the fluorescence microscope ECLIPSE TE300 (Nikon, Japan), or the

[52] P. Chomczynski and N. Sacchi, *Anal. Biochem.* **162**, 156–159 (1987).
[53] A. M. Wobus, G. Kaomei, J. Shan, M. C. Wellner, J. Rohwedel, G. Ji, B. Fleischmann, H. A. Katus, J. Hescheler, and W. M. Franz, *J. Mol. Cell Cardiol.* **29**, 1525–1539 (1997).

TABLE I
PRIMERS FOR RT-PCR ANALYSIS

Gene	Primer sequence (Forward/Reverse)	Annealing temperature	No. of cycles	Product size
Pancreatic markers:				
Glut-2	5′-TTCGGCTATGACATCGGTGTG 5′-AGCTGAGGCCAGCAATCTGAC	60°C	40	556 bp
IAPP	5′-TGATATTGCTGCCTCGGACC 5′-GGAGGACTGGACCAAGGTTG	65°C	40	233 bp
Insulin	5′-CCCTGCTGGCCCTGCTCTT 5′-ATGCTGGTGCAGCACTGA	60°C	40	270 bp
Isl-1	5′-GTTTGTACGGGATCAAATGC 5′-ATGCTGCGTTTCTTGTCCTT	60°C	35	514 bp
Ngn3	5′-TGGCGCCTCATCCCTTGGATG 5′-AGTCACCCACTTCTGCTTCG	60°C	40	157 bp
Pax4	5′-ACCAGAGCTTGCACTGGACT 5′-CCCATTTCAGCTTCTCTTGC	60°C	40	300 bp
Pax6	5′-TCACAGCGGAGTGAATCAG 5′-CCCAAGCAAAGATGGAAG	58°C	35	332 bp
Pdx1	5′-CTTTCCCGTGGATGAAATCC 5′-GTCAAGTTCAACATCACTGCC	60°C	45	230 bp
Shh	5′-TGAGGACGGCCATCATTCAG 5′-CTCCAGCGTCTCGATCACGT	59°C	45	173 bp
Hepatic markers:				
albumin	5′-GTCTTAGTGAGGTGGAGCAT 5′-ACTACAGCACTTGGTAACAT	58°C	35	569 bp
α-1-antitrypsin	5′-CAATGGCTCTTTGCTCAACA 5′-AGTGGACCTGGGCTAACCTT	63°C	30	518 bp
α-fetoprotein	5′-CACTGCTGCAACTCTTCGTA 5′-CTTTGGACCCTCTTCTGTGA	58°C	35	301 bp
HNF3β	5′-GCGGGTGCGGCCAGTAG 5′-GCTGTGGTGATGTTGCTGCTCG	63°C	40	378 bp
Housekeeping gene:				
β-tubulin	5′-TCACTGTGCCTGAACTTACC 5′-GGAACATAGCCGTAAACTGC	60°C	28	317 bp

confocal laser scanning microscope (CLSM) LSM-410 (Carl Zeiss, Jena, Germany) using the following excitation lines/barrier filters: 364 nm/450-490BP (Hoechst 33342), 488 nm/510-525BP (FITC/DTAF), 543 nm/570LP (Cy3).

Semi-Quantitative Determination of Immunofluorescence Signals

Quantification of immunofluorescence signals was performed by two alternative methods depending on the cell culture status. Cells growing in

TABLE II
SELECTED PRIMARY ANTIBODIES TO CHARACTERIZE NESTIN+ PROGENITOR CELLS, PANCREATIC AND HEPATIC CELL TYPES

Primary antibody	Dilution	Company	Fixation
Progenitor cells:			
mouse anti-nestin (clone rat 401)	1:3	Development Studies Hybridoma Bank, IA, USA	4% PFA
Pancreatic markers:			
mouse anti-insulin	1:40	Sigma	4% PFA
rabbit anti-glucagon	1:40	Dako Corporation, CA, USA	4% PFA
rabbit anti-somatostatin	1:40	Dako	4% PFA
rabbit anti-PP	1:40	Dako	4% PFA
Hepatic markers:			
goat anti-α-fetoprotein	1:100	Santa Cruz Biotechnol, USA	Met:Ac (7:3)
goat anti-amylase	1:100	Santa Cruz	Met:Ac (7:3)
goat anti-dipeptidyl peptidase IV	1:100	Santa Cruz	Met:Ac (7:3)
sheep anti-albumin	1:100	Serotec, USA	Met:Ac (7:3)
rabbit anti-α1-antitrypsin	1:100	Sigma	Met:Ac (7:3)
mouse anti-cytokeratin 18 (clone KS-B17.2)	1:100	Sigma	Met:Ac (7:3)
mouse anti-cytokeratin 14 (clone CKB1)	1:100	Sigma	Met:Ac (7:3)
mouse anti-cytokeratin 19	1:100	Chemicon, Hofheim, Germany	Met:Ac (7:3)
mouse anti-glutamine synthetase	1:100	BD Transduction Laboratories, USA	Met:Ac (7:3)

TABLE III
FLUORESCENCE-LABELED SECONDARY ANTIBODIES USED FOR IMMUNOFLUORESCENCE ANALYSIS

Secondary antibody	Dilution	Company
Cy3™-conjugated goat anti-mouse IgG	1:600	
Cy3™-conjugated goat anti-rabbit IgG	1:600	Jackson
Cy3™-conjugated donkey anti-goat IgG	1:600	ImmunoResearch
Cy3™-conjugated goat anti-mouse IgM	1:600	Laboratories,
DTAF-conjugated goat anti-rabbit IgG	1:100	Dianova, Hamburg,
FITC-conjugated donkey anti-sheep IgG	1:100	Germany
FITC-conjugated goat anti-mouse IgG	1:100	

monolayer may be analyzed by direct determination of immunolabeled cells (percentage values), whereas, for cells growing in multilayered clusters, the "labeling index" technique is proposed.

1. *Determination of percentage values of Hoechst-labeled cells:* Cells were analyzed for immunofluorescence signals and the percentage number of immuno-positive cells relative to a total number of ($n = 1000$) Hoechst 33342-labeled cells is given.
2. *Estimation of the "labeling index"*:[33] For cells growing in clusters, immunofluorescence analysis was performed using the inverted fluorescence microscope ECLIPSE TE300 (Nikon, Japan) equipped with a 3CCD Color Video Camera DXC-9100P (Sony, Japan) and the LUCIA M-Version 3.52a software (LIM, Czech Rep.) For each sample, at least 20 randomly, but representative selected pictures were analyzed for the "area fraction" value, which is the ratio of the immunopositive signal area to the measured area. To discriminate the immunopositive signal from background fluorescence, the pictures were binarized with the specific threshold fluorescence values.

Immunohistochemistry (IHC) of Spheroids

Spheroids were collected by sedimentation, washed two times with PBS, fixed in Bouin's solution (75 ml picric acid, 25 ml 4% formaldehyde, 5 ml acetic acid) for 2 hr at RT, washed twice in 70% ethanol and dehydrated in graded ethanol. Fixed spheroids were embedded in paraffin, sectioned at 5 μm slices and mounted on silanized slides using conventional techniques. Slides were deparaffinized at 60°C for 2 hr, rehydrated in xylene (5 min, RT), isopropanol (5 min, RT), processed through graded ethanol (96, 80, 70 and 50%, 5 min each, RT), washed in Aqua dest. (1×, 3 min, RT) and in PBS (3×, 5 min, RT). Immunohistochemistry was performed as described in section "Immunofluorescence analysis" beginning with BSA blocking.

Insulin ELISA

ES-derived cells differentiated into the pancreatic lineage were washed PBS (5×) and preincubated in freshly prepared KRBH (Krebs' Ringer Bicarbonate Hepes) buffer containing 118 m*M* sodium chloride, 4.7 m*M* potassium chloride, 1.1 m*M* potassium dihydrogen phosphate, 25 m*M* sodium hydrogen carbonate (all from Carl Roth GmbH & Co, Karlsruhe, Germany), 3.4 m*M* calcium chloride (Sigma), 2.5 m*M* magnesium sulfate (Merck), 10 m*M* Hepes and 2 mg/ml bovine serum albumin supplemented with 2.5 m*M* glucose (all from Gibco) for 90 min at 37°C.

To estimate glucose-induced insulin secretion, the buffer was replaced by 27.7 m*M* glucose and alternatively with 5.5 m*M* glucose and 10 μM tolbutamide dissolved in KRBH buffer for 15 min at 37°C. The control was incubated in KRBH buffer supplemented with 5.5 m*M* glucose. The supernatant was collected and stored at −20°C for determination of insulin release.

Cells were dissociated by 0.2% trypsin : 0.02% EDTA in PBS (1 : 1) for 3 min and collected by centrifugation. Proteins were extracted from the cells with acid ethanol (1 *M* hydrochloric acid : absolute ethanol 1 : 9) and overnight incubation at 4°C, followed by cell sonification, and stored at −20°C for the determination of total cellular insulin and protein content, respectively. The insulin enzyme-linked immunosorbent assay (ELISA, Mercodia AB, Sweden) was performed according to manufacturer recommendations.

The total protein content was determined by the protein Bradford assay according to manufacturer recommendations (Bio-Rad Laboratories GmbH, Munchen, Germany). Released insulin levels are presented as ratio of released insulin per 15 min and intracellular insulin content. The intracellular insulin level is given as ng insulin per mg protein (see Ref. 33).

Albumin ELISA

ES-derived cells were washed in PBS (5×) and incubated at 37°C in "hepatic differentiation medium" in the absence of BSA/FCS for 24 hr. The supernatant was collected and stored at −20°C for measurement of albumin release.

To determine the total cell number, cells were dissociated by treatment with 0.2% trypsin: 0.02% EDTA (1 : 1) in PBS.

For the estimation of albumin synthesis, the quantitative enzyme-linked immunoassay (Albumin ELISA, Bethyl Laboratories, INC, Montgomery, USA) was performed according to manufacturer recommendations.

Results

By applying the "nestin+ cell selection" protocol followed by specific differentiation induction in the presence of specific differentiation factors, insulin- (Fig. 1A) and albumin- (Fig. 1B) producing cells were generated at high efficiency.[33]

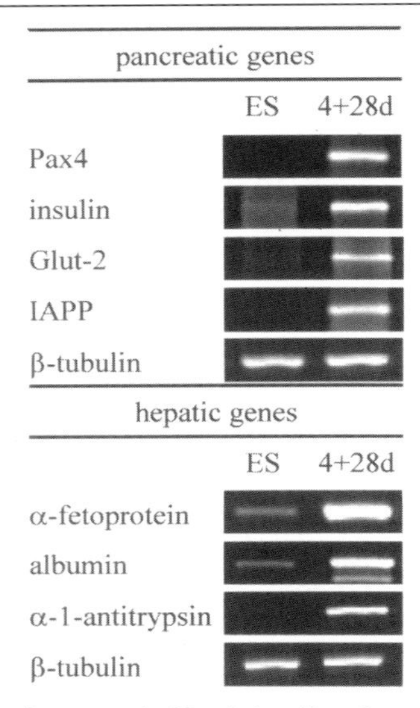

FIG. 2. RT-PCR analysis of pancreatic [Pax4, insulin, glucose transporter-2 (Glut-2), islet amyloid polypeptide (IAPP)] and hepatic [albumin, α-fetoprotein, α-1-antitrypsin] genes. mRNA levels of target genes in undifferentiated ES cells and in cells cultivated according to the "nestin + cell selection" protocol and differentiated into pancreatic and hepatic cell types were analyzed at stage 4 + 28 d (= 4 + 8 + 20 d). The housekeeping gene β-tubulin was used as internal standard.

Differentiation of Pancreatic Cells

During differentiation into the pancreatic lineage, genes specifically expressed in β cells including Pax4, insulin, glucose transporter-2 (Glut-2) and islet amyloid polypeptide (IAPP) were significantly upregulated at stage 4 + 28 d (= 4 + 8 + 20 d) in comparison to undifferentiated ES cells (Fig. 2). The number of pancreatic endocrine cells amounted to a "labeling index" of 0.15 (insulin) and 0.05 (glucagon), respectively, corresponding to about 20–25% insulin- and 5–10% glucagon-producing cells (Fig. 3A). The number of insulin-positive cells increased with differentiation time from Day 4 + 8 + 13 to Day 4 + 8 + 20, whereas no significant changes were detected in the number of glucagon-expressing cells (Fig. 3A). Intracellular insulin levels analyzed by ELISA amounted to 20.7 ng insulin/mg protein at stage 4 + 8 + 20 d.[33] The cells released insulin in response to glucose.[33]

Cells generated according to the "nestin + cell selection" protocol for 4 + 8 + 20 days followed by 10 days of histotypic "Spinner" culture showed a significant accumulation of insulin in spheroids (Fig. 4B). The intracellular insulin levels amounted to 297.4 ± 18.0 ng/mg protein.[33]

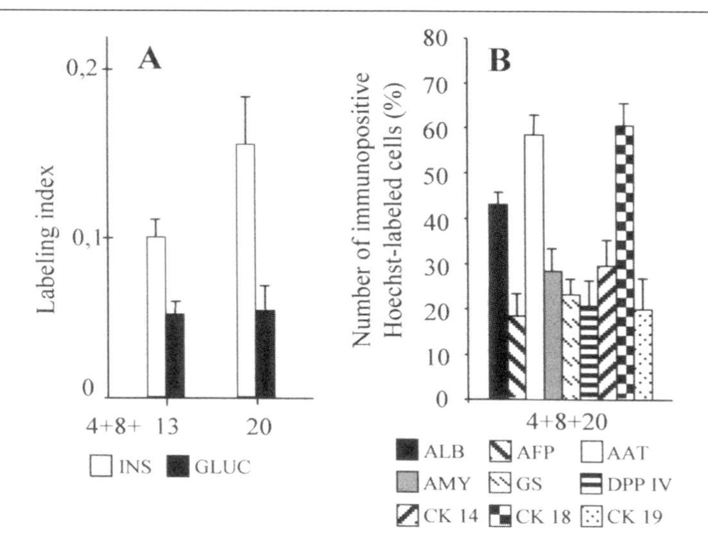

FIG. 3. Semi-quantitative imaging analysis of pancreatic and hepatic proteins in ES-derived cells analyzed for the "labeling index" at days 4 + 8 + 13 and 4 + 8 + 20 (A), and the determination of percentage values of Hoechst 33342-labeled cells expressing hepatic proteins (B) at stage 4 + 8 + 20 d. Each value represents mean ± SEM. Abbreviations: INS, insulin; GLUC, glucagon; ALB, albumin; AFP, α-fetoprotein; AAT, α-1-antitrypsin; AMY, amylase; GS, glutamine synthetase; DPP IV, dipeptidyl peptidase IV; CK 14, cytokeratin 14; CK 18, cytokeratin 18; CK 19, cytokeratin 19. (A) $n = 3$ experiments with 2 parallels, (B) $n = 13$ experiments.

Differentiation of Hepatic Cells

After differentiation of ES cells via the "nestin + cell selection" protocol and differentiation induction into the hepatic lineage (Fig. 1B), liver-specific genes encoding α-fetoprotein, albumin and α-1-antitrypsin were significantly upregulated at stage 4 + 28 d (= 4 + 8 + 20 d) in comparison to undifferentiated ES cells (Fig. 2). Liver-specific proteins were efficiently expressed as determined by immunofluorescence analysis. Albumin (ALB) was produced in up to 45%, α-fetoprotein (AFP) in about 20%, α-1-antitrypsin (AAT) and cytokeratin 18 (CK 18) in nearly 60% of Hoechst 33342-labeled cells. Amylase (AMY), glutamine synthetase (GS), dipeptidyl peptidase IV (DPP IV), cytokeratin 14 (CK 14) and cytokeratin 19 (CK 19) were detected in about 20 to 30% of the total number of Hoechst 33342-labeled cells (Fig. 3B). All these proteins were determined *in vivo* to be specific for the different hepatic phenotypes including hepatocytes, biliary duct epithelium, hepatoblasts and oval cells (for review see Ref. 15). Cells generated according to the "nestin + cell selection" protocol for 4 + 8 + 14 days followed by 10 d of histotypic "Spinner" cultivation showed a significant accumulation of albumin (Fig. 4C) in spheroids.

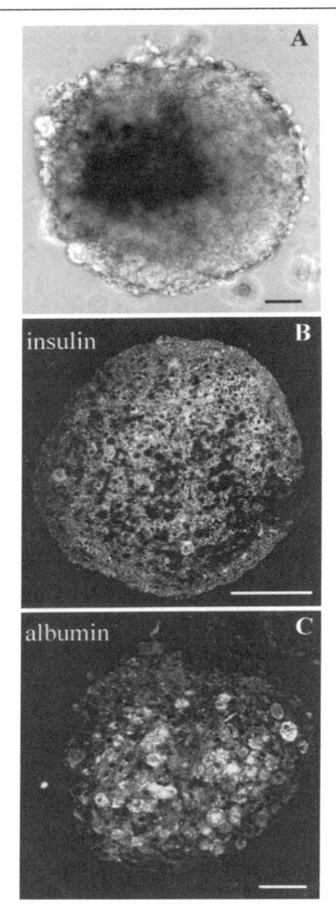

Fig. 4. Phase contrast (A) and immunofluorescence micrographs of insulin (B)- and albumin (C)-labeled cells (white) in ES-derived spheroids generated via the "nestin+ cell selection" protocol, specific differentiation induction and 10 days maturation in the "spinner" culture system. Bar = 50 μm.

Summary

Here, we present efficient strategies to differentiate ES cells either into pancreatic or into hepatic cell types. We recommend a strategy to select nestin+ cells, an early progenitor cell type with high developmental plasticity, followed by differentiation induction with specific growth and extracellular matrix factors into pancreatic and hepatic cell types. Cells differentiating via nestin+ cells into the pancreatic and hepatic lineage expressed tissue-specific genes. Proteins characteristic for mature endocrine pancreatic or hepatic cells were synthesized and released. Further, a histotypic "spinner" culture system was introduced to generate mature insulin- and albumin-producing cells at high efficiency.

Acknowledgments

We want to thank Mrs. S. Sommerfeld, O. Weiß and K. Meier for excellent technical assistance. The financial support by the Deutsche Forschungsgemeinschaft (DFG, WO 503/3-2), DeveloGen AG, Göttingen and Fonds der Chemischen Industrie (FCI), Germany, is gratefully acknowledged.

[22] Generating CNS Neurons from Embryonic, Fetal, and Adult Stem Cells

By JONG-HOON KIM, DAVID PANCHISION, RAJA KITTAPPA, and RON MCKAY

Introduction

In vitro culturing of neuroepithelial stem cells has become an indispensable tool for studying the mechanisms controlling proliferation, mitotic arrest and lineage commitment of cells in the nervous system.[1] Recent progress in this field has improved the possibility of treating patients by transplanting new cells that can replace those lost through trauma or disease. However, any potential clinical use of these cells requires systematic methods to enrich for the cell of interest and demonstrate that these cells show functions that will assist in understanding and treating the disease.

Embryonic stem (ES) cells are pluripotent and immortal cells derived from the inner cell mass of preimplantation blastocysts. They can proliferate extensively and have ability to differentiate into endodermal, mesodermal and ectodermal derivatives. The most important benefit of using ES cells as donor cells is the relative ease of genetic engineering, which permits the enrichment or purification for specific cell types by selectable marker. Recently, we have developed efficient methods of generating central nervous system (CNS) stem cells and their derivatives from ES cells.[2–5]

[1] D. M. Panchision and R. D. McKay, *Curr. Opin. Genet. Dev.* **12**, 478–487 (2002).

[2] S. Okabe, K. Forsberg-Nilsson, A. C. Spiro, M. Segal, and R. D. McKay, *Mech. Dev.* **59**, 89–102 (1996).

[3] O. Brustle, K. N. Jones, and R. D. Learish, *et al., Science* **285**, 754–756 (1999).

[4] S. H. Lee, N. Lumelsky, L. Studer, J. M. Auerbach, and R. D. McKay, *et al., Nat. Biotechnol.* **18**, 675–679 (2000).

[5] J. H. Kim, J. M. Auerbach, and J. A. Rodriguez-Gomez, *et al., Nature* **418**, 50–56 (2002).

0076-6879/2003 $35.00

Primary fetal and adult CNS stem cells are also extensively self-renewing and are multipotent for neuronal and glial fates.[6] CNS stem cells selectively express the intermediate filament Nestin,[7] which remains the principal marker for distinguishing CNS stem cells from their differentiated derivatives[6] or from ES cells.[4] The methods of culturing primary CNS stem cells are included in this chapter, as they have formed the basis for all subsequent methods of generating and comparing CNS derivatives from ES cells.

Derivation of Dopaminergic Neurons from Mouse Embryonic Stem Cells

ES cells can be cultured in a 6-stage protocol that amalgamates ES cell differentiation methods with a standard CNS stem cell culturing method. In this section, we introduce the complete protocol for the isolation of a highly enriched population of midbrain dopamine (DA) transmitting neurons from genetically enhanced ES cells. DA neurons have attracted great interest because the degeneration of these cells in the substantia nigra causes Parkinson's disease (PD).[8] Thus, ES cells might serve as a source of replacement DA neurons in Parkinson's disease. *Nurr*1, an orphan nuclear receptor, is specifically required for the induction of midbrain DA neurons during neurogenesis and for the maintenance of DA phenotype in adulthood.[9] We have shown that *Nurr1*-transduced ES cells generate a higher proportion of DA neurons than wild type ES cells.[4,5] These neurons express typical markers for midbrain DA neurons, including tyrosine hydroxylase (TH). They also release high levels of DA upon depolarization and provide functional recovery after grafting into 6-hydroxydopamine-lesioned rats.[5]

Procedure Outline

Stage 1: Expansion of undifferentiated ES cells on gelatin-coated tissue culture surface with LIF.

Stage 2: Transfection of undifferentiated ES cells with a *Nurr1*-containing plasmid and selection of transfected clones.

[6] K. K. Johe, T. G. Hazel, T. Muller, M. M. Dugich-Djordjevic, and R. D. McKay, *Genes Dev.* **10**, 3129–3140 (1996).

[7] U. Lendahl, L. B. Zimmerman, and R. D. McKay, *Cell* **60**, 585–595 (1990).

[8] M. Gerlach, H. Braak, and A. Hartmann, *J. Neurol.* **249** (Suppl 3), 333–335 (2002).

[9] O. Saucedo-Cardenas, J. D. Quintana-Hau, W. D. Le, *et al., Proc. Natl. Acad. Sci. USA* **95**, 4013–4018 (1998).

Stage 3: Generation of EBs in suspension culture for 4 days in KO-ES medium.

Stage 4: Selection of Nestin$^+$ cells for 6–8 days in ITSFn medium from EBs plated on tissue culture surface.

Stage 5: Expansion of En1$^+$ midbrain precursor cells for 4 days in N2 medium containing bFGF/Shh-N/FGF-8/Ascorbic acid.

Stage 6: Differentiation to TH$^+$ midbrain DA neurons by withdrawing the growth factors from N2 medium containing ascorbic acid.

Materials and Reagents

Equipment and Supplies

Tissue culture cabinet
Incubator, humidified with controlled CO_2 and preferably O_2
Heated water bath at 37°C
Tabletop centrifuge (200g = 1000 rpm in a Sorvall RT6000 with H1000B rotor)
Gene Pulser electroporator (BioRad)
Cold storage (4°C and −20°C)
Hemocytometer (Merrifeld)
Bacterial Petri Dishes (Falcon #351029)
10-cm Culture Dishes (Falcon # 353003)
24-well Culture Plates (Corning Costar #3526)
6-well Culture Plates (Nuclon #152795)
Sterile pipettes (Falcon or other suppliers)
Cryotubes.

Growth Media

Note. If diluting solutions from concentrated stocks or powder, use water certified as tissue-culture grade and mycoplasm-free. Add sterile water directly into the supplied vial of lyophilized powder. Alternatively, sterilize solutions reconstituted with purified (e.g., Milli-Q) water by filtering them through 0.2 μm filters into clean, sterile bottles.

KO-ES medium (containing 15% v/v serum)

1. Mix the following to make 500 ml medium:
   425 ml Knockout-DMEM (Invitrogen #10829-018)
   75 ml ES Cell Qualified Fetal Bovine Serum (Invitrogen # 16141-061)

5 ml MEM Non-Essential AA Solution (Invitrogen #11140-050)
5 ml Penicillin–Streptomycin–Glutamine (Invitrogen #10378-016)
0.5 ml 2-mercaptoethanol (Invitrogen #21985-023)

2. Adjust pH to 7.2 and filter-sterilize; store at 4°C.

ITS medium (DMEM/F12 with ITS supplements)

1. Mix the following powdered reagents in purified water to make 500 ml medium:
   6.0 g DMEM/F12, 1 : 1 (Invitrogen 12500-039)
   1.2 g $NaHCO_3$
   0.775 g Glucose (Sigma #G-6152)
   0.0365 g Glutamine (Sigma #G-5763)
   2.5 mg Insulin (Intergen 4501-01), dissolved in 5 ml of 10 m*M* NaOH and added into the medium
   25 mg Apo-transferrin (Sigma T-2036)
   30 μl selenium (500 μM stock, Sigma S-5261)
2. Adjust pH to 7.2 and filter-sterilize.
3. Add 5 ml penicillin/streptomycin stock antibiotic (LifeTechnologies 15070-089).

Note. If the insulin does not readily dissolve in dilute NaOH it means that exposure to air has lowered the pH; prepare fresh dilute NaOH from stock. Media can be prepared from powder provided that all essential ingredients are present. Always protect media from light exposure by wrapping in foil and storing in a darkened 4°C refrigeration unit. Long-term exposure to air will raise the pH of media even during storage; this can be limited by storing in "air-tight" 50 ml aliquots and warming to 37°C as needed. Discard prepared media that is more than two or three weeks old.

N2 medium (DMEM/F12 with N2 supplements)

1. Start with one 500 ml bottle of 1 : 1 DMEM/F12 medium (Mediatech 10090-CV).
2. Remove 5 ml of medium into a tube, add 50 mg apo-transferrin (Sigma T-2036) into this aliquot and vortex; filter transferrin solution through a 0.22 μm syringe filter (Millipore SLGV-025LS) back into the original 500 ml bottle of medium.
3. Dissolve 12.5 mg insulin (Intergen 4501-01) in 5 ml of 10 m*M* NaOH and add:
   50 μl putrescine (1 *M* stock, Sigma P-5780)
   30 μl selenium (500 μM stock, Sigma S-5261)
   100 μl progesterone (100 μM stock, Sigma P-7556)

Filter this through a 0.22 μm syringe filter into the original 500 ml bottle of medium.

4. Add 5 ml penicillin/streptomycin stock antibiotic (LifeTechnologies 15070-089).

Note. This is similar to the formulation described by Bottenstein and Sato (1979). Use the same precautions as described for ITS medium.

Neurobasal medium (with B27 supplements or FBS)

1. Start with one 500 ml bottle of Neurobasal medium (Invitrogen 21103-049).
2. Add one of the following:
   a. B27 supplements (Invitrogen 17504-044) diluted 1 : 50.
   b. Fetal bovine serum (Invitrogen # 16141-061 or other) diluted to 5% final.

Other Solutions

Phosphate buffered saline (PBS), sterile, mycoplasm-free, 1X (Invitrogen #10010-023)

Trypsin/EDTA (Invitrogen #25300-054).

0.2% Trypan Blue solution.

DMSO or other freezing media.

Gelatin (Sigma G2500), 0.1% Solution. Dissolve 1 g gelatin in 1000 ml purified water, autoclave for 30 min in liquid cycle and let cool; store at 4°C and add directly to dishes as needed.

Poly-L-ornithine (Sigma P-3655). Reconstitute an entire bottle in sterile water to a stock concentration of 3.75 mg/ml and store at −20°C. Dilute in purified water to a working concentration of 15 μg/ml. Sterile-filter into 1 liter bottles, store at 4°C and add directly to dishes as needed.

Bovine Fibronectin (Sigma F-4759). Reconstitute as instructed with sterile water to a stock concentration of 1 mg/ml. Store vial at 4°C and warm to room temperature before using. Stock aliquots can be frozen but repeated freeze-thawing is not recommended. It is critical that lyophilized fibronectin is not agitated during reconstitution as this can create an insoluble aggregate. Fibronectin quality can vary considerably and some sources are sub-optimal for the adhesion of stem cells; poor fibronectin quality or expired lots commonly lead to cell lifting and death.

Growth Factors and Supplements

Note. Reconstitute lyophilized powders as instructed. This is generally done by adding sterile, mycoplasma-free 1X PBS containing 0.1% BSA (fraction V, Sigma) directly to the vial. Aliquot and store at

−20°C. Aliquots can be kept at 4°C for several weeks, but repeated freeze-thawing is not recommended.

Basic Fibroblast Growth Factor (bFGF, R&D Systems 233-FB). Make 10 μg/ml stock.

Fibroblast Growth Factor 8b (FGF8b, R&D Systems 423-F8). Make 25 μg/ml stock.

Sonic hedgehog (Shh, R&D Systems 461-SH). Make 100 μg/ml stock.

Ascorbic Acid (Sigma A-4403). Make 100 mg/ml stock with water.

GDNF (R&D Systems 212-GD-010)

LIF or ESGRO (Chemicon ESG1106)

Mitomycin C (Sigma M-0503)

G418 (Geneticin selective antibiotic, Invitrogen 11811-023)

Preparing Coated Tissue Culture Dishes for Expansion of Nestin^{+} Stem Cells

1. Tissue culture dishes must be coated with poly-L-ornithine followed by fibronectin before they can be used to culture neuroepithelial stem cells. Add 5–7 ml of 15 μg/ml poly-L-ornithine solution to 10 cm tissue culture dishes and incubate several hours to overnight at 37°C.
2. Aspirate the solution and add enough 1X PBS to cover the surface of the dish (5–7 ml per 10 cm dish). Allow the dishes to incubate about 1 hr, then aspirate the solution and repeat the PBS wash two additional times. Note that residual poly-L-ornithine is toxic to cells.
3. Dilute the 1 mg/ml bovine fibronectin stock into 1X PBS to make a 1 μg/ml solution. Aspirate PBS wash from dishes and add fibronectin/PBS to dishes; incubate 1 hr to overnight at 37°C. Aspirate off fibronectin/PBS and store in 1X PBS if needed, or aspirate fibronectin/PBS just prior to plating cells. Note that adhesion of cells correlates directly with the duration of fibronectin coating; short durations are sufficient for plates to be passaged while longer durations are preferred for culturing cells longer than three days in the same dish. Discard dishes stored in PBS if they have not been used within 3–4 days.

Procedure for Generation of Dopaminergic Neurons from ES cells

Stage 1: Culture of Undifferentiated ES Cells without Feeder Layer

1. Coat 10 cm dishes with 0.1% gelatin for 30 min, then replace with 10 ml medium; place the dishes in an incubator until they are required.
2. Seed 3×10^6 ES cells on 0.1% gelatin-coated 10 cm tissue culture dish in ES medium containing 1400 units/ml LIF.

3. Place the dishes in a tissue culture incubator; change the medium every day with LIF; grow cells up to 80% confluence.

Stage 2: Transfection of ES Cells with Nurr1 Construct

1. Change medium on ES cells 2 hr prior to electroporation.
2. Coat 10 cm dishes with 0.1% gelatin for 30 min, then replace with 10 ml medium; place the dishes in an incubator until they are required.
3. Aspirate medium and wash with 1X PBS two times.
4. Harvest ES cells in log phase growth by trypsinization; for best result ES cells should have been passed no more than 3 days before trypsinization. After trypsinization add 3 ml cold medium and triturate well to single cell suspension.
5. Pellet the cells by centrifugation in a table top centrifuge at approximately 200*g* (1000 r.p.m.).
6. Resuspend the cell pellet in 5 ml of ice-cold PBS; determine the cell density and dilute with 1X PBS to a density of 2×10^7 cells in 0.8 ml.
7. Add 25 μg linearized *Nurr1*-construct and place in a chilled 0.4 cm electroporation cuvette and mix gently.
8. Electroporate in Bio-Rad Gene Pulser at 0.25 kV, 500 μF; the time constant will be between 4 and 7; place cuvette on ice for 15 min.
9. Plate the electroporated cells on three 10 cm dishes (with *neo*-resistant feeder cells) in prewarmed medium containing LIF without selection. After 24 hr post-electroporation, change the medium and add 200 μg/ml G418.
10. Continue the selection on G418 up to 10 days until colonies grow to an apparent size amenable to picking; change medium daily with LIF and G418.
11. Remove medium and wash with 1X PBS two times.
12. Pick each colony into a 96-well plate (round bottom) containing 45 μl of trypsin solution per well and incubate at 37°C for 5 min; use a pipette to disperse colonies by pipetting up and down in a circular motion; be careful not to make a bubble.
13. Using a multi-channel pipette, transfer dispersed cells into gelatinized 96-well plate containing 150 μl of ES cell medium containing LIF; incubate at 37°C for 2–3 days.
14. Remove medium and wash with 1X PBS two times.
15. Add 40 μl of trypsin solution, incubate at 37°C for 5 min and add 70 μl of ES medium; use multi-channel pipette to disperse cells.
16. Transfer 70 μl into 96-well plate containing 70 μl of cold 2X freezing medium. Wrap plate with parafilm and place on dry ice for 20 min, then transfer into −80°C. Store the frozen cells until the evaluation of each clone's *Nurr1* expression level is completed.

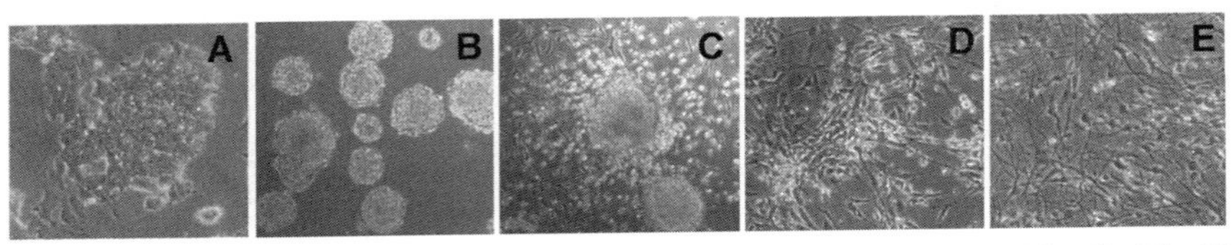

FIG. 1. Procedure for induction of midbrain dopaminergic neurons from ES cells. The ES cells are taken through 5 steps or stages. In stage 1 (A), undifferentiated ES cells are cultured for 5 days in the presence of 15% fetal calf serum (FCS) on gelatin-coated tissue culture dishes in the presence of LIF (1400 U/ml.). In stage 2 (B), embryoid bodies (EBs) are generated in the presence of FCS for 4 days in the presence or absence of LIF (1000 U/ml.). In stage 3 (C), the EBs are plated into ITS medium (fibronectin, 5 μg/ml) where, over 10 days, Nestin$^+$ cells migrate from the cell aggregates. In stage 4 (D), these Nestin$^+$ cells are resuspended and expanded for 4 days in N2 medium containing bFGF, Shh, and FGF8. In stage 5 (E), the medium is changed into N2 medium without bFGF, Shh, or FGF8. These cells differentiate efficiently into neurons and astrocytes over a 2 week period. (See Color Insert.)

17. Transfer remaining cells to gelatinized 24-well plate containing ES medium; grow about 3 days and check expression level of *Nurr1* by Northern or Western Blot analysis. Choose roughly 5 colonies that show the highest level of *Nurr1* expression.
18. Thaw the clone selected for *Nurr1* expression and plate on 6-well plate containing neo-resistant feeder layer. Grow cells for 3 days in the presence of G418
19. Trypsinize cells and transfer to 10 cm dish containing *neo*-resistant feeder; grow up to sub-confluent; make frozen cell stocks for each clone of *Nurr1*-transfected ES cells (3×10^6 cells/cryotube).

Stage 3: Embryoid Body Formation

1. Plate *Nurr1*-transfected ES cells (*Nurr1*-ES cells) on gelatinized 10 cm dishes in ES cell medium with LIF and G418. An initial plating density of 3×10^6 cells per 10 cm dish is reasonable.
2. Grow cells for 3 days with daily medium changes; after 3 days of culture they usually reach sub-confluence (Fig. 1A).
3. After washing cells twice with 1X PBS, trypsinze cells and resuspend pellet in 5 ml ES medium with LIF.
4. Plate cells on bacterial petri dish at 2×10^6 cells per 10 cm dish in ES medium containing 1,000 units/ml of LIF. Note that LIF treatment increases the frequency of Engrailed-1 (En1) expression, marking midbrain precursor cells, and reduces the expression of Pdx1 (Fig. 2A), an early regulator of the differentiation of cells in the endocrine pancreas throughout the ES cell differentiation protocol. LIF also strongly promotes neuronal differentiation during stage 5 (Fig. 2B).

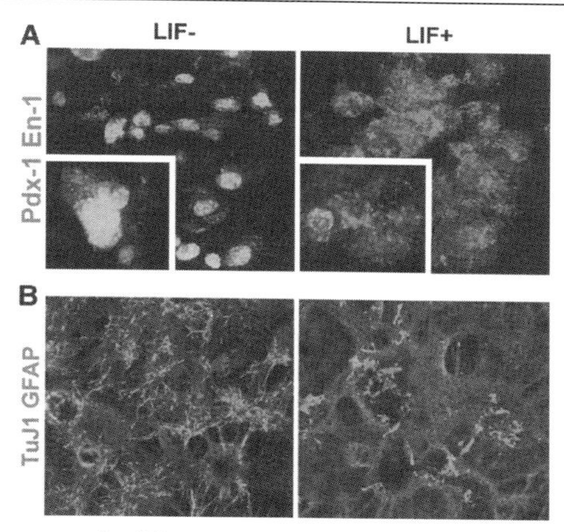

FIG. 2. Neural and pancreatic differentiation of ES cells. Embryoid bodies were generated in the presence (LIF^{+}) or absence (LIF^{-}) of LIF (1000 U/ml) and differentiated. Double-immunostaining for PDX-1/En-1 (A, day 3 in stage 4) and TuJ1/GFAP (B, day 8 in stage 5). Note that LIF treatment in stage 2 (EB formation) increases the neuronal ($TuJ1^{+}$ cells, green) and decreases the astrocytic ($GFAP^{+}$, red) population. LIF treatment efficiently enhances midbrain precursor cells ($En\text{-}1^{+}$ cells, red) and negatively regulates pancreatic precursor cells ($PDX\text{-}1^{+}$ cells, green). (See Color Insert.)

5. Grow the cells for 4 days with medium changes every 2 days; after 2 days of culture, cells will form small clusters. Transfer the small embryoid bodies (EBs) to 15 ml conical tube and let stand until the EBs settle to the bottom. Remove medium by aspiration and replace with fresh ES medium with LIF. Take care not to aspirate EBs. Transfer the EBs to original dishes and after 4 days of culture they form simple EBs (Fig. 1B).

Stage 4: Selection of $Nestin^{+}$ Cells

1. Transfer EBs in bacterial culture dishes to 15 ml tube.
2. Stand until the EBs settle down (it takes 5–10 min depending on the size of EBs) and remove half of medium. Add 5 ml new ES medium without LIF to original dishes and rinse the dishes with the medium and add to 15 ml tube containing EBs.
3. Transfer the EBs to tissue culture dishes. Allow EBs to adhere to tissue culture dishes overnight.
4. After 24 hr, change the medium to ITS medium containing fibronectin (50 μl of fibronectin stock/10 ml medium, final concentration = 5 μg/ml). Note that fibronectin tends to clump when agitated; don't vortex and carefully mix in ITS medium by gently inverting.

5. Feed the cells for 8–10 days in ITS medium containing 5 μg/ml fibronectin; change medium every 2 days. Duration of culture period in ITS medium depends on the mouse strain that is used to establish ES cell line and the passage number. Usually longer culture (up to 12 days) in ITS medium generates more Nestin$^+$ cells. During the first 3 days, a large number of cells detach from the plate and lyse. The remaining cells change their morphology from compacted epithelial cells to small, elongated cells expressing Nestin (Fig. 1C).

Stage 5: Expansion of En1$^+$ Midbrain Precursor Cells

1. Remove medium from the plates and wash cells twice with PBS.
2. Trypsinze cells for 5 min and neutralize with serum containing medium; during the selection of Nestin$^+$ cells, many neural precursors migrated out of EB.
3. Using pipette disperse the cells up and down gently. The EB contains mesoderm, endoderm and non-neural ectoderm; therefore do not dissociate the EB completely.
4. Transfer dispersed cells and EB clumps to 15 ml tube and stand until EB clumps settle to the bottom of tube (about 5 min).
5. Take supernatant containing suspended cells and spin down cells at 1000 rpm for 5 min. Resuspend the pellets in N2 medium and discard the settled EB clumps from step 4.
6. Count cells and plate the cells on poly-L-ornithine/fibronectin-coated plate in N2 medium containing 20 ng/ml bFGF, 500 ng/ml SHH and 100 ng/ml FGF8b. We recommend plating cells at high density as follows:

   24-well plate (200 mm^2): 3–4$\times 10^5$ cells in 0.4 ml of medium
   6-well plate (962 mm^2): 1.5–2$\times 10^6$ cells in 2.5 ml of medium
   6-cm dish (2827 mm^2): 4.5–5.5$\times 10^6$ cells in 4 ml of medium
   10 cm dish (7854 mm^2): 1.2–1.6$\times 10^7$ cells in 10 ml of medium

7. Feed the cells for 4–6 days in N2 medium containing bFGF/SHH/FGF8b. Add bFGF daily to prevent differentiation and replace the medium with growth factors every other day. After 2 days of expansion the cultures will be nearly homogeneous for Nestin$^+$ neural precursor cells ($\sim$95% of the total cells express Nestin; Fig. 1D); after 4 days the culture can be passaged to new dishes to further expand neural precursors. Note that the proportion of TH$^+$ cells will decrease after successive passages for unknown reasons; ascorbic acid appears to partially maintain the proportion of TH$^+$ cells after passaging.

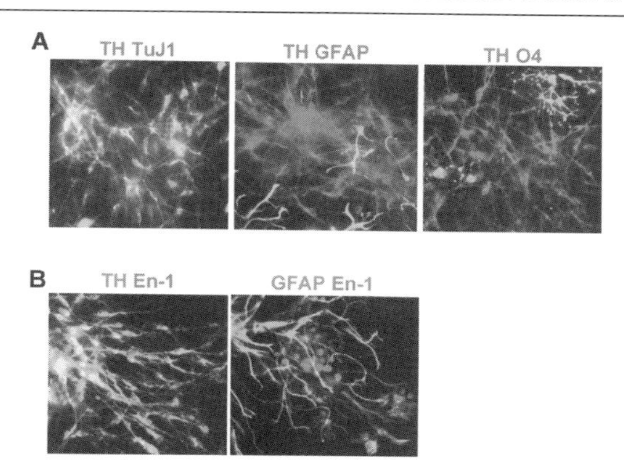

FIG. 3. Differentiation into midbrain TH neurons. (A) Co-staining with TH and markers for neurons (TuJ1), astrocytes (GFAP), and oligodendrocytes (O4) shows that only neurons express TH. (B) Most TH^+ neurons express a midbrain-specific marker En-1 (TH, green; En-1, red). Some $GFAP^+$ astrocytes also express En-1 (GFAP, green; En-1, red). (See Color Insert.)

Stage 6: Differentiation to Midbrain Dopaminergic Neurons

1. Induce differentiation of expanded $En1^+$ cells by withdrawing bFGF, SHH and FGF8b from N2 medium (Fig. 1E).
2. Feed the cells in N2 medium containing 200 μg/ml ascorbic acid and 20 ng/ml GDNF for at least 6 days. Ascorbic acid enhances survival and promotes differentiation.
3. Change N2 medium containing ascorbic acid/GDNF every 2 days.
   a. After 6 days of differentiation, >70% of total cell population expresses β-III Tubulin (TuJ1), 10–15% expresses GFAP and <2% expresses O4 (Fig. 3A). About 80% of $TuJ1^+$ cells and 20% of $GFAP^+$ cells expresses TH and En-1 (Fig. 3B).
   b. Additional growth factors or chemicals can be tested in stage 4 and/or 5 to change fate, to enhance proliferation or to promote differentiation.
   c. It is necessary to use Neurobasal medium containing B27 or 5% serum for long-term culture.
   d. The DA neurons generated from *Nurr1*-ES cells release a high amount of DA (Fig. 4), express GDNF receptor GDNF-Rα1 and c-Ret (Fig. 5A), DA transporter (Fig. 5B), and synaptophysin (Fig. 5C).
   e. For the functional test of the DA neurons generated from *Nurr1*-ES cells, these cells can be grafted into the striatum of rats lesioned with the neurotoxin 6-hydroxydopamine (6-OHDA) causing cell death of the DA neurons in the

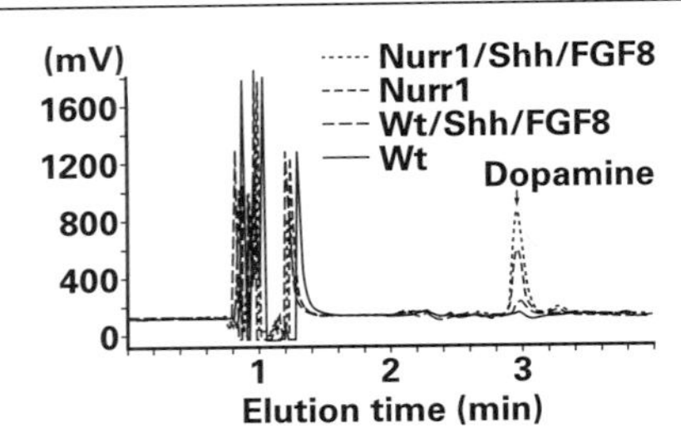

FIG. 4. RP-HPLC determination for dopamine release of stage 5 ES cells after depolarization in HBSS (56 m*M* KCl) for 15 min. The dopamine released by depolarization was markedly elevated in the cultures of stage 5 Nurr1-ES, compared to wt-ES cells.

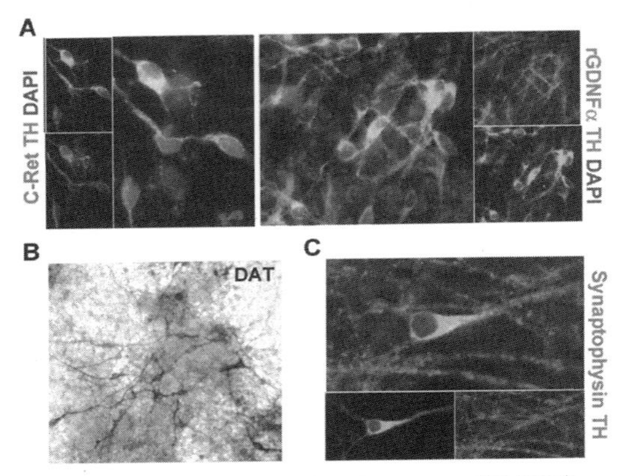

FIG. 5. Characteristics of Nurr1 ES-derived TH neurons. (A) TH^+ neurons derived from Nurr1-transfected ES cells express GDNF receptor c-Ret (c-Ret, red; TH, green; DAPI, blue) and rGDNFα (rGDNFα, red; TH, green; DAPI, blue), dopamine transporter, DAT (B: brown), and synaptophysin (C: synaptophysin, red; TH, green). (See Color Insert.)

substantia nigra (Fig. 6). It is crucial to transplant the cells 3 days after withdrawal of bFGF (day 3 of stage 5). The residual undifferentiated cells may cause tumor formation upon transplantation before stage 5 and dissociation of cells for grafting after long-term differentiation results in cell death caused by ruptured many thin dendrites of matured neurons.

Preparing Cultures of Rodent Fetal Neuroepithelial Stem Cells

Several different methods have been developed to culture neural precursors. Cells can be grown as suspended aggregate called neurospheres[10]

[10] A. L. Vescovi, B. A. Reynolds, D. D. Fraser, and S. Weiss, *Neuron* **11**, 951–966 (1993).

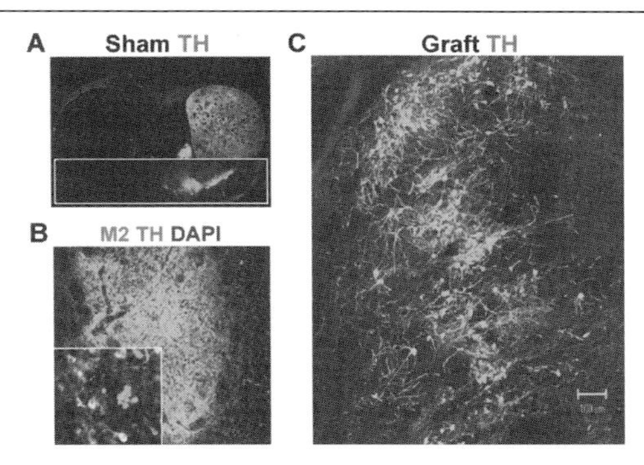

FIG. 6. Integration of Nurr1 ES cell-derived TH neuron into the striatum of hemisparkinsonian rat. (A) Low power photomicrographs showing TH^+ neurons in the striatum, substantia nigra, and ventral tegmental area. Note that the lesioned side of the brain contains no positive cells. (B) Photomicrographs of the graft in the striatum. All grafts were easily detected by staining mouse-specific surface antigen, M2 (green). Many of the $M2^+$ grafted cells also expressed TH (inset: M2, green; TH, red). (C) The dorsal striatum of grafted animals showed discrete grafts containing TH^+ neurons (scale bar, 100 μm). (See Color Insert.)

or on a surface matrix in monolayer cultures.[6,11,12] Cell proliferation can be driven with basic FGF (bFGF) or epidermal growth factor (EGF) alone or in combination. Additionally, cultures can be grown in a fully defined medium or with undefined supplements such as serum or conditioned medium. Although many approaches have been successfully used to generate neural precursors, differences in these protocols may affect the properties of the precursor cells and the efficiency with which they are maintained. This should be kept in mind when interpreting results from experiments using different protocols.

This protocol has been optimized for the production and maintenance of nearly homogeneous monolayer cultures of rat multipotent precursors, as indicated by clonal analysis.[6] These cells readily proliferate in a fully defined medium containing a mitogen and differentiate rapidly upon mitogen withdrawal to yield reproducible proportions of neurons, astrocytes, and oligodendrocytes. These cells can be used to analyze the mechanisms that regulate stem cell proliferation, lineage choices and the maturation of newly-differentiated progeny.[13–20] The monolayer design

[11] T. J. Kilpatrick and P. F. Bartlett, *Neuron* **10**, 255–265 (1993).
[12] A. A. Davis and S. Temple, *Nature* **372**, 263–266 (1994).
[13] P. Rajan and R. D. McKay, *J. Neurosci.* **18**, 3620–3629 (1998).
[14] L. Studer, V. Tabar, and R. D. McKay, *Nat. Neurosci.* **1**, 290–295 (1998).
[15] K. Forsberg-Nilsson, T. N. Behar, M. Afrakhte, J. L. Barker, and R. D. McKay, *J. Neurosci. Res.* **53**, 521–530 (1998).
[16] M. Molne, L. Studer, V. Tabar, Y. T. Ting, M. V. Eiden, and R. D. McKay, *J. Neurosci. Res.* **59**, 301–311 (2000).

of the cultures greatly facilitates the visualization of individual cells and counting of cell numbers. Neurosphere cultures can be generated with minor modifications (see Appendix). This protocol can also be adapted to the culture of mouse neuroepithelial stem cells (see Appendix). The isolation procedures for fetal versus adult stem cells are sufficiently different as to merit separate sections, while the culturing methods are essentially the same.

During the initial culture period most differentiated neurons die due to the absence of appropriate trophic support. The proliferating cells are then harvested in a buffered, nonenzymatic solution, leaving behind any surviving neurons and the few contaminating mesenchymal cells. The passaged cells consist of a nearly homogeneous population of multipotent precursors and can be maintained as such for at least four to five passages.

Procedure Outline

Step A: Dissect fetal CNS tissue of interest in HBSS (without sodium bicarbonate).

Step B: Mechanically dissociate tissue in HBSS to a single cell suspension.

Step C: Culture cells in fully defined N2 medium containing bFGF as a mitogen.

Step D: Lift and dissociate cells with HBSS (with sodium bicarbonate) and passage into new dish; continue bFGF expansion as needed.

Step E: Differentiate cells by withdrawing bFGF and culturing in N2 medium alone.

Materials and Reagents (in addition to those described in ES cell section)

Equipment, Surgical Instruments and Supplies

Dissecting microscope with light source
Inverted light microscope
1 Tungsten needle (adapted from Conrad *et al.*, 1993[21]).
Cut 20–30 mm pieces of 0.01'' diameter tungsten wire with a sharp wire cutter. With power OFF, connect a variable auto-transformer (Variac) to power lines with alligator clip ends. Attach metal part of

[17] C. Vicario-Abejon, C. Collin, P. Tsoulfas, and R. D. McKay, *Eur. J. Neurosci.* **12**, 677–688 (2000).
[18] R. Y. Tsai and R. D. McKay, *J. Neurosci.* **20**, 3725–3735 (2000).
[19] Y. Sun, M. Nadal-Vicens, and S. Misono, *et al., Cell* **104**, 365–376 (2001).
[20] D. M. Panchision, J. M. Pickel, and L. Studer, *et al., Genes Dev.* **15**, 2094–2110 (2001).
[21] G. W. Conrad, J. A. Bee, S. M. Roche, and M. A. Teillet, *J. Neurosci. Methods* **50**, 123–127 (1993).

needle holder to one electrical lead and a metal spatula to the other lead; keep spatula immersed in a 3 *M* KOH solution. Wear insulating gloves and do not allow leads to touch. Place tungsten wire in needle holder and apply tip of tungsten wire to the KOH solution. Dial up current until bubbles form from tungsten wire, indicating that tip is being sharpened by electrolysis. Periodically monitor progress under dissecting scope. Use in needle holder during microdissection; needles can be bent to provide a comfortable working angle. In lieu of a tungsten needle, one may use a 25 G needle attached to a 1 ml syringe.

1 Needle holder (Ted Pella)
1 Small scissors
1 Microdissecting scissors (BRI 11-1020)
Fine micro-forceps, e.g., Dumont Biologie #4, 0.05×0.02 mm (BRI 10-1320)
Blunt micro-forceps, e.g., Dumont #2, 0.34×0.14 mm (BRI 10-1405)
1 Curved spatula
1 Instrument tray (BRI 24-1300)
Cell lifters (Costar)

Solutions (most described in ES cell section)

N2 medium
bFGF
Poly-L-ornithine
Fibronectin
PBS, 1X
0.2% Trypan Blue solution

Hank's Balanced Salt Solution, HEPES-buffered with glucose, 1X

1. Mix the following in a 1 liter cylinder:
   800 ml milliQ water
   100 ml 10X Ca^{2+}/Mg^{2+} free HBSS (Invitrogen 14185-052).
   3.70 g $NaHCO_3$ (Sigma S-5761) if needed
   3.90 g HEPES (Sigma H-0763)
   1.55 g glucose (Sigma G6152)
2. Adjust pH to 7.1 or 7.2 with 1 N HCl.
3. Add milli-Q water to 1 liter.
4. Filter-sterilize and store at 4°C.

The glucose promotes survival of tissue during lengthy dissections. The addition of sodium bicarbonate provides pH buffering in a 5% CO_2 incubator environment but rapidly becomes basic at room temperature and atmosphere. Thus, it is recommended that HBSS for passaging cells at 5% CO_2 be prepared with sodium bicarbonate and that separate HBSS for tissue dissection be prepared without sodium bicarbonate.

Preparing Coated Tissue Culture Dishes for Expansion of Neuroepithelial (Nestin$^+$) Cells

For monolayer culture, use the same fibronectin coating protocol as described above for ES cell culture. For neurosphere culture, use uncoated tissue culture dishes.

Dissecting Tissue from the Rat Fetus

1. Dissection does not have to be done under a laminar flow hood; however, the surgical field should be clean and the instruments disinfected with 70% ethanol or autoclaved. To obtain fetal tissue (i.e., 1–20 days postcoitum and prior to birth), euthanize a pregnant female rodent by carbon dioxide asphyxiation. Death of the pregnant female can be ensured by decapitation, bilateral pneumothorax by stab incision. The abdomen of the mother should be sanitized with ethanol, and the skin and abdominal muscles cut and deflected using sterile instruments. The uterus containing the embryos (10–16 for a pregnant Sprague–Dawley rat) will be separated from the attached connective tissue and fat, detached from the cervix and placed in a 10 cm dish containing HBSS (without sodium bicarbonate). Several rinses can be used to remove excess blood.
2. Place the dish containing the uterus under a dissecting microscope. Cut open the uterus along its length to expose embryos in their yolk sacs. Using blunt forceps, remove each embryo from its yolk sac. Use a curved spatula to transfer each embryo to a new dish containing HBSS. Euthanize the embryo by severing the base of the hindbrain with sharp sterile scissors or (for embryos younger than E15) sharp forceps. It is recommended that embryos and isolated tissue be kept on ice when not being dissected, especially during lengthy dissections of multiple embryos or CNS regions.
3. At this point the method of dissection will vary according to the age of the embryo and the region of the nervous system to be isolated. We will focus on the isolation of telencephalic tissue.

    a. For older rat embryos (E13.5 and older), use one pair of blunt forceps to hold the embryo in place and another pair to gently peel the cranium away from the telencephalon, starting from an incision made at about the midbrain. With experience the cranium can be pulled away in one sheet in an anterior-directed motion. Use blunt forceps to roll brain away from the base of the skull (severing the optic nerves) and transfer isolated brain to a new dish containing HBSS.

b. For younger rat embryos (E12.5 and younger), blunt forceps can be eased under the cranial tissue just above the eyes and pulled upward toward the dorsal midline to loosen cranial tissue. Remove cranial flap using fine forceps or the tungsten needle. Use blunt forceps to roll brain away from the base of the skull and transfer isolated brain to a new dish containing HBSS.

4. Turn the brain ventral-side-up and use a tungsten needle to slice the telencephalon bilaterally into two symmetric hemispheres that will resemble "conch shells." In older embryos this may be obscured by medial diencephalic tissue that can be removed. The "conch shell" can be partially unfurled by slicing away the olfactory lobe with a tungsten needle.
   a. One edge of the "conch shell" will be the medial and lateral ganglionic eminences, two thickened humps that are more pronounced in older embryos. These can be isolated separately using a tungsten needle. The meninges (the fine reddish network of blood vessels that covers the forebrain) can be easily removed with fine forceps in older embryos.
   b. The other lip of the "conch shell" is the choroid plexus epithelium; adjacent to this is the hippocampus. The hippocampus is isolated using a tungsten needle by (1) slicing along the flexure forming the boundary with the cortex and (2) slicing away the choroid plexus epithelium on the medial side. It is extremely difficult to get a clean dissection of early choroid plexus epithelium because the meninges is firmly attached to this region and some cranial tissue is often attached to the dorsal midline.
   c. Lateral to the ganglionic eminences is the cortex. Isolate the cortex by using a tungsten needle to trim away the ganglionic eminences, then trim along the more dorsal flexure point that forms the boundary of the hippocampus. The meninges are easy to peel away in the oldest embryos but are harder to peel away before E15; instead, they can often be removed during the later dissociation step.
5. Collect isolated tissue into a 35 mm dish containing HBSS, placed on ice.

Plating and Culturing Rat Fetal Stem Cells

1. Perform all subsequent steps in a laminar flow hood to prevent contamination. Using a sterile pasteur pipette or 1000 μl tip, consolidate the tissue from all embryos in a sterile 15 ml centrifuge tube and pellet the tissue by centrifugation in a tabletop centrifuge at approximately 200*g* (1000 r.p.m). In this and subsequent steps that involve transferring pieces of tissue, always maintain the pipette in a vertical orientation to prevent the tissue from adhering to the sides of the pipette. Do not exceed the recommended speed for centrifugation, as this decreases the yield of viable cells. Aspirate the HBSS from the tube, taking care not to dislodge the pellet.
2. Add 1 ml fresh sterile HBSS (without sodium bicarbonate) to the tissue pellet and dissociate the cells by pipetting slowly up and down using a 1000 μl pipette with a sterile tip. Allow the pipette tip to rest gently against the bottom of the centrifuge tube in order to assist in dispersing the tissue. Pass the cells no more than 8 times through the pipetman in order to maintain the viability of the cells. Allow undissociated tissue to settle to the bottom of the tube; with limited trituration the meninges of older embryos should stay intact and also settle to the bottom of the tube.
3. Transfer the upper suspended layer of cells to a new 15 ml tube and repeat the previous step with any remaining chunks of tissue.
4. Add approximately 8 ml sterile HBSS to the dispersed suspension and slowly pipette the cells 10–20 times to disperse any remaining cell aggregates. Pellet the cells by centrifugation as above.
5. Aspirate the supernatant, taking care not to disturb the cell pellet. Resuspend the cells in 5–10 ml N2 medium by pipetting gently until the pellet is completely resuspended. Measure the cell recovery and viability by mixing a small aliquot 1 : 1 with a solution of 0.2% trypan blue, a vital dye, and counting the live and dead cells on a hemacytometer. One can generally expect 70–80% of the cells to exclude trypan blue (indicating survival) after this acute dissociation. The yield of cells varies with the region dissected and the gestational age of the embryo.
6. Plate $1.0–1.5 \times 10^6$ cells per precoated 10 cm dish in 10 ml of N2 medium containing 10 ng/ml bFGF as a mitogen (see Table I, acute density). Place the dishes in a tissue culture incubator; add fresh bFGF daily and replace the medium every other day. During this initial expansion the cultures will be heterogeneous, containing differentiated and dying neurons as well as emerging colonies of proliferating precursors. After 5 days the cultures should be 50–75%

TABLE I
FETAL STEM CELL PLATING AND FEEDING RECOMMENDATIONS

Dish size			Plating to achieve density of				Media volume	
Diam. (mm)	Area (cm^2)	Ratio	18,000/cm^2 acute (P0)	10,200/cm^2 expansion	354/cm^2 colony	35/cm^2 clonal	Plating (ml)	Feeding (ml)
150	176.71	2.25	3.2×10^6	1.8×10^6	62,555	6256	23.0	16.0
100	78.54	1.00	1.4×10^6	8.0×10^5	27,803	2780	10.0	7.0
60	28.27	0.36	5.1×10^5	2.9×10^5	10,008	1001	4.0	2.0
35	9.62	0.12	1.7×10^5	9.8×10^4	3405	341	1.5	1.0
25	4.91	0.06	8.8×10^4	5.0×10^4	1738	174	1.0	0.5
15	1.77	0.02	3.2×10^4	1.8×10^4	627	63	0.5	0.3

confluent and can then be passaged to new dishes, a process that selects against differentiated cells.

Passaging Cells

1. After an initial 5 days in culture, harvest the cells for passage to new coated dishes in order to remove undesired cells. Wash cells 2–3 times with 5 ml HBSS (with sodium bicarbonate), then incubate cells 15 min at 37°C in 7–10 ml HBSS. This treatment facilitates removal of the cells without the need for enzymes. Using a 10 ml pipette, gently spray the HBSS over the surface of the dish to wash off lightly adherent cells. Though generally not necessary, any remaining cells can be scraped off of the surface of the dish with a cell lifter. Pipette cells gently to disperse any clumps and pellet by centrifugation as above.
2. Aspirate the supernatant and resuspend the cells in 5–10 ml N2 medium, then count cells and assess the viability as above. For continued expansion, inoculate coated dishes at 8×10^5 per 10 cm dish in N2 medium containing 10 ng/ml bFGF. For experiments, other plating densities can be chosen based on parameters shown in Table I.
3. Expansion cultures should be passaged every 3 days from first passage onward. At this point the culture is relatively homogeneous and consists almost exclusively of Nestin$^+$ multipotent precursors (Fig. 7). Add fresh bFGF daily and replace the medium every other day.
4. In order to induce differentiation of these cells to neurons and glia, remove the medium containing bFGF and replace it with N2 lacking

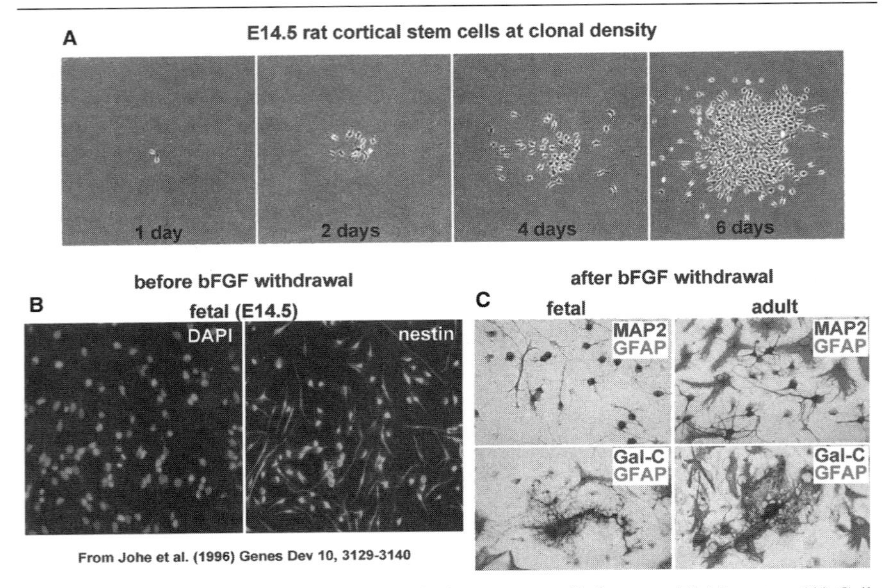

FIG. 7. Proliferation and differentiation of primary stem cells from rat E14.5 cortex. (A) Cells plated at 30 cells/cm^2 generate distinct clones after expansion in bFGF. (B) Cells in proliferating clone uniformly express the intermediate filament Nestin. (C) After 7 days in the absence of bFGF, cells have differentiated into MAP2$^+$ neurons, GFAP$^+$ astrocytes, and Gal-C$^+$ oligodendrocytes. Both fetal and adult stem cell clones give rise to reproducible proportions of all three derivatives. (See Color Insert.)

bFGF. Allow 7 to 10 days of differentiation to permit the expression of many differentiation markers.[6] Additional factors can be added during the expansion or differentiation phase as needed.[6,13–20]

Appendix for Fetal Stem Cell Culture

Critical Parameters

The developmental stage of the embryonic tissue is the major determinant of the yield of proliferating precursors. For example, an average litter of E14.5 rat embryos will yield $5–10 \times 10^6$ cortical cells upon dissociation. The first passage 4–5 days later will typically yield $2–4 \times 10^7$ viable cells. This corresponds to an average doubling time of 18 hr. For the rat telencephalon, proliferating precursors are most abundant in the ganglionic eminence until E14, cortex until E15 and hippocampus until E16.[22] The precursor populations decline substantially after these ages.

[22] S. A. Bayer and A. Altman, *in* "The Rat Nervous System" (G. Paxinos,, ed.), Vol. 2, pp. 1041–1098. Academic Press, San Diego, CA, 1995.

Likewise, the ratio of neurons versus glia that are generated from multipotent precursors shifts as development proceeds to favor the generation of glia.[1]

It is necessary to replenish the bFGF in these cultures daily in order to prevent differentiation of the precursors. Cells will also differentiate prematurely if they become contact inhibited, and for this reason we passage the cells when they are 50–70% confluent. Enzymes such as trypsin are omitted from the passaging step so as to create a selection against more adherent cells such as differentiated astrocytes. It is crucial to avoid exposure of the cells to serum at any stage of precursor culture. Treatment with fetal calf serum rapidly induces differentiation of the multipotent precursors to predominantly astrocytic cells. This effect is prominent even with transient exposure of the cells to less than 0.5% serum.

Clonal density plating can be done at first or subsequent passages (Table I, Fig. 7). Plating efficiency (the proportion of viable plated cells that survive to divide at least once) decreases dramatically below a density of about 350 cells/cm^2. Retroviral lineage analysis[6] of cells in mass culture suggests that this cell death is stochastic rather than selecting for a sub-population of cells. Our experience with rat cells plated at a clonal density of 3 to 35 cells/cm^2 and marked within several hours show a plating efficiency of roughly 10%. This translates to about 20 to 300 clones per 10-cm dish. In contrast, colony density plating (mass culturing where distinct nonclonal colonies can still be visualized) has a plating efficiency of more than 90%.

Transfection efficiency in these cells is much lower than with many cell lines. For example, E14.5 cortical cells transfect with an efficiency of about 5 to 15% using polyamine reagents such as Lipofectamine Plus (Invitrogen). For various reasons, transfection efficiency is better with middle passage cells (around passage 3) in moderate sized clusters. Despite these minor limitations, transfection is routinely performed with these cultures.[13,19,20]

Modifications for Culturing Mouse Fetal Neuroepithelial Stem Cells

The standard protocol described above is optimized for the efficient expansion of rat neuroepithelial stem cells and is generally too stringent for efficient expansion of mouse stem cells. This protocol works if mouse stem cells are plated at densities much higher than those used for rat stem cells, but extra care must be taken to prevent density-dependent differentiation. As a result mouse cells are often passaged more frequently than rat stem cells.

The survival and expansion of mouse stem cells is greatly enhanced by two modifications to the standard protocol: the doubling of bFGF

TABLE II
MODIFICATIONS TO STANDARD PROTOCOL REQUIRED FOR MOUSE CELL CULTURING

Species age	Rat E14.5	Mouse E13.0
bFGF	10 ng/ml once daily	20 ng/ml once daily
O_2 requirement	21% O_2	5% O_2
P0 plating for expansion	1.5×10^6 per 10 cm dish	2.5×10^6 per 10 cm dish
Duration to next passage	5 days	4 to 5 days
P1 plating for expansion	0.8×10^6 per 10 cm dish	1.0×10^6 per 10 cm dish
Duration to next passage	3 days	2 to 3 days

concentration to 20 ng/ml (added daily) and the growth of the cells in a low (e.g., 5%) oxygen incubator environment. The lower oxygen concentration probably minimizes oxidative stress on the cells. We have not found an antioxidant supplement that duplicates the effect of lowered oxygen on these cells. Mouse cortical stem cells can be successfully expanded at low or even clonal density (e.g., 3×10^3 cells per 10 cm dish) under these modified conditions but still grow less robustly than rat stem cells. The addition of B27 supplement as previously described[23] does enhance survival but changes the cells morphologically and may promote differentiation. The addition of N-acetyl-cysteine[23] to our media formulation has no beneficial effect and actually inhibits adhesion and proliferation of these cultures.

Table II summarizes the modifications for dissecting and culturing cortical stem cells of mouse compared to the equivalent developmental stage of rat. Cells at subsequent passages should be cultured exactly as at P1.

Modifications for Neurosphere Culture

Neurosphere cultures are an alternative to monolayer cultures and have the benefit of requiring less labor. Some investigators also like the fact that cell growth occurs on endogenously secreted matrix of other cells rather than an exogenously added substrate. While neurosphere cultures are significantly more difficult to analyze for cell behavior (e.g., fate choice, cell cycle) than monolayer cultures, they are sufficient for assaying simple clone formation.

Neurosphere cultures can be generated with three modifications to the standard protocol described above. First, cells are plated on

[23] X. Qian, Q. Shen, and S. K. Goderie, *et al. Neuron* **28**, 69–80 (2000).

uncoated tissue culture plates rather than matrix-coated plates; the cells will either stay in suspension or will adhere only briefly to the plate. Second, the plates are not disturbed for 5–6 days after plating, as this will disrupt the formation of neurospheres. Third, because the cultures are left unattended for several days, a higher mitogen dose must be given at the time of plating (usually about 100 ng/ml bFGF). It has been suggested that the close proximity of cells promotes proliferation through the secretion of paracrine factors.

Preparing Cultures of Rat Adult Neuroepithelial Stem Cells

Neural precursors are removed from the subventricular zone (SVZ) of the adult rodent central nervous system (CNS) by dissection, dissociated to a single-cell suspension, and plated on coated tissue culture dishes in media containing the mitogen basic fibroblast growth factor (bFGF). During the initial culture period most differentiated neurons die due to the absence of appropriate trophic support while some glia survive. The proliferating cells are then harvested in a buffered solution, leaving behind any surviving neurons and most of the more adherent glial cells. The passaged cells consist of an enriched population of multipotent precursors and can be maintained as such for about one to two passages.

Materials and Reagents (specific to adult dissection)

Trypsin (10,000 BAEE units/mg; Worthington LS003703)
Hyaluronidase (3000 units/mg; Worthington LS005474)
Trypsin inhibitor (7000 BAEE units/mg; Life Technologies 17075-011).
Deoxyribonuclease (Worthington LS002139)

Dissecting Tissue and Plating Cells from the Rat Adult Subventricular Zone

1. Sacrifice five 200–250 g female rats, one at a time, remove brain and place in cold sterile HBSS (without sodium bicarbonate, with 10 mM glucose). Tissue should be kept on ice.
2. Using a razor blade, make coronal sections beginning at the olfactory bulb. These should be as thin as possible; 1–2 mm is reasonable. Under a dissecting microscope, carefully tease off the ventricular tissue with a tungsten needle, including tissue lateral and medial to the ventricle.
3. Mince the tissue with the tungsten needle and combine the pieces in 10 ml HBSS plus 1.5 mg/ml trypsin (10,000 BAEE units/mg; Worthington LS003703) plus 0.35 mg/ml hyaluronidase (3000 units/mg; Worthington LS005474). Agitate the suspension

gently at 34°C for 30–40 min (try using a hybridization oven and taping the tube to the rotating rod). Perform subsequent steps under a laminar flow hood.

4. Pellet the cells and aspirate the supernatant with a pipet-aid (do not use a vacuum because a DNA clot forms during this step, increasing the likelihood that you will aspirate the cells under vacuum). Wash the cells twice with 10 ml DMEM/F12 plus 0.05% DNase (Worthington LS002139) plus 0.7 mg/ml trypsin inhibitor (7000 BAEE units/mg; Life Technologies 17075-011).
5. Wash once with HBSS, then pellet cells at 200*g* (1000 rpm).
6. Resuspend the cells in 2 ml HBSS and triturate 10 times slowly with a 1000 μl pipette tip, then let any chunks settle. Transfer the supernatant to another tube and repeat the trituration of the remaining tissue pieces as necessary.
7. Pellet the cells and resuspend in N2 medium containing 10 ng/ml bFGF. The low cell number and abundant debris preclude accurate cell counting, so simply plate cells into one 10 cm dish per animal dissected (10 ml medium per plate). Replace with fresh N2 medium containing 10 ng/ml bFGF every day; expand for 7 to 8 days before passaging as described above. After 7–8 days, the plates should contain large and well-defined colonies. These can then be passaged and cultured in the same manner as fetal cells.

Appendix for Adult Stem Cell Culture

There is considerable debris in adult cultures that seems to inhibit the proliferation of the cells. Dividing cells will be difficult to see for a couple of days after plating. Ficoll gradient centrifugation[24] may clean up the preps but in our experience will substantially decrease the recovery of cells. However, by simply changing the medium daily (rather than every second day) the debris will clear within a week and improve stem cell proliferation significantly.

Glial contamination is a minor but consistent problem in adult stem cell preps. For this reason cells should be plated at low density after passaging so that individual cells can be marked for clonal analysis. Clones arising from multipotent stem cells have a characteristic morphology (phase-bright and bipolar) that distinguishes them from colonies of more differentiated glia (elongated sickle-shaped cells). Experiments are optimally performed at

[24] T. D. Palmer, E. A. Markakis, A. R. Willhoite, F. Safar, and F. H. Gage, *J. Neurosci.* **19,** 8487–8497 (1999).

passage 1, since glial contamination becomes more pervasive with continued culture.

Cultures derived from SVZ in the most rostral regions of the telencephalon (near the olfactory bulb) give the highest ratio of stem cells versus glia. This ratio decreases caudally and most of the cells lining the ventricles near the hippocampus gave rise to astrocytes in culture, so this tissue should be avoided.

[23] Defined Conditions for Neural Commitment and Differentiation

By QI-LONG YING and AUSTIN G. SMITH

Introduction

Pluripotent mouse embryonic stem (ES) cells can be expanded in culture indefinitely while retaining the capacity to produce seemingly every type of fetal and adult cell.[1] ES cell differentiation *in vitro* is thought to recapitulate *in vivo* developmental programs[2] and generation of various apparently fully specified and functional cell types has been described.[3] However, we cannot yet claim an ability to "direct" ES cell differentiation. Current methods are empirical with outcomes that are invariably heterogeneous and often poorly reproducible. Some progress has been made with controlling intermediate stages of lineage progression,[4,5] but mastering the full sequence of steps necessary for efficient generation of any particular terminally differentiated phenotype remains elusive. There are two major challenges: first to understand and manipulate lineage choices; second to develop culture conditions that support the viability and maturation of progenitor and terminal phenotypes *in vitro*. Our laboratory has begun to investigate these issues in the context of neural differentiation.[5a]

The most widely used method to trigger neural development from ES cells is cell aggregation in suspension culture followed by treatment with retinoic acid. In suspension culture ES cells form

[1] A. G. Smith, *Annu. Rev. Cell. Dev. Biol.* **17**, 435–462 (2001).
[2] G. M. Keller, *Curr. Opin. Cell. Biol.* **7**, 862–869 (1995).
[3] A. M. Wobus and K. R. Boheler, *Cells Tissues Organs* **165**, 129–130 (1999).
[4] S. H. Lee, N. Lumelsky, L. Studer, J. M. Auerbach, and R. D. McKay, *Nat. Biotechnol.* **18**, 675–679 (2000).
[5] H. Wichterle, I. Lieberam, J. A. Porter, and T. M. Jessell, *Cell* **110**, 385–397 (2002).
[5a] A. G. Smith, Converting ES Cells into Neurons, *in* "Research and Perspectives in Endocrinology." Springer-Verlag, in press.

0076-6879/2003 $35.00

multicellular multi-differentiated structures called embryoid bodies.[6] Neural derivatives are present only at low frequency in embryoid bodies generated in serum-containing medium, but their proportion increases dramatically after addition of retinoic acid.[7] Regardless of the concentration or duration or retinoic acid treatment, however, the final cultures are always a heterogeneous mixture of various cell types. Several strategies have been developed to purify or enrich neuroectodermal precursors or more mature neuronal or glial phenotypes from embryoid bodies. These include the introduction of a transgene marker conferring drug resistance and/or cell-sorting capacity specifically to neural lineage cells,[8,9] immunopanning for neural antigens to select neuronal or glial restricted progenitors,[10] and a combination of growth factor stimulation and differential adhesion and proliferation in minimal media.[11,12] Although these techniques are effective, the primary process of neural determination remains unexplained and relatively inefficient.

It is difficult to dissect and manipulate differentiation within embryoid bodies because they are multicellular agglomerations of extraembryonic endoderm and definitive ectodermal, mesodermal and endodermal derivatives.[2] Furthermore, retinoic acid has pleiotropic actions—it induces other lineages[13] and affects positional specification[5]—and the route by which it influences neural commitment is obscure. Several groups have developed methods in which treatment with retinoic acid is avoided.[14–17]

[6] T. C. Doetschman, H. Eistetter, M. Katz, W. Schmidt, and R. Kemler, *J. Embryol. Exp. Morphol.* **87**, 27–45 (1985).

[7] G. Bain, D. Kitchens, M. Yao, J. E. Huettner, and D. I. Gottlieb, *Dev. Biol.* **168**, 342–357 (1995).

[8] M. Li, L. Pevny, R. Lovell-Badge, and A. Smith, *Curr. Biol.* **8**, 971–974 (1998).

[9] N. Lenka, Z. J. Lu, P. Sasse, J. Hescheler, and B. K. Fleischmann, *J. Cell Sci.* **115**, 1471–1485 (2002).

[10] T. Mujtaba, D. R. Piper, A. Kalyani, A. K. Groves, M. T. Lucero, and M. S. Rao, *Dev. Biol.* **214**, 113–127 (1999).

[11] S. Liu, Y. Qu, T. J. Stewart, M. J. Howard, S. Chakrabortty, T. F. Holekamp, and J. W. McDonald, *Proc. Natl. Acad. Sci. USA* **97**, 6126–6131 (2000).

[12] S. Okabe, K. Forsberg-Nilsson, A. C. Spiro, M. Segal, and R. D. McKay, *Mech. Dev.* **59**, 89–102 (1996).

[13] C. Dani, A. G. Smith, S. Dessolin, P. Leroy, L. Staccini, P. Villageois, C. Darimont, and G. Ailhaud, *J. Cell Sci.* **110**, 1279–1285 (1997).

[14] V. Tropepe, S. Hitoshi, C. Sirard, T. W. Mak, J. Rossant, and D. van der Kooy, *Neuron* **30**, 65–78 (2001).

[15] M. V. Wiles and B. M. Johansson, *Exp. Cell. Res.* **247**, 241–248 (1999).

[16] H. Kawasaki, K. Mizuseki, S. Nishikawa, S. Kaneko, Y. Kuwana, S. Nakanishi, S. I. Nishikawa, and Y. Sasai, *Neuron* **28**, 31–40 (2000).

[17] J. Rathjen, B. P. Haines, K. M. Hudson, A. Nesci, S. Dunn, and P. D. Rathjen, *Development* **129**, 2649–2661 (2002).

Neural cells appear in embryoid bodies in the absence of serum[14,15] or presence of conditioned medium extracts.[17] They can also be obtained at high frequency upon coculture with a particular stromal cell line, PA6,[16] an effect ascribed to an unidentified stromal cell-derived inducing activity (SDIA). In all cases, however, the mechanism of neural commitment remains elusive. Data from one study indicates that neural cells can be derived from individual ES cells when placed in suspension in the absence of serum.[14] However, the interpretation that neural specification arises by default[18] is challenged by the low frequency of this event (1 in 1000 cells).

We sought to develop a simple system that would allow direct observation, analysis, and manipulation of the process of neural specification without the confounding influences of cell aggregation, coculture, uncharacterized media constituents, or cell selection. We also wished to avoid retinoic acid because this is likely to restrict the regional identity of neural precursors.[5] Here, we describe defined conditions for conversion of ES cells to neural fates in monolayer culture.

Cells, Reagents, and Experimental Protocols

ES Cells

Parental ES cell lines are germline competent CGR8[19] and E14Tg2a[20] derived from 129/Ola mice. Genetically manipulated ES cells are:

46C: generated by gene targeting in E14TG2a.[21] The open reading frame of the *Sox1* gene is replaced with the coding sequence for enhanced green fluorescent protein (GFP) and an internal ribosome entry site (IRES)-linked puromycin resistance gene. In embryos generated from 46C ES cells GFP is expressed in neuroepithelial cells throughout the neuraxis and in the lens but in no other tissue. The fidelity of *Sox1*-GFP expression is maintained *in vitro*. Consequently this cell line is a valuable tool for monitoring and quantitating the acquisition of neural identity by ES cells.[21,22]

[18] I. Munoz-Sanjuan and A. H. Brivanlou, *Nat. Rev. Neurosci.* **3**, 271–280 (2002).

[19] P. Mountford, B. Zevnik, A. Duwel, J. Nichols, M. Li, C. Dani, M. Robertson, I. Chambers, and A. Smith, *Proc. Natl. Acad. Sci. USA* **91**, 4303–4307 (1994).

[20] M. Hooper, K. Hardy, A. Handyside, S. Hunter, and M. Monk, *Nature* **326**, 292–295 (1987).

[21] Q. L. Ying, M. Stavridis, D. Griffiths, Meng Li, and A. Smith, *Nat. Biotechnol.* **21,** 183–186 (2003).

[22] J. Aubert, H. Dunstan, I. Chambers, and A. Smith, *Nat. Biotechnol.* **20**, 1240–1245 (2002).

[23] N. Billon, C. Jolicoeur, Q. L. Ying, A. Smith, and M. Raff, *J. Cell Sci.* **115**, 3657–3665 (2002).

OS25: generated from E14TG2a by sequential gene targeting.[23] First a bifunctional *βgeo* cassette was inserted into *Sox2* locus[8] by homologous recombination. One of the correctly targeted clones was then subjected to second round gene targeting in which the counter-selectable *hygromycin-thymidine kinase* (*hy-tk*) fusion gene[24] was integrated into the *Oct4* (*Pou5f1*) locus. The *Sox2* targeting construct and flanking probes were generously provided by Silvia Nicolis and Robin Lovell-Badge. The *Oct4* construct was generated by replacing *βgeo* in the *Oct4βgeo* targeting vector[19] with the *hytk* fusion gene and was provided by Hitoshi Niwa. OS25 cells can be positively selected for ES cells and neural precursors (Sox2-expressing) and then subjected to negative selection with Ganciclovir for specific elimination of undifferentiated ES cells (Oct4-expressing).

Oct4GiP: derived from strain 129Ola mice carrying an *Oct4-GFPiresPac* transgene.[25] GFP and puromycin resistance are expressed exclusively in undifferentiated ES cells under the direction of regulatory sequences of the mouse *Oct4* gene. GFP expression facilitates monitoring of undifferentiated ES cells and selection in puromycin allows ES cells to be propagated as pure cultures of undifferentiated stem cells.

TK23: generated from E14TG2a. The GFP coding sequence was targeted into the *Tau* locus by homologous recombination. Kerry Tucker and Yves Barde kindly provided the *Tau* targeting vector. GFP expression in these cells serves as a reporter of neuronal differentiation.[26]

Culture Medium

ES cells are routinely propagated without feeders on gelatin-coated plastic in Glasgow modification of Eagle's medium (GMEM) containing 10% fetal calf serum (FCS), 2-mercaptoethanol, and leukaemia inhibitory factor.[27]

For neural differentiation we have found empirically that a mixed formulation of basal media and supplements provides optimum cell viability and efficient neural differentiation. This formulation, which we call N2B27, is as follows:

DMEM/F12 (50/50) medium supplemented with modified N2 (25 μg/ml insulin, 100 μg/ml apo-transferrin, 6 ng/ml progesterone, 16 μg/ml putrescine, 30 n*M* sodium selenite) plus 50 μg/ml bovine serum albumin fraction V, combined 1 : 1 with Neurobasal™ medium supplemented with B27.

[24] S. D. Lupton, L. L. Brunton, V. A. Kalberg, and R. W. Overell, *Mol. Cell. Biol.* **11**, 3374–3378 (1991).

[25] Q. L. Ying, J. Nichols, E. P. Evans, and A. G. Smith, *Nature* **416**, 545–548 (2002).

[26] K. L. Tucker, M. Meyer, and Y. A. Barde, *Nat. Neurosci.* **4**, 29–37 (2001).

[27] A. G. Smith, *J. Tiss. Cult. Meth.* **13**, 89–94 (1991).

Basal media and B27 supplement are from Gibco (Paisley, UK). N2 is prepared from individual constituents.

Basic Protocol for Monolayer Neural Differentiation

Prior to initiating differentiation ES cells are plated at relatively high density ($2–3 \times 10^6$ per T25 flask) and cultured for 24 hr in standard ES cell medium containing LIF. To start monolayer differentiation, undifferentiated ES cells are dissociated using 0.025% trypsin solution (0.025% trypsin, 1.3 m*M* EDTA, 0.1% chicken serum, in PBS) at 37°C. The trypsin is neutralized with serum containing medium and ES cells are spun down and resuspended directly in N2B27 medium. ES cells are then plated onto 0.1% gelatin-coated tissue culture plastic at a density of $0.5–1.5 \times 10^4/\text{cm}^2$. Typically, 3×10^5 ES cells in 3 ml N2B27 medium are plated per 60 mm tissue culture dish (Nunc). Thereafter medium is changed every other day. Under this monoculture condition in the absence of LIF, ES cells lose pluripotent status and predominantly commit to a neural fate over a 4–5 day period. Subsequently neurons appear and increasingly populate the cultures. Neuronal differentiation is most efficient if the cells are replated at the neural precursor stage (see the following subsection).

Monitoring Monolayer Differentiation Using 46C ES Cells

To the trained eye neural precursors are morphologically distinguishable from ES cells. The cells are closely apposed but with distinct intercellular boundaries. The cell bodies are opaque and nuclei difficult to discern, in contrast to ES cells which have very prominent nuclei (Fig. 1). Neural precursors can also be identified by fixation and immunostaining for Sox1 or for nestin, although nestin is less specific since it is also present in somitic and pancreatic cells. The vital reporter provided by *Sox1*-GFP in 46C cells is a useful adjunct to the above approaches because it allows unambiguous visualization of living neural precursors (Fig. 1). The GFP fluorescence from the knock-in allele is readily seen using a standard fluorescence microscope in medium lacking serum (which quenches the fluorescent signal). If the cultures are transferred to PBS for observation, the rosette organization of the neural precursor clusters becomes more obvious (Fig. 2).

FACS Quantitation of Neural Conversion

Use of 46C cells also allows reliable quantitation of neural precursors by flow cytometry. First, cells are trypsinized and washed in serum-containing medium. The cells are then spun down and resuspended in FACS buffer

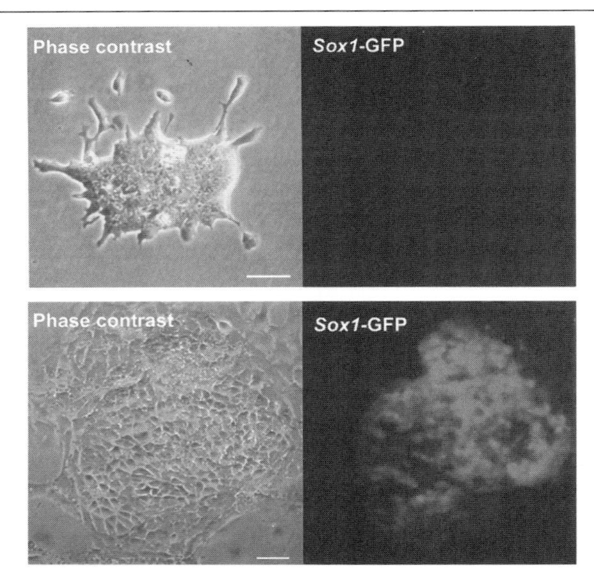

FIG. 1. Phase contrast and fluorescent images of monolayer differentiation of *Sox1*-GFP ES cells. Upper panels show a colony of undifferentiated 46C ES cells cultured in medium containing serum and LIF. Lower panels show a colony cultured in GMEM/10% FCS medium plus LIF for 2 days then transferred to N2B27 for 5 days to induce neural differentiation. Bar: 50 μm. (See Color Insert.)

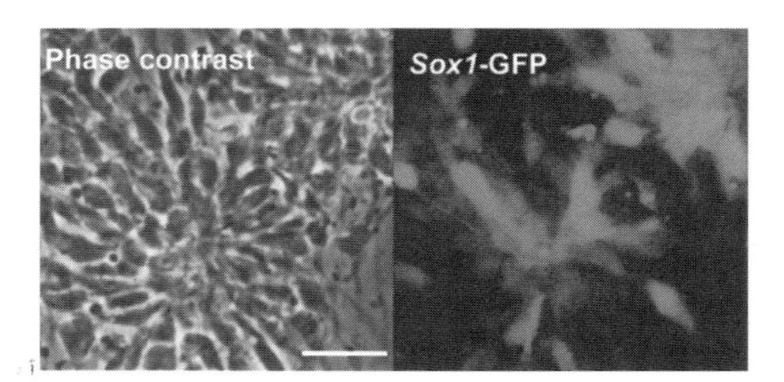

FIG. 2. Expression of *Sox1*-GFP in rosette of neural precursors derived from 46C ES cells after 7 days in monolayer differentiation. Bar: 50 μm. (See Color Insert.)

(PBS with 4% FCS) and the total cell numbers from each dish are counted. In our analyses we used a FACS Calibur flow cytometer (Becton Dickinson) and CellQuest software. Gates are set at 10 units of fluorescence, which excludes >99% of undifferentiated ES cells. 10,000 events are scanned. Cell debris and dead cells are excluded from the analysis based on electronic gates set using forward scatter (size) and side scatter (cell complexity) criteria. All the settings are determined at the start of the experiment using undifferentiated 46C ES cells as negative control and ES cells ubiquitously expressing GFP[28] as positive control. The settings are then saved and used throughout the experiment to ensure consistency of data acquisition.

In the first two days after plating, very few cells become positive for *Sox1*-GFP. Then on day 3 and day 4, the proportions of *Sox1*-GFP positive cells rise steeply to around 75%. From day 4 to day 8, the proportion of *Sox1*-GFP positive cells remains relatively stable around 70–80%. Thereafter, progressive differentiation into neurons and glia is accompanied by downregulation of *Sox1*.[29] Most importantly, although there is some compromise of initial plating efficiency under these serum-free conditions, subsequent viability is high and cell numbers increase exponentially throughout the culture period without any lag period.[21] Therefore, this protocol results in quantitative conversion of ES cells to a neural phenotype.

Monitoring the Monolayer Differentiation Process

An advantage of the monolayer differentiation system over embryoid body methods is that the whole process can be observed at the level of individual cells or colonies without intervention in the differentiation procedure.

Lower cell plating density facilitates the observation and recording of individual cells or colonies. At such densities it is advisable to plate in serum-containing ES cell medium plus LIF to optimize cell attachment. Cultures are maintained overnight and then switched to N2B27 without LIF to start differentiation and recording. If 46C ES cells are used, the emergence of *Sox1*-GFP positive neural precursors can be visualized during the transition from undifferentiated ES cells to neural lineages. The basic method is as below:

46C ES cells are plated at 5×10^3 to 5×10^4 cells per gelatin-coated 60 mm tissue culture dish in GMEM/10% FCS medium plus LIF. After overnight culture, medium is replaced with N2B27. Phase contrast and GFP fluorescence images of the same fields may be captured using a suitable microscope and imaging package (e.g., Openlab Software) at regular time intervals. Medium is renewed every other day. At the final time point, cells are fixed and immunostained for other neural markers (i.e., nestin, βIII-tubulin) to confirm the final differentiation status of the culture.

[28] T. Pratt, L. Sharp, J. Nichols, D. J. Price, and J. O. Mason, *Dev. Biol.* **228**, 19–28 (2000).

[29] L. H. Pevny, S. Sockanathan, M. Placzek, and R. Lovell-Badge, *Development* **125**, 1967–1978 (1998).

These analyses show that neural conversion occurs throughout the culture and is not accompanied by major cell death.[21] This is consistent with bulk conversion of ES cells to a neural phenotype rather than selection and amplification of a rare sub-population. This can be confirmed by clonal anaysis.

Plating and Differentiation Efficiency at Clonal Density

The monolayer protocol allows clonal assays to gauge the efficiency of differentiation of individual ES cells. Dissociated ES cells are plated onto 0.1% gelatin-coated 6-well plates (Iwaki) at the density of 1–5×10^3 cells/well and cultured in serum-containing ES cell medium (3 ml/well) plus LIF. At this density more than 90% of colonies derive from single cells. After 2 days, medium is replaced with serum-free neural differentiation medium N2B27. Medium is renewed with fresh N2B27 every other day. After 5–7 days culture in N2B27 medium, total colonies and colonies containing *Sox1*-GFP$^+$ and/or nestin$^+$ neural precursors may be counted. Alternatively, neural colonies can also be identified and counted by their characteristic morphology as shown in Fig. 1. The proportions of *Sox1*-GFP positive cell populations in individual wells may also be determined by FACS analysis. Typically by 5 days in N2B27 medium, more than 90% of total colonies contain *Sox1*-GFP positive neural precursors. Due to the plating in serum these colonies usually also contain nonneural cells around the periphery. Nonetheless more than 50% of total cell population are *Sox1*-GFP positive by FACS. At later time points the proportion of neuron and/or astrocyte containing colonies can be scored using *Tau*-GFP or antibody markers such as Tuj1 or GFAP.

Monitoring Neuronal Differentiation

In order to monitor and quantitate accurately the production of neurons during monolayer differentiation we engineered another ES cell line, TK23, in which GFP is integrated into the *Tau* locus and is therefore expressed only in cells of a neuronal phenotype.[26] The production of neurons during monolayer differentiation can be observed and quantified by FACS as described for neural precursors using *Sox1*-GFP positive cells. *Tau*-GFP positive cells appear 4–5 days after monolayer differentiation and their proportion increases steadily thereafter. Up to 30% of TK23 cells activate *Tau*-GFP expression after 12 days in monolayer differentiation. Many of the remaining cells are neural precursors as evident by nestin antibody staining (or *Sox1*-GFP if using 46C cells). The frequency of neuronal differentiation is promoted if the cells are dissociated and replated. This is probably due to a combination of factors. Without replating, cell density becomes very high which may increase the potential for autocrine growth factors to suppress differentiation and sustain neural precursors in an undifferentiated state. Cell–cell contacts may also inhibit neuronal differentiation. In addition, gelatin is not the optimal substratum for neuronal differentiation. Dissociation and replating the cells will dilute the concentration of autocrine factors, disrupt cell–cell contacts, and provide the opportunity to alter the substrate.

Protocol for Efficient Derivation of Neurons via Monolayer Differentiation

Monolayer differentiation is initiated as described earlier. On day 8, cells are dissociated using 0.025% trypsin, washed in serum containing medium and collected by gentle centrifugation. The cells are resuspended in N2B27 medium and replated onto fibronectin- or PDL-laminin-coated plastic at a density of $0.5–1.5 \times 10^4/cm^2$. Addition of FGF-2 (10–20 ng/ml) improves cell viability on replating. Medium is renewed every 3–4 days. The production of neurons is examined by *Tau*-GFP (for TK23) or immunostaining for neuronal markers (e.g., βIII-tubulin), combined with nuclear counter staining to facilitate cell number counting. Mature neuronal morphology with extensive arborization becomes apparent on extended culture in N2B27 medium. Mature neuronal cells can be maintained under these conditions for 3–4 weeks. It should be noted, however, that there is considerable cell death during neuronal maturation, possibly because N2B27 does not contain specific neurotrophic factors.

Protocol for Generation of Tyrosine Hydroxylase (TH)-Positive Cells from Monolayer Culture

Many of the neurons that persist in N2B27 express gamma-aminobutyric acid (GABA). In contrast, relatively few tyrosine hydroxylase positive cells are seen. However, exposure during differentiation to the ventral midbrain patterning factors FGF8 and Sonic Hedgehog[30] results in appearance of significant numbers of tyrosine hydroxylase immunoreactive neurons.[21] The protocol is based on that described by Lee *et al.*[4] for embryoid body derived neural precursors:

[30] W. Ye, K. Shimamura, J. L. Rubenstein, M. A. Hynes, and A. Rosenthal, *Cell* **93**, 755–766 (1998).

Four days after initiation of monolayer differentiation, cells are dissociated and replated onto PDL-laminin-coated plastic in N2B27 medium supplemented with FGF2 (20 ng/ml), Sonic Hedgehog (400 ng/ml) and FGF8 (100 ng/ml) (R&D System). Two days later, medium is changed to N2B27 plus FGF2 (10 ng/ml). Cells are cultured for another 2 days before fixation and immunostaining for TH or analysis for other mesencephalic dopaminergic markers.

Purification and Propagation of Neural Precursors from Monolayer Differentiation

As with embryoid body methods or PA6 induction, the population of neural precursors obtained by the monolayer protocol is not pure. Elimination of residual ES cells and contaminating nonneural cells is therefore a significant issue, in particular for expression profiling studies and transplantation experiments. A reliable strategy for purifying neural precursors is the use of lineage selection.[8]

Purification of Sox1-GFP Positive Neural Precursors by FACS

In the case of 46C cells, FACS purification can be used to separate *Sox1*-GFP positive neural precursors from undifferentiated ES cells, differentiated neurons and glia, and any contaminating nonneural cells. In our studies, we have used a Cytomation MoFlo flow cytometer, but the basic protocol should be applicable to other machines.

Monolayer differentiation is conducted as described earlier. After 4–6 days, 46C cells are dissociated and resuspended in FACS buffer. The cell suspension is passed through a cell-strainer (35 μm pore size) and stored on ice. *Sox1*-GFP positive and negative populations are sorted separately into tubes containing N2B27 medium. To keep the sorted cell samples free from contamination, all steps are performed using aseptic technique. To monitor the purity of FACS sorting, reverse transcription PCR analysis may be used to detect the expression of *Sox1* and *Oct4* mRNAs.[21] Presence or absence of ES cells can be confirmed by plating aliquots of sorted cells in ES cell medium plus LIF and monitoring formation of undifferentiated colonies.

To determine the neural differentiation potential of sorted cell populations, *Sox1*-GFP positive and negative cells are replated onto PDL-laminin or fibronectin-coated plastics in N2B27 medium. The cells are cultured for a period of time before being fixed and immunostained for neural markers. Neurons and glia are obtained at high efficiency from GFP positive, but not the GFP negative population. Cells with overt neuronal morphology that express the pan-neuronal marker βIII-tubulin become apparent within 2–3 days of plating *Sox1*-GFP sorted cells. Astrocytes are evident only 10 days or longer after plating but their proportions increase with time or on addition of serum.

Purification of Sox1-GFP Positive Neural Precursors by Puromycin Selection

An alternative to FACS purification that can be applied with 46C cells is selection with puromycin. The puromycin resistance gene, *pac*, is coupled to *Sox1*-GFP via an internal ribosome entry site and is therefore coexpressed with GFP. An advantage of puromycin selection is that nonresistant cells are rapidly eliminated so that "clean" populations can be obtained by brief exposure.

To purify neural precursors via puromycin selection, monolayer differentiation is carried out with 46C cells as described above. After 6–8 days puromycin is added at a concentration of 0.5 μg/ml. (The effective dose for selective killing of nonexpressing cells in N2B27, 0.5 μg/ml, is lower than in serum, 1.5 μg/ml.) Cells may be treated *in situ* with puromycin or may first be dissociated and replated on gelatin-coated dishes. Puromycin selection is maintained for at least 48 hr but for no longer than 96 hr. We have observed that in the absence of serum exposure to puromycin for more than 4 days kills *pac* expressing as well as nonexpressing cells. Four days selection effectively eliminates ES cells and other cell types without reducing the neural precursor population.

Generation of Purified Sox2-Expressing Neural Precursors Using OS25 ES Cells

A further option is to purify *Sox2* positive, *Oct4*-negative neural precursors. In this case monolayer differentiation is carried out with OS25 cells. On day 6–8, cells are dissociated and replated onto gelatin-coated dishes in N2B27 medium plus FGF-2 (20 ng/ml), G418 (100 μg/ml) and Ganciclovir (2.5 μM). Medium is renewed after 48 hr. After 4 days in selection, G418 and Ganciclovir are removed. Four days in Ganciclovir is sufficient to eliminate ES cells and longer treatment should be avoided to minimize "by-stander" killing.

The pure neural precursors cells resulting from any of the above procedures may be expanded either in suspension (neurospheres) or adherent culture in the presence of mitogens FGF-2 and/or EGF. Alternatively they can be induced to differentiate into neurons and glia simply by plating onto PDL-laminin or fibronectin-coated plastic in reduced or no FGF-2. The absence of residual ES cells also means that these populations are suitable for transplantation into foetal or postnatal brain.

Trouble-Shooting Monolayer Differentiation

The monolayer protocol is a straightforward method for neural differentiation of mouse ES cells. However, due to variation between cell lines, reagents and laboratory practices, achieving consistent high efficiency neural differentiation may prove challenging. Below are some comments and tips on tackling problems that may arise during monolayer differentiation.

ES Cells

To date we have tested 15 different ES cell lines, derived from three independent primary ES cell isolates, and all generate neural cells efficiently. Therefore, we consider that the phenomenon of monoculture neural commitment is a generic property of ES cells. It should be noted, however, that we do not use feeder cells in routine ES cell propagation. We anticipate that the presence of feeders will impede neural differentiation. Therefore, we recommend that ES cells established on feeders should be passaged without feeders and adapted to culture on gelatin before initiating this protocol. We have observed some variations regarding neural differentiation efficiency between different ES cell lines, and even within the same ES cell line at different passage numbers, requiring adaptation of seeding densities (see below). We also noticed that two ES cell lines with abnormal karyotypes did not differentiate well under these conditions.

Substrates

The use of gelatin appears to be a crucial factor in this procedure. Prolonged gelatin coating time (>45 min at room temperature) may improve plating efficiency. Poly-D-lysine (PDL)-laminin and fibronectin appear to give better cell attachment, but they promote nonneural differentiation. Unusually, we observed that an aneuploid ES cell line plated very poorly on gelatin but did attach and undergo neural differentiation when plated on PDL-laminin-coated plastic. It is possible that cell attachment properties may be vulnerable to genetic change in culture, which will then alter differentiation behavior.

Cell Plating Density

Under serum-free culture conditions, high plating density assists cell survival but neural differentiation becomes less efficient, while low cell plating density leads to reduced cell attachment and viability. For most ES cell lines we have tested so far, the best neural induction is achieved when cells are plated at a density between $0.5 \times 10^4/\text{cm}^2$ and $1.5 \times 10^4/\text{cm}^2$. The optimum plating density for each ES cell line needs to be determined empirically, however.

Serum

Although neural differentiation is conducted under serum-free culture conditions, the serum in the previous ES cell culture medium may influence the outcome. We have observed that prior exposure to some batches of serum reduces the efficiency of subsequent neural differentiation.

Cell Death

Cell death is a common problem when cells are transferred directly from serum-containing medium to serum-free medium. This may appear quite dramatic initially as up to 50% of cells may fail to attach, but abundant viable and proliferative cells should be apparent upon medium change. To improve cell viability in this monolayer differentiation condition, you can:

1. Add L-glutamate (0.5–1 m*M*) and β-mercaptoethanol (0.1 m*M*) to N2B27 medium. This improves cell survival without affecting neural differentiation.
2. Adjust the relative proportion of DMEM/F12-N2 and Neurobasal-B27 in N2B27 medium. Increasing the proportion of Neurobasal-B27 in N2B27 improves cell attachment and viability, while reducing the efficiency of neural induction. Increasing the proportion of DMEM/F12-N2, in contrast, may lower viability, but neural and especially neuronal differentiation are promoted.
3. Add LIF for the first day of monolayer differentiation. This will increase cell viability, but also delay neural differentiation.
4. Increase cell plating density.

Finally, during neuronal maturation many cells die, presumably due to inadequate trophic support. FGF-2 can enhance survival and there is clearly scope to investigate the ability of factors such as BDNF and GDNF to enhance neuronal viability at these later stages.

Role of Components in N2B27 and Extrinsic and Intrinsic Factors

Using this protocol, ES cells convert efficiently into neural precursors in monoculture in serum-free N2B27 medium without any other added factors. To investigate minimal conditions for neural determination of ES cells,

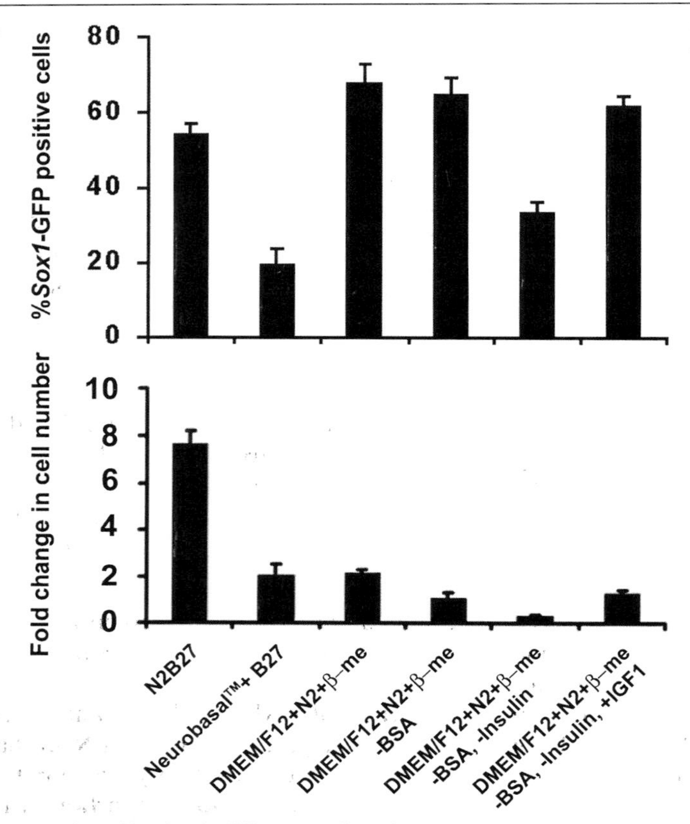

FIG. 3. Neural specification in different media and supplements. Data are averaged from two independent experiments performed in triplicate. For method in detail see text.

requirements for components in the N2B27 medium have been evaluated using 46C ES cells (Fig. 3).

Exponentially growing 46C ES cells were dissociated and plated onto 0.1% gelatin-coated 6-well plates at a density of 1.5×10^5 cells/well in serum-free media with different supplements. 0.1 m*M* mercaptoethanol was added when B27 was omitted. Medium was renewed every other day. On day 5, cells were dissociated using 0.025% trypsin solution and resuspended in FACS buffer. Total cell numbers for each well were counted and the proportions of *Sox1*-GFP positive cells quantified by FACS.

Data summarized in Fig. 3 show that Neurobasal™ medium and B27 enhance cell viability and/or proliferation but do not promote neural conversion. Of the protein components in N2 supplement, transferrin is essential for cell viability whilst BSA assists with cell attachment but is dispensable. Insulin can be substituted by insulin-like growth factor (IGF-1). In the absence of insulin/IGF-1 and BSA, cell plating efficiency and viability are significantly reduced but *Sox1*-GFP cells are still generated, although at lower frequency. The finding that neural cells are generated in medium in which the only protein component is transferrin establishes that neural specification of ES cells does not require exogenous inductive stimuli, although may be augmented by added survival factors, most importantly insulin/IGF-1. This does not mean that neural determination from ES cells is a simple default process, however. On the contrary FGF signaling appears to be essential, but this is provided by autocrine production of FGF4, and perhaps other FGFs.[21]

Summary

The efficiency of monolayer differentiation establishes that commitment of ES cells to a neural fate needs neither multicellular aggregation nor extrinsic inducers. The entire process by which pluripotent ES cells acquire neural specification can be visualized and recorded at the level of individual colonies. Furthermore this simple culture system is amenable to cellular and molecular dissection, promising to yield new insights into the mechanism underlying neural determination in mammals and perhaps to deliver the goal of "directed" homogeneous differentiation of ES cells.

Acknowledgment

We are indebted to Marios Stavridis, Meng Li, Dean Griffiths and Katherine Rennie for their contributions to these studies. This work was supported by the International Human Frontiers Science Program Organisation, and by the Medical Research Council and the Biotechnology and Biological Sciences Research Council of the United Kingdom.

[24] Development of Melanocytes from ES Cells

By TAKAHIRO KUNISADA, TOSHIYUKI YAMANE, HITOMI AOKI, NAOKO YOSHIMURA, KATSUHIKO ISHIZAKI, and TSUTOMU MOTOHASHI

Introduction

Melanocytes or their precursors melanoblasts are derived from the neural crest. After extensive migration, they reside in the skin, inner ear, and uveal tract as highly dendritic, heavily pigmented cells and are generally located in the epidermal basal cell layer of these areas, including

0076-6879/2003 $35.00

hair follicles.[1] Genetic observations indicate that steel factor (SLF, also known as MGF, SCF)[2] and endothelin 3 (ET3)[3] are essential factors for the early stages of melanocyte development.

This chapter describes a procedure for inducing the differentiation of the melanocyte lineage from embryonic stem (ES) cells.[4] We utilize a coculture system of ES cells with a bone-marrow-derived stromal cell line, ST2 originally developed for the induction of hematopoietic cells.[5] After inoculating ES cells onto an ST2 layer and regularly changing the culture medium, colonies containing melanocytes are induced from single ES cells. Since ST2 cells are derived from an albino mouse,[4] any pigmented cells that appear in this system are derived from ES cells. To maintain good efficiencies for melanocyte induction, quality of the ST2 cells and ES cells are both important. Especially, certain cell lines are never found to generate melanocytes even if the cell line differentiates into hematopoietic lineages. If appropriate ES cell lines are chosen and ST2 cells are prepared properly, you can constantly and efficiently induce melanocytes.

After inoculating undifferentiated ES cells onto plates preseeded with ST2 cells and culturing in the presence of dexamethasone and other supplements, melanocytes are generated within the colonies after 12–13 days of differentiation. The number of melanocytes continues to increase for up to 3 weeks after plating the ES cells. It should be noted that these colonies are composed of highly heterogeneous populations that include cardiac muscle cells, hematopoietic cells, endothelial cells and so on. Markers for melanoblasts, committed melanocyte precursors, are first detected on day 6 of differentiation. Generation of melanocytes from ES cells depends on steel factor and is promoted by ET3. More precise analysis suggested that steel factor-dependent melanoblasts are generated from day 6 to 12 of differentiation. Provided that the ES cell lines

[1] J. J. Nordlund, R. E. Boissy, W. Hearing, R. A. King, and J.-P. Ortonne, eds.), "The Pigmentary System." Oxford University Press, New York, NY, 1998.

[2] S. J. Galli, K. M. Zsebo, and E. N. Geissler, The Kit ligand, stem cell factor, *Adv. Immunol.* **55**, 1–96 (1993).

[3] A. G. Baynash, K. Hosoda, A. Giaid, J. A. Richardson, N. Emoto, R. E. Hammer, and M. Yanagisawa, Interaction of endothelin-3 with endothelin-B receptor is essential for development of epidermal melanocytes and enteric neurons, *Cell* **79**, 1277–1285 (1994).

[4] T. Yamane, S. A. Hayashi, M. Mizoguchi, H. Yamazaki, and T. Kunisada, Derivation of melanocytes from embryonic stem cells in culture, *Dev. Dyn.* **216**, 450–458 (1999).

[5] M. Ogawa, S. Nishikawa, K. Ikuta, F. Yamamura, M. Naito, K. Takahashi, and S.-I. Nishikawa, B cell ontogeny in murine embryo studied by a culture system with the monolayer of a stromal cell clone, ST2: B cell progenitor develops first in the embryonal body rather than in the yolk sac, *EMBO J.* **7**, 1337–1343 (1988).

are derived from the 3.5 days postcoitum blastocysts, the time course of *in vitro* differentiation of melanocytes from ES cells corresponds well to that of neural crest-derived melanocyte development *in vivo*.[4]

Instead of conventional gene knockout approach, ES culture system is fast to test whether or not the gene inactivated in ES cell is indispensable for neural crest or melanocyte development. Taking this advantage, many candidate genes possible to affect melanocytes or neural crest development could be functionally verified. In addition, factors that may affect melanocytes could be tested by using this culture system.

Materials

Media and reagents were prepared as follows. Medium for the maintenance of ES cells (ESM): Dulbecco's Modified Eagle's Medium (DMEM: Gibco BRL, Grand Island, NY; Cat. No. 12800-017) was supplemented with 15% fetal calf serum (FCS), 2 m*M* L-glutamine (Gibco BRL; Cat. No. 25030-081), 1× nonessential amino acids (Gibco BRL; Cat. No. 11140-050), 0.1 m*M* 2-mercaptoethanol (2-ME), 1000 units/ml recombinant leukemia inhibitory factor (LIF) or equivalent amounts of a culture supernatant from CHO cells producing LIF, 50 U/ml streptomycin, and 50 μg/ml penicillin. ESM should be used within one month. Medium for embryonic fibroblasts (EFM): DMEM supplemented with 10% FCS, 50 U/ml streptomycin, and 50 μg/ml penicillin. 0.1% gelatin solution: Dissolve 0.5 g of gelatin (Sigma Chemical Co., St Louis, MO; Cat. No. G2500) into 500 ml distilled water by autoclaving. Store at room temperature. Mitomycin C (Sigma Chemical Co.; Cat. No. M0503): For x100 solution, dissolve the powder into sterile distilled water, and adjust the concentration to 1 mg/ml just before use. x100 solution could be stored at −70°C without refreezing for one month. Medium for the maintenance of ST2: RPMI-1640 (Gibco BRL Cat. No. 31800-022) was supplemented with 5% FCS, 50 μ*M* 2-ME, 50 U/ml streptomycin, and 50 μg/ml penicillin. Medium for the differentiation of ES cells on ST2 (DiM): α-Minimum Essential Medium (α-MEM: Gibco BRL Cat. No. 11900-024) was supplemented with 10% FCS, 50 U/ml streptomycin, and 50 μg/ml penicillin. Dexamethasone (Dex: Sigma Chemical Co., St Louis, MO; Cat. No. D4902): Prepare original stocks at 10^{-2} *M* in ethanol and store at −70°C. Make 10^{-3} *M* working solution in ethanol. This is stable for three months at 4°C. Human recombinant basic fibroblast growth factor (bFGF: R&D System Inc., Minneapolis, MN; Cat. No. 233-FB): Prepare 200 n*M* stock with PBS containing 0.1% bovine serum albumin, 1 m*M* DTT and store the aliquots in −70°C. Store at 4°C after thawing and use within one month. Cholera toxin (CT: Sigma Chemical Co.; Cat. No. C8052): Dissolve at 50 nM in distilled water and store the aliquots at −70°C. Store at 4°C after thawing and use within one month. Endothelin 3 (ET3: Peptide Institute, Inc., Osaka, Japan; Cat. No. 4199-s): Dissolve at 0.1 mM in 0.1% acetic acid and store the aliquots at −70°C. Stable for one month when thawed and stored at 4°C. For the maintenance of ES cells, choose a good lot of serum that supports the growth of ES cells well, does not generate differentiated cells, and has high plating efficiency. The culture at the clonal density without the addition of exogenous LIF would allow the selection of such a lot. The batches of serum suitable for the maintenance of ES cells are also available from several commercial suppliers. Other generally used materials were purchased from Sigma Chemical Co., Wako Pure Chemical (Osaka, Japan) and Gibco BRL.

Methods

For each manipulation, prewarm the medium to 37°C. The cultures are maintained at 37°C with 5% CO_2 in a humidified incubator.

Preparation of Embryonic Fibroblast Cells Treated with Mitomycin C for Feeder Cells

We ordinarily maintain ES cells on mitomycin C-treated embryonic fibroblasts. Prepare E12.5 to 14.5 embryos from pregnant C57BL6 mouse and remove head and visceral tissues by fine forceps in DMEM supplemented with 50 U/ml streptomycin, and 50 μg/ml penicillin. After changing the medium several times, 6–10 embryos were minced into small pieces by scissors and transferred to 10 ml of 0.2% trypsin/1 mM EDTA and incubated for 15 min at 37°C in 15 ml conical tube with occasional mixing. Dissociate the digested embryos by pipetting up and down. After allowing it to settle for 2 min, supernatant containing well dissociated cells was transferred to 50 ml conical tube and 40 ml of DMEM supplemented with 10% FCS, 50 U/ml streptomycin, and 50 μg/ml penicillin (EFM) was added. After centrifugation at 200g for 5 min, sedimented cells were dissolved in EFM. 8×10^6 cells were seeded in 10 cm tissue culture dish with 10 ml of EFM.

After one to two days, cells become just confluent and then passage 1 : 4 to new culture dishes. For harvesting the cells for passage, we usually wash dishes two times by PBS and treat cells with 1 ml of 0.1% trypsin/1 mM EDTA for 2 min at 37°C then add 9 ml EFM. When the cells are just confluent (usually takes one to two days, and never leave the cells in a confluent state for more than one day), harvest cells by trypsinization, wash two times with EFM and dissolve the cells with EFM containing 10%

DMSO. We usually adjust 5–6 $\times 10^6$ cells/ml and 1 ml of the cell suspension was transferred to freezing vials and stored in liquid nitrogen tank.

For the preparation of mitomycin C-treated embryonic fibroblasts, vials containing cells were quickly thawed in 37°C water bath, washed once with EFM and 5–6 $\times 10^6$ cells were transferred to 10 cm tissue culture dish supplemented with 10 ml EFM. After reaching just confluent, cells were treated with 10 μg/ml mitomycin C for 2.5 hr. If necessary, cells could be passaged 1 : 4 before mitomycin C treatment. After washing the cells well, seed 10^6 or 3×10^6 cells into 60 mm dish or 100 mm dish, respectively. It is also possible to freeze mitomycin C-treated cells. In this case, mitomycin C-treated cells were harvested and frozen as previously described and 6×10^6 cells thawed were used for a 100 mm dish or three 60 mm dishes. Dishes should be coated with gelatin by covering with 0.1% gelatin solution at 37°C for 30 min or at room temperature for 2 hr. Remove the gelatin solution just before the plating of embryonic fibroblasts.

Preparation of Undifferentiated ES Cells

The day before, prepare mitomycin C-treated confluent embryonic fibroblasts on gelatin-coated dishes according to the previous section. Thaw frozen ES cells and transfer them on mitomycin C-treated embryonic fibroblasts. Usually, 5–6 $\times 10^6$ ES cells were inoculated per 6 cm dish or 1–1.2 $\times 10^7$ cells per 10 cm dish with 4 ml or 10 ml of ESM. Grow them up to a subconfluent state with daily change of ESM. Time to reach a subconfluent state may vary according to the condition of frozen ES stock. Replace the ESM 2–4 hr before the passage. Wash three times with PBS and trypsinize the culture with 0.25% trypsin/ 0.5 m*M* EDTA for 5 min at 37°C. Add medium and dissociate the cell clump by pipetting up and down. Centrifuge at 120*g* for 5 min at 4°C, aspirate the supernatant, and resuspend the cell pellet in ESM. During this, count the number of ES cells.

Replate them onto the freshly prepared mitomycin C-treated embryonic fibroblasts. For 60 mm or 100 mm dishes, inoculate 10^6 or 3×10^6 ES cells, respectively. Change the medium of the ES cell cultures with fresh medium the next day of passage. Two days after the passage, the cultures must reach to a subconfluent state. Harvest ES cells by the above-mentioned manner and resuspend in ESM on ice for further use or passage them. The passage numbers of ES cells should be kept as small as possible. In this respect it is advisable to prepare sufficient amounts of young stocks and recover the stocks when needed. To prepare frozen ES cell stocks, dissolve harvested ES cells in ESM and place them on ice. Then, add the same volume of separately prepared, ice chilled ESM containing 20% DMSO and transfer

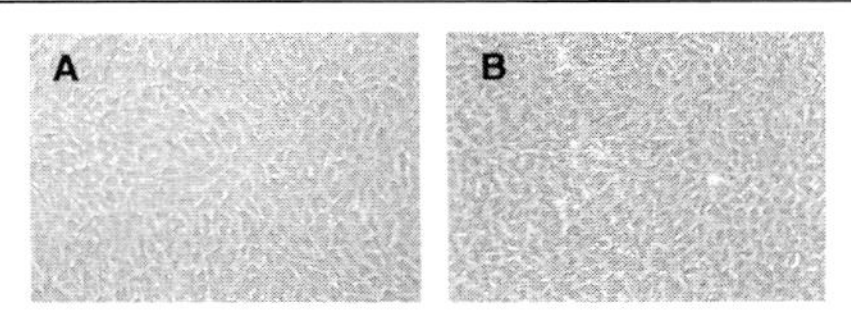

FIG. 1. Appearance of ST2 cells ready for induction (A). After 21 days of culture even without ES cells, ST2 cells remain contact-inhibited single-cell layer (B).

them to ice cooled vials. 6×10^6 ES cells per ml per tube is good for preservation.

Maintenance and Preparation of ST2 Stromal Cells

ST2 cells are maintained in the RPMI 1640 media supplemented with 5% FCS and 50 μM 2-ME. ST2 cultures grown long periods sometimes change in appearance (i.e., changing to a more dendritic shape or senescent appearance). Discard such cultures, and use freshly thawed young ST2 cell stocks. Figure 1A shows the appearance of normal ST2 ready for induction. ST2 cells remain the same in appearance after 21 days of culture (Fig. 1B). Trypsinize the confluent cells in a dish or flask with 0.05% trypsin/0.5 mM EDTA at 37°C for 1–2 min. Add medium, dissociate the cells by pipetting, centrifuge them at 200g for 5 min, and split them 1 : 4. Maintain by regularly passaging them every 3 or 4 days.

To prepare feeder layers in 6-well plates, we ordinarily seed cells of a confluent 100 mm dish into four 6-well plates. Three to four days later, the cells reach confluency and are ready to use for the differentiation of ES cells. It is also possible to seed at higher or lower density, and then culture the ST2 cells a shorter or longer time, until the ST2 cells are in a confluent state when the ES cells are inoculated. Neither irradiation nor treatment with mitomycin C is needed.

Induction of Differentiation into Melanocytes

Prepare confluent ST2 feeder layers in 6-well plates as described in the previous section. Grow ES cells up to a subconfluent state as described and replace the medium of the ES cell culture with fresh medium 2–4 hr before inducing the differentiation. Do not use ES cells that have just been thawed from frozen stocks. Use ES cells that have been cultured for at least 1 day after thawing. Wash the ES cell culture for three times with PBS and trypsinize with 0.25% trypsin / 0.5 mM EDTA for 5 min at 37°C. Add DiM and dissociate the cell clump by pipetting up and down. Centrifuge at 120g for 5 min at 4°C, aspirate the supernatant, and suspend the cell pellet in

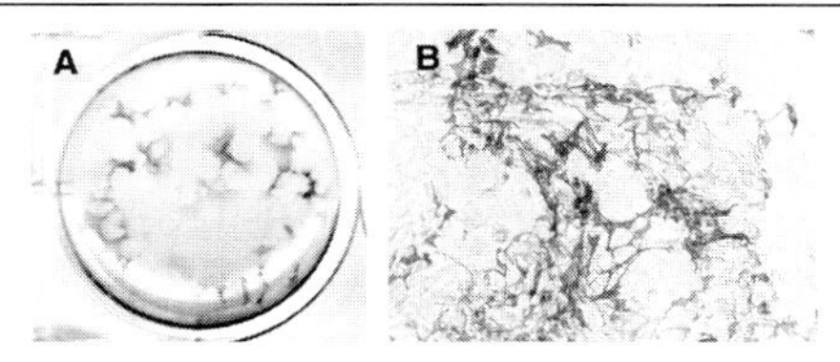

FIG. 2. Appearance of melanocytes induced from ES cells. After 21 days, melanocytes appear as clonal populations in the colonies (A) and the melanocytes are dendritic and heavily pigmented (B), typical for cutaneous melanocytes.

DiM. Count the ES cell number and dilute to an appropriate cell density (usually 200–1000 per well) with DiM. If you want to remove the embryonic fibroblasts, replate the ES cells on gelatin-coated dishes and incubate for 30 min in the culture medium, then collect the nonadherent cells by pipetting and count them. However, it does not cause any problem to directly seed the mixture of ES cells and embryonic fibroblasts onto the ST2 layer.

Aspirate the medium from the plates of ST2 feeder layer. Dispense 2 ml of the ES cell suspension into each well. Supplement with 10^{-7} *M* Dex, 20 p*M* bFGF, and 10 p*M* CT (final concentration). Dex is most important, while bFGF and CT are not so critical. If you remove Dex, the number of melanocytes generated is reduced by two orders of magnitude. You need not add steel factor because ST2 cells produce a sufficient amount of this cytokine. Other factors, 50 n*M* 12-*O*-tetradecanoyl phorbol acetate (TPA), 10^{-8} *M* 1α,25-dihydroxyvitamin D_3, and 100 ng/ml endothelia 3, also increase the number of melanocytes in this culture system. In case you need larger number of melanocytes, the addition of endothelin 3 is most effective.

Incubate cultures at 37°C with 5% CO_2 in an incubator. Just after the cultures are started, the growth of ST2 cells may be transiently promoted and their appearance may become fibroblastic and more tightly packed due to the effects of bFGF and CT, but there is no need to worry as this occurs normally and ST2 cells will settle down after a while. Change the culture medium every 2 or 3 days, and perform the culture for up to 3 weeks of differentiation. Do not scratch the ST2 layer when you aspirate the medium and leave cells for a long time out of the incubator, because these cause detachment of the ST2 feeder from the edge of the culture plates. Adipocytes are frequently generated from ST2 layers. This occurs normally in well-maintained cultures. After 2 or 3 days of differentiation, ES cell clusters derived from single ES cells are discernible. These colonies grow by piling up until around day 12 of differentiation. Pigmented melanocytes appear on day 12–13 and the number increases until 3 weeks of differentiation as shown in Fig. 2. Further cultivation only augments the level of pigmentation of each cell. To count the melanocytes, see the next section.

Counting of the Number of Melanocytes

As shown in Fig. 2, dendritic mature melanocytes containing melanin granules in their cytoplasm will be observed within and around the ES-derived colonies by the above-mentioned culture. An average of about 20,000 to 50,000 melanocytes (more than 100,000 when endothelin 3 was supplied) would be generated per well. As the culture described here is not a two-dimensional one, and melanocytes overlap with each other in the colony, it is necessary to dissociate colonies into single-cell suspensions as described below in order to count melanocytes. If the numbers of melanocytes are small enough to count without dissociation, score the number of melanocytes after the fixation of the culture with 10% formalin/PBS for 10 min on ice.

Aspirate the medium from the culture, and wash three times with PBS. Add 400 μl of 0.25% trypsin/0.5 m*M* EDTA. Incubate at 37°C for 10 min. Dissociate the colonies by pipetting up and down vigorously, and then return the cultures to the incubator and incubate for a further 5 min. Add 100 μl of FCS to inactivate trypsin, and dissociate the clumps into a single-cell suspension by pipetting vigorously. Count the number of pigmented cells within a 3 mm^2 region of hemocytometer. They will be easily detected by the observation without the phase-contrast. If the number of pigmented cells is small, count replicate aliquots so that reliable data is obtained.

Comments

At least two ES cell lines D3[6] and J1,[7] gave similar results using our protocol. It is also possible to use ES cells that are adapted to feeder-independence. ST2 (RCB0224) cells have been registered in the RIKEN Cell Bank (http://www.rtc.riken.go.jp/CELL/HTML/RIKEN_Cell_Bank.html). If the cells do not function well, please contact us.

There is batch-related variation of FCS. About half the batches of serum are not good for the maintenance of ST2 for as long as 3 weeks in our experience. Also, the plating efficiency of ES cells and the number of melanocytes generated on ST2 varies according to the lot of FCS. Check the batches of serum by seeding ES cells in the range from 100 cells to 2000 cells per well of 6-well plates. Select serum that is good for

[6] T. C. Doetschman, H. Eistetter, M. Katz, W. Schmidt, and R. Kemler, The in vitro development of blastocyst-derived embryonic stem cell lines: formation of visceral yolk sac, blood islands and myocardium, *J. Embryol. Exp. Morphol.* **87**, 27–45 (1985).

[7] E. Li, T. H. Bestor, and R. Jaenisch, Targeted mutation of the DNA methyltransferase gene results in embryonic lethality, *Cell* **69**, 915–926 (1992).

the maintenance of ST2 and efficiently supports the generation of melanocytes. The efficiency with which ES cells form colonies on the ST2 feeder layers changes according to the lot of serum from less than 1% to about 40%. It is recommended to seed ES cells so that about 80 colonies are generated per well in the 6-well plate. It should be noted that plating efficiency is not correlated with the ability to support melanocyte formation.

Seeding at higher density reduces the efficiency of melanocyte induction. Although inoculation at a lower density does not reduce the efficiency of melanocyte induction, the results of such experiments fluctuate from well to well. The plating efficiency also varies according to the ES line you use. Carry out preliminary experiments to check the plating efficiency using your lines and serum.

Acknowledgment

We thank Drs. Shin-Ichi Hayashi and Hidetoshi Yamazaki (Tottori University) for collaborative work.

Section II

Differentiation of Mouse Embryonic Germ Cells

[25] Isolation and Culture of Embryonic Germ Cells

By MARIA P. DE MIGUEL and PETER J. DONOVAN

Introduction

In mammals, three types of pluripotent stem cells have been identified and isolated into culture (see Ref. 1 for review). Embryonic stem (ES) cells are derived from the inner cell mass of the preimplantation blastocyst.[2,3] Embryonic germ (EG) cells are derived from primordial germ cells (PGCs) of the postimplantation embryo.[4,5] Embryonal carcinoma (EC) cells are derived from testicular tumors observed in young adult males but also likely derive from PGCs.[6] These pluripotent stem cells share two important properties. First, they can be maintained indefinitely in culture as an essentially immortal cell line. Second, they are capable of giving rise to every cell type in the body.[1] These features make such cells potentially important tools for the treatment of human disease since differentiated derivatives of pluripotent stem cells could be used to replace damaged or diseased cells via transplantation. Pluripotent stem cells and their derivatives will also likely generate important information about embryonic development. Because EC cells are derived from tumors they typically are aneuploid and have proven to be less useful than ES or EG cells.[1] This chapter outlines the protocols developed for culture of mouse PGCs and their conversion into EG cells but which may be applicable to the culture of the same cells from a variety of different species including birds and humans. Many of the standard techniques used for the culture of ES cells, such as culture medium, feeding regimen, subculture technique, etc., can be used for the culture of EG cells. The major difference lies in the initial isolation of the cells. EG cell lines have been derived from both mouse and human embryos using the same culture

[1] P. J. Donovan and J. Gearhart, The end of the beginning for pluripotent stem cells, *Nature* **414**, 92–97 (2001).

[2] M. J. Evans and M. H. Kaufman, Establishment in culture of pluripotential cells from mouse embryos, *Nature* **292**, 154–156 (1981).

[3] G. R. Martin, Isolation of a pluripotent cell line from early mouse embryos cultured in medium conditioned by teratocarcinoma stem cells, *Proc. Natl. Acad. Sci. USA* **78**, 7634–7638 (1981).

[4] Y. Matsui, K. Zsebo, and B. L. Hogan, Derivation of pluripotential embryonic stem cells from murine primordial germ cells in culture, *Cell* **70**, 841–847 (1992).

[5] J. L. Resnick, L. S. Bixler, L. Cheng, and P. J. Donovan, Long-term proliferation of mouse primordial germ cells in culture, *Nature* **359**, 550–551 (1992).

[6] L. C. Stevens, Origin of testicular teratomas from primordial germ cells in mice, *J. Natl. Cancer Inst.* **38**, 549–552 (1967).

0076-6879/2003 $35.00

conditions.[4,5,7] These observations suggest that some of the factors regulating PGC growth and differentiation have been conserved during evolution and, therefore, that the techniques for isolating EG cells described here may be applicable to other mammalian and nonmammalian species.

Materials

Buffers and Solutions

1. PBS: Phosphate buffered saline, without Ca^{+2} and Mg^{+2}, pH 7.0 (Gibco/Invitrogen).
2. Trypsin/EDTA: HBSS containing 0.05% trypsin and 0.53 m*M* EDTA.
3. Fast Red/Naphtol phosphate solution: Make up a 1 mg/ml solution of Fast red TR salt (Sigma, stored at −20°C) in dH_2O. Add 40 μl/ml Naphtol AS-MX phosphate (Sigma, stored at 4°C). Both reagents must be very fresh. Use immediately.

Culture Media

1. Basic culture medium for PGCs: DMEM (Dulbecco's modified Eagle's medium) high glucose, supplemented with 15% Fetal Calf Serum (Hyclone) (for serum batch testing, see Note 1), 2 m*M* glutamine (Gibco/InVitrogen), 5 U/ml penicillin–streptomycin (Gibco/BRL), and 1 m*M* Na^+ pyruvate (Sigma).
2. Growth factors for EG cell derivation and culture: mSCF (Genzyme) 10 ng/ml, mLIF (Chemicon) 1000 U/ml, hbFGF (Gibco/Invitrogen) 1 ng/ml and Forskolin (Sigma) 100 μM.
3. Basic culture medium for STO cells: DMEM high glucose, supplemented with 10% Fetal Calf Serum (Hyclone), 2 m*M* glutamine (Gibco/Invitrogen), 5 U/ml penicillin–streptomycin (Gibco/Invitrogen), and 1 m*M* Na^+ pyruvate (Sigma).
4. Freezing medium for EG cells: 9 ml of Basic culture medium for PGCs (see above) plus 1 ml of dimethyl sulfoxide (DMSO). Mix thoroughly and keep on ice prior to use.

Fixatives

1. 4% Paraformaldehyde: 4 g paraformaldehyde in 100 ml PBS. To dissolve the paraformaldehyde, preheat the PBS at 90°C, and add

[7] M. J. Shamblott *et al.*, Derivation of pluripotent stem cells from cultured human primordial germ cells, *Proc. Natl. Acad. Sci. USA* **95**, 13726–13731 (1998).

NaOH dropwise slowly until the solution turns completely clear. Let it cool down before use. Make it fresh every time.

2. acid–ethanol: 1 ml acetic acid and 19 ml ethanol precooled at −20°C.

Methods

PGCs Isolation and Culture

Embryo Dissection

a. *8.5 days post coitum (dpc):* Separate each implantation site cutting the uterus between them, very near to each embryo in order to allow the deciduum to protrude (Fig. 1). Put the dissected implantation sites in a petri dish with ice cold PBS. Make a slit in the protruding decidua with a pair of fine forceps (Dumont #5). With fine forceps apply pressure to the other side and base of the decidua (see Fig. 1) and the embryo should pop out of the slit in the deciduum. Flatten out the embryo, and remove the posterior third, including the caudal

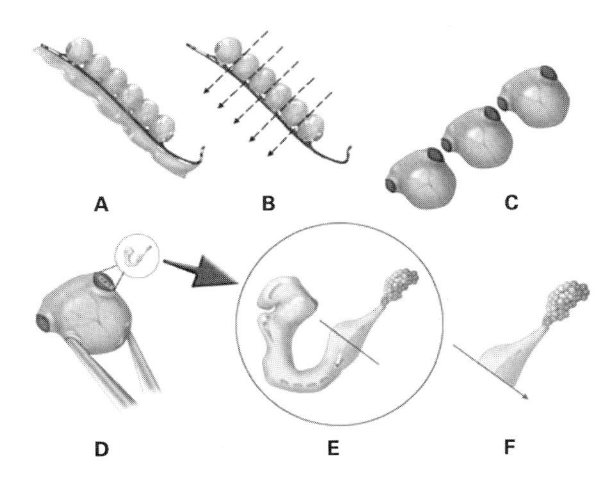

FIG. 1. Dissection of 8.5 dpc embryos. The presence of a vaginal plug is denoted as 0.5 dpc. Remove the uterine horns from timed-mated animals. Cut the uterine horns into individual implantation sites, cutting as close to the deciduum as possible (A–C). Make a slit or incision in the deciduum that is protruding from one side of the implantation site (D). It may be necessary to tear the uterine muscle wall in order to do this. With a fine pair of forceps squeeze the implantation site from the base and the opposite side of to where the incision has been made (E). The embryos will usually pop out of the incision. Again using fine forceps, cut off the caudal half of the embryo (F). The germ cells are found at the end of the caudal end of the primitive streak close to the allantoic diverticulum. Collect either individual fragments or pool fragments for trypsinization.

end of the primitive streak and allantois. The PGCs are localized at the junction of the primitive streak and allantois in the hindgut diverticulum (Fig. 1). Collect the embryo fragments in PBS and keep on ice until time for trypsinization.

b. *10.5 days post coitum:* Separate each implantation site as for 8.5 dpc embryos. Dissect the embryo free of the placenta and amnion, and cut it in half, just below the fore limbs (Fig. 2). Keep the caudal half of the embryo. Remove the viscera including the heart, liver, and intestines being careful not to remove the genital ridges. The genital ridges lie at about the level of the developing hind limbs. Cut down both sides of the ventral body wall with forceps, following the line of the spine and peel away the ventral body wall (Fig. 2). Be careful not to pull out the developing gut because at this age the gonads can be torn out along with it. Separate the gonads, mesonephros, and part of the aorta fragment with forceps. Holding the fragment with the fine forceps, pull it briefly out of the PBS and then quickly replace it

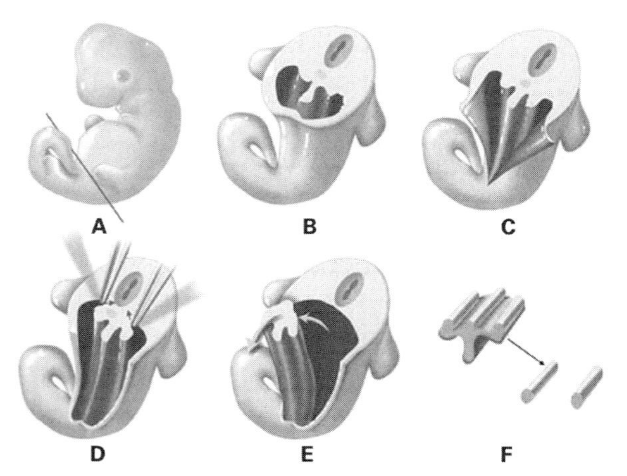

FIG. 2. Dissection of 10.5 dpc embryos. Remove the embryo from the implantation site and decidua. With a pair of forceps, cut the embryo in half just below the fore limbs (A–B). Discard the upper portion and carefully remove the viscera from the lower portion. Take special care when removing the gut since the gonads can sometimes be removed when removing the gut. Tear the embryo down the mildline (C). The genital ridges or gonad anlagen lie on the dorsal body wall either side of the hindgut mesentery. Lie the embryo on its back and place the tines of a pair of forceps either side of the genital ridges (D). Pinch the tines together and then gently lift the forceps upwards and towards the tail end of the embryo. The genital ridges, hindgut and descending aorta will come out of the embryo together (E). Briefly lift the fragment out of the PBS and the blood will drain out of the descending aorta. It is not necessary to dissect the gonad from the other structure at this stage of embryogenesis. The dissection of gonads from older embryos follows essentially the same method. At later stages in embryogenesis it is preferable to separate the gonad from the associated structure, which can be easily done using fine forceps (F).

in the PBS. This serves to drain the blood out of the descending aorta and removes a potential source of contaminating cells. Place the fragments in a tube with PBS on ice. At this stage of embryogenesis it is not necessary to dissect the genital ridges away from the descending aorta and gut. For older embryos, see Note 2.

c. *Enzymatic digestion and homogenization:* Remove as much of the PBS from the tube as possible. Wash once with 1–2 ml of the trypsin/EDTA solution and then aspirate, being careful not to aspirate the tissue. Add another 1 ml of trypsin/EDTA and incubate the embryo fragments or gonads in the solution for 8 min at 37°C. Remove the trypsin/EDTA solution gently, again being careful not to aspirate the tissue. Add the appropriate volume of PGC basic culture medium and pipette slowly up and down with a micropipette tip (200 μl) until a single-cell suspension is obtained (about 25 strokes). The final dilution volume of the mixture should be calculated based on the number of PGCs to be plated per well. The number of PGCs isolated is estimated from the known numbers of PGCs present in the embryo (see Note 2). We aim to add the PGCs in a small volume (10 μl) of medium to avoid dilution of medium and growth factors already present in the feeder layer cultures. It is possible to dissociate embryo fragments or genital ridges individually. This is most easily accomplished by placing the individual fragments in individual wells of a 96-well plate and reducing the volume of trypsin/EDTA accordingly. Once the fragments have been dissociated, dilute the mixture with basic culture medium and aliquot this mixture onto preplated feeder layers of irradiated STO cells (see below).

d. *Culture:* PGCs are routinely cultured on confluent monolayers of STO cells (but see Note 3) at 37°C in an atmosphere of 95% air and 5% CO_2. STO cells themselves can be maintained in medium containing 10% FCS and should be split prior to confluence in order to prevent overgrowth and selection for transformed cells. To prepare feeder layers, coat the plates with 0.1% gelatin for 1 hr. Aspirate the gelatin and wash the wells once in PBS. Trypsinize subconfluent plates of STO cells to produce a single-cell suspension. Plate them at a density of 10×10^4 cells/cm^2. Typically the feeder layers are plated in 200 μl in a 96-well plate or 0.5 ml in a 24-well plate. Allow the STO cells to settle down and attach and the next morning, γ-irradiate them (5000 rads) to induce cell cycle arrest (but see Note 4). After irradiation, immediately remove the culture medium and refeed the STO cells with PGC basic culture medium plus growth factors (see "Materials" section). Plate PGCs on top of the STO feeder layer. We aim to plate approximately 100–200 PGCs per well of a 24-well plate, or 50 PGCs per well of a 96-well plate. This is the equivalent of three 8.5 dpc embryos worth of PGCs per well of a 24-well plate, or 75 PGCs/cm^2. For larger or smaller wells, scale the amount of cells plated up or down accordingly. Change the medium every day by gentle aspiration of 2/3 of medium and refeeding with fresh medium. If using a 96-well plate, avoid the use of the outer wells as they can dry up more quickly if the incubator is not properly humidified. Fill the outer wells with sterile distilled water.

e. Identification of PGCs and EG cells
PGC identification: There are several techniques for identification of PGCs in culture but the simplest method is alkaline phosphatase histochemistry: Wash the cultures in PBS without Ca^{+2} and Mg^{+2} (prewarmed at 37°C) and then fix the cultures in 4% paraformaldehyde in PBS (pH 7.4) for 30 min at room temperature (RT). Wash the cultures three times in PBS, once in distilled water, and then incubate them in freshly made Fast Red/Naphtol phosphate solution for 30 min. After staining wash them again in distilled water; PGCs will be stained red. We recommend counting the cells within a few days of staining, otherwise the cell morphology deteriorates. PGCs can also be identified in culture using monoclonal antibodies such as anti-SSEA-1.[8] The anti-SSEA-1 monoclonal antibody can be obtained from the Developmental Studies Hybridoma Bank (http://www.uiowa.edu/~dshbwww/info.html). EG cells also express alkaline phosphatase and SSEA-1 and there is currently no easily available marker that distinguishes between PGCs and EG cells. Therefore, the same markers that are used to identify PGCs are also used to identify EG cells. The most straightforward way to distinguish between PGCs and EG cells is in their growth properties; PGCs are mortal and will survive for only about 7 days in culture, whereas EG cells are immortal and can be maintained indefinitely.

Cell Proliferation

Analysis of cell proliferation can be carried out in all types of cultures by BrdU incubation and anti-BrdU immunocytochemical detection using a cell proliferation kit (Amersham), following the

[8] P. J. Donovan, D. Stott, L. A. Cairns, J. Heasman, and C. C. Wylie, Migratory and postmigratory mouse primordial germ cells behave differently in culture, *Cell* **44**, 831–838 (1986).

manufacturer's instructions. In brief, incubate the cultures with BrdU at 1 : 1000 dilution in culture medium for 1 hr at 37°C. Wash the cultures several times in PBS and then fix them in acid : ethanol for 10 min exactly. Timing is important, because longer fixation will destroy alkaline phosphatase activity. Rehydrate the cultures in PBS and add the anti-BrdU antibody, for 1 hr at RT. Then wash in PBS, incubate with the second antibody (peroxidase anti-mouse IgG) for 30 min, wash again in PBS and develop with DAB. Proliferative rates can be obtained by counting labeled versus unlabeled cells. BrdU-labeled PGCs can be identified by double staining for BrdU and, afterwards, alkaline phosphatase. Make sure to wash the cultures in dH_2O for at least 20 min before alkaline phosphatase staining. Monitor the alkaline phosphatase histochemical reaction to ensure that the alkaline phosphatase precipitate does not obscure the BrdU precipitate.

Derivation of EG Cells

To derive EG cells, PGC cultures are maintained for up to 9 days in basic culture medium supplemented with leukemia inhibitory factor (LIF) and basic Fibroblast Growth Factor (bFGF) (but see Note 5). By 7–9 days of primary culture large colonies of EG cells should be growing on the feeder layer. These colonies of cells can then subcultured by trypsinization onto fresh, mitotically inactive, feeder layers. EG cell colonies are very similar in morphology to colonies of ES cells. EG colonies can be visualized by fixing the cultures and staining for alkaline phosphatase or by observing living cultures under a microscope equipped with phase contrast or Hoffman modulation contrast optics. Alternatively, colonies of EG cells can be visualized at low power under a dissecting microscope by adjusting the reflecting mirror under the baseplate. This will achieve a pseudo-Nomarski image. Aspirate the medium from the well of a 96-well plate and wash the culture with 200 μl of trypsin/EDTA mixture. Immediately aspirate the trypsin/EDTA mixture and add 100 μl of fresh trypsin/EDTA. Incubate the culture at 37°C in a humidified CO_2 incubator for 5–10 min. Remove the plate from the incubator and add 100 μl of PGC basic culture medium to neutralize the action of the trypsin. Using a pipette equipped with a 100 μl pipette tip, pipette the solution up and down for approximately 100 strokes. Monitor the extent of cell disruption under an inverted microscope. The aim is to disrupt the colonies of EG cells into a single-cell suspension. Clumps of EG cells will differentiate if passaged. Two types of cells should be seen under the microscope, large cells, which are STO cells and embryonic fibroblasts, and smaller cells that are EG cells. At the early stages of EG cell derivation it may be difficult to see the small numbers of EG cells present.

Place the contents of the well onto a single well of a 24-well plate that has been recently refed with 0.5 ml of fresh medium. Replenish the medium every 24 hr by removing as much of the existing medium as possible and adding fresh medium—0.5 ml per well of a 24-well plate. After 3–5 days the EG cells should have expanded and the well should be covered in large numbers of colonies of EG cells.

Passaging Colonies of EG Cells

EG cells can be split again by trypsinization and expanded or frozen down. Prepare feeder cells ahead of time. Remove as much of the medium as possible by aspiration and then wash the well with 0.5 ml of trypsin/EDTA mixture. Immediately aspirate the trypsin/EDTA and add 0.5 ml fresh trypsin/EDTA. Incubate at 37°C for 5–10 min in a CO_2 incubator. Towards the end of 5 min monitor the extent of trypsinization under an inverted microscope. Excessive trypsinization will cause a loss of cell viability. Add an equal volume of complete medium to neutralize the trypsin and pipette up and down 50–100 times with a pipettor equipped with a 100 μl pipette tip to break clumps into a single-cell suspension. If necessary, break up clumps into a single-cell suspension using a Pasteur pipette which has had its tip flame-polished to reduce the diameter. The contents of a single well of a 24-well plate can be placed onto multiple wells of a 24-well plate or onto a single 35 mm tissue culture dish.

Picking and Passaging Single EG Colonies

Single colonies of EG cells can be isolated and expanded. Prepare a 96-well plate with feeder cells 24 hr ahead of time and irradiate it on the morning of use. Prepare another 96-well plate with 50 μl of trypsin/EDTA per well and place it at 37°C in a CO_2 incubator. Remove the medium from a 35 or 100 mm dish of EG cells. Wash the plate with 1 or 2 ml of trypsin/EDTA. Immediately aspirate the trypsin/EDTA mixture and add another 1 or 2 ml. Leave at RT and monitor on an inverted microscope or on a dissecting microscope. After 5–10 min the STO cell feeder layer will begin to fall apart but the colonies of EG cells, being more compact and resistant to the effects of trypsin, will remain intact. Under a dissecting microscope pick individual colonies using a pipettor equipped with a disposable 10 μl pipette tip. Place the isolated colony into a well of the 96-well plate containing 50 μl of trypsin/EDTA. Pick as many colonies as possible in a short period of time (so as to avoid over trypsinization of the first-picked colonies) and place the 96-well plate into the incubator for 5 min. Remove the plate from the incubator and add 50 μl of complete medium. Using a pipettor equipped with a disposable 100 μl tip pipette the mixture up and down until the clumps are disrupted and a single-cell suspension is achieved. Repeat for each well remembering to replace the pipette tip between wells. Place the contents of each well into a single well of the 96-well plate containing feeder cells. Feed daily until the cells have expanded sufficiently that they can be passaged onto a larger plate.

Freezing and Thawing EG Cells

Once EG cells have been established they can be expanded and then frozen in liquid nitrogen. Expand the cells onto 10 cm plates and allow them to grow to 75% confluence. Prepare 10 ml of freezing medium prior to trypsinization of the EG cells. Aspirate the medium from the plate and wash with 2 ml of trypsin/EDTA mixture. Immediately aspirate the trypsin/EDTA mixture and add another 2 ml of trypsin/EDTA. Allow the dish to sit for 1 min at RT. Aspirate the trypsin/EDTA and place the culture dish into the incubator. Note, that although aspiration removes most of the trypsin/EDTA, there is enough liquid present to cover the cells. In a properly humidified incubator the cells will not dry out. Incubate the dish at 37°C for 5–10 min and monitor at intervals to follow the extent of cell dissociation. Remove the dish from the incubator and add 10 ml of complete medium. Pipette up and down to dissociate the cells into a single-cell suspension. If necessary, use a flame-polished Pasteur pipette to break up the clumps of EG cells into a single-cell suspension. Pellet the cells by centrifugation (400*g* for 5 min). While the cells are spinning label cryotubes and place them in the tissue culture hood. Aspirate the medium from the centrifuge tube and resuspend the cells in 10 ml of freezing medium. Immediately place 1 ml aliquots of the cells into the cryotubes, seal the tubes and begin cryopreservation. There are many techniques that can be used to freeze cells. We place the tubes upright inside a plastic beaker that has been completely lined with two layers of paper hand towels. Cover the vials with more paper towels and place the beaker in a −80°C freezer overnight. Cells can be stored for several weeks at −80°C but for longer storage they should be transferred to liquid nitrogen. Prior to thawing, prepare plates of irradiated feeder cells to plate the EG cells onto. Make one 10 cm plate per vial of cells to be thawed. To thaw cells prepare 20 ml of complete medium per vial to be thawed and then remove the vial directly from the liquid nitrogen into a 37°C water bath. Exercise caution when thawing vials as they can explode! Wear eye goggles. Agitate the vial to accelerate the thawing. When thawed, remove the contents of the vial into a 10 ml conical tube and add 10 ml of complete medium. Mix thoroughly. Centrifuge the cells (400*g* for 5 min) and then aspirate the medium. Resuspend the cells in 10 ml of complete medium and place the cells onto a 10 cm dish preplated with STO feeder cells. After 3–4 days colonies of EG cells should become apparent on the dish.

Notes

1. The serum batch used for cell culture may be especially critical for the growth of PGCs and the derivation of EG cells. We recommend testing batches of serum before purchasing to determine which serum batch is best suited for growing these cell types. For historical purposes we have tested batches of serum for their ability to support the growth of three cell types, F9 EC cells, PCC4 EC cells and TG-1 Hybridomas. These cells are plated at low density in wells of a 24-well plate with each row containing a medium made with different batch of serum. The cells are allowed to grow to confluence and the growth rate determined. This can be done quantitatively, or more easily, by monitoring the color of the medium. Cells that are growing quickly will exhaust the medium more rapidly and the color of the medium will change to a golden yellow color. In addition, it is advisable to test serum batches for toxicity by growing cells in medium with 30% FCS.
2. Dissection of older embryos (11.5–13.5 dpc) is carried out in basically the same way as the 10.5 dpc embryos, with the expected increase in embryo size associated with a higher yield of PGCs. At later stages in embryogenesis it is preferable to separate the gonad from the associated structure which can be easily done using fine forceps. The number of PGCs per embryo is approximately 50 PGCs/embryo at 8.5 dpc, 1×10^3 PGCs/embryo at 10.5 dpc, 5×10^3 PGCs/embryo at 11.5 dpc, 10×10^3 PGCs/embryo at 12.5 dpc, and 35×10^3 PGCs/embryo at 13.5 dpc. There may be some strain variation. We routinely use F1 hybrid strains (e.g., B6C3F1 or B6D2F1) since they have relatively large litter sizes. We do not routinely isolate PGCs from 9.5 dpc embryos because at this stage the developing genital ridges and mesonephros are difficult to isolate. Any possible advantage that might be gained from the greater cell numbers with respect to 8.5 dpc embryos is far outweighed by the difficulty and, therefore, the increased time of dissection.
3. We routinely use STO cells as feeder cells for growing PGCs and EG cells. However, other feeder layers can and have been used to derive EG cells. Notably, Sl^{m220} cells have been used by many investigators to derive EG lines.[4] Derived from the bone marrow of *Steel* mutant

mice that do not express SCF, Sl^{m220} cells have been engineered to express a membrane-bound form of SCF.[9]

4. STO cells and other fibroblasts can be inhibited from transiting through the cell cycle by Mitomycin C rather than irradiation. However, we do not recommend that method, since in our hands PGCs do not grow well on feeder layers that are inactivated with Mitomycin C.
5. Some methods for deriving EG cells also include the addition of recombinant soluble stem cell factor (mSCF (Genzyme) 10 ng/ml),[4] although we have not found that to be necessary. The derivation of human EG cells also required the addition of forskolin (Sigma; 100 μm), a cAMP agonist and one of the most potent PGC mitogens.[7] Pay careful attention to the recommended storage conditions for growth factors since they can be easily inactivated. We suggest that growth factors be aliquoted and stored at the temperature suggested by the manufacturer.

[9] D. Toksoz *et al.*, Support of human hematopoiesis in long-term bone marrow cultures by murine stromal cells selectively expressing the membrane-bound and secreted forms of the human homolog of the steel gene product, stem cell factor, *Proc. Natl. Acad. Sci. USA* **89**, 7350–7354 (1992).

Section III

Gene Discovery by Manipulation of Mouse Embryonic Stem Cells

[26] Gene Trap Mutagenesis in Embryonic Stem Cells

By WEISHENG V. CHEN and PHILIPPE SORIANO

Introduction

Draft sequences of human and mouse genomes are now available in public databases awaiting functional annotation. Although the power of bioinformatics in *ab initio* gene prediction and functional deduction has been well demonstrated, its limitations are also apparent and the inferred functional cues need to be validated in a biological context, preferentially within a whole organism. The laboratory mouse has long been the premier mammalian model system for dissecting *in vivo* gene functions because of its small size, large litter, short generation time and most importantly, a plethora of powerful tools available for genetic intervention. Most remarkably, with gene targeting in embryonic stem cells,[1,2] virtually any desired mutation can be created for any given gene to obtain information regarding its functions in resulting mutant mice. Moreover, conditional gene alterations with site-directed recombination and transcriptional transactivation systems are now in common use to switch gene activity spatially or temporally,[3] allowing functional aspects of many genes elusive with straight knockouts to be revealed. However, since knowledge of target sites is absolutely required *a priori*, only known genes discovered by independent studies can be approached by this technique. Besides, construction of targeting vectors and screening of targeted clones are rather time-consuming. Although improved methods[4,5] have been developed to speed up the process, mutating every mouse gene with targeted mutagenesis alone would be a formidable and lengthy mission.

To accelerate the rate of gene discovery and functional validation, a number of random mutagenesis schemes have been employed in the mouse. Forward genetics mutagenesis screens with *N*-ethyl-*N*-nitrosourea (ENU) represent one of the major efforts. A potent DNA alkylating agent, ENU primarily introduces point mutations in mouse germline, resulting in a wide spectrum of mutation types including loss-of-function, gain-of-function,

[1] K. R. Thomas and M. R. Capecchi, *Cell* **51**, 503 (1987).
[2] T. Doetschman, R. G. Gregg, N. Maeda, M. L. Hooper, D. W. Melton, S. Thompson, and O. Smithies, *Nature* **330**, 576 (1987).
[3] M. Lewandoski, *Nat. Rev. Genet.* **2**, 743 (2001).
[4] P. O. Angrand, N. Daigle, F. van der Hoeven, H. R. Scholer, and A. F. Stewart, *Nucleic Acids Res.* **27**, e16 (1999).
[5] P. Zhang, M. Z. Li, and S. J. Elledge, *Nat. Genet.* **30**, 31 (2002).

0076-6879/2003 $35.00

hypomorphic, neomorphic, and antimorphic alleles.[6,7] Strategies for both dominant and recessive screens, as well as chromosome/region/allelic-specific and sensitized screens have been described, and systematic large-scale screens are being performed in several centers world-wide. Mutant generation by this approach is almost effortless, but identifying the responsible genes by positional cloning remains the bottleneck. Although promising gene-driven approaches by prescanning mutations in mutagenized ES cells[8] or in mutant mouse DNA archives[9] have been suggested, they are only applicable to known, but not novel genes.

Another major mutagenesis scheme is gene trapping in mouse embryonic stem cells. Gene trapping evolved from insertional mutagenesis in transgenic mice,[10] where a transgene introduced by retroviral infection or pronuclear injection is used both as a mutagen to randomly disrupt endogenous genes, and as a molecular landmark to facilitate the cloning. However, since there is no selection for mutation-causing insertions, the occurrence of observable phenotypes is relatively low, which promoted the pursuit for improved vectors with a higher mutagenic capacity. Taking advantage of position effects, enhancer traps that contain a reporter gene driven by a minimal promoter were then designed to enrich for intragenic insertions by capturing endogenous *cis*-acting elements, and to follow gene expression with simple reporter assays.[11,12] By extrapolating the same principle of regulatory element complementation, a variety of gene trap vectors containing a reporter/selectable marker transgene cassette devoid of promoter or polyadenylation signal sequences were soon devised to specifically target exonic or intronic sequences and arrest endogenous gene transcription.[13–16] Gene trap events are isolated in embryonic stem cells with

[6] M. J. Justice, J. K. Noveroske, J. S. Weber, B. Zheng, and A. Bradley, *Hum. Mol. Genet.* **8**, 1955 (1999).

[7] R. Balling, *Annu. Rev. Genomics Hum. Genet.* **2**, 463 (2001).

[8] Y. Chen, D. Yee, K. Dains, A. Chatterjee, J. Cavalcoli, E. Schneider, J. Om, R. P. Woychik, and T. Magnuson, *Nat. Genet.* **24**, 314 (2000).

[9] E. L. Coghill, A. Hugill, N. Parkinson, C. Davison, P. Glenister, S. Clements, J. Hunter, R. D. Cox, and S. D. Brown, *Nat. Genet.* **30**, 255 (2002).

[10] T. Gridley, P. Soriano, and R. Jaenisch, *Trends Genet.* **3**, 162 (1987).

[11] N. D. Allen, D. G. Cran, S. C. Barton, S. Hettle, W. Reik, and M. A. Surani, *Nature* **333**, 852 (1988).

[12] R. Kothary, S. Clapoff, A. Brown, R. Campbell, A. Peterson, and J. Rossant, *Nature* **335**, 435 (1988).

[13] A. Gossler, A. L. Joyner, J. Rossant, and W. C. Skarnes, *Science* **244**, 463 (1989).

[14] G. Friedrich and P. Soriano, *Genes Dev.* **5**, 1513 (1991).

[15] H. von Melchner, J. V. DeGregori, H. Rayburn, S. Reddy, C. Friedel, and H. E. Ruley, *Genes Dev.* **6**, 919 (1992).

[16] H. Niwa, K. Araki, S. Kimura, S. Taniguchi, S. Wakasugi, and K. Yamamura, *J. Biochem. (Tokyo)* **113**, 343 (1993).

the use of a selectable marker, and mutant mice can be derived following chimera production and germline transmission. Although generalization is difficult to make since the mutagenicity of different vectors varies, nearly 60% gene trap insertions described in the literature led to overt embryonic or adult phenotypes, a frequency comparable to that of targeted mutagenesis.[17] Together, the readily obtainable sequence, expression, and phenotype information on randomly mutated genes make gene trap mutagenesis a potent tool for functional genomics studies. This chapter describes the methods utilized in our laboratory to generate and characterize gene trap mutations, and strategies for gene trap mutagenesis screens are briefly discussed.

ROSA Series Retroviral Gene Traps

Although plasmid-based gene trap constructs can be conveniently delivered into ES cells by electroporation, multiple copy insertions, concatemers, and deletions of the gene traps as well as rearrangements of integration sites are frequently observed, which may complicate the analysis of gene trap events. In contrast, gene traps mediated by retroviral vectors integrate cleanly without causing rearrangements of host flanking sequences except for a duplication of a few nucleotides, and single copy insertion can be ensured by infecting cells at low multiplicity. For these reasons, retroviral gene traps have been employed predominantly in our laboratory to transduce ES cells and obtain gene-trapped clones. These gene traps are based on self-inactivating (SIN) recombination-incompetent Moloney murine leukemia virus (Mo-MuLV) retroviral vectors that harbor a deletion of the viral enhancer alone or together with promoter sequences in the U3 region of 3′ LTR. Upon reverse transcription, these changes are reproduced in both LTRs and result in an inactive provirus, so that potential influence of viral regulatory elements on transgene or endogenous gene expression is significantly reduced.[18] An additional advantage of retroviral gene traps is that retrovirus-mediated integration tends to occur close to DNase I hypersensitive sites frequently found at the 5′ end of genes,[19,20] which increases the likelihood of generating null alleles.

All the retroviral gene traps developed in our laboratory are promoter traps, which function through a promoterless reporter/selectable marker

[17] W. L. Stanford, J. B. Cohn, and S. P. Cordes, *Nat. Rev. Genet.* **2**, 756 (2001).
[18] P. Soriano, G. Friedrich, and P. Lawinger, *J. Virol.* **65**, 2314 (1991).
[19] S. Vijaya, D. L. Steffen, and H. L. Robinson, *J. Virol.* **60**, 683 (1986).
[20] H. Rohdewohld, H. Weiher, W. Reik, R. Jaenisch, and M. Breindl, *J. Virol.* **61**, 336 (1987).

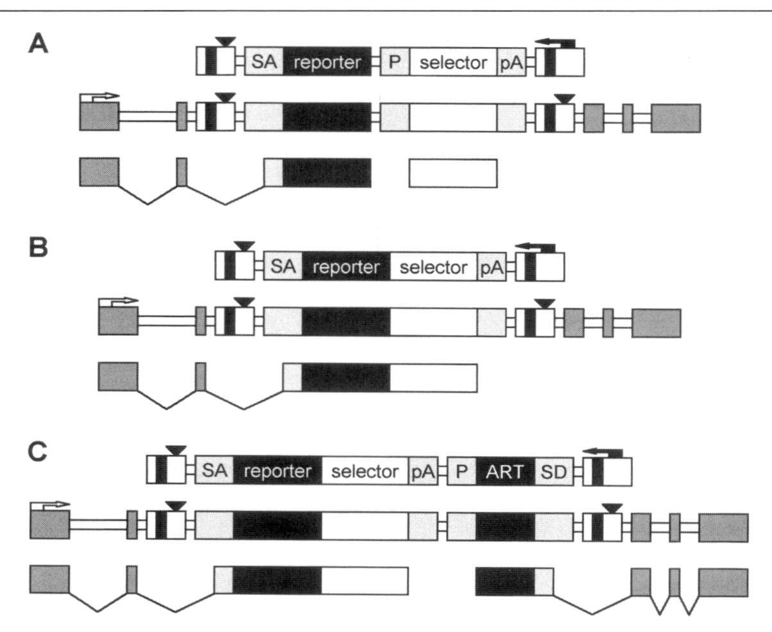

FIG. 1. ROSA gene trap vectors. Vector, provirus, and transcripts are illustrated sequentially for each vector type. All ROSA vectors function through a promoterless reporter or a reporter-selectable marker (selector) fusion downstream of a splice acceptor (SA) to arrest the endogenous transcription of trapped genes. The gene trap elements are placed in reverse orientation relative to viral transcription (indicated by arrow) to avoid interference with packaging. Viral enhancer/promoter is deleted in the U3 region of 3′ LTR (indicated by inverted triangle) and the change is reproduced in both LTRs in the provirus. Type A vectors contain an independent selectable marker cassette regulated by an internal promoter (P) and a polyadenylation signal (pA). Vectors of this type trap genes regardless of their expression levels in ES cells, but most drug-resistant clones are derived from intergenic insertions. Type B vectors utilize a reporter-selectable marker fusion instead to increase trapping efficiency. Type C vectors are modified from type B by including a 3′ exon trapping cassette containing an internal promoter-driving artificial exon (ART) followed by a splice donor (SD) to facilitate the cloning of trapped cDNA flanks. Note that only the exon portion of splice acceptor and splice donor is included in fusion transcripts.

gene downstream of a splice acceptor. Because a polyadenylation signal sequence is included to terminate gene transcription, the gene trap elements are inserted between the LTRs in reverse orientation relative to viral transcription to avoid interference with virus packaging (Fig. 1). These vectors and resulting mutant lines are thus referred to as ROSA (reverse-orientation-splice-acceptor). Alternatively, a promoterless reporter/selectable marker gene can be inserted into the U3 region of 3′ LTR to construct retroviral exon trap vectors.[15,21] When the ROSA gene trap is integrated into an intron of an endogenous gene in the correct orientation, 5′ exons are forced to splice into the promoterless reporter and the endogenous transcription is arrested. A Kozak-optimized ATG[22] is placed in front of the reporter coding sequence to serve as either an initiator codon or an internal methionine in the fusion protein, so that both the 5′ UTR and the protein coding region of endogenous genes can be targeted. The splice acceptor used in ROSA vectors is a short consensus sequence derived from the intron 1/exon 2 boundary of type II adenovirus major late transcript. To date, all ROSA gene trap insertions that have been characterized in our laboratory are null mutations and "splicing-around" the gene trap insertion has not been observed, indicating that the splice acceptor is of sufficient strength to override endogenous splicing in most cases. Nonetheless, no gene trap design can prevent splicing bypass in every gene trap event, considering the complicated mechanisms governing splice site recognition and selection, and the fact that a significant portion of mammalian genes are alternatively spliced.[23,24]

The prototype of ROSA vectors (Fig. 1A) contains a separate selectable marker cassette. Although such vectors can be used to trap genes regardless of expression levels, the majority of drug resistant clones are derived from intergenic insertions. To increase trapping efficiency, ROSA vectors with a reporter-selectable marker fusion cassette were developed (Fig. 1B). Because capturing an endogenous promoter is absolutely required to activate the bifunctional fusion, essentially every resistant clone obtained with these vectors represents a *bona fide* gene trap event. However, since only 5′ exons are included in the fusion transcript and most trapping events obtained with these vectors occur at the 5′ end of genes,[14] cloned cDNA flanking sequences are often too short to be informative with the present 3′-enriched NR and EST databases. This problem can be solved by incorporating a parasitic 3′ exon trapping cassette consisting of an internal promoter-driven artificial exon followed by a splice donor (Fig. 1C). The splice donor used in the 3′ exon trapping cassette is derived from the exon 1/intron 1 boundary of the type II adenovirus major late transcript, which has been shown to mediate efficient splicing into 3′ exons, thereby allowing longer and more informative cDNA flanking sequences to be cloned. It should be noted that vectors of this type differ from polyA traps which utilize a selectable

[21] G. G. Hicks, E. G. Shi, X. M. Li, C. H. Li, M. Pawlak, and H. E. Ruley, *Nat. Genet.* **16**, 388 (1997).

[22] M. Kozak, *J. Cell Biol.* **108**, 229 (1989).

[23] A. J. Lopez, *Annu. Rev. Genet.* **32**, 279 (1998).

[24] B. Modrek and C. Lee, *Nat. Genet.* **30**, 13 (2002).

marker-containing 3′ exon trapping cassette to derive gene-trapped clones.[16,25–27] Unlike promoter traps which only target 5′ UTRs or coding sequences, polyA traps can be activated regardless of the insertion site within a gene. Therefore, although both expressing and nonexpressing genes can be tagged with polyA traps, these vectors are prone to create more hypomorphic and neutral mutations, given that mammalian mRNAs typically have a long 3′ UTR (over 30% of entire length in average for human protein-coding genes[28]).

Retroviral gene trap vectors are introduced into a packaging cell line such as GP + E86[29] to obtain stable virus-producing cells.[18,30] Briefly, vectors are linearized by cutting in the plasmid backbone, purified by phenol/chloroform extraction and dissolved at 1 μg/μl in TE (pH 7.5) after precipitation. Ten microgram linearized plasmid is used to transfect 1×10^6 GP + E86 cells in 0.8 ml phosphate-buffered saline (PBS) by electroporation. After selecting in presence of the appropriate antibiotic for 10–12 days, resistant clones are pooled, expanded and frozen in batch. To collect infectious viral particles, virus-producing cells are grown to near confluence and overlaid with fresh medium. After 16 hr, virus-containing medium is filtered through a 0.22 μm filter to remove cells and debris before being used for infection. To determine titer, serial 10-fold dilutions of the viral stock are made in medium containing 4 μg/ml polybrene (Hexadimethrin bromide, Sigma) to infect log phase 3T3 cells or ES cells in 6-well dishes. Cells are selected in antibiotic-containing medium until colonies can be counted macroscopically by staining with 0.5% methylene blue in 50% methanol. Note that the observed titer of gene trap retroviruses is lower than the actual titer, which can be extrapolated based on the trapping efficiency.[18]

Derivation of Gene-trapped ES Cell Clones

ES Cell Culture

Maintenance of the pluripotency of ES cells in culture is of paramount importance for gene trap experiments. Unlike targeted mutagenesis where multiple cell lines harboring the identical mutation can be established, gene

[25] M. Yoshida, T. Yagi, Y. Furuta, K. Takayanagi, R. Kominami, N. Takeda, T. Tokunaga, J. Chiba, Y. Ikawa, and S. Aizawa, *Transgenic Res.* **4**, 277 (1995).

[26] B. P. Zambrowicz, A. Imamoto, S. Fiering, L. A. Herzenberg, W. G. Kerr, and P. Soriano, *Proc. Natl. Acad. Sci. USA* **94**, 3789 (1997).

[27] M. Salminen, B. I. Meyer, and P. Gruss, *Dev. Dyn.* **212**, 326 (1998).

[28] International Human Genome Sequencing Consortium, *Nature* **409**, 860 (2001).

[29] D. Markowitz, S. Goff, and A. Bank, *J. Virol.* **62**, 1120 (1988).

[30] G. Friedrich and P. Soriano, *Methods Enzymol.* **225**, 681 (1993).

trap events are unique and failure of germline transmission would nullify all the upstream efforts in identifying the desired clones. Strict procedures should be followed for tissue culture reagent preparations. Phosphate-buffered saline (PBS; 138 m*M* NaCl, 2.67 m*M* KCl, 8.1 m*M* Na_2HPO_4, 1.47 m*M* KH_2PO_4, pH 7.2), 0.1% gelatin, and DMEM (Invitrogen, Cat. No. 12100-061) should be made with freshly drawn Milli-Q-purified water (Millipore) and stored in glass containers with absolutely no trace of detergents or preferably disposable tissue culture ware. Buffer DMEM with 2.2 g/liter of sodium bicarbonate, instead of 3.7 g/liter as recommended by the manufacturer, to allow proper pH equilibration in a 5% CO_2 incubator. Osmolality of each new batch should be measured to ensure a range between 280 and 300 mmol/kg. Medium stored over 1 month should be replenished with L-glutamine to a final concentration of 2 m*M*.

Embryonic stem cells are grown in DMEM with 15% fetal bovine serum (FBS, HyClone), 50 I.U./ml penicillin, 50 μg/ml streptomycin, and 0.1 m*M* 2-mercaptoethanol on mitotically inactive feeder cell layers. Several ES-verified serum lots should be tested with the ES cells and feeders of choice for highest plating efficiency (typically 30–35%). AK7.1 ES cells derived from 129S4 mice have been employed primarily in our gene trap experiments in conjunction with SNL 76/7 feeder cells,[31] which express leukemia inhibitory factor (LIF) and neomycin phosphotransferase (neo). If desired, SNL sublines expressing other selection markers can be engineered by transfection. Alternatively, primary mouse embryonic fibroblast cells (PMEFs) carrying the selection marker of choice can be used as feeders. Feeder cells must be mitotically arrested by mitomycin C (Sigma, Cat. No. M-4287) treatment (10 μg/ml for 2 hr) or γ-irradiation (4000 rads total) before plating at 5×10^4 cells/cm^2 onto tissue culture plates precoated with 0.1% gelatin. Feeder plates can be used within a period of a few weeks, but the quality drops over time. Medium on rapidly growing ES cells should be replaced whenever an obvious change in color to golden yellow is observed. ES cells are cultured until just subconfluent, at which time they are washed with PBS, overlaid with 0.25% trypsin–EDTA (Invitrogen, Cat. No. 25200-072) and incubated for 5 min at 37° in 5% CO_2. Add growth medium to stop trypsinization, and reduce cell clumps to the level of single cells by vigorous pipetting. Disposable polyethylene transfer pipettes (Fisher Scientific, Cat. No. 13-711-9C) are most convenient for this procedure. ES cells are usually split at 1 : 5–1 : 10 during passage to maintain a high density. Insufficient digestion and triturating, as well as growing ES cells at low densities or overgrowth

[31] A. P. McMahon and A. Bradley, *Cell* **62**, 1073 (1990).

tend to trigger spontaneous differentiation and thus should be avoided. Minor differentiation can often be controlled following successive passages since differentiated cells tend to be outgrown by stem cells. For freezing, centrifuge trypsinized ES cells at 1000 rpm for 5 min and resuspend the pellet in growth medium. Count cell numbers and adjust to the desired density (usually 1×10^7 cells/ml), add an equal volume of 2× freezing medium (20% DMSO and 40% FBS in DMEM) and aliquot 0.5 ml into each cryotube. Leave cryotubes in a Styrofoam container overnight at −70° before transferring into liquid nitrogen for permanent storage. To thaw cells, incubate the cryotube in a 37° water bath till completely thawed, wash cells once in growth medium by centrifugation, resuspend in fresh medium and plate onto a premade feeder plate.

Retroviral Infection and Electroporation

Low passage ES cells and freshly made feeder plates should be used to derive gene-trapped ES cell clones. For retroviral infection, thaw ES cells and virus producing cells and grow to subconfluence. Trypsinize ES cell culture the evening before infection, and seed 1×10^6 cells onto each 100 mm feeder plate. Meanwhile, refeed virus producing cells with fresh ES cell medium. After 16 hr, virus containing medium is collected, filtered and diluted in ES cell medium, and polybrene is added to a final concentration of 4 μg/ml. Dilution ratio is determined by prior titration on ES cells with the same batch of virus producing cells to ensure an amount sufficient to obtain 200–400 resistant clones per plate. This represents a multiplicity of infection (M.O.I.) significantly lower than one virus per cell, and therefore only one provirus is present in each clone in most cases. Infect ES cells for 6 hr before replacing the infection medium with fresh ES cell medium for recovery.

Alternatively, electroporation can be used to deliver gene trap constructs into ES cells if desired. After trypsinization, reduce cell clumps into single cell suspension and count cell numbers. Approximately 1×10^7 cells are pelleted at 1000 rpm for 5 min, resuspended in 0.8 ml PBS and transferred into a 4 mm gap width electroporation cuvette (Eppendorf, Cat. No. 4307-000-623). Add 25 μg linearized plasmids into cell suspension, mix gently by pipetting and remove any resulting air bubbles. Cells are electroporated immediately at 250 V/500 μF using a Gene Pulser unit (Bio-Rad) equipped with a capacitance extender. This setting usually gives rise to a time constant between 5 and 7 msec; unusual values may indicate problems such as improperly made PBS, incorrect cell density, or electroporation failure. Cells are transferred into fresh ES cell medium, aliquoted into several premade feeder plates and allowed to grow overnight.

Antibiotic selection is applied about 20 hr after infection or electroporation. For G418 selection, the medium is supplemented with 175 μg/ml (active ingredient) Geneticin (Invitrogen). Cells are kept under selection for 10–12 days, with daily medium change for the first 5 days and when needed thereafter.

Picking, Synchronizing, Freezing, and Thawing ES Cell Clones

Drug-resistant ES cell clones are macroscopically visible 6 or 7 days after selection and should be harvested after 10 days. Discard growth medium by aspiration, wash cells once with PBS, and overlay the plate with 10 ml PBS. At the same time, load 25 μl 0.25% trypsin–EDTA into each well of a U-bottom 96-well plate using a multiple channel pipettor. Place the plate containing ES cell clones onto a stereoscope, preferentially situated inside a tissue culture hood to maintain sterility, and identify individual clones under magnification. With a mouth pipette (100–200 μm inner diameter, hand-pulled from a 1.0 mm glass capillary), an undifferentiated ES cell clone well-separated from neighboring ones is dislodged from feeder layer by applying gentle suction and transferred into a well of the prepared 96-well plate containing trypsin solution. Each clone is picked with a new pipette to avoid cross-contamination. Alternatively, clones can be picked with disposable pipette tips, with the volume control of micropippettor set at 10 μl to allow minimum carry over of PBS. After 96 clones are picked, place the plate in a 37°, 5% CO_2 incubator for 10 min, and then 100 μl ES cell medium is added into each well. Completely disrupt cell clumps by pipetting up and down for 30–40 times using a multiple channel pipettor, and the resulting single cell suspensions are transferred into corresponding wells of a premade 96-well feeder plate with 100 μl medium per well. After overnight growth, ES cells are refed with fresh medium (200 μl/well) and allowed to grow for another 2–3 days until medium color in most wells turns golden yellow.

The expanded cultures of picked clones vary vastly in the number, size, and morphology of ES cell colonies formed. To enrich for quality clones and facilitate downstream applications, a preselection and synchronization procedure is performed. Examine each well under microscope, and wells either undergoing pronounced differentiation or containing too few cells are marked for discarding. Aspirate growth medium with a multiple channel aspirator, wash cells once with 100 μl PBS, and then add 25 μl 0.25% trypsin–EDTA into each well. After incubation for 10 min at 37° in 5% CO_2, cell clumps in unmarked wells are disrupted by vigorous pipetting, and the entire cell suspension in each well is transferred individually into new 96-well feeder plates as described above. ES cell colonies in most wells reach subconfluence within 24–36 hr and are ready for freezing at this point. The expanded clones are frozen in U-bottom 96-well plates or preferentially 96-format microtubes (Continental Lab Products, Cat. No. 2600.mini), which allow individual clones to be retrieved without thawing the whole plate. Two sets of frozen stocks are usually made for security. After trypsinization, 150 μl growth medium is added into each well, and ES cells are completed dissociated and transferred in a volume of 50 μl into each corresponding well or microtube preloaded with 50 μl 2× freezing medium. Immediately pipette a few times to mix well. After all the 96 clones are transferred, 50 μl paraffin oil is added into each well or microtube to prevent evaporation, and both sets are sealed with parafilm and transferred to a $-70°$ freezer in a Styrofoam container. For long term storage, the frozen stocks must be kept at $-135°$ or lower in a low-temperature freezer. We use an isothermal vapor storage system (Custom Biogenic Systems) to avoid direct contact of cells with liquid nitrogen. The remaining cells in the original plate can be transferred into a new feeder plate for further culture.

To thaw cells, place individual microtubes onto a 96-format rack, cover with lid, and incubate in a 37° water bath. Thawed cells in each microtube are resuspended by pipetting and transferred into one well of a 4- or 24-well feeder plate loaded with 1 ml of growth medium. On the following day, replace the oil and DMSO containing medium with fresh medium. Frozen plates can be thawed in a 37°, 5% CO_2 incubator for 10–15 min, and the cells are resuspended and transferred into a new 96-well feeder plate with 200 μl growth medium per well. After incubating for 3 hr at 37° in 5% CO_2, carefully aspirate medium without dislodging the attached cells and add 200 μl fresh medium into each well. The thawed clones must be synchronized as described above before further expansion or freezing. Although 100% recovery is usually achieved, repeated freezing and thawing in 96-well format should be avoided since cell viability and pluripotency might be compromised under these conditions.

Cloning cDNA and Genomic DNA Sequences Flanking Gene Trap Insertion Sites

Since trapped genes are tagged with the introduced gene traps, genomic DNA or cDNA flanks can be readily isolated with a variety of cloning methods. In many cases, the identity or functional clues of trapped genes can be revealed by database searching and bioinformatics analysis. Besides, the cloned sequences can be used to derive probes or to design PCR primers for genotyping or expression studies.

LA PCR System

A modified long and accurate (LA) PCR system[32] is employed for cloning both cDNA and genomic DNA flanks in our laboratory. The robustness and high fidelity of the LA PCR system is achieved primarily by using a highly active and thermostable form of Taq polymerase devoid of 5′ exonuclease activity (KlenTaq1, Ab Peptides) in combination with a proofreading enzyme (*Pfu*, Stratagene). Functionally equivalent enzyme mixes are available in many commercial kits at much higher prices. Besides, the use of a high pH buffer (PC2; 50 m*M* Tris–HCl pH 9.1, 16 m*M* ammonium sulfate, 3.5 m*M* $MgCl_2$, and 150 μg/ml BSA) together with betaine (1.3 *M*) and DMSO (1.3%) also contribute to the overall yield and quality by reducing DNA hydrolysis and melting down high GC contents. KlenTaq1 tolerates a broad range of magnesium concentrations and optimization is generally not necessary. An all-purpose formula is designed to amplify fragments up to 2 kb in a 20 μl reaction, which usually yields sufficient products for most downstream applications. For each reaction, mix 10 μl diluted templates with 10 μl 2× PCR premix containing 5.2 μl 5 *M* betaine, 2.0 μl 10× PC2 buffer, 1.0 μl *Redi*load (Research Genetics), 0.5 μl 10 m*M* dNTPs, 0.5 μl 10 μM forward primer, 0.5 μl 10 μM reverse primer, 0.26 μl DMSO, and 0.04 μl LA16 enzyme mix (KlenTaq1: *Pfu* = 15 : 1, v/v). Prepare reactions on ice and transfer to a hot-lid thermocycler preheated to 94° before commencing cycling. These PCR settings were optimized with PTC-100 (MJ Research) and GeneAmp 9700 (Applied Biosystems) and should be tested if other thermocyclers are used.

Cloning cDNA Flanks with 5′ RACE and 3′ RACE

Since fusion transcripts are generated between endogenous exons and the gene trap elements upon splicing, 5′ or 3′ flanking cDNA sequences can be cloned with 5′ RACE (for all ROSA vectors) or 3′ RACE (for type C ROSA vectors), respectively (Fig. 2A and B). The RACE protocols provided below are based on a described anchoring oligo system,[33] except that the original anchoring primers (Q0 and Q1) are extended for higher Tm (QA and QB) to reduce background amplification (Table I). All gene-specific primers are designed to have high Tm (around 70°) and a touchdown PCR module[34] is used to further increase amplification specificity. An intrinsic problem with RACE is that many templates of

[32] W. M. Barnes, *Proc. Natl. Acad. Sci. USA* **91**, 2216 (1994).

[33] M. A. Frohman, *Methods Enzymol.* **218**, 340 (1993).

[34] R. H. Don, P. T. Cox, B. J. Wainwright, K. Baker, and J. S. Mattick, *Nucleic Acids Res.* **19**, 4008 (1991).

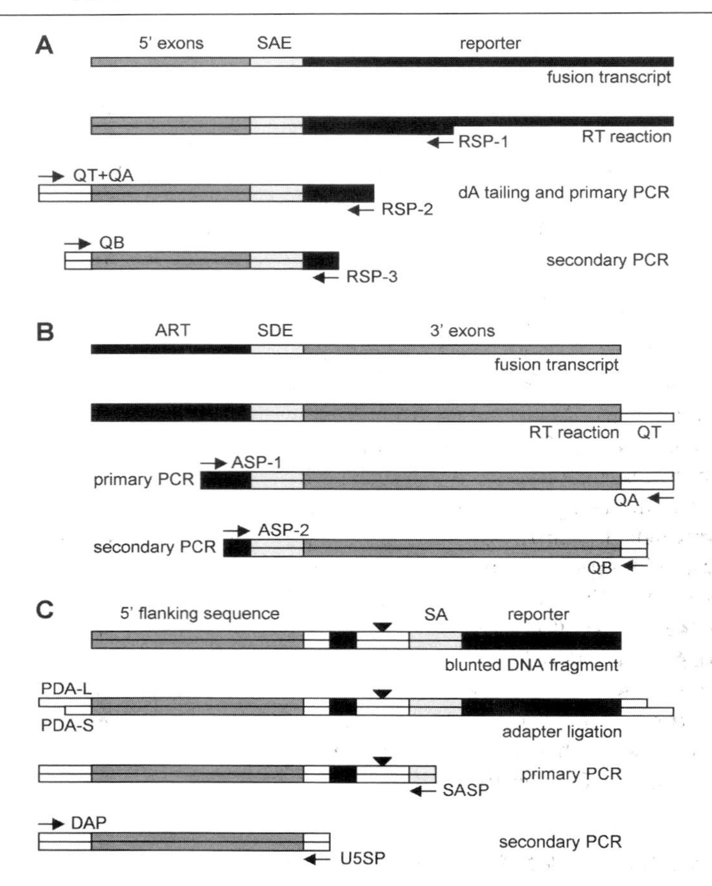

FIG. 2. Schematic representations of 5′ RACE (A), 3′ RACE (B) and anchoring PCR (C) procedures for cloning cDNA and genomic DNA sequences flanking gene trap insertion sites.

different sizes are available for amplification, primarily due to problems with reverse transcription such as premature synthesis (especially for 5′ RACE) and low stringency of primer binding (especially for 3′ RACE), which occasionally leads to failure in amplification or complicated banding. These problems can be alleviated by optimizing reverse transcription and PCR conditions, as well as adjusting template input by using different dilution ratios in primary PCR reactions.

Total RNA is usually isolated with TRIzol reagent (Invitrogen) following instructions provided by the manufacturer and dissolved in H_2O at a final concentration of 0.1–0.5 $\mu g/\mu l$. For reverse transcription, incubate 10 μl RNA at 65° for 10 min and then chill on ice. Mix denatured RNA with

TABLE I
ANCHORING OLIGO SYSTEMS FOR RACE AND ANCHORING PCR

Oligo	Sequence
QT	CCAGTGAGCAGAGTGACGAGGACTCGAGCTCAAGC TTTTTTTTTTTTTTTTT
QA	CCAGTGAGCAGAGTGACGAGGAC
QB	GACGAGGACTCGAGCTCAAGC
PDA-L	AGCAGCGAACTCAGTACAACAACTCTCCGACCTCT CACCGAGT
PDA-S	ACTCGGTGA
DAP	AGCAGCGAACTCAGTACAACA

2× RT mix containing 4.0 μl 5× first strand buffer (375 mM KCl, 15 mM $MgCl_2$, 250 mM Tris–HCl, pH 8.3), 2.0 μl 100 mM DTT, 1.0 μl 10 mM dNTPs, 1.0 μl 10 μM cDNA synthesis oligo, 1.0 μl 40 U/μl RNaseOUT RNase inhibitor (Invitrogen) and 1.0 μl 200U/μl SuperScript II RT (Invitrogen). Transfer reactions to thermocycler preheated to 45° and incubate for 45 min at 45°, 15 min at 50°, 15 min at 55°, and 15 min at 75°. The step-up incubation temperatures are designed to allow the reverse transcriptase to pass through difficult secondary structures while balancing the loss of enzyme activity. The anchoring oligo QT (Table I) is used to prime first strand synthesis for 3′ RACE. For 5′ RACE, a reporter-specific primer (RSP-1) is used instead of a general dT oligo to enrich for fusion cDNAs. Following synthesis, add 1 μl RNase A (10 μg/μl) and incubate at 37° for 20 min to digest RNA.

For 5′ RACE (Fig. 2A), purify the synthesized first strand cDNA with QIAEXII (Qiagen) or an equivalent product and elute in 20 μl 10 mM Tris–HCl (pH 8.0). Mix 10 μl purified cDNA with 2× tailing mix containing 4.0 μl 5× reaction buffer, 4.0 μl 1 mM dATP, 1.2 μl 25 mM $CoCl_2$ and 0.8 μl 25 U/μl terminal transferase (Roche). Incubate at 37° for 10 min followed by heat inactivation at 75° for 15 min. Make 1 : 5, 1 : 25, 1 : 125 dilutions of the dA-tailed cDNA products and set up primary PCR reactions with a second reporter-specific primer (RSP-2) and anchoring primer QA, as well as an additional 0.1 μl 10 μM anchoring oligo QT. After incubating at 94° for 3 min, the reactions are cycled for eight times with denaturing at 94° for 30 sec, annealing at 72°–65° (−1°/cycle) for 1 min, and extension at 72° for 5 min. The 5 min extension is designed to reduce amplification bias towards short products during initial cycles. The reactions are cycled for additional 32 times with 94° for 30 sec, 64° for 1 min, and 72° for 3 min per cycle. Primary PCR products are diluted at 1 : 100 and amplified with a third nested reporter-specific primer (RSP-3) and anchoring primer QB with the same program. For detection, run 10 μl nested PCR products on 1% agarose gel and transfer to membrane. Hybridize with a splice acceptor exon-specific primer (SAESP) to verify bands or to detect products invisible with ethidium bromide staining. A splice acceptor intron-specific primer (SAISP) can be used to detect products originating from genomic DNA or pre-mRNA contamination in RNA samples or splicing bypass. Desired products are gel-purified from the rest of 10 μl PCR reaction, and cloned to TA vectors according to standard procedures prior to sequencing.

For 3′ RACE (Fig. 2B), first strand cDNA is directly diluted without purification and tailing. The dilution ratios, reaction mixes and cycling conditions are essentially the same as for 5′ RACE, except that QT is not included in the primary PCR reaction and artificial exon-specific primers ASP-1 and ASP-2 are used to prime the primary and nested PCR reactions, respectively. 3′ RACE usually gives rise to distinct bands that can be directly sequenced after purification. If desired, a splice donor intron-specific primer (SDISP) can be used to detect contaminated or bypassed products.

Cloning Genomic DNA Flanks with Anchoring PCR

The following anchoring PCR protocol for cloning 5′ genomic DNA flanking sequences is modified from a genotyping method designed to identify the scrambler mutation[35] and is compatible with all existing ROSA vectors. The procedures involve directional ligation of blunt-ended genomic DNA fragments with a pseudo-double-stranded adapter (PDA) generated by annealing a pair of complementary primers varying in length, followed by PCR with vector-specific and adapter primers (Fig. 2C and Table I). We found this method more straightforward and efficient than inverse PCR which requires much fine-tuning for circularizing ligation conditions.

Isolate high molecular weight genomic DNA with standard procedures. Digest 1 μg DNA with 20 U of at least two different blunt-end enzymes (such as Hpa I and Dra I) separately in 20 μl TER (TE, pH 8.0 with 10 μg/ml RNase A) at 37° overnight. Purify digested DNA with QIAEXII or an equivalent product and elute in 30 μl 10 m*M* Tris–HCl (pH 8.0). Mix 15 μl purified DNA fragments with 1 μl 10 μ*M* PDA-L primer, 1 μl 10 μ*M* PDA-S primer, 2 μl 10× ligation buffer and 1 μl T4 DNA ligase (Roche) and incubate at 16° overnight. Primary PCR reactions are prepared by mixing 10 μl ligation products with 10 μl 2× PCR premix containing a splice acceptor-specific primer (SASP). Since there are sufficient carry-overs of adaptor primers in the ligation products, it is unnecessary to

[35] N. Usman, V. Tarabykin, and P. Gruss, *Brain Res. Protoc.* **5**, 243 (2000).

supply them in the 2× PCR premix and the volume is brought to 10 μl with 0.5 μl H_2O. After denaturing at 94° for 3 min, the reactions are cycled 30 times at 94° for 30 sec, 60° for 30 sec, and 72° for 3 min. For nested PCR, 1 : 100 diluted primary PCR products are amplified with a LTR U5-specific primer (U5SP) and a distal adapter primer (DAP) with the same program. A distinct band is usually obtained with at least one of the DNA digests, which can be sequenced directly after purification. If desired, gene-specific primers can be derived from the sequence obtained to clone the 3′ genomic DNA flanks using the same ligation products.

Monitoring Expression of Trapped Genes by Reporter Assays

A promoterless reporter gene is typically included in gene trap constructs to allow easy assessment of the expression of trapped genes. The *Escherichia coli lacZ* gene encoding β-galactosidase (βgal) is most commonly used. βgal expressing cells can be detected *in situ* in fixed samples by a sensitive enzymatic assay with 5-bromo-4-chloro-3-indolyl-β-D-galactopyranoside (X-Gal), or isolated by fluorescence activated cell sorting (FACS) after staining with fluorescein di-β-D-galactopyranoside (FDG). In addition, βgal activity can be quantified with *o*-nitrophenyl-β-D-galactoside (ONPG) or chlorophenol red-β-D-galactopyranoside (CPRG). Recently, other reporters such as green fluorescent protein (GFP), β-lactamase, and human placental alkaline phosphatase (PLAP) have been introduced to visualize gene expression in live cells and/or to render higher sensitivity for detection.[36–40] Binary gene trapping systems using site-directed recombinases have also been developed to conditionally activate reporter gene expression in pre-engineered indicator cell lines or mouse embryos.[41,42]

For X-Gal staining, cells, embryos, and tissues are fixed at room temperature in freshly made fixative solution containing 2% formaldehyde (from a 37% stock) and 0.2% glutaraldehyde (from a 25%

[36] M. Whitney, E. Rockenstein, G. Cantin, T. Knapp, G. Zlokarnik, P. Sanders, K. Durick, F. F. Craig, and P. A. Negulescu, *Nat. Biotechnol.* **16**, 1329 (1998).

[37] J. W. Xiong, R. Battaglino, A. Leahy, and H. Stuhlmann, *Dev. Dyn.* **212**, 181 (1998).

[38] Y. Ishida and P. Leder, *Nucleic Acids Res.* **27**, e35 (1999).

[39] E. Medico, G. Gambarotta, A. Gentile, P. M. Comoglio, and P. Soriano, *Nat. Biotechnol.* **19**, 579 (2001).

[40] P. A. Leighton, K. J. Mitchell, L. V. Goodrich, X. Lu, K. Pinson, P. Scherz, W. C. Skarnes, and M. Tessier-Lavigne, *Nature* **410**, 174 (2001).

[41] A. P. Russ, C. Friedel, K. Ballas, U. Kalina, D. Zahn, K. Strebhardt, and H. von Melchner, *Proc. Natl. Acad. Sci. USA* **93**, 15279 (1996).

[42] L. Vallier, J. Mancip, S. Markossian, A. Lukaszewicz, C. Dehay, D. Metzger, P. Chambon, J. Samarut, and P. Savatier, *Proc. Natl. Acad. Sci. USA* **98**, 2467 (2001).

stock) in PBS. Adjust fixation time for different samples, using 5 min for cells and preimplantation embryos, 15 min for E7.5–E8.5 embryos, 30 min for E9.5–E10.5 embryos, 60 min for E11.5–E12.5 embryos, and 90 min for E13.5–E14.5 embryos. For fetus and adult tissues, an intracardial perfusion procedure[43] is recommended to improve fixative penetration. After fixation, samples are washed in PBS (up to 20 min with two changes for larger samples) and then incubated in staining solution containing 5 m*M* potassium ferricyanide, 5 m*M* potassium ferrocyanide, 2 m*M* $MgCl_2$, 0.01% sodium deoxycholate, 0.02% nonidet P-40, and 1 mg/ml X-Gal. Staining solution without X-Gal can be prepared in advance and stored in the dark at room temperature, and X-Gal is added just before use. X-Gal stock (40 mg/ml) is made in dimethyl formamide (DMF) instead of dimethyl sulfoxide (DMSO) to prevent freezing and stored in small aliquots at -20°. Samples are usually stained overnight at room temperature to reduce endogenous βgal activity, and wild-type controls should always be included to differentiate nonspecific staining. For better staining solution penetration, a sagittal section is made with a razor blade for embryos at E13.5 or later stages before staining, and vibratome or cryostat sections can be made for staining fetus or adult tissues. After staining, samples are washed with several changes of 70% ethanol to inactivate βgal activity and can be stored indefinitely in 70% ethanol. To enhance staining intensity and visualize internal structures, embryos can be cleared with methyl salicylate or HistoClear after complete dehydration in 100% ethanol.

It should be noted that reporter gene expression does not always faithfully reveal the expression levels and patterns of the trapped genes, and conflicts have been reported in several cases.[44–46] This could be due to the interruption or deprivation of endogenous regulatory elements in introns or 5′/3′ UTRs, the difference in transcript or protein stability, or the loss of tissue-specific splicing isoforms. Therefore, it is advisable to verify endogenous gene expression with other methods such as *in situ* hybridization and/or immunohistochemistry.

Production and Analysis of Gene Trap Mouse Strains

Mice should be housed in SPF environment with good ventilation and controlled temperature and light cycles. ES cells derived from the 129S4 strain (agouti, *AA*) are injected into C57BL/6J (black, nonagouti, *aa*)

[43] C. Bonnerot and J. F. Nicolas, *Methods Enzymol.* **225**, 451 (1993).

[44] J. M. Deng and R. R. Behringer, *Transgenic Res.* **4**, 264 (1995).

[45] A. K. Voss, T. Thomas, and P. Gruss, *Dev. Dyn.* **212**, 171 (1998).

[46] W. Shawlot, J. M. Deng, L. E. Fohn, and R. R. Behringer, *Transgenic Res.* **7**, 95 (1998).

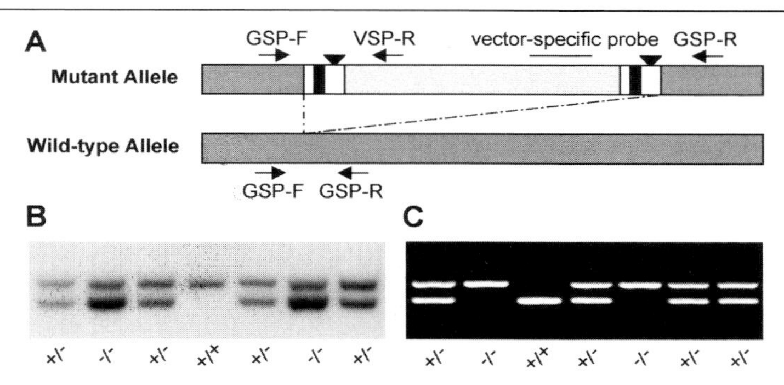

FIG. 3. Genotyping methods. (A) Schematic representations of wild-type and mutant alleles. Relative locations of primers and the vector-specific probe are indicated. (B) Quantitative Southern hybridization with a vector-specific probe and an internal control probe derived from an unrelated single-copy gene. Genotypes can be determined by comparing the relative intensity of transgene band (lower) to that of control band (upper). (C) Genotyping with 3-primer PCR. The wild-type band (lower) and mutant band (upper) are amplified with GSP-F/GSP-R and GSP-F/VSP-R primer pairs, respectively.

blastocysts and transferred into the uterus of 2.5 dpc F1 (C57BL/6J × CBA/J) pseudopregnant recipients. Chimeras are recognized by the presence of agouti fur a week after birth and coat color chimerism is assessed. High percentage male chimeras are bred to wild-type C57BL/6J females, and germline transmission is indicated by agouti offspring (*Aa*). Low percentage chimeras and female chimeras can also transmit mutations through germline on occasion. Since genetic modifiers may profoundly affect the penetrance and expressivity of phenotypes,[47] it is desirable to analyze mutations in homogeneous genetic backgrounds. Once germline transmission is ascertained, the chimera is crossed to 129S4 females to transmit the mutation into a congenic 129S4 background. Mutations can also be passed into a congenic C57BL/6J or any other genetic background after backcrossing for 10 or more generations. For practical reasons, most studies are performed on a hybrid 129S4 × C57BL/6J background while maintaining a minimum number of congenic breeding pairs, which should be expanded for further analysis whenever a clear difference in phenotypes is observed.

Heterozygous mice obtained from chimera breeding are identified by PCR with vector-specific primers. Mating pairs are set up and their offspring screened for homozygotes before weaning. Genotypes can be conveniently determined by quantitative Southern hybridization (Fig. 3A and B). DNA samples are digested with an appropriate enzyme and hybridized with a vector-specific probe and an unrelated probe of a single copy gene as loading control. Homozygotes can be differentiated from heterozygotes by the

[47] J. H. Nadeau, *Nat. Rev. Genet.* **2**, 165 (2001).

higher relative intensity of the transgene band, which is absent in wild-type littermates. However, ambiguous results are sometimes obtained with this method. For the most accurate genotyping, it is recommended to clone genomic DNA flanking sequences and derive specific probes or PCR primers. PCR-based methods are preferred especially for genotyping early embryos from which limited materials can be obtained. A 3-primer PCR system is routinely used in our laboratory to genotype both gene-targeted and gene-trapped mutants, which includes a gene-specific forward primer 5′ to insertion site (GSP-F), a gene-specific reverse primer 3′ to insertion site (GSP-R), and a vector-specific reverse primer (VSP-R). The wild-type band and mutant band are amplified with GSP-F/GSP-R and GSP-F/VSP-R primer pairs, respectively (Fig. 3A and C).

Viable homozygous mice can be screened for adult phenotypes such as reproductive, metabolic, morphological, immunological, neurological, or behavioral abnormalities. If no homozygous pups are obtained in 3–4 litters, the mutation is likely to result in embryonic or perinatal lethality. Timed matings are then set up and embryos retrieved at different stages for genotyping and phenotyping. Choose E10.5–E12.5 as a starting point, and then trace up or down to other stages according to the genotypes and phenotypes obtained.

Gene Trap Mutagenesis Screens

The three types of information attainable with gene trap mutagenesis—mutant phenotype, trapped sequence, and reporter expression—have all been exploited, alone or in combination, to select gene trap events of interest for further functional analysis. Forward genetics phenotype-driven screens are the most straightforward and reliable in that they provide direct evidence on the functions of trapped genes without presumptions. However, these *in vivo* screens are very labor-intensive and time-consuming, and require vast resources for mouse housing and husbandry. On the contrary, sequence-driven screens are intrinsically high-throughput. With automated cloning procedures, over 10,000 sequence-indexed ES cell clones have been generated and made freely available to public by combined efforts in several major centers, and an International Gene Trap Mutagenesis Consortium has been established aiming at achieving saturating mutagenesis in the next 5 years. Despite the remarkable promise in a long run, a caveat of these *in silico* screens lies in that only known genes are accessible, while the vast majority of trapped clones represent either ESTs or novel genes, and little information regarding their potential functions can be extracted from the sequence tags. A balanced alternative is to conduct expression-driven screens, which preselect gene trapped ES cell clones *in vitro* prior to generating mice without a bias towards known genes. Popular expression-driven screen strategies include differentiation trapping that selects genes expressed in certain lineage by identifying reporter-expressing cell types derived from ES cell differentiation, induction trapping that selects genes responsive to exogenous stimuli by following reporter expression changes upon treatment, and compartment trapping that selects genes encoding secreted or nuclear proteins by examining compartment specificity of fusion proteins.[17]

Nevertheless, the reliability, efficiency and spectrum of these expression-driven screens are severely limited by the reliance on reporter gene expression in ES cells. Although using alternative cell types could alleviate this problem, the mutation of interest would have to be reproduced in ES cells after gene identification when *in vivo* functional analysis is desired. To circumvent the restraint, we have designed a new expression-driven screening platform by combining microarray technology[48] with gene trapping. With a specifically-designed gene trap and a high-throughput RACE procedure, 3′ cDNA flanks of trapped genes are cloned to construct a spotted cDNA array. The gene trap array thus permits direct monitoring of the endogenous expression of trapped genes rather than the expression of tagged reporters in any cell type, and the screening scope is far extended with the broad applications of microarrays. For example, tissue-specific genes,[49] secreted proteins,[50] transcription factor targets,[51] genes responsive to external stimuli,[52–54] and genes changed in mutant cells or tissues[55,56] can all be identified with this unified screening platform. A major advantage of the gene trap array over conventional microarrays is that mutant mice of any gene identified after screening can be directly derived without

[48] D. J. Lockhart and E. A. Winzeler, *Nature* **405**, 827 (2000).

[49] T. S. Tanaka, S. A. Jaradat, M. K. Lim, G. J. Kargul, X. Wang, M. J. Grahovac, S. Pantano, Y. Sano, Y. Piao, R. Nagaraja, H. Doi, W. H., 3rd, Becker, K. G. Wood, and M. S. Ko, *Proc. Natl. Acad. Sci. USA* **97**, 9127 (2000).

[50] M. Diehn, M. B. Eisen, D. Botstein, and P. O. Brown, *Nat. Genet.* **25**, 58 (2000).

[51] D. A. Bergstrom, B. H. Penn, A. Strand, R. L. Perry, M. A. Rudnicki, and S. J. Tapscott, *Mol. Cell* **9**, 587 (2002).

[52] V. R. Iyer, M. B. Eisen, D. T. Ross, G. Schuler, T. Moore, J. C. Lee, J. M. Trent, L. M. Staudt, J., Jr., Boguski, M. S. Hudson, D. Lashkari, D. Shalon, D. Botstein, and P. O. Brown, *Science* **283**, 83 (1999).

[53] D. Fambrough, K. McClure, A. Kazlauskas, and E. S. Lander, *Cell* **97**, 727 (1999).

[54] G. G. McGill, M. Horstmann, H. R. Widlund, J. Du, G. Motyckova, E. K. Nishimura, Y. L. Lin, S. Ramaswamy, W. Avery, H. F. Ding, S. A. Jordan, I. J. Jackson, S. J. Korsmeyer, T. R. Golub, and D. E. Fisher, *Cell* **109**, 707 (2002).

[55] C. M. Simbulan-Rosenthal, D. H. Ly, D. S. Rosenthal, G. Konopka, R. Luo, Z. Q. Wang, P. G. Schultz, and M. E. Smulson, *Proc. Natl. Acad. Sci. USA* **97**, 11274 (2000).

[56] F. J. Livesey, T. Furukawa, M. A. Steffen, G. M. Church, and C. L. Cepko, *Curr. Biol.* **10**, 301 (2000).

engineering knockout constructs *de novo*, while allowing novel genes currently unavailable elsewhere to be discovered and studied.

Conventional gene trap vectors are designed to create loss-of-function alleles. Now, novel vectors are being developed in our laboratory and others to generate gain-of-function alleles, which would enable screens for genetic interactors or protein binding partners of genes of interest. On the other hand, site-directed recombination systems are being exploited extensively to devise exchangeable gene traps, which permit gene trap insertions to be replaced with any desired transgene cassette for further functional studies or making useful mouse models.[57–59] Together with the improved method for mutant mouse generation,[60] these new developments herald exciting prospects for functional genomics in the future.

Acknowledgments

We thank our colleagues for critical comments on the manuscript. This work is supported by NIH grant HD24875 to P. S.

[57] K. Araki, T. Imaizumi, T. Sekimoto, K. Yoshinobu, J. Yoshimuta, M. Akizuki, K. Miura, M. Araki, and K. Yamamura, *Cell. Mol. Biol.* **45**, 737 (1999).

[58] N. Hardouin and A. Nagy, *Genesis* **26**, 245 (2000).

[59] K. Araki, M. Araki, and K. Yamamura, *Nucleic Acids Res.* **30**, e103 (2002).

[60] K. Eggan, A. Rode, I. Jentsch, C. Samuel, T. Hennek, H. Tintrup, B. Zevnik, J. Erwin, J. Loring, L. Jackson-Grusby, M. R. Speicher, R. Kuehn, and R. Jaenisch, *Nat. Biotechnol.* **20**, 455 (2002).

[27] Gene Trap Vector Screen for Developmental Genes in Differentiating ES Cells

By HEIDI STUHLMANN

Introduction

A powerful approach to screen for genes involved in developmental processes is to generate mutations by random insertion of gene trap vectors into mouse embryonic stem (ES) cells. Insertion of foreign DNA elements into coding or regulatory regions of the genome frequently disrupts or alters expression of flanking genes. The insertion serves to "tag" the endogenous genes, allowing for subsequent cloning. The availability of mouse ES cells, together with the development of "entrapment vectors," has made it possible to efficiently introduce insertional mutations in ES cells and to screen for candidate genes in ES cells and embryos derived from the ES

0076-6879/2003 $35.00

cells. ES cells are undifferentiated, pluripotent cells derived from the inner cell mass (ICM) of blastocysts.[1–3] When introduced into morula- or blastocyst-stage embryos, they contribute to the formation of somatic tissues as well as functional germ cells.[4,5] Furthermore, they can be induced to differentiate *in vitro* into embryoid bodies (EBs), either spontaneously by removal of the feeder layer of mouse embryonic fibroblasts (MEFs) and leukemia inhibitory factor (LIF), or directed by removal of differentiation inhibitors and addition of specific growth factors and cytokines. Examples of specific ES differentiation protocols are found elsewhere in this volume. Upon differentiation, ES cells give rise to a variety of different cell lineages, including parietal and visceral endoderm, mesoderm, cardiac and skeletal myocytes, hepatocytes, hematopoietic and endothelial cells, dendritic cells, neurons, and glial cells.[5–19] The process of EB formation *in vitro* closely mimics early postimplantation development, as has been demonstrated by

[1] F. A. Brook and R. L. Gardner, *Proc. Natl. Acad. Sci. USA* **94**, 5709–5712 (1997).
[2] M. J. Evans and M. H. Kaufman, *Nature* **292**, 154–156 (1981).
[3] G. R. Martin, *Proc. Natl. Acad. Sci. USA* **78**, 7634–7638 (1981).
[4] A. Bradley, M. Evans, M. H. Kaufman, and E. Robertson, *Nature* **309**, 255–256 (1984).
[5] A. Nagy, J. Rossant, R. Nagy, W. Abramow-Newerly, and J. C. Roder, *Proc. Natl. Acad. Sci. USA* **90**, 8424–8428 (1993).
[6] G. Bain, D. Kitchens, M. Yao, J. E. Huettner, and D. I. Gottlieb, *Dev. Biol.* **168**, 342–357 (1995).
[7] V. L. Bautch, W. L. Stanford, R. Rapoport, S. Russell, R. S. Byrum, and T. A. Futch, *Dev. Dyn.* **205**, 1–12 (1996).
[8] K. Choi, M. Kennedy, A. Kazarov, J. C. Papadimitriou, and G. Keller, *Development* **125**, 725–732 (1998).
[9] T. C. Doetschman, H. Eistetter, M. Katz, W. Schmidt, and R. Kemler, *J. Embryol. Exp. Morph.* **87**, 27–45 (1985).
[10] K. Eto, R. Murphy, S. W. Kerrigan, A. Bertoni, H. Stuhlmann, T. Nakano, A. D. Leavitt, and S. J. Shattil, *Proc. Natl. Acad. Sci. USA* **99**, 12819–12824 (2002).
[11] A. Fraichard, O. Chassande, G. Bilbaut, C. Dehay, P. Savatier, and J. Samarut, *J. Cell Sci.* **108**, 3181–3188 (1995).
[12] G. Keller, M. Kennedy, T. Papayannopoulou, and M. V. Wiles, *Mol. Cell. Biol.* **13**, 473–486 (1993).
[13] M. H. Lindenbaum and F. Grosveld, *Genes Dev.* **4**, 2075–2085 (1990).
[14] V. A. Maltsev, J. Rohwedel, J. Hescheler, and A. M. Wobus, *Mech. Dev.* **44**, 41–50 (1993).
[15] T. Nakano, H. Kodama, and T. Honjo, *Science* **265**, 1098–1101 (1994).
[16] J. Rohwedel, V. Maltsev, E. Bober, H. H. Arnold, J. Hescheler, and A. M. Wobus, *Dev. Biol.* **164**, 87–101 (1994).
[17] Senju, S., Hirata, S., Matsuyoshi, H., Masuda, M., Uemura, Y., Araki, K., Yamamura, K.L., and Nishimura, Y. (2002) *Blood* online prepublication.
[18] C. Strübing, G. Ahnert-Hilger, J. Shan, B. Wiedemann, J. Hescheler, and A. M. Wobus, *Mech. Dev.* **53**, 275–287 (1995).
[19] R. Wang, R. Clark, and V. L. Bautch, *Development* **114**, 303–316 (1992).

the temporal appearance of differentiated cell lineages and expression of stage- and lineage-specific markers.

The essential feature of entrapment vectors is a promoter-less reporter gene, usually the *E. coli lacZ* gene, whose expression becomes dependent on transcription initiated from cis-acting regulatory sequences of a targeted cellular gene. Therefore, candidate genes can be identified based upon the pattern of reporter gene expression rather than a mutant phenotype (reviewed in Refs. 20, 21, 22). Three types of entrapment vectors have been previously developed, enhancer trap, promoter trap, and gene trap vectors that contain a splice acceptor site positioned 5′ to the reporter gene, allowing for the generation of fusion transcripts of the reporter gene to upstream exon sequences upon integration into an intron.[23–27] Entrapped endogenous genes can be rapidly cloned by PCR-based methods, such as inverse PCR or 5′RACE, or by cloning of genomic DNA flanking the vector insertions. New vector designs include polyadenylation site vectors to efficiently trap genes that are not expressed in undifferentiated ES cells,[28] and secretory-trap vectors to identify secreted and transmembrane proteins expressed in ES cells.[29] Retrovirus- or plasmid-based entrapment vectors have been used in mutagenesis screens, expression- and induction-trapping screens, or sequence-based screens in ES cells and embryos derived from the ES cells. Several consortia presently perform large-scale gene trap projects and provide ES cell lines and insertion site sequences as public resources [University of Manitoba Institute of Cell Biology (http://www.escells.ca); The Gene Trap Project of the German Human Genome Project (http://tikus.gsf.de); The BayGenomics Gene Trap Project (http://baygenomics.ucsf.edu); The CMHD Gene Trap Project (http://www.cmhd.ca)]. In addition, specialized smaller screens have been performed by a number of individual groups. An excellent review by

[20] A. Gossler and J. Zachgo, *in* "Gene Targeting: A Practical Approach" (A. L. Joyner, ed.), pp. 181–227. Oxford University Press, New York, 1993.

[21] W. L. Stanford, J. B. Cohn, and S. P. Cordes, *Nat. Rev. Genet.* **2**, 756–768 (2001).

[22] I. J. Jackson, *Nat. Genet.* **28**, 198–200 (2001).

[23] G. Friedrich and P. Soriano, *Genes Dev.* **5**, 1513–1523 (1991).

[24] A. Gossler, A. L. Joyner, J. Rossant, and W. C. Skarnes, *Science* **244**, 463–465 (1989).

[25] R. Korn, M. Schoor, H. Neuhaus, U. Henseling, R. Soininen, J. Zachgo, and A. Gossler, *Mech. Dev.* **39**, 95–109 (1992).

[26] W. Skarnes, B. A. Auerbach, and A. L. Joyner, *Genes Dev.* **6**, 903–918 (1992).

[27] H. von Melchner, J. V. DeGregori, H. Rayburn, S. Reddy, C. Friedel, and H. E. Ruley, *Genes Dev.* **6**, 919–927 (1992).

[28] M. Salminen, B. I. Meyer, and P. Gruss, *Dev. Dyn.* **212**, 326–333 (1998).

[29] W. C. Skarnes, J. E. Moss, S. M. Hurtley, and R. S. P. Beddington, *Proc. Natl. Acad. Sci. USA* **92**, 6592–6596 (1995).

Stanford *et al.* (2001)[21] provides detailed information on entrapment vectors and different approaches for screening.

Here, we describe an expression screen to identify genes that are expressed at specific stages of mouse development by using retroviral entrapment vectors. ES cell clones with entrapment vector integration into putative developmentally regulated genes can be selected by virtue of their stage-specific reporter gene expression, and by colocalization with stage- and lineage-specific markers upon *in vitro* differentiation into embryoid bodies (Fig. 1). Selected clones can be further examined *in vivo*, either by generating transgenic lines or "transient" chimeric embryos. Finally, genomic integration sites can be cloned and used to screen stage-specific embryonic cDNA libraries. This approach allows us to efficiently identify and preselect *in vitro* for insertions into genes that are restricted in their temporal and spatial distribution in the postimplantation embryo and that may not be expressed in pluripotent embryonal cells.[30,31]

This chapter focuses on the strengths of the *in vitro* differentiation screen to preselect for entrapment integrations into developmentally regulated genes (Fig. 1). First, we discuss infection of ES cells with retroviral entrapment vectors. Second, screening for reporter gene expression during EB differentiation will be described. The third section provides protocols for colocalizing reporter gene expression and developmental markers. The protocols described in this chapter use commercially available reagents, they are reproducible and efficient.

Retroviral Entrapment Vector Infection of ES Cells

Bifunctional Retroviral Vectors

The design and use of retroviral- or plasmid-based entrapment vectors has been reviewed in detail elsewhere.[20,21] We chose to use bifunctional retroviral entrapment vectors with an alkaline phosphatase (AP) reporter gene (Fig. 2) to identify genes that are differentially expressed during early mouse development. Retrovirus vectors efficiently infect ES cells and stably integrate into their genomic DNA. Other advantages of retroviral vectors are that infection at a low multiplicity leads to single copy insertions, their conservative integration via their long terminal repeat sequences, and their propensity to integrate into 5′ untranslated or first intron genomic

[30] J.-W. Xiong, R. Battaglino, A. Leahy, and H. Stuhlmann, *Dev. Dyn.* **212**, 181–197 (1998).
[31] A. Leahy, J.-W. Xiong, F. Kuhnert, and H. Stuhlmann, *J. Exp. Zool.* **284**, 67–81 (1999).

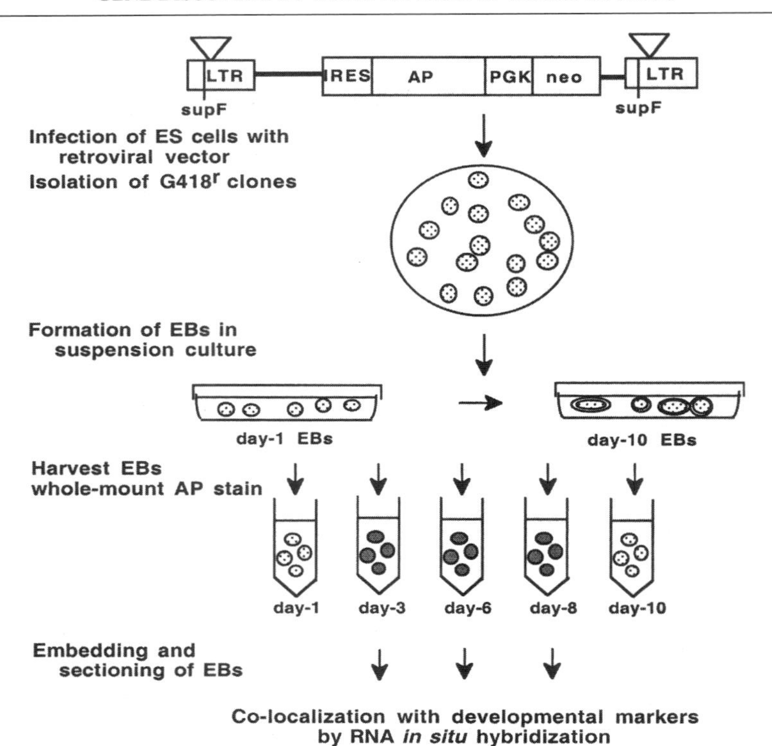

FIG. 1. Schematic overview of a genetic screen for developmentally regulated genes. ES cells are infected with a retroviral entrapment vector (depicted is PT-IRES-AP) and integration events are selected for in the presence of G418. Following selection and isolation of single colonies, ES cell clones are differentiated *in vitro*, and EBs are screened at multiple days of differentiation for temporally and spatially regulated AP gene expression. AP-stained EBs are embedded, sectioned and *in situ* hybridization is performed to assay for colocalization with developmental marker genes.

sequences. The human placental AP gene was chosen as a reporter, because we found significantly less variability in its expression in ES cells and EBs as compared to *lacZ* (R. Battaglino and H. Stuhlmann, unpublished). In order to allow for selection of cells carrying the integrated entrapment vector independent of AP reporter expression, the neo^R gene was transcribed from an internal phosphoglycerokinase (*pgk-1*) promoter that is functional in a variety of undifferentiated and differentiated cells.

The entrapment vectors (Fig. 2) contain 5′ and 3′ Moloney murine leukemia virus- (Mo-MLV-) derived sequences, including the two LTRs and about 1-kb of *gag* sequences that are important for efficient packaging of the retroviral genome. The plasmid constructs contain a 300-bp deletion of

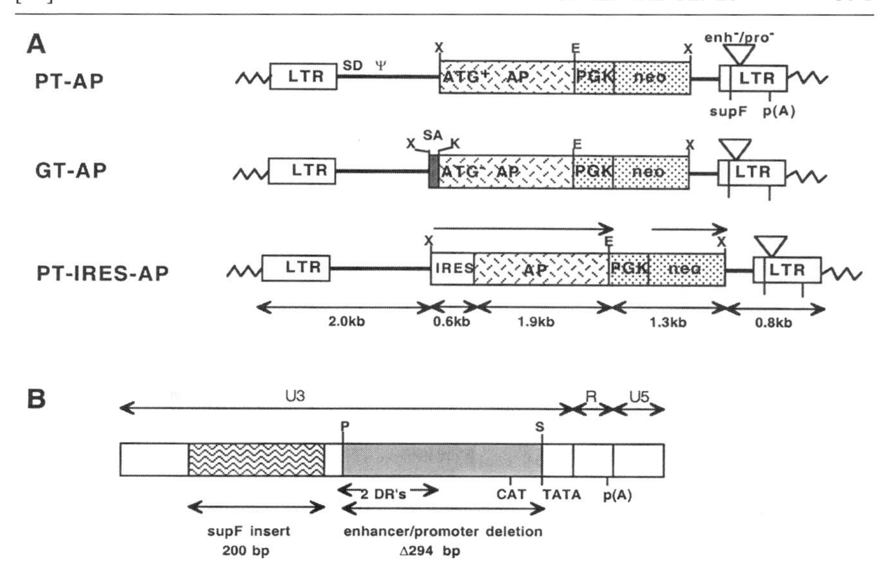

FIG. 2. Structure of bifunctional AP entrapment vectors. (A) The constructs contain a human placental AP reporter gene. PT-AP contains a 1.9-kb AP cDNA fragment including its ATG. GT-AP contains a 170-bp XhoI-KpnI fragment with the Mo-MLV *env* splice acceptor site (SA) and the AP cDNA with a deletion of its ATG. PT-IRES-AP contains a 0.6 kb IRES fragment from EMCV and the AP cDNA. All constructs contain a *neo* expression cassette with a *pgk-1* promoter and the neo^R gene. The vectors contain 5′ Mo-MLV sequences including the 5′LTR, the packaging site ψ, and 1-kb of gag sequences. The 3′ end of the vectors contains Mo-MLV sequences including the 3′LTR with the 200-bp *E. coli supF* gene inserted in U3. The 3′LTR also contains a 300-bp deletion of retroviral enhancer and promoter sequences. The relative position of the *supF* insert, the deletion of enhancer and promoter (enh⁻/pro⁻), and the polyadenylation [p(A)] sequence are indicated. (B) The structure of 3′ Mo-MLV LTR in AP entrapment vectors is schematically shown. Indicated are the positions of the 200-bp *E. coli supF* insert, enhancer sequences (direct repeats; DR), CAT and TATA transcriptional sequences, and the polyadenylation sequence p(A) in the LTR. The enhancer and promoter sequences are deleted between the PvuII (P) and the SacI (S) sites.

enhancer and promoter sequences in the 3′ LTR. Following infection and reverse transcription the deletion is copied into both LTRs in the provirus. Expression of the AP reporter gene requires initiation of transcription from a promoter of an endogenous gene 5′ to the retroviral integration site and termination at the retroviral polyadenylation sequence in the 3′ LTR. In addition, the LTRs contain an insert of a bacterial suppresser tRNA gene, *sup F*, to allow cloning of proviral and flanking genomic host sequences in a suitable bacterial host. With the PT-AP (promoter trap-AP) vector, integration close to a cis-acting cellular regulatory sequence and in the sense orientation leads to read-through transcription into the retroviral genome. Depending on the position of the first ATG in the cellular gene relative to the initiator ATG in AP, either a chimeric protein with the AP at its C-terminus, or an authentic AP protein will be made. In the GT-AP (gene trap-AP) construct, the initiator ATG in AP is deleted, and a 170-bp fragment containing the splice acceptor site from the Mo-MLV *env* mRNA is inserted upstream of the AP gene. Integration of the entrapment vector within an intron of a transcribed cellular gene generates a spliced fusion transcript and, if in-frame, leads to expression of a chimeric AP protein. The construct PT-IRES-AP contains a 600-bp internal ribosomal entry site (IRES) element from encephalomyocarditis virus (EMCV) upstream of the AP gene. Integration of the PT-IRES-AP vector into a transcribed gene leads to cap-independent translation of an authentic AP reporter protein from a fusion transcript, independent of the reading frame.

Virus producing lines are generated by electroporation of retroviral entrapment vector constructs into retrovirus packaging lines; these are genetically engineered to provide helper virus function in trans but are incapable to produce wild-type virus. In our hands, all three entrapment vectors gave rise to stable producer lines with comparable titers, between 2 to 5×10^4/ml on NIH3T3 fibroblasts, and 0.5 to 1×10^3/ml on ES cells, respectively.

Propagation and Maintenance of ES Cells

We have successfully used the ES cell lines R1[5] and W4[32] for retroviral infection, *in vitro* differentiation, and ES cell chimera formation. Both are derived from the mouse inbred strain 129Sv, they have been shown to contribute with high efficiency to the germ line upon blastocyst injection and to chimera formation by morula-ES cell aggregation, and they maintain these characteristics at high passage numbers. We found that conditions for optimal growth and differentiation are very similar for the R1 and W4 cell lines. Stocks of ES cells at low passage numbers are kept in liquid nitrogen, and cells are propagated for fewer than five passages prior to an experiment. To retain the cells in an undifferentiated state, both lines require growth on primary mouse embryonic fibroblast feeders (MEFs) and the presence of LIF in the culture medium. Equally important is the quality of the fetal bovine serum (FBS). We routinely test several lots of serum for maintenance of the undifferentiated phenotype upon at least five serial passages, cloning

[32] W. Auerbach, J. H. Dunmore, V. Fairchild-Huntress, Q. Fang, A. B. Auerbach, D. Huszar, and A. L. Joyner, *BioTechniques* **29**, 1024–1032 (2000).

efficiency, and cell toxicity at a serum concentration of 30%. We have obtained very good success with ES cell-qualified FBS from HyClone and Gemini.

Culture Medium and Solutions

All reagents for culturing ES cells and MEFs should be "tissue culture grade" or "tissue culture tested." Preferably, only sterile tissue culture grade plastic ware and pipettes should be used. For culture of ES cells, we use Dulbecco's modified Eagle's medium (DMEM) with high glucose and glutamine (Mediatech Cellgro) supplemented with 15% heat-inactivated fetal bovine serum (FBS, HyClone or Gemini), 20 m*M* HEPES (50X stock, pH 7.3, GIBCO), 0.1 m*M* nonessential amino acids (100X stock, GIBCO), 0.1 m*M* β-mercaptoethanol (1000X stock, GIBCO), penicillin and streptomycin (100X stock, GIBCO), and 500 u/ml LIF (GIBCO). ES cells are cultured on gelatin-coated plastic plates or flasks (0.7% solution; gelatin from porcine skin, approximately 300 bloom, Sigma No. G1890). Cells are passaged every 2–3 days. For this, dishes are rinsed twice with phosphate buffered saline (PBS), and cells are carefully dissociated with Trypsin–EDTA (0.05% solution, GIBCO) to obtain a single cell suspension. Trypsin digestion is stopped by adding an equal amount of culture medium, and cells are transferred to a new plate at a dilution 1 : 5 or 1 : 10. Culture medium for MEFs is DMEM supplemented with 10% heat-inactivated FBS, 20 m*M* HEPES, and penicillin and streptomycin.

Preparation of Mouse Embryonic Fibroblast Feeder Cells

We routinely use the transgenic line DR-4 as a donor for multiple drug-resistant MEFs. The DR-4 mice bear selectable genes neo^R, $puro^R$, hyg^R, and a deletion of the X-linked *Hprt* gene,[33] which confers resistance to G418, puromycin, hygromycin, and 6-thioguanine. DR-4 mice can be purchased from The Jackson Laboratory (Stock Tg(DR4)1Jae, Stock No. 003208). MEFs are isolated from mouse embryos at day 13.5 or 14.5 of gestation. For maintenance of ES cells in the absence of drug selection, MEFs from CD-1 embryos can be used. Each preparation of primary MEFs has to be tested for mycoplasma before expansion and coculture with ES cells. We are routinely using a PCR-based and an ELISA-based mycoplasma detection kit, both of which detect the most common mycoplasma species found in contaminated mammalian cell culture systems.

[33] K. L. Tucker, Y. Wang, J. Dausman, and R. Jaenisch, *Nucleic Acids Res.* **25**, 3745–3746 (1997).

Embryos are dissected from their implantation sites and transferred to a dish with sterile PBS and rinsed once in PBS. Under a stereomicroscope in a sterile hood, heads and soft tissues (liver, intestine, kidneys, lung, and heart) are removed with watchmaker forceps. The embryo carcasses from one litter (6–10 embryos) are placed into a fresh 10 cm dish, and just enough Trypsin/EDTA is added to cover them (about 5 ml for 10 embryos). The carcasses are minced into fine pieces by using sterile scalpels, and incubated for 30 min in the 37°C incubator. MEF culture medium is added, and the digested tissues are dissociated by repeated cycles of vigorous pipetting and collecting the supernatant cell suspension. Cells from combined supernatants are plated into T165 flasks (10 flasks for 10 embryo carcasses) in MEF culture medium. Cells reach near confluence usually 2 days after plating, at which time they are frozen (2 vials per embryo) at −80°C and transferred to liquid N_2 1 day later.

To expand primary MEFs, frozen cells from one vial are quickly thawed at 37°C in MEF culture medium and divided into three T165 flasks. Cells are expanded on 150 mm plates in intervals of 2–3 days two more times at a dilution of 1 : 3. Cells from the final passage (total of 27×150 mm plates) are harvested upon reaching complete confluence, pelleted and resuspended in 45 ml ice-cold MEF culture medium. The cells are mitotically inactivated using a γ-irradiation source. Dimethyl sulfoxide (DMSO, 5 ml) is added to the cell suspension, and cells are frozen down in aliquots (0.2 ml aliquots for T25 flasks feeder layer, 0.5 ml for T75 flask or 100 mm dish, and 1.5 ml for 3× T75 flasks or 3× 100 mm dishes, respectively). The MEF feeder cells can be stored at −80°C for several months. It is crucial that the MEFs are actively dividing during their expansion and that they do not reach senescence. In our hands, the expansion protocol described above generates reproducibly MEFs that support maintenance of undifferentiated ES cell cultures.

Culture of ES Cells on MEFs

ES cells are routinely cultured in ES medium on mitotically inactivated MEF feeder layers in gelatin-coated flasks. A fresh vial of inactivated MEFs is thawed out for each ES cell passage. The MEFs can be plated either before passage of ES cells, or at the same time. Culture medium is changed every day.

To thaw ES cells, a frozen vial is quickly warmed in a 37°C water bath. Cells are transferred to a 15-ml conical tube filled with 5 ml ES medium and pelleted for 5 min at 1000 rpm. Cells are resuspended in 1 ml ES medium and transferred into a gelatin-coated T25 flask with MEF cells and 5 ml ES cell medium. Culture medium is changed after ES cells have attached in order to remove dead cells (usually 2–3 hr after thawing cells).

ES cells should be passaged when they are about 80% confluent. It is crucial to dissociate cells well to achieve a single cell suspension (otherwise the ES cells will spontaneously differentiate), and to seed the cells at appropriate densities. For regular passages, the following protocol is used: Cells in a T25 flask are rinsed 2× and trypsinized with 0.5 ml Trypsin–EDTA for 2 min at 37°C. The cells are detached from the bottom by knocking against the flask, and incubated for another 2–4 min at 37°C. After adding 1.5 ml ES cell medium, cells are completely dissociated by pipetting several times with a 1-ml pipette against the bottom of the flask. An aliquot of the suspension is counted, and 1.5×10^6 cells are added to a new T25 gelatin-coated flask with a MEF feeder layer. Alternatively, ES cells are plated in a new flask at a 1 : 5 or 1 : 10 dilution. The cells should be passaged every 2–3 days. ES cells are frozen down from ~80% confluent cultures at 5×10^6 cells/ml in 0.5-ml aliquots per freezing vial. For this, cells are washed, dissociated with Trypsin–EDTA as described above, and counted. Cells are pelleted in a 15 ml conical tube and resuspended in ES cell culture medium supplemented with 10% (v/v) DMSO at a concentration of 5×10^6/ml. The cell suspension is frozen in 0.5-ml aliquots over night at −80°C and transferred to liquid N_2 the next day.

Retroviral Entrapment Vector Infection and Selection of ES Cells

Titration of Virus Supernatant on ES Cells

Our experience and that of other groups is that infectious titers from retroviral producer lines are between 10- and 50-fold lower on ES cells than on NIH 3T3 fibroblasts. Therefore, virus titers should be determined on ES cells. Virus supernatant from producer lines is freshly harvested and passed through a 0.4-μm syringe filter to remove cells. Alternatively, supernatant can be stored as frozen stocks in aliquots at −80°C. Since infectious titers drop to about 50% during each freezing/thawing cycle, we only thaw virus supernatant once. Serial dilutions (10^0 to 10^{-4}) are made in PBS supplemented with 3% FBS and 4-μg/ml polybrene in an ice-water bath. Two milliliters from each dilution are added to a 100 mm dish (plated the day before with 0.5×10^6 ES cells in the presence of 4-μg/ml polybrene). ES cell medium (8 ml) is added 4 hr later, and selection medium (culture medium supplemented with 0.35-mg/ml G418) is added after 36-hr incubation. Seven days later, cells are fixed with 100% methanol and stained with GIEMSA (1:10 dilution in H_2O). To determine entrapment efficiencies, plates infected in parallel with virus dilutions (10^0 and 10^{-1}) are selected for G418 resistance, and clones are stained 7 days later for AP activity (see below). Under a stereomicroscope, the total number of G418-resistant colonies, and the number of AP positive colonies per plate are counted.

Infectious titers are scored as number of G418-resistant colony forming units per ml. The entrapment efficiencies are determined as percentage of AP-positive clones out of the total number of G418-resistant clones. In our hands, the trapping frequency with the PT-AP virus was 1–3% in fibroblasts and in ES cells, and similar frequencies were found with the GT-AP virus. The intensity of the purple AP stain varied somewhat among positive clones but was similar in most cells within one particular clone, consistent with the notion that many entrapment events lead to the expression of AP fusion proteins with varying enzymatic activities. Significantly higher entrapment efficiencies (5–13% in NIH 3T3 cells, and 2–5% in ES cells) were obtained when the IRES element was included upstream of the AP reporter gene (PT-IRES-AP), consistent with the notion that expression of a functional AP gene is independent of the translational reading frame. Histochemical staining for AP was uniformly strong among different clones as well as among cells within the same clone, consistent with the notion that the reporter gene is expressed as an authentic rather than a fusion AP protein.

Because the entrapment efficiencies vary between 1 and 5% in ES cells, large numbers of G418-resistant colonies have to be screened. To obtain between 100 and 500 of G418-resistant ES cell clones from one 100-mm dish, we use two alternative protocols for retrovirus infection, as outlined below.

Retrovirus Infection with Virus Supernatants

This protocol is being used if: (i) the virus titers on ES cells are at least 10^2–10^3/ml; and (ii) single proviral integrations per infected ES cell are desired. Twelve to 18-hr before virus infection, $0.5–1 \times 10^6$ ES cells are plated per 100-mm gelatin-coated dish with feeder cells in ES cell culture medium, and in the presence of 4-μg/ml polybrene. For a typical experiment in which several hundred colonies are to be isolated, we use four to six 100-mm dishes. The next day, aspirate medium and infect cells with 2-ml of virus supernatant (fresh harvested or frozen stocks, sterile filtrated through 0.4-μm filters). After 4-hr of incubation at 37°C, 8 ml of culture medium is added. Thirty-six hours later, selection medium (ES cell culture medium supplemented with 0.35 mg/ml G418 [dry powder, GIBCO; stock solution 50 mg/ml]) is added. Medium is changed every day, and colonies are picked 7–8 days later (see below).

Retrovirus Infection by Cocultivation

This protocol describes retrovirus infection of ES cells by cocultivation with virus-producing cells. This method is being used if: (i) the virus titers are very low ($< 10^2$/ml), and/or (ii) drug selection of infected ES cells is not possible and infection of a high percentage of ES cells should be achieved. One 100-mm dish with virus producer cells at 70–80% confluence is treated with 5-ml of 10-μg/ml mitomycin C in culture medium for 3 hr at 37°C. The cells are carefully washed 3× with PBS to remove residual mitomycin C, trypsinized, and plated on 100-mm gelatinized dishes with fresh feeder cells in ES cells medium supplemented with 4-μg/ml polybrene. ES cells are trypsinized, counted, and pelleted as described above. Between $0.5–1 \times 10^6$ ES cells are seeded for cocultivation into one 100-mm dish. Cells are cocultivated for 48 hr, and fresh medium supplemented with polybrene is added every day. The cells are trypsinized and plated at a 1:100 dilution into new dishes in ES medium supplemented with G418 and fresh MEFs added. Medium is changed every day, and colonies are isolated 7–8 days later.

Isolation of ES Cell Clones and Preparation of Master Plates

At the time of their isolation, ES colonies should have a smooth, rounded "hunched-up" shape, with no differentiating cells visible on the borders and no dead cells in the center. Cells from a single ES cell clone are divided into two 96-well plates, one of which will be frozen as a master plate, and the second will be used for expansion and expression screening during *in vitro* differentiation.

The following plates are prepared for the isolation of clones: (i) 96-well plates with flat-bottom wells, gelatin-coated, with MEFs added (use one 0.2-ml frozen aliquot per 96-well plate) in 200-μl ES cell culture medium per well. Duplicate plates for each clone are prepared. We routinely isolate 400 clones per day; thus eight 96-well plates are needed. (ii) 96-well plates with V-shaped bottom wells, filled with 30-μl Trypsin–EDTA each. Colonies are picked in a laminal flow hood under a stereomicroscope. Using a 20-μl pipetman, scrape with the tip in a circle around the ES cell clone to disrupt the layer of MEFs, carefully scoop up and then suck up the ES cell clone in a small volume. Transfer to a V-shaped 96-well plate filled with Trypsin–EDTA. Continue to pick one row or two rows of 8 clones each. It is important not to let the clones incubate in trypsin for extended periods of time. Using a multichannel-pipettor with eight tips attached, cells from one row of eight wells are dissociated by pipetting 20–30 times up and down. Culture medium (75 μl) is added, and cell suspension (50 μl each) is plated into the prepared duplicate 96-well plates. Clones are grown until they reach 50–80% confluence. Cells are frozen down either from both plates or, alternatively, cells are frozen from one plate and cells from the duplicate plate are immediately expanded to induce EB formation as described in the following section.

To freeze cells in 96-well plates, cells are washed twice with PBS and dissociated in 20-μl Trypsin–EDTA using the multichannel-pipettor. Eighty microliters of freezing solution (70-μl culture medium plus 10-μl DMSO) is mixed with the cell suspension. The plates are sealed with parafilm, placed in a freezing bag and frozen at −80°C. To thaw clones quickly, 200-μl warm medium is added to 96-well plates, and plates are placed in the 37°C incubator. After cells are attached, medium is replaced with a freshly prepared suspension of MEFs in 200-μl culture medium per well. Cells are expanded from 96-well plates to 24-well plates and then to T25 flasks.

Screen for Differential AP Reporter Expression During *In Vitro* Differentiation of ES Cells

Design of Differential Expression Screen

To identify clones that contain entrapment vector insertions into developmentally regulated genes, we designed a screen in which candidate clones are selected by virtue of their stage-specific reporter gene expression upon *in vitro* differentiation into embryoid bodies. Studying the onset of expression of a panel of germ layer and cell lineage marker genes allowed us to determine the time course of differentiation during the formation of embryoid bodies (Table I). Thus, stages equivalent to embryogenesis between implantation and the beginning of gastrulation (4.5–6.5 dpc) occur within the first two days of EB differentiation. Between days 3 and 5, EBs contain cell lineages found in embryos during gastrulation at 6.5–7.0 dpc, and after day-6 in culture, EBs are equivalent to early organogenesis-stage embryos (7.5 dpc).[31]

The screen described here eliminates a bias for insertions into genes that are transcriptional active in undifferentiated ES cells. It also reduces the number of candidate clones to be analyzed later in developing embryos. In a screen of 2400 ES cell clones, 41 (1.7%) displayed a differential reporter gene expression during EB formation. Strikingly, the majority of these insertions (76%) were in genes that showed alkaline phosphatase expression during *in vitro* differentiation but not in undifferentiated ES cells[30] (Xiong *et al.*, 1998[30]). A crucial aspect of this *in vitro* differentiation screen is that the temporal and spatial reporter gene expression during EB formation is a good predictor for its expression in the embryo. Thus, we found that clones with restricted reporter gene expression *in vitro* exhibited similar temporally and spatially restricted AP expression *in vivo*.[30,31]

TABLE I

SUMMARY OF MARKER GENE EXPRESSION IN EMBRYOID BODIES

Probe	Days in suspension culture										*In vivo* expression (dpc)
	1	2	3	4	5	6	7	8	9	10	
oct-3	+++	+++	+++	+++	++	++	+	+	+	+	• Embryonic ectoderm 0.5–8.5 dpc
Fgf-5	+	+++	+++	+++	+++	++	+	+	+	+	• Embryonic ectoderm and distal embryo proper 5.25–7.5 dpc
GATA-4	+	+	+	+	+	++	++	+++	+++	+++	• Primitive endoderm, visceral and parietal endoderm Endodermal derivatives: endocardium, gut, gonad 5.0 dpc—adult
nodal	–	+	+	++	++	+++	++	++	+	+	• Embryonic ectoderm, primitive endoderm, anterior primitive streak 6.25–7.5 dpc
Brachyury	–	–	++	+++	+++	+++	++	+	+	+	• Mesoderm and notochord 6.5–8.5 dpc
flk-1	–	–	++	++	+++	+++	++	++	++	+	• Vasculature 7.0 dpc—adult
Nkx-2.5	–	–	–	–	–	++	++	++	++	++	• Cardiogenic mesoderm 7.5 dpc—adult
EKLF	–	–	–	–	–	+	+	++	++	++	• Blood cells of yolk sac and hepatic primordia 7.5 dpc—adult
Msx3	–	–	–	–	–	–	–	+	+++	+++	• Neural tube 8.5–16.5 dpc

Stages of development in EBs equivalent to:

Postimplantation ⟶ (days 1–10)

Gastrulation ⟶ (days 3–5)

Early organogenesis ⟶ (days 6–10)

The protocol provided below is used for screening a large number of ES cell clones with random retroviral entrapment vector insertions. Prior to inducing spontaneous differentiation, single ES cell clones are pooled in groups of 8–10. Since the trapping efficiency with our entrapment vectors is between 1 and 5%, one expects that about 1 out of two to ten pools will contain one positive clone. Once a positive pool has been identified, individual clones from this pool are rescreened.

Culture and Assay Conditions

Embryoid Body Culture Conditions

Embryoid bodies are cultured in suspension in DMEM (containing high glucose and glutamine) supplemented with 10% heat-inactivated FBS (Gemini or Atlas), 20 m*M* HEPES (50X stock, GIBCO), and penicillin and streptomycin (100X stock, GIBCO). Suspension cultures are grown in bacterial dishes (LabTech, Nunc). It is crucial to test serum samples for their ability to support good EB formation. In our experience, serum that is optimal for undifferentiated ES cell cultures often is not ideal for *in vitro* differentiation. Criteria for a good serum are formation of intact, spherical EBs of uniform size with smooth borders, absence of sticking to the bacterial surface, and absence of cell death. We have had good success with FBS from Gemini and Atlas.

Alkaline Phosphatase Activity Assay

This protocol has been optimized from a previously published procedure[34] to assay for human placental AP activity in ES cells, EBs and embryos. AP expression is detected as a strong, dark purple histochemical stain. Uninfected mouse ES cells and embryoid bodies exhibit a high level of endogenous AP activity, even in the presence of Levamisole, a specific inhibitor for rodent APs. Heat treatment at 70°C for 30 min in the presence of $MgCl_2$ prior to staining with BCIP/NBT reduces the nonspecific activity to a faint pink background staining while not interfering with the dark purple stain from the human placental AP activity. In our experience, using the BM Purple AP substrate (Boehringer Mannheim/Roche) substantially reduces the background from endogenous AP activities and is preferred over BCIP/NBT. However, when used in combination with RNA *in situ* analysis, the strong purple precipitate of the BM Purple AP substrate can interfere with detection of silver grains. If this proves to be a problem, the samples

[34] S. C. Fields-Berry, A. L. Halliday, and C. L. Cepko, *Proc. Natl. Acad. Sci. USA* **89**, 693–697 (1992).

should either be stained for a shorter time, or BCIP/NBT should be used instead.

Cells, EBs, or embryos are washed 1× in PBS at room temperature and fixed for 5 min at room temperature in fixation solution. To prepare the fixation solution, 2-g solid paraformaldehyde is dissolved in the fume hood on a heat stirrer for 5–10 min in 100-ml 0.1 *M* PIPES (pH 6.9), 2 m*M* $MgCl_2$, 1.25 m*M* EGTA, pH 8.0, and subsequently cooled on ice. The fixation solution should be freshly prepared. Samples are washed twice for 5 min each in PBS containing 10 m*M* $MgCl_2$. To inactivate endogenous AP activities, samples are heat-treated for 30 min at 70°C in PBS containing 10 m*M* $MgCl_2$. Samples are washed for 10 min at room temperature in AP buffer (100 m*M* Tris–HCl, pH 9.5; 100 m*M* NaCl; 10 m*M* $MgCl_2$) and stained in AP staining solution (0.1-mg/ml BCIP; 1-mg/ml NBT; 0.24-mg/ml Levamisole (SIGMA) in AP buffer) or with BM Purple AP substrate. Samples should be stained for at least 4 hr or overnight at room temperature in the dark and are subsequently washed 1× in PBS.

Differentiation of ES Cells into Embryoid Bodies

Individual ES cell clones are expanded from 96-well plates to 24-well plates. About 2 days later, when cells reach 30–50% confluence, clones are pooled in groups of 8–10 and induced to spontaneously differentiate into embryoid bodies. About 1/4 of cells from each clone are replated in the original 24-well plate in the presence of fresh MEFs, and they are fixed and stained for AP activity 1–2 days later when small clumps of undifferentiated ES cells have formed. Pooled cells are plated in 60-mm gelatin-coated plates in the absence of MEFs in ES cell culture medium without LIF. After 2 days in culture, small ES cell clumps are lifted off the plates by gentle trypsinization and transferred with a wide-mouth pipette into suspension culture in EB culture medium. It is important to incubate in trypsin for the shortest time possible to avoid breaking up of the embryoid bodies. Every 48 hr half of the culture supernatant is replaced with fresh culture medium. This is achieved by tilting the plates to let EBs settle down on one side of the plate. After 4 days, "simple embryoid bodies" form, consisting of an outer layer of endodermal cells, a basal lamina and an inner layer of columnar, ectodermal cells. The simple EBs are kept for another 6 days in suspension culture, where they further differentiate into complex "cystic embryoid bodies" containing visible cavitations, pockets of primitive blood islands and rhythmically contracting cardiomyocytes. In parallel, day-4 simple EBs are replated on gelatinized tissue culture plates, where they reattach within 24 hr. Outgrowth of differentiated cells with various phenotypes, including fibroblasts, endothelial-like cells, neurons, contracting cardiomyocytes, differentiated cell types with an uncharacterized phenotype, as well as pockets of undifferentiated cells, are observed during the next 6 days. Aliquots of EBs at day 4, day 7, and day 10 in suspension culture are harvested, fixed and histochemically stained for AP activity. Day-6 cultures of replated EBs are assayed for AP activity as well. Once positive pools are identified, individual clones from these pools are expanded from the frozen master plates and analyzed for AP activity in undifferentiated ES cells, upon EB formation, and in cultures of replated EBs. Only those ES cell clones should be scored positive that display reproducible AP expression in at least 50% of all EBs present in each culture sample. Positive ES cell clones should be retested for their AP expression pattern, expanded and frozen in aliquots for further analysis. Examples for AP staining in EBs from ES cell clones that were infected with retroviral entrapment vectors are depicted in Fig. 3.

Colocalization of AP Reporter Gene and Developmental Marker Expression

Typically, characterization of entrapment events has been dependent on the generation of transgenic lines where reporter gene expression can be examined *in vivo*, or mutagenicity can be assessed. However, the time-consuming and expensive nature of *in vivo* analysis limits its practicality for large-scale entrapment screens. Here, we describe a method for characterizing the tissue specificity of reporter gene expression *in vitro* in order to preselect candidate ES cell clones for generating transgenic lines. By colocalizing reporter gene expression with RNA *in situ* hybridization signals from a panel of developmental marker genes (listed in Table I) in EBs, information on the cell lineage specificity of entrapment vector insertions can be obtained. In order for this approach to be successful, colocalization has to fulfill three criteria: (i) localization of *in situ* hybridization signals in EBs should be temporally and spatially defined. Leahy *et al.*[31] have shown this to be true. (ii) It should be possible to examine histochemical staining for AP and *in situ* hybridization with riboprobes on the same section of EBs. We found that AP staining prior to *in situ* hybridization did not interfere with the sensitivity of hybridization. However, a strong AP precipitate can sometimes interfere with signal intensity by "masking" the silver grains. To minimize this problem, we recommend either reducing the time for staining, or using BCIP/NBT as a substrate instead of BM Purple (see above). (iii) Colocalization in EBs should accurately reflect colocalization in the embryo. Below, proof of principle is provided for one particular entrapment insertion identified in our screen.

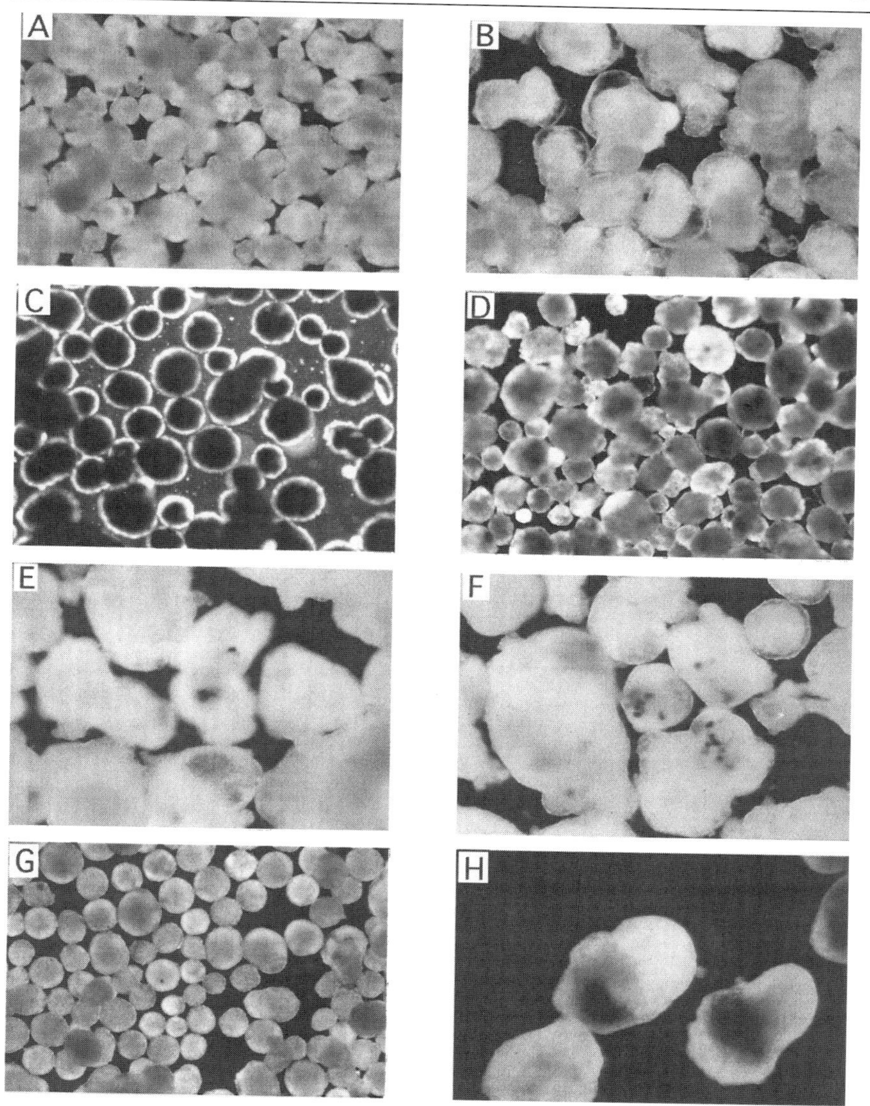

FIG. 3. Differential AP reporter gene expression in EBs. ES cells were infected with entrapment vectors and differentiated *in vitro* to spontaneously form EBs in suspension culture. Aliquots were harvested and stained for AP activity. (A, B) Day-4 and -7 EBs from uninfected ES cells. (C, D) Day-4 and -7 EBs from a clone infected with PT-AP. (E, F) Day-7 and -10 EBs from a clone infected with GT-AP. (G, H) Day-4 and -7 EBs from a clone infected with PT-IRES-AP. (See Color Insert.)

RNA in situ *Hybridization*

Procedures used for embedding, sectioning, and mRNA *in situ* hybridization are essentially as described previously.[31,35] Embryoid bodies are fixed in 4% paraformaldehyde and embedded in paraffin at 60°C overnight following a series of dehydrating washes (from 50 to 100% ethanol, final wash for 30 min in xylene). Embedded material is sectioned at 6 μm, deparaffinized in xylene, rehydrated, and digested with Proteinase K (Sigma) for 7.5 min. Sections are treated with acetic anhydride in 0.01 *M* triethanolamine and dehydrated prior to a 16-hr hybridization with ^{35}S-radiolabeled antisense mRNA probes. [α-^{35}S] UTP methionine (specific activity 1250 Ci/mmol) is obtained from NEN/Dupont (Cambridge, USA). The final probe concentration during hybridization should be 35 cpm/ml in formamide hybridization buffer (50% deionized formamide, 0.3 *M* NaCl, 20 m*M* Tris–HCl pH 7.4, 5 m*M* EDTA, 10 m*M* $NaH_2PO_4 \times H_2O$ pH 8.0, 10% dextran sulfate, 1× Denhardt's, 0.5-mg/ml total yeast RNA, 10 m*M* DTT). Posthybridization washes of increasing stringency (50% formamide, 2× SSC, 0.1× SSC) and 1 hr RNase A treatment at 37°C are included to reduce background. Slides are dehydrated through a series of ethanol/NH_4OH washes and dried for 1 hr. Slides are dipped in Kodak NBT-2 emulsion, dried over-night and exposed for 5 days to 2 weeks at 4°C. Sections are counter-stained with toluidine blue (Sigma), unless they were stained before as whole-mount for AP, dehydrated and coverslipped.

Colocalization: Proof of Principle

For colocalization analysis, ES cell clones are induced to spontaneously differentiate in suspension culture. Embryoid bodies are harvested between days 4 and 10, fixed and histochemically stained as whole-mounts for AP (see above). *In situ* hybridization is performed on serial sections of paraffin-embedded EBs, using the panel of developmental markers listed in Table I. As an example and proof of principle, we will discuss our results with ES clone 1–13 that contains an insertion of the PT-IRES-AP virus (Fig. 4). Clone 1–13 displayed spotty AP reporter gene expression in embryos, varying between day 4 and 10 in culture. Expression of the AP reporter in embryos derived from clone 1–13 was detected in the visceral yolk sac and developing vascular structures.[30] AP reporter expression at various stages of EB differentiation was found to colocalize with signal from the endothelial marker *flk-1*, but not with other developmental markers, including *Brachyury* T (Fig. 4). Furthermore, colocalization of AP reporter activity

[35] D. Sassoon and N. Rosenthal, *Methods Enzymol.* **225**, 384–404 (1993).

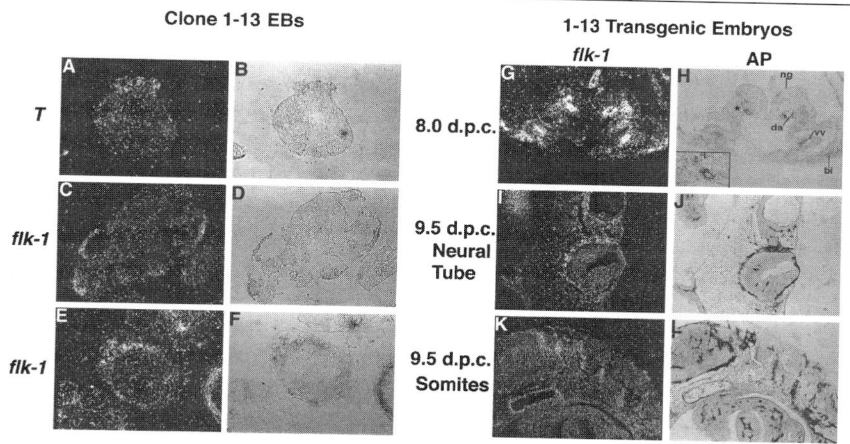

FIG. 4. Colocalization of *flk-1* and AP reporter gene expression in EBs and transgenic mice derived from ES cell clone 1–13. *In situ* hybridization on sectioned, AP stained clone 1–13 EBs (A–F) and clone 1–13-derived transgenic embryos (G–L). (A, B) *Brachyury* T riboprobe did not colocalize with AP reporter gene expression. Riboprobes corresponding to *flk-1* colocalized with AP reporter gene expression in (C, D) day-4 EBs; (E, F) day-9 EBs; (G, H) 8.0 dpc embryo (transverse section), insert showing higher magnification of the dorsal aorta (dorsal aorta, da; neural groove, ng; vitteline vein, vv; blood island, bi); (I, J) neural tube of 9.5 dpc embryo; (K, L) somites of 9.5 dpc embryo. Dark field images of A, C, E, G, I, and K correspond to bright field images of B, D, F, H, J, and L, respectively. (See Color Insert.)

and *flk-1 in situ* hybridization signal was also observed in sections of 8.0 and 9.5 dpc embryos derived from clone 1–13 (Fig. 4). Thus, reporter gene expression during *in vitro* differentiation accurately reflected expression during embryogenesis, and its colocalization with *flk-1* was predictive of expression *in vivo* in the endothelial lineage.

Conclusions

The genetic screen described in this chapter combines an approach that is unbiased for gene expression in ES cells with an *in vitro* screen in EBs to identify candidate genes with restricted expression during early postimplantation development as well as in particular cell lineages. We have used this approach successfully to isolate previously unidentified genes with lineage restricted expression (Refs. 30, 36, 37; our unpublished results). This general approach is versatile and can be adapted easily for different purposes. For example, different stages of embryonic development could be targeted by using a different time frame during *in vitro* differentiation for the

[36] J. W. Xiong, A. Leahy, H. H. Lee, and H. Stuhlmann *Dev. Biol.* **206**, 123–141 (1999).
[37] J. W. Xiong, A. Leahy, and H. Stuhlmann, *Mech. Dev.* **86**, 183–191 (1999).

screen. Second, conditions that induce differentiation of ES cells into particular cell lineages could be applied in order to enrich for insertions into genes that are specifically expressed in this lineage (e.g., Refs. 38, 39). Finally, it might be possible to modify the approach to allow for functional screening. For example, one might select for ablation of a particular cell lineage or cell type if a Herpes virus simplex thymidine kinase (TK) gene or a Diphtheria toxin gene is used as a reporter gene.

Acknowledgments

This work was supported by NIH grants HD31534 and HL65738, and AHA Grant-in-Aid #9950585N to H. S.

[38] L. M. Forrester, A. Nagy, M. Sam, A. Watt, L. Stevenson, A. Bernstein, A. L. Joyner, and W. Wurst, *Proc. Natl. Acad. Sci. USA* **93**, 1677–1682 (1996).

[39] W. L. Stanford, G. Caruana, K. A. Vallis, M. Inamdar, M. Hidaka, V. L. Bautch, and A. Bernstein, *Blood* **92**, 4622–4631 (1998).

[28] Gene-Based Chemical Mutagenesis in Mouse Embryonic Stem Cells

By YIJING CHEN, JAY L. VIVIAN, and TERRY MAGNUSON

Introduction

Assigning specific functions to all genes identified through genome sequencing efforts has taken on an unprecedented urgency in the postgenome era. Analyses of mutants identified in systematic mutagenesis screens provide genetic evidence crucial for deciphering the normal *in vivo* functions of the mutated genes. Several genome-wide and regional phenotype-based whole animal screens have been initiated in the U.S., Europe and Australia to achieve the goal of generating a comprehensive collection of mutants for all mouse genes. The fact that usually only one mutation is identified per gene and, perhaps more importantly, the fact that once a phenotype has been identified, the arduous task of genetic mapping, positional cloning and gene validation must be undertaken before the gene underlying a particular mutant trait is identified, have motivated the development of alternative gene-based mutagenesis strategies utilizing totipotent mouse embryonic stem (ES) cells. ES cells can be grown in large numbers, are amenable to a variety of genome modifications and can contribute to the germ-line to allow for functional assessment of the genetic perturbation in the whole organism. The ability to create, characterize and maintain mutations in cell culture presents an attractive alternative to expand the current mouse mutant collection.

Gene targeting and gene-trap-based insertional mutagenesis are two widely adopted techniques for generating mouse mutants using ES cells. In both cases, severe loss-of-function mutations often arise as a consequence of gross disruption of the genomic locus. In contrary, alkylating agents, such as *N*-ethyl-*N*-nitrosourea (ENU), mainly cause single base substitutions. ENU has been shown to be the most potent mouse germ-line mutagen capable of generating allelic series of mutations including partial and complete loss-of-function as well as gain-of-function mutations critical for underscoring a full range of gene functions. We have taken advantage of the mutagenic properties of ENU and the experimental malleability of ES cells to develop an ES cell-based chemical mutagenesis method that would allow for efficient generation of a large allelic series of mutations in any genes of interest. In brief, ES cells are treated with ENU and individual mutagenized clones are isolated and cultured in duplicate. One set of samples is cryopreserved as an archived library. Nucleic acids are isolated from the other set for screening of ENU-induced mutations in genes of interest. Clones harboring desired mutations are recovered from the library and injected into blastocysts to achieve germ-line transmission, thereby enabling the assessment of the functional significance of the ENU-induced mutations *in vivo*.

The feasibility of such a gene-based mutagenesis approach was first demonstrated using selectable genes in ES cells.[1] We have since generated a cryopreserved library composed of 2060 clonal ENU-mutagenized ES cell lines. A gene-based screen of this mutant ES cell library identified a total of 29 mutations in *Smad2* and *Smad4*, two nonselectable genes of the transforming growth factor beta (TGFβ) superfamily of signaling molecules.[2] Five of the 13 karyotypically normal ES cell lines carrying nonsilent mutations have been passed through the germ-line. Whole animal phenotypic analyses of some of these alleles provided evidence for novel developmental processes mediated by these components of TGFβ signaling, confirming the utility of nonnull alleles created by chemical mutagens. The accurately assessed mutation load of the ES cell library (Table I) indicates that it is a valuable resource for developing mouse lines for genetic and functional studies. In this chapter, we discuss the technical details of a

[1] Y. Chen, D. Yee, K. Dains, A. Chatterjee, J. Cavalcoli, E. Schneider, J. Om, R. P. Woychik, and T. Magnuson, *Nat. Genet.* **24**, 314 (2000).

[2] J. L. Vivian, Y. Chen, D. Yee, E. Schneider, and T. Magnuson, *Proc. Natl. Acad. Sci. USA*, **99**, 15542 (2002).

TABLE I
MUTATION FREQUENCIES OBSERVED IN GENE-BASED SCREENS

Organism	Mutagen	bp Screened (Mb)	Sequence variations (Kb)	References
Mouse ES cells	ENU	13.46	1/464	2
Mouse	ENU	0.37	1/62	4
Mouse	ENU	9.5	1/1900	5
Zebrafish	ENU	7.3	1/482	6
Drosophila	EMS	3.34	1/209	7
Arabidopsis	EMS	~2	1/153	8

bp, base pairs.

gene-based screen in chemically mutagenized ES cells for generating allelic series of mutations in genes of interest.

Methods

Preparation of ENU

ENU (Sigma Isopac) solution is prepared similarly to what is described for whole animal mutagenesis.[3] Briefly, on the same day of the mutagenesis, the ENU crystal in the Isopac container is dissolved in an equal volume mixture of 95% ethanol and the PCS buffer (50 m*M* sodium citrate, 100 m*M* sodium phosphate dibasic, pH 5.0) in a well-ventilated hood. The concentration of the ENU solution is determined by spectrophotometry based on the observation that a 1 mg/ml solution gives an OD_{398} reading of 0.72. ENU solutions and contaminated labware are soaked in 1 N NaOH solution overnight to inactivate any residual ENU activity. The ENU solution is diluted to the desired concentration in the ES cell growth medium (see below) and filter-sterilized before adding to cells. The excess ENU is inactivated and discarded appropriately. In a series of pilot experiments, a reasonable balance between mutation frequency and germ-line competency was observed in ES cells treated with 0.2 mg/ml of ENU (Y. Chen, unpublished results).

[3] M. J. Justice, D. A. Carpenter, J. Favor, A. Neuhauser-Klaus, M. Hrabe de Angelis, D. Soewarto, A. Moser, S. Cordes, D. Miller, V. Chapman, J. S. Weber, E. M. Rinchik, P. R. Hunsicker, W. L. Russell, and V. C. Bode, *Mamm. Genome* **11**, 484 (2000).
[4] D. R. Beier, *Mamm. Genome* **11**, 594 (2000).
[5] E. L. Coghill, A. Hugill, N. Parkinson, C. Davison, P. Glenister, S. Clements, J. Hunter, R. D. Cox, and S. D. Brown, *Nat. Genet.* **30**, 255 (2002).
[6] E. Wienholds, S. Schulte-Merker, B. Walderich, and R. H. Plasterk, *Science* **297**, 99 (2002).
[7] A. Bentley, B. MacLennan, J. Calvo, and C. R. Dearolf, *Genetics* **156**, 1169 (2000).
[8] C. M. McCallum, L. Comai, E. A. Greene, and S. Henikoff, *Nat. Biotechnol.* **18**, 455 (2000).

Mutagenizing ES Cells

Several ES cell lines have been used in chemical mutagenesis.[1,9] CT129/Sv ES cells[10] are used to generate the archived mutant library. Culturing conditions described in this section are optimized for this cell line. For routine maintenance, the cells are cultured on a layer of primary mouse embryonic fibroblast feeders at 37°C in a humidified incubator with 5% CO_2. The culture medium consists of MEM-α, 15% fetal calf serum, 0.1 mM β-mercaptoethanol and 1000 units/ml of leukemia inhibitory factor. Upon reaching confluency, passaging is done by treating cells with Trypsin–EDTA solution (Invitrogen, 0.05% trypsin, 0.53 mM EDTA) for 3–4 min.

For mutagenesis, the ES cells are passaged onto feeder-free, gelatin-coated plates at least once. The day before ENU treatment, the ES cells are plated at a density of 2×10^6 per gelatin-coated 100 mm plate. After overnight growth, the regular growth medium is replaced with growth medium supplemented with appropriate concentration of ENU. The treatment typically lasts for 2 hr at 37°C. At the end of the treatment, the cells are rinsed three times with MEM-α to remove residual ENU. The mutagenized cells are dissociated with trypsin solution to a single-cell suspension and plated onto gelatin-coated plates at an appropriate density to allow for the formation of around 200 surviving colonies per plate. The overall clone viability post ENU treatment is calculated by dividing the number of colonies formed by the number of mutagenized cells plated. The viability is then adjusted with plating efficiency. To estimate the plating efficiency in a given experiment, we plate 1×10^3 nonmutagenized cells and calculate the percentage of surviving cells that form colonies. The typical plating efficiency for CT129/Sv cells under normal growth conditions is around 20–30%. For an ENU dose of 0.2 mg/ml, the survival rate adjusted for plating efficiency is typically around 10%.

Picking and Cryopreservation of Mutagenized Clones

The growth medium is changed everyday after the ENU treatment. Cell proliferation is often retarded and extensive cell death is evident 2–3 days post ENU treatment. The colonies formed from mutagenized cells are hand-picked into 96-well plates when the colonies reach a size of 0.7–1 mm in diameter after 7–8 days. A significant number of colonies exhibit signs of

[9] R. J. Munroe, R. A. Bergstrom, Q. Y. Zheng, R. Smith, S. W. John, K. J. Schimenti, B. J. Libby, V. L. Browning, and J. C. Schimenti, *Nat. Genet.* **24**, 318 (2000).

[10] D. W. Threadgill, D. Yee, A. Matin, J. H. Nadeau, and T. Magnuson, *Mamm. Genome* **8**, 390 (1997).

differentiation, presumably due to the mutagenic and cytotoxic effects of ENU. Colonies with suboptimal morphology are avoided during the hand-picking process.

The picked colonies are grown in regular growth medium for about three more days after which they are dissociated with trypsin solution and expanded through replica plating into 96-well plates. Once the replica plated samples reach confluency, typically in 2–3 days, one set is cryopreserved in MEM-α supplemented with 10% fetal calf serum and 10% DMSO. The cells are archived in the 1.2 ml Micro Titertubes (Out Patient Services) arrayed in a 96-well format to allow for individual thawing and easy localization of the clones. All clones are stored in the vapor phase of a liquid nitrogen tank.

The mutagenic efficiency of any given ENU treatment of ES cells can be monitored through the induced loss-of-function mutation frequency in selectable genes. For example, when male ES cells are mutagenized, one set of replicate plated sample can be subjected to 6-thioguanine selection to reveal loss-of-function mutations in the X-linked hypoxanthine phosphoribosyl transferase (*Hprt*) gene.

Gene-based Mutation Screening by Heteroduplex Analysis

Mutation scanning of ENU-induced sequence variations is performed through the analysis of nucleic acids isolated from one set of replica plated samples. Total RNA is isolated using the Rneasy 96 kit (Qiagen) according to the manufacturer's protocol. The DNase digest step is omitted to allow for copurification of the genomic DNA. From one confluent 96-well of ES cells, approximately 2 μg of total RNA can be recovered. About 0.2 μg of total RNA is used as a template for reverse transcription (RT) reactions in a volume of 25 μl to generate cDNAs for PCR amplification of genes expressed in ES cells. The same RT reaction mix can also be used directly as a genomic DNA template for amplification of genes not expressed in ES cells.

ENU-induced mutations can be detected using a variety of techniques, such as Single Strand Conformation Polymorphism (SSCP) analysis and direct sequencing of PCR products. We employ the Denaturing High Performance Liquid Chromatography (DHPLC)-based heteroduplex analysis technology due to its sensitivity and relative cost-effectiveness. Mutation scanning by DHPLC involves subjecting PCR products to ion-pair reverse-phase liquid chromatography in a column containing alkylated nonporous particles. Under conditions of partial heat denaturing within a linear acetonitrile gradient, heteroduplexes that form in PCR samples having internal sequence variations display reduced column retention time relative to their homoduplex counterparts.[11] The DHPLC analysis of the PCR products amplified from our mutagenized ES cell clones is performed on the automated WAVE fragment analysis system (Transgenomic, Inc.) using UV absorbance for detection. PCR products are denatured and annealed after PCR in the thermocycler prior to DHPLC analysis. As the sensitivity of heteroduplex detection is heavily influenced by the melting behavior of PCR fragments, primers spanning regions of even melting temperature and in the range of 250–600 bp are preferred. Computational algorithms are available for predicting the optimum temperature for analysis (http://insertion.stanford.edu/melt.html). For inexperienced users, the DHPLC conditions for each amplicon can be verified by analysis of point mutations created by *in vitro* mutagenesis. Some examples of heteroduplex elution profiles are shown in Fig. 1. ES cell clones harboring candidate mutations are thawed and expanded, and total RNAs are isolated for RT-PCR. Presence of the mutation is confirmed by heteroduplex analysis. Direct sequencing of the PCR product identifies the base change of each mutation.

The determination of an optimum partial denaturing temperature for a given PCR product is key to the DHPLC-based mutation detection technology. We have combined the computational predictions (WAVEMAKER software, Transgenomic, Inc.) and empirical examination of the elution profile over a temperature gradient[11] to determine the optimum partial denaturing temperature at which the analysis is performed. Although multiplexing with DHPLC is possible with fluorescently labeled PCR products, we typically analyze one sample at a time using conventional unlabeled primers. Under optimal conditions, a WAVE DHPLC system with a Dual-Plate configuration can analyze 192 samples in 24 hr. Recently, a new mutation detection technology called temperature gradient capillary electrophoresis (TGCE) has been developed[12] (SpectruMedix). TGCE detects the presence of heteroduplexes in PCR products by electrophoresis through a capillary array that is exposed to a thermal gradient covering all possible melting temperatures for different regions of a given DNA fragment. The ability to analyze samples over a temperature gradient maximizes the mutation detection sensitivity. TGCE also allows simultaneous analysis of samples, thereby offering a significant improvement in the throughput of sample analysis. With a configuration of 96 capillary arrays, 96 samples can be analyzed in as little as 1 hr. TGCE can potentially become a higher throughput mutation detection alternative for large-scale gene-based screens.

[11] W. Xiao and P. J. Oefner, *Hum. Mutat.* **17**, 439 (2001).

[12] Q. Li, Z. Liu, H. Monroe, and C. T. Culiat, *Electrophoresis* **23**, 1499 (2002).

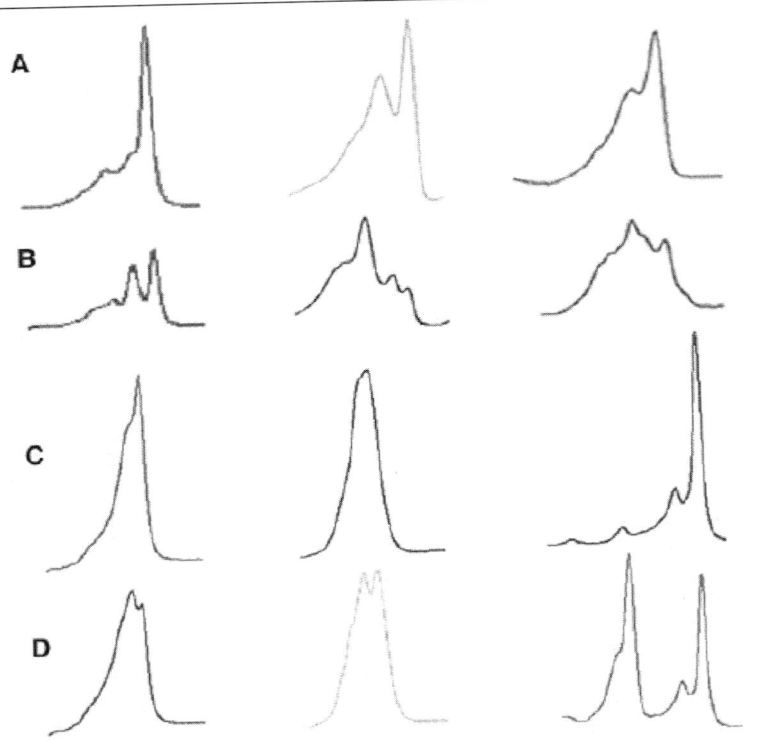

FIG. 1. Representative chromatograms of DHPLC heteroduplex analysis. Rows A and C are typical chromatograms of wild-type PCR products. Rows B and D are chromatograms of the corresponding heteroduplexes-containing PCR products. Heteroduplexes are detected as additional peaks in the elution profile. All samples shown contain single nucleotide substitutions except for the last sample in Row D, where the mutant product harbors a 138-bp deletion that resulted in the dramatic leftward shift in the elution profile.

Generation of Mice Carrying ENU-induced Mutations

ES cell clones carrying confirmed point mutations are thawed from their multi-well array and expanded on primary fibroblast feeders. Since ENU-induced lesions only affect one strand of the DNA double helix, subsequent DNA replication and mitosis result in mosaic colonies composed of both wild-type and heterozygous cells (Fig. 2). A subcloning procedure is performed to isolate a pure population of cells heterozygous for the ENU-induced mutation. For subcloning, trypsin-dissociated single-cells from each ES cell line are picked by mouth pipette using a microscope and seeded sparsely onto feeders. After 7–8 days, at least 24 single-cell colonies on these

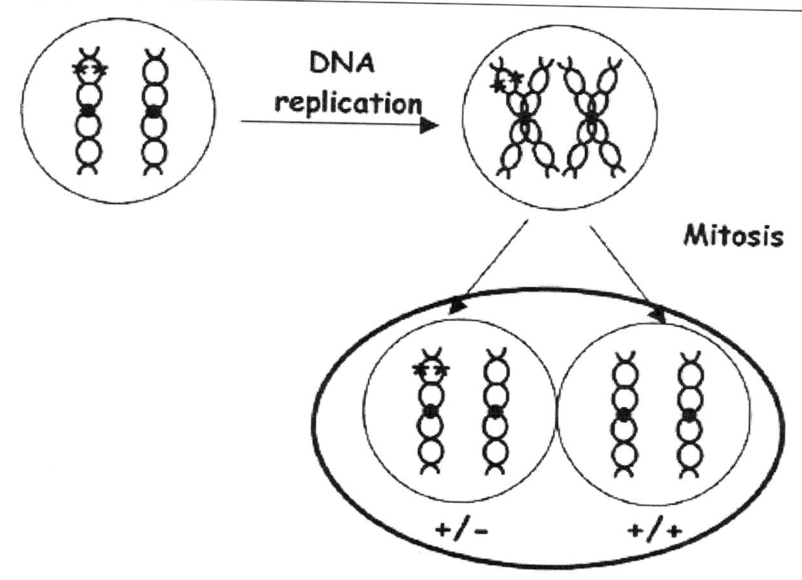

Fig. 2. Mosaic ES cell colonies form following ENU treatment. Circles represent ES cells. The helical structure within each circle symbolizes a double-stranded DNA molecule. Asterisks represent ENU-induced sequence variations. The large oval represents a mosaic colony containing both heterozygous mutant cells (+/−) and wild-type (+/+) cells with respect to a specific ENU-induced sequence variation.

plates are picked into a multi-well plate and subsequently expanded for cryopreservation and nucleic acid isolation. PCR products are analyzed by DHPLC. Up to three single-cell clones carrying the heterozygous mutations are then pooled and prepared for injection into blastocysts. In cases where all subclones of a given line are identified as carrying the mutation the original parental line is processed for injection. In rare occasions, clones containing the mutation are lost following the expansion and subcloning process.

Blastocyst injections for production of chimeric mice are performed following standard protocols. Donor blastocysts are of C57BL/6 origin. Resulting chimeras are mated to wild-type C57BL/6 or Black Swiss mice for testing germ-line transmission. Germ-line transmission is assessed by coat color and by genotyping with genomic DNA from tail biopsy. Genotyping is performed by heteroduplex analysis of PCR products amplifying the genomic region spanning the point mutation. Mouse stocks are maintained in the heterozygous state by crossing heterozygous mice to wild-type mice of the desired background for multiple generations.

Various strategies can be employed to characterize the phenotype(s) of ENU-induced alleles. If loss-of-function alleles via gene targeting or other means are available for the gene under investigation, they can be crossed to the ENU-induced alleles immediately after germ-line transmission to assess the strength of the ENU-induced alleles. Otherwise, complementation tests can be performed among multiple ENU-induced alleles to categorize the alleles based on the range of phenotypes observed. To study an ENU-induced mutation in its homozygous state, it is desirable to cross the mutation onto a wild-type background for several generations before intercrossing to avoid potential complications resulting from extragenic mutations. For example, a four-generation outcross to a wild-type strain will result in the replacement of 93.75% of mutagenized genome, essentially eliminating most of the extragenic mutations that are not closely linked to the mutation of interest. Based on the results of the *Smad2* and *Smad4* screen, our ES library on average contains one ENU-induced sequence variation per 464 Kb of DNA (Table I). Given this mutation frequency, the average distribution of nonsilent coding sequence variations in this ES cell library is approximately 1 per 13 centimorgans, assuming a 1600-centimorgan, 2.7 Gb haploid mouse genome of which 3% is coding. Therefore, the likelihood of obtaining closely linked coding sequence variations is small. Gene-based mutagenesis in whole animals has been carried out in several organisms (Table I). It is worth noting that a similar mutation frequency of 1 sequence variation per 482 Kb was observed in a gene-based screen in zebrafish exposed to a dose of ENU typically used in phenotype-driven screens.[6]

Concluding Remarks

Gene-based screens using chemically mutagenized ES cells allow investigators to proceed rapidly from gene sequence to gene function in a relatively cost-effective manner. Attempts have been made to conduct similar gene-based screens using genomic DNA of ENU-mutagenized mice,[4,5] however, highly variable mutation frequencies were reported, presumably due to variations in mutagen dose, sample size and effectiveness of mutation detection method. As a result, the relative efficiency of such gene-based approaches in mice remains to be determined.

It is worth pointing out that chemical mutagenesis in ES cells has broad applications beyond the realm of gene-based screens. For example, chemically mutagenized ES cells have been used in phenotype-based screens,[9] where a broad spectrum of intriguing whole animal phenotypes is recovered. In addition, ES cell-based mutagenesis opens the possibility of conducting phenotype screens directly in ES cells, bypassing the mouse until after a desirable phenotype is uncovered.[13] The versatility of ES cell-based mutagenesis makes it a powerful addition to the existing tools that are contributing to the functional annotation of the mammalian genome.

[13] Y. Chen, J. Schimenti, and T. Magnuson, *Mamm. Genome* **11**, 598 (2000).

Section IV

Differentation of Monkey and Human Embryonic Stem Cells

[29] Growth and Differentiation of Cynomolgus Monkey ES Cells

By HIROFUMI SUEMORI and NORIO NAKATSUJI

Introduction

Human embryonic stem (ES) cell lines provide great potential and expectation for cell therapy and regenerative medicine, because many types of human cells can be produced by the unlimited proliferation and differentiation of stem cells in culture. Therefore, it is important to obtain reliable methods to maintain proliferation of stem cells, and to devise various methods of inducing differentiation into useful cell populations.

In 1995, nonhuman primate ES cell lines were established from blastocysts of the rhesus monkey and marmoset.[1–3] There were several differences between mouse and monkey ES cells. First, addition of LIF in the culture medium produced no effects in the maintenance of primate stem cells. Second, the shape of monkey ES cell colonies was flatter than the compacted and domed mouse ES cell colonies. Third, expression of a few stem cell markers were different. Although alkaline phosphatase activity was detected, SSEA-1 antigen, which is expressed strongly in mouse ES cells, was not expressed in the primate lines. Instead, SSEA-3 and SSEA-4 antigens were expressed in monkey ES cells.

In 1998, human ES cell lines were established from blastocysts, which had been produced but not used in the clinical treatment of infertility.[4] Human ES cell lines showed very similar characteristics to monkey ES cells. Since then, other groups have reported establishment of human ES cell lines.[5]

[1] J. A. Thomson, J. Kalishman, T. G. Golos, M. Durning, C. P. Harris, R. A. Becker, and J. P. Hearn, Isolation of a primate embryonic stem cell line, *Proc. Natl. Acad. Sci. USA* **92**, 7844–7848 (1995).

[2] J. A. Thomson, J. Kalishmanm, T. G. Golos, M. Durning, C. P. Harris, and J. P. Hearn, *Biol. Reprod.* **55**, 254 (1996).

[3] J. A. Thomson and V. S. Marshall, *Curr. Top. Dev. Biol.* **38**, 133 (1998).

[4] J. A. Thomson, J. Itskovitz-Eldor, S. S. Shapiro, M. A. Waknitz, J. J. Swiergiel, V. S. Marshall, and J. M. Jones, *Science* **282**, 1145 (1998).

[5] B. E. Reubinoff, M. F. Pera, C. Y. Fong, A. Trounson, and A. Bongso, *Nat. Biotech.* **18**, 399 (2000).

0076-6879/2003 $35.00

There have been reports of the differentiation of several useful cell types from these human ES cell lines.[6–8]

Significance of Monkey ES Cell Lines

Before clinical application of cell therapy using human ES cells, the effectiveness and safety of cell transplantation needs to be tested, using animal models. Nonhuman primate models are necessary in addition to rodent models for preclinical assessment, because they are closer to humans in phylogeny, body sizes and physiology. For example, the possibility of tumorgenesis caused by transplanted cells should be evaluated using not only the rodent models but also monkey models which would have a much higher physiological significance to clinical human treatments. Another important aspect to evaluate is the immuno-rejection after transplantation of allogenic ES-derived cells. Such immunological responses are more adequately predicted by using nonhuman primate models rather than other animal models.

Also, sizes and structures of various organs and tissues are very important for actual cell transplantation. For example, the brain is one of the prime targets for cell therapy, and the brain structure of monkeys is most similar to humans. Also, assessment of cell therapy of other anticipated targets such as the liver and eyes requires the proper sizes of these organs to simulate transplantation methods, number of engrafted cells, and physiological effects caused by the transplanted cells.

The rhesus and cynomolgus monkeys are macaques belonging to the old world monkeys, which are closely related to humans. They are bred as experimental animals and widely used for medical research. Also, various disease models are available for research purposes, such as a Parkinson's disease model in macaques. For these reasons, ES cell lines of nonhuman primates are valuable and indispensable tools for preclinical research of cell therapy.

Establishment of Cynomolgus ES Cell Lines

We established ES cell lines (Fig. 1A, B) from cynomolgus monkey (*Macaca fascicularis*) blastocysts produced by *in vitro* fertilization (IVF) or

[6] S. Assady, G. Maor, M. Amit, J. Itskovitz-Eldor, K. L. Skorecki, and M. Tzukerman, *Diabetes* **50**, 1691 (2001).

[7] D. S. Kaufman, E. T. Hanson, R. L. Lewis, R. Auerbach, and J. A. Thomson, *Proc. Natl. Acad. Sci. USA* **98**, 10716 (2001).

[8] C. Xu, S. Police, N. Rao, and M. K. Carpenter, *Circ. Res.* **91**, 501.

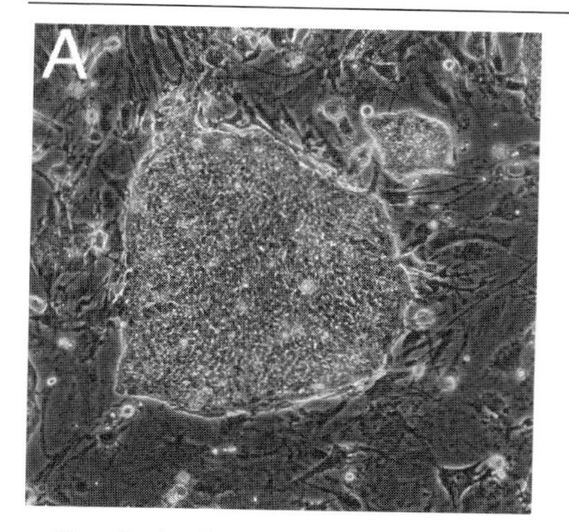

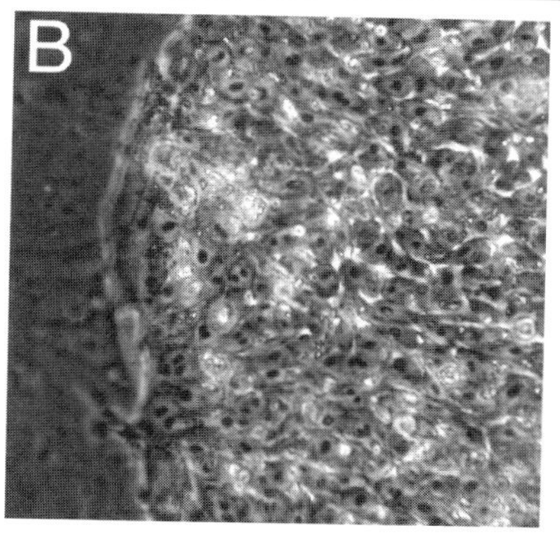

FIG. 1. A phase contrast micrograph of an undifferentiated cynomolgus monkey ES cell colony (A), and a higher magnification view (B). Note the high nucleus/cytoplasm ratio and prominent nucleoli. (See Color Insert.)

intracytoplasmic sperm injection (ICSI).[9] We examined the expression of several stem cell markers using cynomolgus ES cells. They expressed alkaline phosphatase activity and the SSEA-4 antigen, but not the SSEA-3 or SSEA-1 antigens. While human and rhesus ES cells were reported to express the SSEA-3 antigen at variable levels among stem cell colonies, cynomolgus ES cells were negative for SSEA-3 in immunostaining.

Cynomolgus ES cells showed extensive spontaneous differentiation when cultured using the standard medium for mouse ES cells with feeder cells and LIF. However, after improvement of the culture methods as described below, these monkey ES cell lines were successfully maintained in an undifferentiated state and with a normal karyotype for prolonged periods of more than 12 months. We obtained cell lines with either the male or female karyotypes.[9]

Improved Methods for Maintenance of Monkey ES Cell Lines

Spontaneous differentiation of stem cells and low efficiency in subculturing have been the major problems hindering the stable maintenance of monkey and human ES cell lines. In such conditions, it was necessary to collect individual colonies for enrichment of undifferentiated stem cells at regular intervals during maintenance of ES cell lines. Also, a relatively unreliable method of mechanical disruption of ES cell colonies

[9] H. Suemori, T. Tada, R. Torii, Y. Hosoi, K. Kobayashi, H. Imahie, Y. Kondo, A. Iritani, and N. Nakatsuji, *Dev. Dyn.* **222**, 273 (2001).

was used to avoid cell damage, instead of a proteinase treatment that would enable uniform dissociation of cell colonies. Recently however, we have succeeded in obtaining significant improvement on these aspects in propagation on cynomolgus ES cell lines.

Culture Medium

Spontaneous appearance of differentiated cells during monkey ES cell culture was reduced remarkably when fetal bovine serum (FBS) was replaced with Knockout Serum Replacement (KSR, Invitrogen) in the culture medium. Although we used FBS samples that had been extensively tested for maintenance of mouse ES cells, it is possible that such FBS still contained various differentiation-inducing factors, which were not present in the KSR. In such serum-free medium, cynomolgus ES cells were maintained in an undifferentiated state for a longer period without periodical collection of the stem cell colonies, which had been necessary when using the FBS medium. In our culture conditions, splitting of the ES cell culture into 3–4 duplicate dishes was possible every 3–4 days.

Subculturing

Similar to other primate ES cell lines, cynomolgus ES cells exhibited a very low plating efficiency when dissociated into single cells with trypsin solution. During subculturing, limited dissociation into cell clusters of 50–100 stem cells were required to enable continued growth. Thus, the standard dissociation procedure for mouse ES cells using trypsin caused excessive damage to the monkey ES cells. Without trypsin however, we could not obtain proper reproducible dissociation. After testing various conditions, we devised an adequate method for efficient subculturing by using 0.25% trypsin supplemented with 1 m*M* $CaCl_2$ and 20% KSR. Presence of Ca^{2+} ions at this concentration slowed down the cell dissociation by trypsin. Also, it may have a protective effect on the cell membrane. Thus, this method enabled well-controlled and reproducible dissociation of the ES cell colonies throughout the whole culture dish for routine efficient subculturing.

Protocol

Preparation of Feeder Cell Layer

Primary culture of mouse embryonic fibroblasts is used as the feeder cell layer. Quality of the feeder layer is very important to maintain the undifferentiated state of ES cells. Early passage (up to 5 passages) of cells should be used to prepare feeder cells. Mitotically inactivated cells are seeded to gelatinized dishes at the density of $1.5–2\times10^4/cm^2$. Feeder cells should be used within 4 days of preparation.

Preparation of Reagents

Culture medium:

DMEM/Ham F-12 1 : 1	400 ml
Non-Essential Amino Acids solution	5 ml
100 m*M* Na Pyruvate	5 ml
200 m*M* L-Glutamine	5 ml
β-Mercaptoethanol	4 μl
Knockout Serum Replacement	100 ml

After mixing, the culture medium can be stored at 4°C for 2 weeks.

Dissociation solution:

2.5% Trypsin Solution	10 ml
Knockout Serum Replacement	20 ml
100 m*M* $CaCl_2$	1 ml
PBS (Ca^{2+} and Mg^{2+} free)	69 ml

Aliquots should be stored at –20°C. They can be thawed once and refreezing should be avoided.

Subculture of ES Cells

ES cells should be subcultured when the cells cover about 50–60% of the surface of feeder layer (Fig. 2A). ES cells should also be subcultured before colonies become multilayered (Fig. 2B).

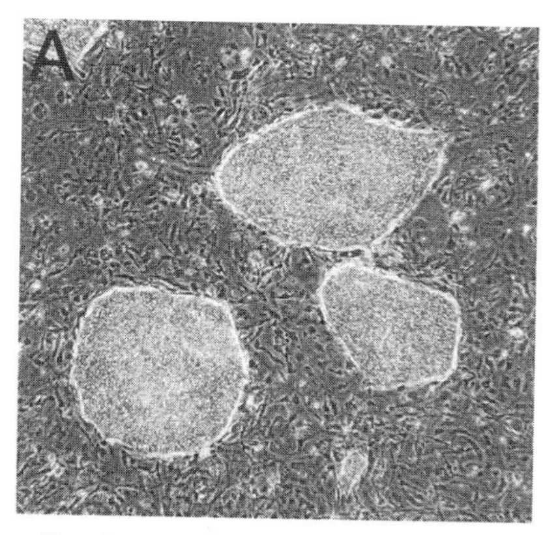

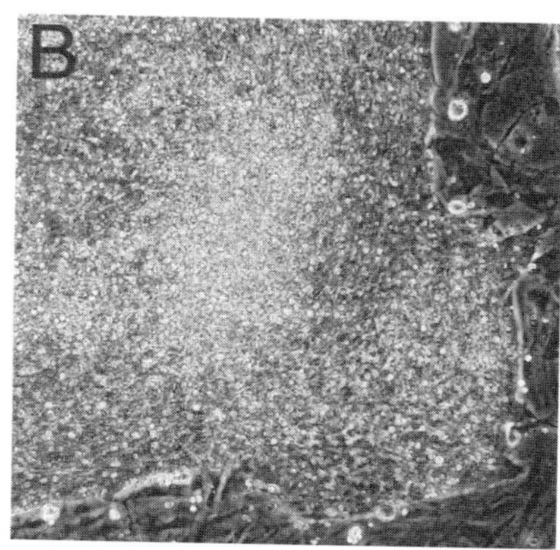

FIG. 2. A low magnification view of a confluent culture of cynomolgus monkey ES cells (A), and a piled-up colony (B). (See Color Insert.)

1. Aspirate medium, and rinse the cells with PBS.
2. Cover the cells with the dissociation solution.
3. Aspirate the dissociation solution.
4. Incubate for 5–6 min in 37°C incubator. Examine cells under the microscope, and confirm that feeder cells show rounded shape and that most colonies of ES cells are partially detached from the dish (Fig. 3A, B).
5. Add culture medium, and dissociate ES cell colonies into clusters consisting of 50–100 cells by gentle pipetting (Fig. 4). Excess dissociation will damage cells and result in loss of ES cells. Feeder cells secrete a viscous matrix resistant to trypsin digestion in ES cell culture medium, and sheets of feeder cells remain after pipetting. Do not try to dissociate these sheets into small clumps, which will cause over dissociation of ES cell clusters.
6. Transfer the cell suspension to a 15-ml centrifuge tube.
7. Centrifuge the cells at *ca.* 170*g* for 5 min.
8. Aspirate the supernatant, and resuspend the cell pellet in the culture medium.
9. Dispense the cell suspension to culture dishes with fresh feeder layer.
10. Refeed the medium daily (Fig. 5). Cell population will triple in 3–4 days, and will reach the appropriate density for subculturing.

Differentiation of Cynomolgus Monkey ES Cells

Differentiation of ES Cell In Vitro

When cynomolgus ES cells were allowed to grow to higher densities to induce differentiation, several kinds of differentiated cells were observed.

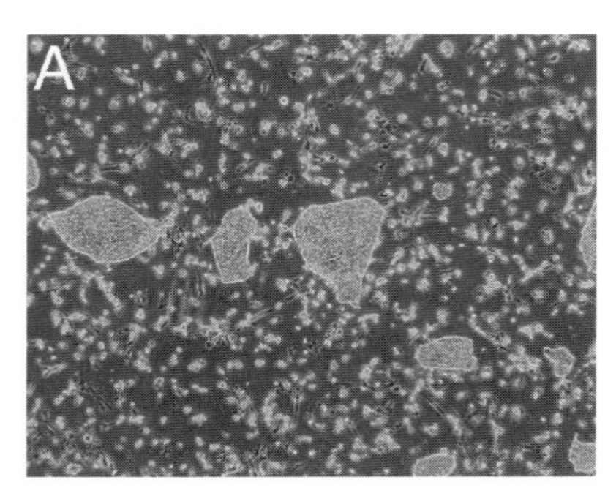

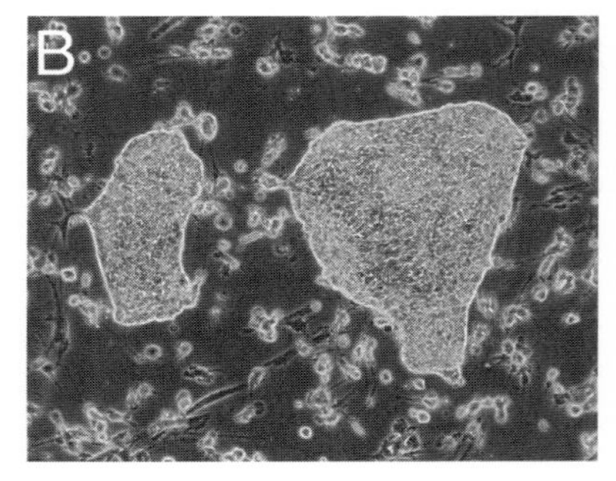

FIG. 3. Appearance of ES cell colonies after trypsin treatment (A), and colonies detaching from the culture dish (B). (See Color Insert.)

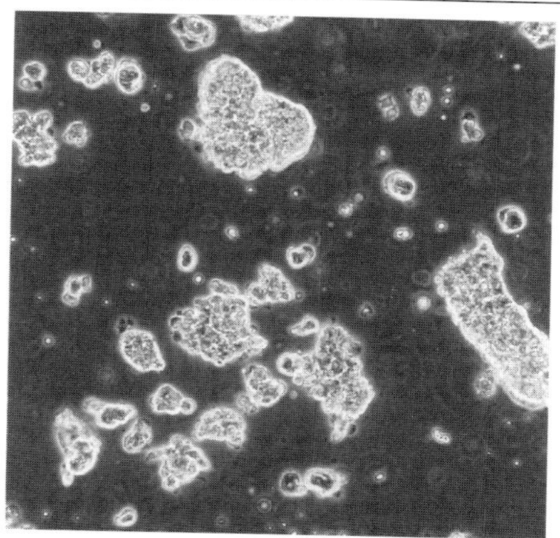

FIG. 4. Appearance of ES cells after dissociation by pipetting. Cells must be kept as clusters of 50–100 cells, or minimum of 15–20 cells, for efficient subculture. (See Color Insert.)

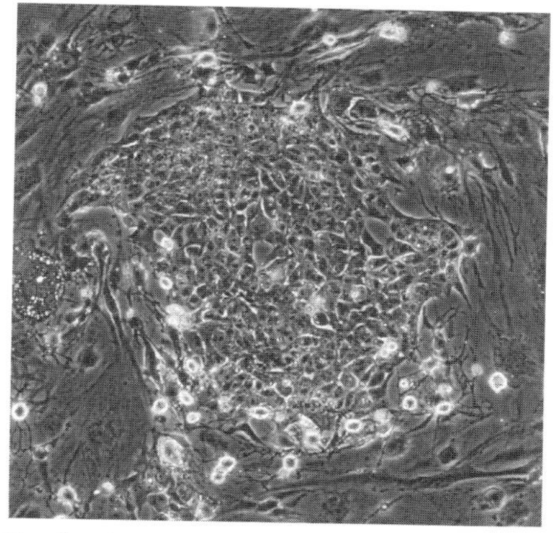

FIG. 5. An ES cell colony one day after subculturing. (See Color Insert.)

These included vesicular epithelia resembling the visceral endoderm or yolk sac, mesenchymal cells that showed outgrowth from cell clumps, and clusters of neurons and pigment cells (Fig. 6). The differentiation potency of cynomolgus ES cells has been further confirmed by production of dopaminergic neurons and retinal pigmented epithelia by using culture conditions to induce such differentiation.[10]

[10] H. Kawasaki, H. Suemori, K. Mizuseki, K. Watanabe, F. Urano, H. Ichinose, M. Haruta, M. Takahashi, K. Yoshikawa, S.-I. Nishikawa, N. Nakatsuji, and Y. Sasai, *Proc. Natl. Acad. Sci. USA* **99**, 1580 (2002).

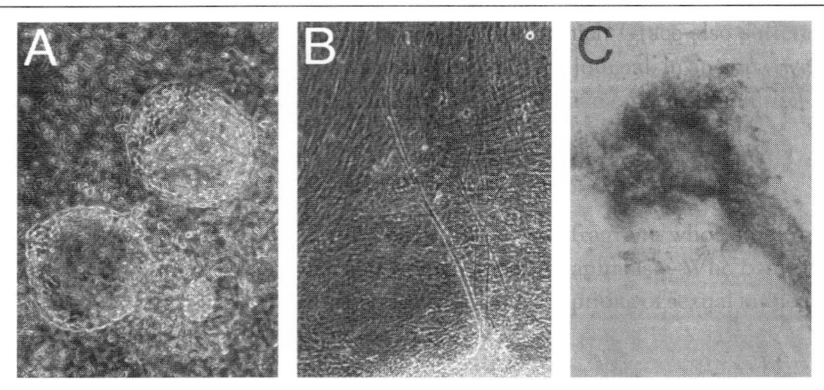

FIG. 6. Cell differentiation *in vitro*. Yolk sac-like vesicular epithelia (A), neuronal cells extending axons (B) and pigmented epithelia (C). (See Color Insert.)

Formation of embryoid bodies (EBs) is an effective method to induce ES cells to differentiate into various cell types. EBs are formed by culturing ES cell aggregates in suspension. Since proliferation of undifferentiated monkey ES cells is dependent on the feeder layer and LIF cannot support their growth, ES cells start differentiation immediately after detachment from the feeder layer. Therefore, undifferentiated stem cells do not proliferate enough to form EBs if the starting cell aggregates are too small. Therefore, it is important to start from larger ES cell aggregates to obtain EBs.

Protocol

1. Reagent: 1 mg/ml collagenase in PBS.
2. Prepare a confluent culture of monkey ES cells. Colonies of about 300 μm in diameter or more are required.
3. Aspirate the medium, and rinse the cells with PBS.
4. Cover the cells with the collagenase solution, and aspirate.
5. Incubate at 37°C for 10 min. Add the medium, and detach colonies from feeder layer by gentle pipetting to avoid dissociation of colonies.
6. Transfer the suspension of colonies to a gelatinized dish, and incubate at 37°C for 30 min to separate ES cell colonies from feeder cells. After attachment of feeder cells to the dish surface, collect floating ES cell aggregates in the medium, and transfer to a new petri dish.
7. ES cell aggregates will form simple EBs in a few days (Fig. 7A). They can be cultured in suspension until apparent cell differentiation is observed. In 2–3 weeks, beating heart muscle and blood islands may

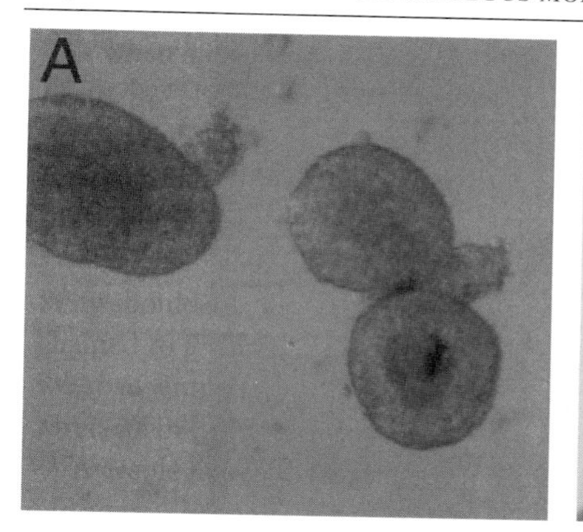

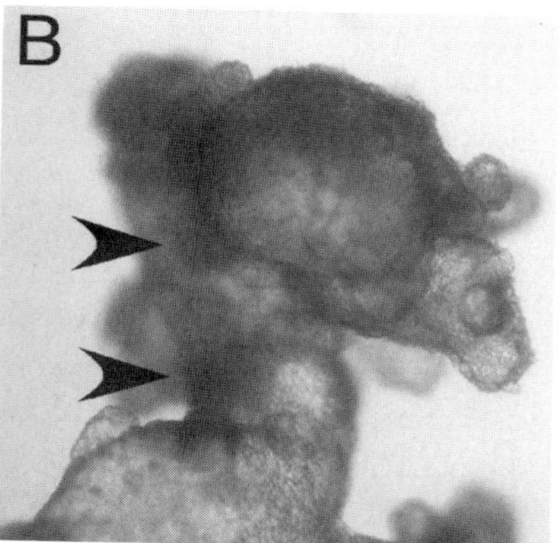

FIG. 7. Simple embryoid bodies (A) and a cystic embryoid body (B) of cynomolgus monkey ES cells. Arrowheads indicate blood island-like structure. (See Color Insert.)

be formed (Fig. 7B). Alternatively, EBs are plated to a tissue culture dish at any time in the suspension culture. EBs get attached to the dish and undergo cell differentiation into various tissues such as neurons and cardiac muscle.

Teratoma Formation in SCID Mice

We transplanted cynomolgus ES cells into SCID mice to produce teratomas. 10^6–10^7 cells (corresponding to the confluent cell layer in a 60- or 100-mm dish) were injected subcutaneously or intraperitoneally to SCID mice. Formation of teratoma becomes recognizable in 2–3 months (Fig. 8A). Histological examination of these teratomas revealed that they contained various tissues derived from all three embryonic germ layers (Fig. 8B). We frequently observed ectodermal tissues containing neurons, glia, glands and epithelia. Mesodermal tissues such as muscle, cartilage and bone were also observed frequently. We also observed typical hair follicles. However, endodermal tissues such as gut epithelium were relatively rare. A columnar epithelium, probably tracheal ciliated epithelium, was the most frequently recognizable putative endodermal tissue. Expression of several tissue-specific proteins was examined to assess differentiation in teratomas by using antibodies prepared for pathological examination of human specimens that recognize tissue-specific antigens. Most tissues expressed typical tissue-specific markers. For example, mature neurons and glia surrounding primitive neuroectoderm expressed the neuron specific enolase (NSE). Glial cells also expressed the glial fibrillary acidic protein (GFAP). Muscle and cartilage, both of which are derived from the mesoderm,

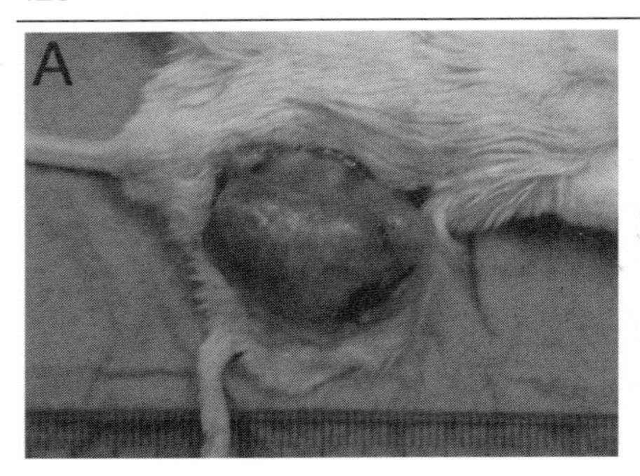

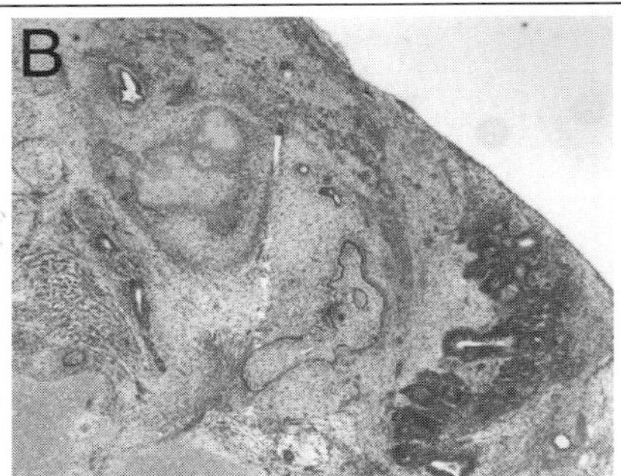

FIG. 8. A teratoma produced by subcutaneous transplantation of cynomolgus ES cells (A), and low magnification view of a histological section of a teratoma (B). (See Color Insert.)

expressed the specific markers desmin and S-100, respectively. As for the endodermal tissues, we found cell clusters expressing alpha-fetoprotein, which may be expressed not only by the liver but also by the yolk sac endoderm.

Prospects of Primate ES Cells

For further progress in basic research and the medical application of primate and human ES cells, we still need to improve many aspects of the manipulation of stem cells. First, it is necessary to propagate ES cells in a completely defined medium, excluding not only the serum but also animal proteins for clinical application. Second, we must find out a way to derive and maintain ES cell lines without a feeder cell layer to avoid unexpected contamination such as by animal retrovirus. To this end, there have been only incomplete trials using human ES cells.[11] Third, and most importantly, we need to devise methods for gene targeting to produce genetically modified monkey and human ES cells. There have been only a few studies of gene transfection in these lines.[12] Such genetic modifications would enable reduction of antigenicity in cell therapy and efficient selection of particularly useful cell types.

In every aspect of ES cell study, nonhuman primate ES cell lines provide important research tools for basic and applied research. In most countries, usage of human ES cells is strictly regulated because of ethical considerations. In such situations, monkey ES cells could provide valuable

[11] C. Xu, M. S. Inokuma, J. Denham, K. Golds, P. Kundu, J. D. Gold, and M. K. Carpenter, *Nat. Biotech.* **19**, 971 (2001).

[12] R. Eiges, M. Schuldiner, M. Drukker, O. Yanuka, J. Itskovitz-Eldor, and N. Benvenisty, *Curr. Biol.* **11**, 514 (2001).

materials for research to advance various aspects of regenerative medicine directly applicable to humans. Finally, they are indispensable for preclinical research using primate models of allogenic cell transplantation therapies to evaluate effectiveness, safety and immunological reaction in physiological conditions similar to the treatment of patients.

Acknowledgment

Our original study included in this article was supported in part by a Research for the Future (RFTF) program of the Japan Society for the Promotion of Science.

[30] Isolation, Characterization, and Differentiation of Human Embryonic Stem Cells

By Martin F. Pera, Adam A. Filipczyk, Susan M. Hawes, and Andrew L. Laslett

Introduction

This chapter will deal with the isolation, characterization and differentiation of human embryonic stem (ES) cell lines from preimplantation blastocysts. The first derivation of human ES cells was reported in 1998,[1] and although there have been a number of anecdotal reports of isolation of new ES cells since then, published data are based only on a few cell lines. Timely progress in this field will depend upon the derivation of new ES cell lines, their proper characterization and comparison with existing isolates, and in-depth analysis of their differentiation under different conditions.

Previous reviews have compared the properties of published human ES cells with those of pluripotent cell lines derived from nonhuman primates, or with human embryonal carcinoma cell lines, and with the counterparts of both cell types in the mouse.[2,3] While these comparisons have given rise to some consensus regarding the canonical primate pluripotent cell phenotype, the data are limited, and the cultures are probably more heterogeneous than the published descriptions imply. Moreover, it is unclear to what extent cell

[1] J. A. Thomson, J. Itskovitz-Eldor, S. S. Shapiro, M. A. Waknitz, J. J. Swiergiel, V. S. Marshall, and J. M. Jones, *Science* **282**, 1145–1147 (1998).
[2] J. A. Thomson and J. S. Odorico, *Trends Biotechnol.* **18**, 53–57 (2000).
[3] M. F. Pera, B. Reubinoff, and A. Trounson, *J. Cell Sci.* **113**, 5–10 (2000).

0076-6879/2003 $35.00

lines isolated in different laboratories show similar growth and differentiation phenotypes. The purpose of this chapter is to help provide some guidelines for human ES cell derivation, and a methodological basis for the characterization and comparison of stem cells and their differentiation in different laboratories.

Isolation of Human ES Cells

Human Embryos

Human ES cells are isolated from preimplantation blastocysts obtained from surplus embryos donated with informed consent by couples undergoing *in vitro* fertilization therapy. Those workers who are planning to derive new ES cell lines should endeavor to meet certain ethical standards if their cell lines are to be of wide use to the research community. Investigators must of course comply with local regulations, but some widely endorsed ethical guidelines might include the following: the protocol for accessing the embryos should be approved by an Institutional Ethics Committee; the embryos should be surplus to clinical requirement and the decision to donate embryos for research should be clearly separated from the production of embryos for treatment; the investigator wishing to derive the ES cell line should not be involved in patient care or in obtaining consent for research use of spare embryos; the patients should give informed consent that includes permission for use in commercial applications where appropriate; there should be no financial inducement to donate the embryo and it should be made clear that the donation will not result in direct medical benefit to the donor. Obviously, it is important to fully document all aspects of the consent, donation and derivation process. Most existing ethical guidelines specify that the ES cell line should be de-identified after derivation (i.e., the name of the cell line or data associated with it should not enable it to be traced to a particular patient).

Prior to embarking on a program of isolation of new ES cell lines, workers should gain experience in the maintenance of primate pluripotent stem cells, by working with established monkey or human ES cell lines, or human embryonal carcinoma cell lines. Mouse ES cells are sufficiently different from primate cells that experience with the former only may not provide adequate preparation to enable efficient derivation and/or culture of human ES cells.

In establishment of human ES cells, our laboratory uses disposable plastic vessels and pipets only. Prior to ES cell establishment, embryos should be cultured to the blastocyst stage of development. ES cell lines have been developed from both fresh and cryopreserved embryos, the latter frozen at an early stage then grown on to produce blastocysts. Usually, embryos are cultured to the blastocyst stage using a two-step sequential culture methodology which employs one type of basic media to the eight cell-stage and a more complex formulation (typically containing high glucose and nonessential amino acids and more similar to conventional tissue culture medium) to the blastocyst stage (for example G1.2 and G2.2 from Vitro Life, Sweden). Where possible high quality blastocysts should be used; there are various morphological grading systems to assess the blastocyst quality and such assessment should be carried out by a clinical embryologist familiar in grading human embryos.

For ES cell isolation, the zona pellucida is removed by treatment with pronase (10 U/ml in serum-free embryo culture medium at 37°C for 2 min). The embryo is then washed in serum-containing culture medium and the inner cell mass is isolated by immunosurgery. While some workers have raised antisera against trophectoderm cell lines for use in this procedure, any antibody that recognizes human cells is suitable. Antisera should be heat inactivated for 30 min at 56°C prior to use to destroy endogenous complement. Titration of antibody and complement may be achieved using spare embryos, or a cultured human cell line, and the minimum concentrations achieving complete cell lysis should be used. Antibody or complement alone should not induce lysis. Antibody incubation is carried out at 37°C for 30 min, followed by addition of baby rabbit or guinea pig complement for 30 min (or less if lysis is complete prior to this time) at 37°C. The lysed trophectoderm is then removed by pipetting of the inner cell mass through a glass capillary drawn out to a diameter slightly larger than the inner cell mass. The inner cell mass is then washed thoroughly in culture medium, and plated out onto a mouse embryonic fibroblast feeder cell layer. The culture medium employed is DMEM (high glucose, low pyruvate formulation) supplemented with 20% foetal calf serum, glutamine, nonessential amino acids (e.g., Invitrogen cat. no. 11140-050), 10 μM beta-mercaptoethanol, ITS (proprietary mixture of insulin selenium and transferrin, Invitrogen cat. no. 41400-045).

Mouse Embryo Fibroblast Feeder Cells

The mouse embryo fibroblasts are isolated from decapitated and eviscerated late midgestation mouse fetuses (E13.5). We have no consistent data to indicate that one strain of mouse is superior to others for derivation of feeder cells. To produce feeder cell stocks, use only well-developed fetuses of normal appearance. Following the removal from the uterus, dissection free from the extraembryonic membranes, and removal of head and viscera, the foetal carcasses are washed in Dulbecco's phosphate buffered saline without calcium or magnesium (PBS-) twice, minced using crossed scalpels, then digested for 5 min in 0.25% trypsin-1 m*M* EDTA in PBS-. Serum-containing culture medium (DMEM high glucose low pyruvate, supplemented with 10% foetal calf serum, glutamine and antibiotics) is added to neutralize the effects of trypsin, the cells are plated out at 2E4/cm^2 and are harvested when confluent; thereafter they are subcultured at a 1 : 5 split ratio and/or cryopreserved. The fibroblasts are used at passage level 2–4. To prepare the fibroblasts for use, the cultures are set up at 2E4/cm^2 and grown for several days until near confluent, at which point they are treated with mitomycin C at 10 μg/ml in serum-containing medium for 2 hr. The treated cells are plated out 24 hr to 5 days prior to ES cell addition at a density of 7.5E4/cm^2 in 1 ml into the central well of organ culture dishes pretreated with gelatin (1% solution in distilled water). Mitomycin C is cytotoxic and carcinogenic; appropriate safety precautions including the use of gloves, gowns, eye protection, and disposal procedures for cytotoxic agents, should be followed when handling this agent. Individual batches of fibroblasts should be tested for their ability to support the growth of established human ES cell lines prior to use. While it is preferable to use freshly prepared fibroblasts for ES cell culture, backup stocks of mitomycin C-treated cells may be cryopreserved in undiluted fetal calf serum using standard techniques and stored in liquid nitrogen for emergency use. Feeder cells may also be prepared by treatment with 75 Gy ionizing radiation to achieve mitotic inactivation. There are no convincing published data to indicate whether there are important differences between mitomycin C and irradiated feeder cells, but where a suitable radiation source is available, irradiation may prove more convenient.

Establishment and Maintenance of ES Cells

The inner cell mass will not resemble established ES cells during the initial phases of culture (Fig. 1A and C), and growth of the colony may not be obvious immediately. Usually by 10–14 days, the ICM will be ready to subcultivate (Fig. 1B and D). Subcultivation is accomplished by mechanical dissection of the growing colony into pieces approximately 0.5 mm^2 (Fig. 1; the nascent ES colonies in the primary cultures in Fig. 1B and D would have been dissected in two) under a dissecting stereomicroscope. Portions of the colony that are overtly differentiated are avoided, and areas of stem cell morphology are sliced into fragments using either a drawn out capillary pipet or a narrow (27–30) gauge syringe needle. The pieces may then be harvested by incubation in a 10 mg/ml solution of dispase (Dispase II, Roche Diagnostics; cat. no. 165 859) in complete culture medium; dispase treatment will remove colony fragments en bloc from the culture dish, after

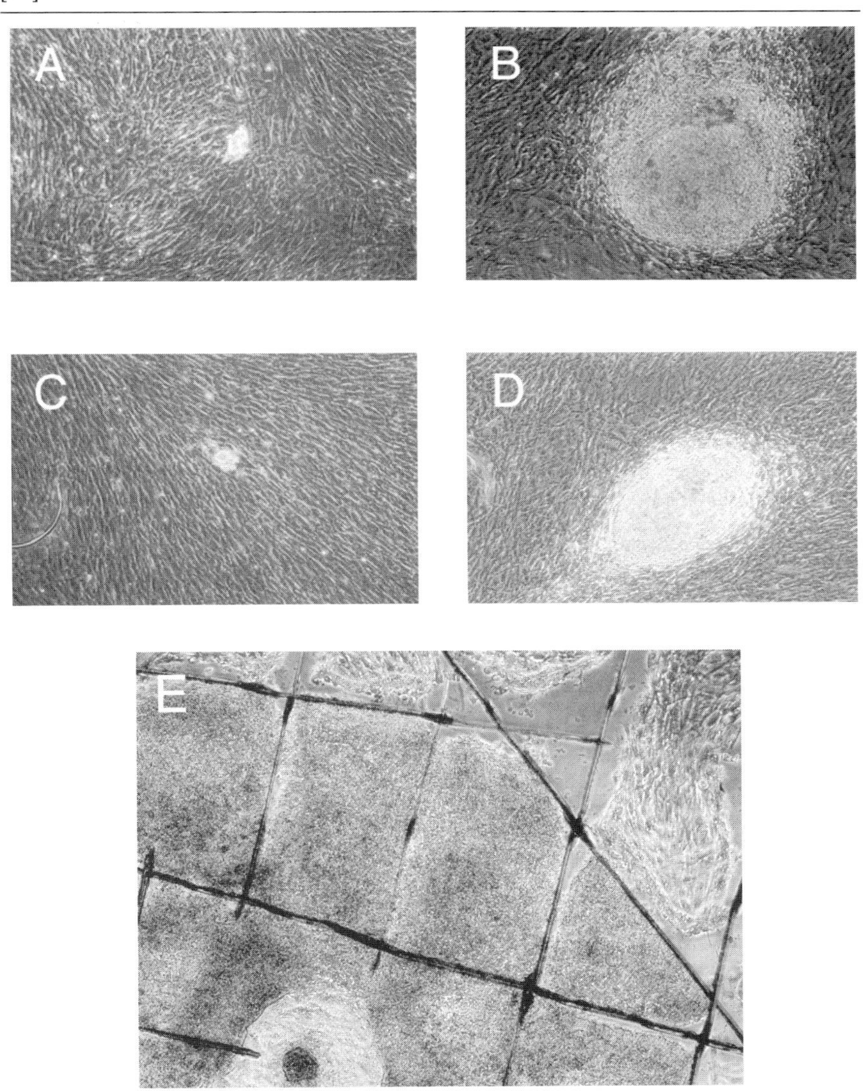

FIG. 1. Derivation and subculture of human ES cells. A, blastocyst from which HES-3 was derived several days after plating; B, HES-3 primary culture 10 days later; C, blastocyst from which HES-4 was derived; D, HES-4 primary culture 10 days later; E, ES colony sliced with drawn out capillary and ready for harvest with dispase. Primary cultures in B and D were successfully subcultured at these stages. In E, healthy areas at the edge of the colony will be selected for subculture, and differentiating area toward the center (bottom left of figure) is avoided. Magnification in A–D, 60×; in E, 90×.

which they are washed in fresh serum-containing medium and replated (between 2 and 8 pieces per dish) onto fresh organ culture dishes pretreated with gelatin and containing feeder cells. Subculture may then be carried out every 7 days.

Once ES cell lines are established, it is possible to serially passage them in a serum-free system that enables subculture in bulk using enzymatic harvest, rather than mechanical dissection.[4] There are several advantages to this methodology: first, it is less labor intensive than passage by dissection under microscopic control, and second, it eliminates the use of serum, which may contain factors that affect ES cell differentiation. The medium used is DMEM:F12 supplemented with 20% Knockout serum replacer and 4 ng/ml FGF-2.[5] The cell monolayer is washed twice with PBS solution at ambient temperature. ES cell colonies are then enzymatically removed by treatment with a thin layer (1–2 ml for a T25 flask) of Collagenase IA (1 mg/ml) (e.g., Sigma Chemical Company C2674) for 5 min, dissolved in serum-free medium, at 37°C and 5% CO_2. Next the surface of the flask is gently and repeatedly rinsed with culture medium. This treatment effectively dissociates feeder cells into a single cell suspension, whilst maintaining ES cell colonies intact; harvest of ES cells is facilitated by rapping the flask gently. The feeder cells are then separated from human ES colonies by allowing the heavier colonies to settle under gravity for 1 min, then immediately aspirating the feeder cells still suspended in the medium. This washing procedure is repeated several times, after which the ES colonies are dispersed into small clumps by trituration using a 200 μl micropipettor tip. The cells are replated on to plastic dishes pretreated with gelatin and containing a feeder cell layer which is approximately one-third the density used in the standard cell culture protocol described above.[5]

Characterization of Human ES Cells

Karyotype

ES cell cultures should be examined regularly for normal G-banded karyotype. This is best performed by a qualified cytogenetics laboratory.

Immunochemical Characterization of Human ES Cells

To date a rather limited range of immunochemical markers has been used to identify human ES cells (Table I). The available antibodies are

[4] M. Amit, M. K. Carpenter, M. S. Inokuma, C. P. Chiu, C. P. Harris, M. A. Waknitz, J. Itskovitz-Eldor, and J. A. Thomson, *Dev. Biol.* **227**, 271–278 (2000).

[5] J. A. Thomson, "Current Protocols in Stem Cell Biology." The Jackson Laboratories, 2002.

TABLE I
MONOCLONAL ANTIBODIES REACTIVE WITH EPITOPES FOUND ON HUMAN ES CELLS

Antibody	Type	Antigen	Source[f]	Reference
MC-631	M IgM	SSEA-3[a]	A	1,6,7
MC-813-70	M IgG3	SSEA-4[b]	A	1,6,7
TRA-1-60	M IgM	KSPG[c]	B	1,6–10
TRA-1-81	M IgM	KSPG[c]	B	1,6–10
GCTM-2	M IgM	KSPG[d]	C	1,6–10
TG343	M IgM	KSPG[d]	C	11
TG30	MIgG2a	25 kDa[e]	C	Pera, unpublished
OCT-4 (C-10)	MIgG2b	Oct-4	D	Pera, unpublished
P1/33/2	M IgG1	CD 9	E	Pera, unpublished

[a]Globo series glycolipid epitope.
[b]Globo series glycolipid epitope.
[c]200 kDa cell surface keratan sulfate/chondroitin sulfate proteoglycan with extensive O-linked carbohydrate. These antibodies react with carbohydrate epitopes.
[d]These antibodies react with the core protein of the proteoglycan.
[e]This reagent reacts with a 25 kDa cell surface protein identical to CD9.
[f]*Sources:* A, Developmental Studies Hybridoma Bank, University of Iowa; B, Chemicon; C, this laboratory; D, Santa Cruz Biotechnology Inc.; E, Dako Corporation.

generally directed against cell surface epitopes which, while not exclusive to ES cells, are relatively limited in their distribution. The epitopes are expressed on pluripotent cell lines and early mammalian embryos at specific stages of development. In most cases, the protein products and their corresponding cDNAs that carry the epitopes recognized by these reagents are not defined. Nevertheless, just as stages of hematopoietic and lymphoid cell differentiation can be dissected by cluster analysis of differentiation antigens, a series of surface antigens characterize primate pluripotent stem cells and their differentiation. The stage-specific embryonic antigens 1, 3, and 4 are globoseries glycolipids recognized by monoclonal antibodies originally raised to distinguish early stages of mouse development. Primate

[6] B. E. Reubinoff, M. F. Pera, C. Y. Fong, A. Trounson, and A. Bongso, *Nat. Biotechnol.* **18**, 399–404 (2000).
[7] P. W. Andrews, J. Casper, I. Damjanov, M. Duggan-Keen, A. Giwercman, J. Hata, A. von Keitz, L. H. Looijenga, J. L. Millan, J. W. Oosterhuis, M. Pera, M. Sawada, H. J. Schmoll, N. E. Skakkebaek, W. van Putten, and P. Stern, *Int. J. Cancer* **66**, 806–816 (1996).
[8] J. K. Henderson, J. S. Draper, H. S. Baillie, S. Fishel, J. A. Thomson, H. Moore, and P. W. Andrews, *Stem Cells* **20**, 329–337 (2002).
[9] G. Badcock, C. Pigott, J. Goepel, and P. W. Andrews, *Cancer Res.* **59**, 4715–4719 (1999).
[10] S. Cooper, M. F. Pera, W. Bennett, and J. T. Finch, *Biochem. J.* **286 (Pt 3)**, 959–966 (1992).
[11] S. Cooper, W. Bennett, J. Andrade, B. E. Reubinoff, J. Thomson, and M. F. Pera, *J. Anat.* **200**, 259–265 (2002).

pluripotent cells express SSEA-3 and SSEA-4 and express SSEA-1 only upon differentiation. Essentially, the reverse is true for mouse ES cells. Also characteristic of human EC cells is the expression of a set of antigens associated with a pericellular matrix keratan sulfate/chondroitin sufate proteoglycan found on the surface of these cells. Keratan and chondroitin sulfate are both types of polylactosamine, and it is known that mouse ES cells carry polylactosamine on their surface, but it is not certain whether the mouse cells express the core protein of the human stem cell proteoglycan. Finally, both mouse and human ES cells express the tetraspanin molecule CD9.

Indirect Immunofluorescence

Human ES cells are transferred intact as clumps or clusters of cells using standard methodology described above to either chamber well slides or multiwell slides placed in rectangular culture dishes with individual compartments to a size of a standard microscope slide (Vivascience quadriPerm dishes, cat. no. IV-76077310). The slides are pretreated with gelatin and where required mouse embryonic feeder cells are added at least 1 day and not more than 5 days prior to ES cell addition. The ES cells are cultivated for any length of time between 1 day to 3–4 weeks. Thereafter the slides are rinsed in PBS- and fixed. For cell surface glycolipid antigens, we prefer 90% acetone : 10% water v/v for 5 min. For intracellular antigens, use either methanol : acetone 1 : 1 v/v or 4% paraformaldehyde in PBS, both at room temperature for 5 min. Methanol acetone slides should be air-dried directly after fixation, whilst paraformaldehyde fixed slides should be rinsed with water prior to air drying. Absolute ethanol is a good all purpose fixative which should be applied for 5 min after which slides should be air-dried. Slides can be stored for at least six months at −20°C.

Antibodies are applied in a humidified atmosphere for 30 min then rinsed with PBS- followed by secondary detection reagents, which may be conjugated to fluorochromes or enzymes. We prefer indirect immunofluorescence as it provides more detailed information about intracellular localization of antigens and enables more clear-cut discrimination between nonspecific background binding and genuine reactivity. After the detection reagent has been applied, the slides are rinsed again in PBS- and then mounted. For fluorescence, it is useful to counterstain nuclei with a DNA binding dye such as Hoechst 33258 (1 μg/ml in PBS- for 30 sec); this enables localization of cells and easy discrimination between human and mouse feeder cells (the latter show a speckled appearance of chromatin after staining with these dyes, whilst human cells show more uniform nuclear staining). There are many proprietary mountants that incorporate anti-bleaching compounds designed to inhibit loss of fluorescence; these are particularly useful when examining cells under high power with high numerical aperture lenses. However, some antifade compounds may interfere with certain fluorochromes such as Alexa Fluor 350. For immunohistochemical detection, we use antibodies conjugated to alkaline phosphatase and detection with fast red TR (an easy-to-use staining kit is available from Sigma Chemical Company, cat. no. F4648) followed by counterstaining with Mayer's hematoxylin (e.g., Dako Corporation cat. no. S3309). The presence of levamisole in the substrate reaction mixture blocks endogenous alkaline phosphatase activity, and the blue counterstain affords a good high contrast image with the fast red.

While the technique of immunostaining is simple, some experience and skill is required to correctly interpret the images that are obtained. It is recommended that wherever possible positive controls be employed and investigators consult the published literature for examples of positive images of certain stains. Workers should consult the literature for examples of staining patterns observed for some commonly used human ES cell antigens. Human embryonal carcinoma cell lines can serve as positive controls for immunostaining protocols.

Flow Cytometry

Flow cytometry may be used to obtain quantitative information on the proportion of cells in an ES culture expressing particular surface markers. Flow cytometry requires dissociation of ES cell colonies into single cell suspensions, and for many ES cell lines this process is often associated with a significant degree of cell killing. Therefore, it is important to monitor cell viability and to bear in mind the possibility that some subpopulations of cell types within an ES culture may be more susceptible to death following dissociation than others. This protocol describes the method of immunostaining cells in single cell suspension for the stem cell marker GCTM-2 and quantitative analysis by flow cytometry.

ES cell colonies harvested as described above are rinsed with PBS- to remove Ca^{+2} and Mg^{+2} ions prior to dissociation, which may be carried out using gentle trypsin/EDTA treatment or by mechanical agitation in cell dissociation buffer (Hanks balanced salt solution with chelating agents, e.g., Invitrogen 13150-016). For trypsin digestion, the PBS wash solution is aspirated from the colonies and 300 μl of 0.05% trypsin/200 μM EDTA in PBS- is added for no longer than a minute at room temperature. Gentle pipetting using a 200-μl pipette helps to break down ES colonies into single cell suspension. Enzyme activity is quenched with 1 ml of serum-containing ES cell medium. It is crucial to trypsinize cells for the least possible time to avoid excessive cell killing. As an alternative to trypsinization, cells may be resuspended in cell dissociation buffer and triturated using a 200-μl micropipet tip. Colonies are readily dissociated into single cells by this procedure.

Cells are then centrifuged for 2 min at 500*g* in a microfuge and resuspended in 300 μl of mouse IgM GCTM-2 antibody supernatant or class matched control antibody for 30 min on ice. Cells are then pelleted for 2 min at 500*g* and washed with 1 ml of wash buffer (containing 10 m*M* Hepes/NaOH, pH 7.4, 140 m*M* NaCl and 5 m*M* $CaCl_2$). After pelleting again for 2 min at 500*g* and removal of the wash buffer, the cells are incubated for 30 min on ice in 100 μl of rabbit anti-mouse Ig-FITC antibody conjugated to fluorescein isothyocianate (DAKO Corporation cat. no. F0261), diluted 1 : 40 in wash buffer. Samples are incubated in the dark to avoid potential bleaching of the fluorochrome.

Cells are washed to free from antibody, as described above and are resuspended in 400 μl of wash buffer. It is desirable to add propidium iodide at 10 μg/ml to the samples and incubate for 10 min at room temperature. This allows the identification of cells with compromised cell membrane integrity, which may comprise a substantial minority of the cell population. With immunolabeling complete, samples were ready for single or dual color flow cytometric analysis.

Immunomagnetic Isolation of Viable ES Cells

As noted above, at least in our study, most ES cell lines are adversely affected by dissociation to single cells using available methodology including nonenzymatic methods. It is however possible to immunosort ES cells separated into small clusters (as in standard subcultivation methodology described above) using magnetic beads, and to recover a high frequency of viable cells after the procedure. With the GCTM-2 antibody, when clusters of cells are isolated, the majority of cells within the cluster are positive for the antibody when stained after separation.

We employ DYNAL (rat anti-mouse IgM, Dynal cat. no. 110.15) beads to use with GCTM-2. Our protocol prearms the magnetic beads with antibody prior to incubation with the cells. The beads (25 μl of bead suspension for up to 4×10^7 cells in a 1 ml sample) are thoroughly resuspended and transferred to a microfuge tube. The tube is placed in a Dynal magnetic particle concentrator (magnet) for 1 min and the buffer is removed. 1.5 ml of PBS- is added to rinse the beads, which are placed in the magnet again for 1 min, after which the beads are resuspended in 50 μl of PBS-. The PBS- is removed, and 500 μl of neat GCTM-2 supernatant is added and incubated at room temperature for 30 min. The beads are washed

TABLE II

PRIMERS USED IN RT-PCR STUDIES OF HUMAN ES CELLS

Gene	Primer Name	Oligonucleotide Sequence (5′ → 3′)	$MgCl_2$ (mM)	Melting temperature (°C)	Size (bp)
Oct4	Oct15	CGT TCT CTT TGG AAA GGT GTT C	1.5	55	309
	Oct26	ACA CTC GGA CCA CGT CTT TC			
FoxD3	GenF480	GCA GAA GAA GCT GAC CCT GA	1.5	55	305
	GenR785	CTG TAA GCG CCG AAG CTC T			
Cripto	CriptoF484	CAG AAC CTG CTG CCT GAA TG	1.5	55	185
	CriptoR668	GTA GAA ATG CCT GAG GAA ACG			
GCNF	GCNF1074	TAC CTG GCA GGA GCT AAT CC	1.5	55	250
	GCNF1321	AGC TGT GAG GCA CTG GTC AG			
Transferrin	TRFF1197	CTG ACC TCA CCT GGG ACA AT	1.5	55	367
	TRF1765	CCA TCA AGG CAC AGC AAC TC			
Vitronectin	VNF24	TTG CAG GCA CTC AGC TAG AA	1.5	55	300
	VNR336	TGT TCA TGG ACA GTG GCA TT			
Alphafeto protein	AFP736	CCA TGT ACA TGA GCA CTG TTG	1.5	55	338
	AFP1173	CTC CAA TAA CTC CTG GTA TCC			
Sonic Hedgehog	SHH	GAG ATG TCT GCT GCT AGT CCT CG	1.5	60	442
	SHHR	GGT CAG ACG TGG TGA TGT CCA CTG			
GATA6	GATA6F	CCA GCA AGC TGC TGT GGT C	1.5	60	571
	GATA6R	CGA CAG CGA GAG CTG TAC TG			
HNF4	HNF4F	GCT TGG TTC TCG TTG AGT GG	1.5	60	462
	HNF4R	CAG GAG CTT ATA GGG CTC AGA C			
HNF3α	HNF3F	GAG TTT ACA GGC TTG TGG CA	1.5	55	400
	HNF3R	GAG GAC AAT TCC TGA GGA T			
Brachyury	BRACHF	GTG ACC AAG AAC GGC AGG AGG	1.5	63	700
	BRACHR	TGT TCC GAT GAG CAT AGG GGC			
Nestin	NEST856F	CAG CTG GCG CAC CTC AAG ATG	1.5	55	208
	NEST1064R	AGG GAA GTT GGG CTC AGG ACT GG			
Pax6	PX6F1368	AAC AGA CAC AGC CCT CAC AAA CA	1.5	55	274
	PX6R1642	CGG GAA CTT GAA CTG GAA CTG AC			
Sox2	SOX2F	GGC AGC TAC AGC ATG ATG CAG GAG CC	1.5	67	130
	SOX2R	TG GTC ATC GAG TTG TAC TGC AGG			

several times in PBS- then washed three times in incubation buffer (HEPES buffered DMEM, e.g., Invitrogen cat. no. 10315, plus 1% foetal calf serum) after which they are resuspended in 50 μl of incubation buffer. The beads are then ready for addition of cells.

Cells may be harvested from serum-containing cultures using dispase, or from serum-free cultures using collagenase, as described above. The cells are dispersed into small clumps, which are harvested by centrifugation at 250*g* for 4 min in a benchtop microfuge, then resuspended in 500 μl incubation buffer. A small aliquot (50 μl) of cells is set aside for antibody staining, to assess the proportion of GCTM-2 positive cells prior to separation. Next, a 50 μl aliquot of GCTM-2-coated beads are added to 450 μl of cells and incubated at 4°C for 30 min with occasional gentle agitation. Following this period of incubation at 4°C, the tube is placed in the magnet for 1 min. The liquid phase, which represents the unbound fraction, is removed and saved for later staining. The beads are rinsed twice in incubation buffer, then finally resuspended in 500 μl incubation buffer and are kept as the bound fraction.

It is not necessary to remove the beads from the cells; cells can reattach and grow readily, even though they remain attached to beads. All fractions should be replated onto slides in the presence of feeder cells to assess the efficacy of the immunomagnetic separation.

Gene Expression in ES Cells

A second means of characterization of stem cells and their differentiated derivatives is by examination of the RNA transcripts that the cells express. There are a number of genes characteristic of primate pluripotent stem cells which show only limited expression in other normal cell types. Some of these molecules are known to be important to the development and differentiation of pluripotent cells in the embryo, while the function of others remains to be elucidated. Examples of genes found in primate pluripotent stem cells include transcription factors such as Oct-4 and FoxD3, and GCNF, as well as growth factors such as cripto or TDGF, and GDF-3 or Vgr-1. Likewise, there are many genes identified through molecular embryological studies that are characteristic of early stages in various particular differentiation lineages, as well as genes whose expression is characteristic of specific types of mature cell. For many genes expressed primarily in the early mammalian embryo, good quality antibodies with known reactivity against human proteins may be lacking. Gene expression at the RNA level is thus widely used to monitor the differentiation status of human ES cell cultures. There are some caveats that should be considered in designing gene expression studies. While some studies of mouse ES cells support the notion that patterns of gene expression in differentiating ES cells reflect the usual sequence of events during embryogenesis, it is probably unwise to assume that this will inevitably be the case, and at any rate the corresponding data for the human embryo will be unavailable in most instances. Another problem in human cell work with genes expressed only early in development may be finding appropriate positive control material to test primers or other probes. Where possible, use of homologous mouse sequences in primer or probe design may circumvent this problem. Whenever cells undergoing analysis have been cultured on mouse fibroblast feeder cells, RNA from these feeder cells should be included as a control. Alternatively, stem cells and differentiating cells from human EC cultures may be useful as positive controls.

While microarray based analyses will play an increasing role in these studies in future, much work to date has relied on RT-PCR. RT-PCR is relatively inexpensive and is particularly suitable for studies in which small numbers of cells are analysed and it is desirable to look at a relatively modest number of multiple differentiation markers. Whatever means of analysis is used, it is important to remember the potential for contamination of samples with minority populations of cells at various stages of differentiation. Table II lists primer pairs that have proven useful in studies of human ES cell differentiation.

RNA Isolation from ES Cells Using Dynalbeads mRNA DIRECT Kit

RT-PCR may be carried out on material from a few cells upwards. In the case of stem cells, usually 5–10 colonies are isolated, rinsed, and placed directly into lysis buffer. Following freezing, the samples are stored at –80°C for future analysis. For differentiation studies, either entire colonies or embryoid bodies may be lysed in a similar fashion to ES cells, or areas of the culture with particular morphologies may be identified under the dissecting microscope and mechanically dissected away from the rest of the culture. Lysis is as above for ES cell colonies. When samples are recovered from frozen storage for analysis, they are spun down at 15,000*g* in a microfuge to remove insoluble material, which is discarded. Poly A+ mRNA is isolated on magnetic beads bearing oligo dT, first strand cDNA synthesis is carried out directly on these magnetic beads and the beads themselves are placed directly into the PCR reaction. The creation of a solid phase cDNA library reduces losses associated with elution of the cDNA from the magnetic beads and thereby enhances the sensitivity of the procedure.

RNA isolation of human ES cells and derivative cells (e.g., neural progenitor cells and neurospheres) is carried out using Dynalbeads mRNA kit according to the manufacturer's instructions (cat. no. 610.02, Dynal, Norway). The Dynalbead method isolates mRNA by incubation of crude cell lysates with magnetic beads conjugated to oligo $(dT)_{25}$ molecules. Use of a Magnetic Particle Concentrator or magnet (Dynal) separates the beads with bound mRNA from other cellular material. ES cells (1–8 colonies or equivalent) are washed twice in warmed PBS- and then twice in buffer containing 0.1 *M* NaCl, 10 m*M* TrisCl, 1 m*M* EDTA pH 7.4, prior to lysis in 300 μl lysis/binding buffer (100 m*M* Tris–HCl pH 7.5, 500 m*M* LiCl, 10 m*M* EDTA, 5 m*M* dithiothreitol, 1% lithium dodecyl sulfate, LiDS). Prior to use, oligo $(dT)_{25}$ conjugated magnetic beads are washed with lysis/binding buffer. The crude cell lysates are then incubated with 20 μl washed beads for 10 min at room temperature. Unbound material is removed following placement of the lysate–bead mixture onto the magnet. Bound material is washed twice, first, with 400 μl buffer containing LiDS (10 m*M* Tris–HCl pH 7.5, 0.15 *M* LiCl, 1 m*M* EDTA, 0.1% LiDS) and secondly, 200 μl buffer without LiDS. Samples are resuspended in 100 ul of buffer without LiDS.

Reverse Transcription

mRNA conjugated to oligo $(dT)_{25}$ magnetic beads is reverse transcribed in reactions containing 4 μl 5X RT buffer, 2 μl 100 m*M* dithiothreitol, 1 μl 10 m*M* mixed dNTPs, 1 μl (40U) RNAseOUT (Invitrogen cat. no. 10777), 1 μl (200 U) Superscript (Invitrogen cat. no. 18064) and water to bring up the volume to 20 μl. Following reverse transcription for 1 hr at 42°C, the enzyme and buffer are removed by washing with water containing RNase out (40 U/20 μl). The beads are then resuspended in 20 μl of water containing RNase out. Evidence that samples are not contaminated with genomic DNA results from negative controls without reverse transcriptase.

Polymerase Chain Reaction

2–5 μl of the RT sample mixture undergoes polymerase chain reaction in 25 μl reactions. Each reaction involves 2.5 μl 10x buffer, 0.5 μl of mixed 10 m*M* dNTPs, 1 μl (1 μ*M*) of each PCR primer, 1.25 units Taq polymerase and 2–5 μl RT reaction. 1–2 m*M* $MgCl_2$ is added to the reaction dependant on specific PCR primers. Mixed sample is then overlayed with oil. Amplification is performed with an initial incubation at 95°C for 5 min, followed by cycles of 95°C for 1 min and 55–62°C for 1 min and 74°C for 1 min, ending with a final incubation at 74°C for 6 min. Amplified PCR products are separated and visualized by 2% (w/v) agarose gel electrophoresis containing ethidium bromide. Sizes of amplification products are estimated by comparison with a 1 kb and 100 bp DNA molecular weight standards (InVitrogen).

ES Cell Differentiation

Xenografts in SCID Mice

The ability of a human ES cell line to form teratomas is the best test of pluripotentiality presently available. In our laboratory we implant cells for testing beneath the testis capsule of SCID mice between 5–6 weeks of age. If obtained from an outside supplier, the animals are given a week on arrival to acclimate to the mouse house. The first part of the procedure, carried out to exteriorize the testis, is similar to the procedure used in vasectomy, which many animal house technicians will be familiar with. Using aseptic technique within a laminar flow hood, the animal is anesthetized with avertin or any suitable general anesthetic. The abdomen is swabbed with alcohol, and a small longitudinal incision is cut through the skin and abdominal wall just below the level of the origin of the hindlimb. At this stage it is easier to perform the remaining procedure under a dissecting microscope. The scrotum of the mouse is squeezed to bring the testes into the abdominal cavity. A piece of fat will be visible inside the incision toward the caudal part of the mouse and just a bit lateral to the midline. This tissue is drawn outside the abdomen with a forceps, and the vas deferens and testis will follow.

The cells should be prepared for inoculation at this stage of the procedure. A small artery clamp is used to gently put traction on the testis, which should be fully outside the body. A 26 gauge needle is used to make a small hole in the testis capsule in a region where there are no major blood vessels. Either a 25 g syringe, or a drawn out capillary pipet is used to introduce the cell inoculum. Either device should be inserted into the testis about halfway, so the cells do not escape. Approximately 50 μl may be introduced before the testis capsule begins to swell. A few seminiferous tubules may herniate out from the incision. We inject as many cells as possible, but we have obtained tumors from say 5–10 ES cell colonies containing perhaps 50,000 cells in total. There is no need to attempt to repair the hole in the testis capsule. After the testis is returned well inside the abdominal cavity, the procedure is repeated with the other testis, then the internal and external incisions are closed separately with 6-0 silk. One suture will be enough for each layer, provided the initial incisions were not excessively large. The mice should be monitored until they gain consciousness, and checked again the following day.

The animals are monitored weekly beginning at around 4 weeks for tumor development. Clinical examination is carried out by bringing the testis down into the scrotum and palpating it. Sometimes, the testis will not come down if it is enlarged or adherent to surrounding tissue, in which case it is necessary to palpate the organ up in the abdomen. Lesions usually become apparent as swellings in about 5 weeks. The tumors are removed, fixed in formalin, and sent for routine histological processing.

Teratomas from human ES cells will contain a variety of tissues, some showing a high degree of histiotypic organization. However, it is not always possible to identify definitively the tissues present in such lesions, because they may be immature. It may be necessary to section through the tumor at multiple levels to identify all existing tissues. Typical lesions will contain muscle, neural tissue, and various forms of epithelia.

In Vitro *Differentiation*

Several approaches have been used to induce differentiation of human ES cells. To date, most studies of human ES cell differentiation have examined cultures undergoing spontaneous differentiation. Human ES cells' spontaneous differentiation has been studied in our laboratory by "*in situ*" differentiation of adherent cells. This is achieved simply by maintaining routine stock cultures in their original dishes with daily medium changes and without renewal of the feeder layer. Over the course of 3–6 weeks, a range of differentiated tissues appears as the cells pile up and form cystic structures and three-dimensional aggregates of varying shape and morphology.

The differentiation of ES cells *in situ* is accompanied by a gradual loss of surface marker and gene expression characteristic of human ES cells. In the adherent cultures, a range of different cell morphologies is observed. It is often difficult to disaggregate these multilayered cell cultures into viable single cell suspensions using conventional enzymatic techniques, but they may be dispersed into clumps by trituration after the harvest by scraping, then subjected to immunoisolation using surface markers and magnetic beads as described above. Alternatively, the differentiating cultures may be studied by indirect immunofluorescence using markers specific to certain lineages, including transcription factors, surface markers, and structural proteins such as intermediate filaments, to help identify cells with distinct morphological features. By associating a particular pattern of marker expression with cell morphology under the phase contrast microscope, it is possible to identify regions within the culture dish that contain cells committed to specific lineages. For example, areas undergoing neural differentiation show expression of polysialylated N-CAM on their surface. If areas of cells undergoing lineage specific differentiation can be identified, it is then possible to use mechanical dissection (with drawn out capillary pipets or narrow gauge syringe needles) under microscopic control, to separate these areas and culture them independently. This approach was successfully used in our laboratory to isolate neural precursors from ES cell cultures.

Others have used embryoid body formation to induce differentiation of ES cells. In early work in our laboratory, embryoid body formation was carried out using hanging drop methodology often employed in mouse ES cell differentiation studies. Because this method required dissociation of the stem cells to single cells prior to hanging drop cultivation, viability was poor. Since then, other groups have found that embryoid bodies may be generated simply by growing clumps of ES cells harvested by mechanical dissection and dispase treatment in suspension under nonadherent conditions. It is not yet possible to tell whether this technique yields a different mix of cell types compared to that found when monolayer cultures are simply allowed to overgrow in their original dishes.

Embryoid bodies may be formed from ES cells grown in serum-containing medium or under serum-free conditions as described above. Colonies about 1 week old are sliced into pieces approximately 0.5 mm^2 in size under a dissecting stereomicrosope, and are then harvested with a 10 mg/ml solution of dispase in serum-containing medium. The pieces are transferred to nonadherent culture vessels in either serum-containing medium or serum-free medium with no growth factors added. Costar Ultra Low Attachment dishes (96 or 24 well) are suitable for this purpose. The embryoid bodies are allowed to grow for periods from several days to weeks at which point they may be analyzed for gene expression by RT-PCR as described above or prepared for immunocytochemical analysis. For the latter, either the structures may be frozen in Tissue-Tek and sectioned in a cryostat, or pelleted, fixed, embedded in wax, and sectioned in a microtome. As with monolayer cultures allowed to overgrow *in situ*, it may be difficult to dissociate the embryoid bodies to single cells enzymatically. Alternatively, the embryoid bodies may be returned to monolayer culture conditions in the presence of serum with or without the addition of feeder cell support.

Growth factor treatment, or coculture with different types of supporting cell, may help to bias ES cell differentiation into particular lineages. In our study, self-grown ES cells under standard conditions are relatively refractory to the action of many growth factors. However, partial induction of spontaneous differentiation by overgrowth *in situ* in monolayer or by embryoid body formation may be combined with factor treatment or transfer to coculture conditions with other cell types, to favor the development of particular lineages.

Once differentiation has been induced, RT-PCR or immunocytochemical techniques may be used to monitor changes in gene expression and cell phenotype. It is also very important ultimately to obtain evidence of differentiated cell function. A discussion of the scope of functional tests available for differentiated cell types is beyond the scope of this chapter, but might include tests *in vitro* (electrophysiology for neurons, glucose-dependent insulin secretion for beta islet cells of the pancreas) as well as transplantation studies *in vivo* into developing tissue or into damaged tissue, with assessment of appropriate integration of cells into the tissue and repair of tissue function following lesion or injury.

Acknowledgment

Work in our laboratory is supported by ES Cell International Pte., the National Health and Medical Research Council, and the Juvenile Diabetes Research Foundation. We thank all the members of our laboratory for their input into this chapter.

[31] Factors Controlling Human Embryonic Stem Cell Differentiation

By MAYA SCHULDINER and NISSIM BENVENISTY

Introduction

Human embryonic stem (ES) cells are pluripotent cell lines derived from the inner cell mass (ICM) of blastocyst stage human embryos.[1,2] These cells possess self-renewal capabilities, which can be preserved through tight regulation of their growth conditions. Such regulation enables these cells to proliferate indefinitely in culture. However, once the cells are allowed to differentiate, they spontaneously develop into various cell types.[3] The differentiation process can be influenced, to some extent, by use of external

[1] J. A. Thomson, J. Itskovitz-Eldor, S. S. Shapiro, M. A. Waknitz, J. J. Swiergiel, V. S. Marshall, and J. M. Jones, *Science* **282**, 1145 (1998).

[2] B. E. Reubinoff, M. F. Pera, C. Y. Fong, A. Trounson, and A. Bongso, *Nat. Biotechnol.* **18**, 399 (2000).

[3] J. Itskovitz-Eldor, M. Schuldiner, D. Karsenti, A. Eden, O. Yanuka, M. Amit, H. Soreq, and N. Benvenisty, *Mol. Med.* **6**, 88 (2000).

0076-6879/2003 $35.00

growth factors.[4] The special properties and human origin of human ES cell lines make them a unique model system for elucidating the processes of early embryogenesis and analyzing the effects of growth factors on early stem cells in the developing human embryo. In addition, they are an invaluable biotechnological and medical tool as they provide an unlimited cell source for generating purified cells for cellular transplantation. More than two decades ago, mouse ES cells were isolated,[5,6] and research conducted on them has provided invaluable protocols and insights into ES cell growth, differentiation and manipulation.[7–10] The work conducted on human ES cells is based on these tools and concentrates on special methods to control differentiation into desired tissues. This review focuses on methods to differentiate human ES cells.

Spontaneous Differentiation of Human ES Cells

ICM cells possess the potential to develop into all embryonic cell types. Obviously, in order for proper differentiation to occur, each cell must have the capacity to respond to an enormous network of signals which must work in tandem to control the cellular fate. These signals include soluble factors, cell surface molecules, and intrinsic control mechanisms such as transcription factors that work parallel to such physical mechanisms as gravity and pressure. Human ES cells are derived from this ICM population and preserve their pluripotent nature. In addition to having the ability to readily differentiate, they are expected to respond similarly to developmental cues. Indeed, in the absence of necessary support from a feeder layer and specific media additions, human ES cells spontaneously differentiate. When differentiating in a monolayer, ES cells have been observed to form mainly extra-embryonic cells. In order to produce a wide range of cellular phenotypes from the three embryonic germ layers, human ES cells were allowed to aggregate into spheroid clumps termed embryoid bodies (EBs).[3]

[4] M. Schuldiner, O. Yanuka, J. Itskovitz-Eldor, D. A. Melton, and N. Benvenisty, *Proc. Natl. Acad. Sci. USA* **97**, 11307 (2000).
[5] M. J. Evans and M. H. Kaufman, *Nature* **292**, 154 (1981).
[6] G. R. Martin, *Proc. Natl. Acad. Sci. USA* **78**, 7634 (1981).
[7] A. Bradley and E. J. Robertson, *in* "Current Topics in Developmental Biology," Vol. 20, p. 357. Yamada Science Foundation and Academic Press, Japan, 1986.
[8] R. H. Lovell-Badge, *in* "Teratocarcinomas and Embryonic Stem Cells: A Practical Approach" (E. J. Robertson, ed.), p. 153. IRL Press, Washington DC, 1987.
[9] E. J. Robertson, *in* "Teratocarcinomas and Embryonic Stem Cells: A Practical Approach" (E. J. Robertson, ed.), p. 71. IRL Press, Washington DC, 1987.
[10] T. Burdon, I. Chambers, C. Stracey, H. Niwa, and A. Smith, *Cell. Tiss. Org.* **165**, 131 (1999).

TABLE I

DIFFERENTIATION POTENTIAL OF HUMAN ES CELLS

Cell type	Differentiation	Functional assay	*In vivo* assay	RNA markers	Protein markers	Ref.
Ectoderm						
Neurons	Spontaneous	—	—	NFL	—	3
	Spontaneous	—	—	Nestin, Pax6, GAD, GABA	NFL/M/H NCAM, Nestin, MAP2, Vimentin, Glutamate, Synaptophysin, βTubulin	2
	Spontaneous/Directed by RA, NGF	—	—	NFH	—	4
	Directed by RA, NGF	—	—	5HTR2A/5A, DDC, DRD1, NFL	NFH	24
	Directed by RA, EGF, bFGF PDGF, IGF-1, NT-3, BDNF & sorting	Calcium imaging, Electrophysiology	—	—	NCAM, Nestin, A2B5, MAP2, TH, GABA, Synaptophysin, Glutamate, Glycine, ChAT	26
	Directed by bFGF, EGF	—	Integration and migration in mice brains	Nestin, Pax6, NFM, NSE,	NCAM, GAD, Vimentin, TH, Nestin, A2B5, NFM/L, MAP2, Synaptophysin, GABA, 5HT, Glutamate,	25
	Directed by bFGF	—	Integration and migration in mice brains	—	Nestin, MAP2, Musashi1, NFH, NCAM, GABA, Glutamate, TH	35
	Directed by transplantation into chick embryos	—	Integration in chick neural tube	—	NFM/H, HNK1, β3Tubulin	18
Astrocytes & Oligodendrocyte	Directed by RA, EGF, PDGF, IGF1, NT3, BDNF & sorting	—	—	—	GFAP	26
	Directed by bFGF, EGF	—	Integration and migration in mice brains	GFAP, MBP, Plp	GFAP, NG2, CNPase, O4	25
	Directed by bFGF	—	Integration and migration in mice brains	—	GFAP, O4	35
Skin	Spontaneous/Directed by EGF	—	—	Keratin	—	4
Adrenal	Directed by RA	—	—	DβH	—	4

Endoderm						
Liver	Spontaneous/Directed by NGF, HGF	—	—	Albumin, α1AT	—	4
Pancreas	Spontaneous/Directed by NGF	—	—	PDX1, Insulin	—	4
	Spontaneous	Glucose sensitivity	—	Insulin, Ngn3 IsGK, Pdx1, Glut1/2	Insulin	13
Mesoderm						
Muscle	Spontaneous	—	—	Enolase	—	4
Bone	Spontaneous	—	—	CMP	—	4
Kidney	Spontaneous/Directed by NGF	—	—	Renin, Kallikrein	—	4
Uro-genital	Spontaneous/Directed by NGF	—	—	WT1	—	4
Cardiomyocytes	Spontaneous	Rhythmic pulsations	—	cActin	—	3
	Spontaneous	Rhythmic pulsations	—	—	Desmin, cActin	2
	Spontaneous/Directed by TGFβ	—	—	cActin	—	4
	Spontaneous	Electrophysiology, Calcium transients	—	GATA4, Nkx2.5, cTN1/T, MLC2A/V cMHC	cMHC, αActinin, Desmin, cTroponin	14
	Directed by feeder layer	Electrophysiology	—	—	αActinin	20
Blood cells	Spontaneous	—	—	ζglobin	—	3
	Spontaneous	—	—	β/δglobin	—	4
	Directed by feeder layer & cytokines & sorting	—	—	Tal1, GATA2, $\alpha/\beta/\gamma$Globin	CD34, CD31, CD45, CD41, CD15, Glycophorin	19
Endothel	Spontaneous & sorting	Formation of blood vessels	Formation of functional vessels *in vivo*	PECAM1, VECad, CD34, GATA2	PECAM1, vWF, VECad,	12

Growth factors: BDNF: brain derived neurotrophic factor; EGF: epidermal growth factor; FGF: fibroblast growth factor; HGF: hepatocyte growth factor; IGF: insulin like growth factor; NGF: nerve growth factor; NT: neurotrophin; PDGF: platelet derived growth factor; RA: retinoic acid; TGF: transforming growth factor.

Molecular markers: 5HTR: 5 hydroxytryptamine (serotonin) receptor; α1AT: alpha 1 antitrypsin; CD: cluster of differentiation; ChAT: choline acetyl transferase; CMP: cartilage matrix protein; CNPase: cyclic nucleotide phosphodiesterase; cTN: cardiac troponin; DDC: dopa de-carboxylase; DR: dopamine receptor; DβH: dopamine β hydroxylase; GABA: gamma amino buteric acid; GAD: glutamic acid decarboxylase; GFAP: glial fibrillary acidic protein; IsGK: islet glucokinase; MAP: microtubule associated protein; MBP: myelin basic protein; MLC/MHC: myosin light/heavy chain; NCAM: neuronal cell adhesion molecule; NFL,M,H: Neurofilament light, medium, heavy chain; NG: neurogenin; Ngn: Noggin; NSE: neuron specific enolase; O4: oligodendrocyte4; PDX: pancreas duodenum homeobox gene; PECAM: platelet endothelial cell adhesion molecule; Plp: proteolipid; Tal: T-cell acute lymphoblastic leukemia; TH: tyrosine hydroxilase; VECad: vascular endothelial cadherin; vWF: vascular Willebrand factor; WT: Wilms tumor.

After several days as EBs various cell types appeared expressing markers of the three embryonic germ layers, and the human cells now responded to a range of extra-cellular signals that effect their differentiation capacity. It has been shown that human ES cells and their differentiated derivatives express a broad spectrum of receptors for growth factors.[4] This is yet another indication of their ability to respond to a wide variety of signals. After formation of EBs, the cells readily differentiate into several cell types[4] such as muscle, bone, kidney, skin, liver, neurons,[2,3] endothel,[12] hematopoietic cells,[3] pancreatic β cells[13] and even functional cardiomyocytes[3,14] (for a detailed description on the range of different cell types, see Table I). This process is not organized and does not lead to the formation of entire organs. Lack of proper gastrulation signals and organizer activity during the aggregation step are most likely responsible for this. Despite the lack of spatial organization, temporal cues do apparently exist as can be seen in mouse ES cells. During EB formation, the differentiating cells sequentially activate tissue specific genes. For example, endothelial specific transcripts[15] or erythrocyte globins[16] are transcribed in a manner that recapitulates their normal timed order of appearance during early embryonic development.

While EB formation is an important factor in influencing the formation of a variety of cell types, it is not necessarily essential for every type of differentiation. Ectodermal cells, such as neurons may form in a monolayer.[2] This is a likely consequence of their being the first differentiated cells to form in the developing embryo and as such probably do not require as many inductive signals.

Protocol for Formation of EBs

Human ES cells are usually grown on a layer of feeder cells, such as arrested mouse embryonic fibroblasts (MEFs) in order to retain their self-renewal and pluripotency properties. It is advisable, before beginning the

[11] C. L. Mummery, A. Feyen, E. Freund, and S. Shen, *Cell Differ. Dev.* **30**, 195 (1990).

[12] S. Levenberg, J. S. Golub, M. Amit, J. Itskovitz-Eldor, and R. Langer, *Proc. Natl. Acad. Sci. USA* **99**, 4391 (2002).

[13] S. Assady, G. Maor, M. Amit, J. Itskovitz-Eldor, K. L. Skorecki, and M. Tzukerman, *Diabetes* **50**, 1691 (2001).

[14] I. Kehat, D. Kenyagin-Karsenti, M. Snir, H. Segev, M. Amit, A. Gepstein, E. Livne, O. Binah, J. Itskovitz-Eldor, and L. Gepstein, *J. Clin. Invest.* **108**, 407 (2001).

[15] D. Vittet, M. H. Prandini, R. Berthier, A. Schweitzer, H. Martin-Sisteron, G. Uzan, and E. Dejana, *Blood* **88**, 3424 (1996).

[16] M. H. Lindenbaum and F. Grosveld, *Genes Dev.* **4**, 2075 (1990).

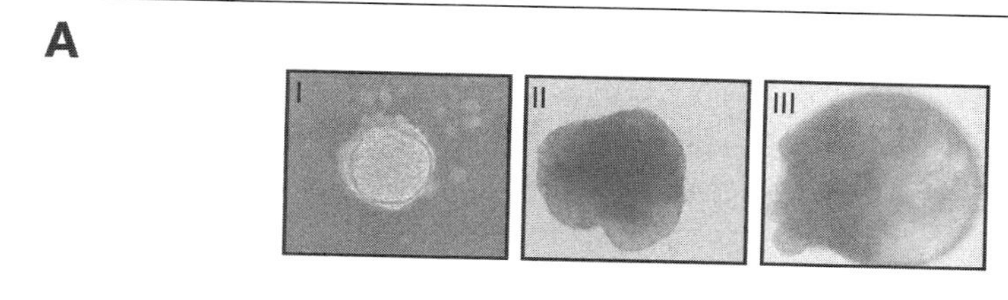

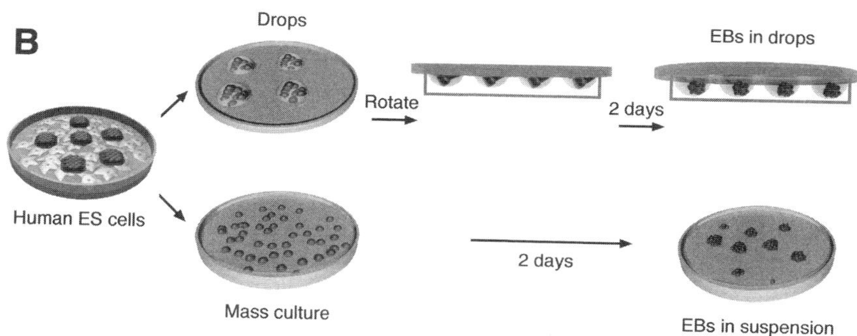

FIG. 1. Formation of embryoid bodies. The three stages of EBs development: I-Simple EB, II-Cavitated EB, III-Cystic EB. Schematic protocol for preparing EBs either in suspension or by use of the "hanging drop" method.

differentiation process, to transfer the cells for a single passage onto a chemical matrix, such as 0.1% gelatin coated plates, in order to reduce the percentage of MEFs present in the culture. Human ES cells are grown in medium that contains: 400 ml KnockOut™ DMEM, 80 ml KnockOut™ SR, 5ml (2 m*M*) glutamine, 50 μl (0.1 m*M*) β-mercaptoethanol, 5 ml (1×) nonessential amino acids stock, 2.5 ml (50 units/ml) penicillin (50 μg/ml) streptomycin, 1 ml (4 ng/ml) basic fibroblast growth factor (bFGF). Upon reaching confluence, cells should be dissociated from the plate using 0.1% Trypsin/1 m*M* EDTA. To form EBs the dissociated cells are placed into petri dishes at a high concentration (approximately 10^7 cells per 10 cm plate). Bacterial petri dishes should be used to avert adherence of the cells to the plate, thus allowing spontaneous aggregation (see Fig. 1B). EBs are grown in the same medium as ES cells but without the addition of bFGF. After 2 days of incubation, clumps of cells form and the medium can be changed for the first time. Medium is changed either by tilting the plate and aspirating approximately half of the media, or by gently transferring the

EBs into conal tubes and allowing the clumps to settle (do not centrifuge). The medium is then aspirated and the EBs returned to the petri dish after being resuspended in new growth media. Media should be changed every 2 days in the manner described above. The EBs develop in three main stages: 2 days following aggregation, small clusters, or simple EBs, are visible (see Fig. 1A-I). A few days later the EBs appear dark in the center as they start cavitating (see Fig. 1A-II). The final stage is characterized by the formation of cystic EBs which possess a fluid filled cavity and harbor many differentiated cell types (see Fig. 1A-III). The different sizes of EBs formed using this method may effect their differentiation status. This obstacle can be overcome by determining a specific EB size using the "hanging drop" method.[17] This procedure involves placing drops of cells onto a tissue culture plate lid. The lid is then placed back onto a phosphate buffered saline (PBS) filled plate to avoid drying up the samples (see Fig. 1B). In this way a specific amount of cells aggregate at the bottom of the drop and form homogeneous EBs. Drops should be 40 μl in size and contain approximately 2000–4000 cells. No more than 25 drops should be placed onto a 10-cm plate lid. This limit will ensure they do not merge. After two days, EBs should be carefully collected and placed in a petri dish. Growth should then be continued according to the first protocol.

Factors Effecting Differentiation of Human ES Cells

In the developing embryo, gastrulation is followed by formation of the neural tube. This initiates a cascade of inductive interactions between different cell types and germ layers and eventually leads to the formation of the complete organism. By placing ES cells in a specific embryonic location advantage may be taken of inductive signals that are important for directing differentiation into a particular tissue. Indeed, when this procedure was performed through injecting human ES cells in the vicinity of the chick neural tube, neuronal differentiation was observed.[18] These signals can be partially reconstructed *in vitro* by growing the cells on specialized feeder layers. As cell lines produced from different tissues secrete a variety of substances into the growth media, they are expected to effect differentiation in a complex manner. Such matrices influence the stem cells not only via secreted growth factors but also through their cell surface proteins. Specialized feeder layers have been used to direct differentiation of human

[17] V. A. Maltsev, J. Rohwedel, J. Hescheler, and A. M. Wobus, *Mech. Dev.* **44**, 41 (1993).
[18] R. S. Goldstein, M. Drukker, B. E. Reubinoff, and N. Benvenisty, *Dev. Dyn.* **225**, 80 (2002).

ES cells to hematopoietic[19] and cardiac muscle cells[20] (see Table I). The drawback of such a system is that feeder layer cells must be eliminated before therapeutic use for transplantation. Another disadvantage is that the actual signals required from the matrix remain unknown. These problems can be partially addressed by substituting the feeder layer with better-defined conditions such as soluble growth factors and chemical matrices.

Soluble growth factors constitute the most studied group of effectors of differentiation. There are approximately 20 families of cytokines found in humans with roughly 100 different factors and their derivatives. The sheer number of factors and possible combinations makes studying their effects very problematic. Furthermore, it is likely that human ES cells themselves secrete growth factors. It is therefore virtually impossible to arrive at a true analysis of the effect of a single factor. Such obstacles not withstanding, single soluble growth factors have indeed been added to differentiating ES cells. The results of this procedure indicated that each factor has a unique effect on differentiation (see Table I). Thus, for example, activin A, inhibited formation of most cell types analyzed, causing differentiation to mainly mesodermal tissues such as skeletal and cardiac muscle. Epidermal growth factor (EGF) allowed formation of both ectodermal and mesodermal tissues and enhanced expression of markers for skin cells in the culture. When neuronal growth factor (NGF) was present differentiation into cells of all three germ layers occurred, with a more pronounced increase in the number of neuronal cells. There was also an increase in expression of molecular markers for liver and pancreas cells derived from the endoderm.[4] Any observed change in the cell types in culture could in principle occur as a result of two separate mechanisms. The first is through ability to effect cellular fate decision so that cells differentiate into the desired path. The second possibility is that the conditions cause positive and negative selection optimizing the growth of a specific cell type already present in the culture. If we knew which mechanism was used by specific factors, we would be able to develop better protocols: growth factors that direct differentiation may work more efficiently if administered at early stages of the differentiation process. In the case of growth factors that work by selection, later administration following primary differentiation may be more effective. In order to administer growth factors optimally, their effective concentration

[19] D. S. Kaufman, E. T. Hanson, R. L. Lewis, R. Auerbach, and J. A. Thomson, *Proc. Natl. Acad. Sci. USA* **98**, 10716 (2001).

[20] C. Mummery, D. Ward, C. E. van den Brink, S. D. Bird, P. A. Doevendans, T. Opthof, A. Brutel de la Riviere, L. Tertoolen, M. van der Heyden, and M. Pera, *J. Anat.* **200**, 233 (2002).

levels should be taken into account. Many of the growth factors have characteristics of morphogens,[21–23] and exert different effects according to their concentration. Thus, several possible influences may emerge from analyses of a range of concentrations. Given these constraints it is hardly surprising that no single factor tested until now has led to the formation of a single cell type culture. Yet, the more complex the protocols and the greater the number of stages or growth factors used, the nearer this goal becomes. For example, neuronal cells are formed spontaneously after aggregation of EBs[3] or when grown in a monolayer.[2] When single growth factors such as retinoic acid (RA) or NGF are added basal neuronal differentiation can be enhanced from approximately 20% to approximately 50%.[24] Methods utilizing manual isolation followed by growth factor combinations such as bFGF and EGF serve to further enhance these numbers.[25] Finally, protocols that comprise several stages of growth factor administration including epidermal growth factor (EGF), basic fibroblast growth factor (bFGF), platelet derived growth factor (PDGF), insulin like growth factor 1(IGF1), neurotrophin 3 (NT3) and brain derived neurotrophic factor (BDNF), allow more homogeneous cultures to arise. These contain close to 100% neuronal cells that have also proved functional[26] (see Table I).

Combining growth factors with chemical matrices may enable selection of desired cell types and may even effect differentiation. The chemical matrices act through binding and activation of different cell surface receptors. Such combinations have been used in mouse ES cells to enhance neuronal numbers in culture by utilizing laminin and polylysine, which are the optimal matrices for obtaining neuronal proliferation in addition to growth factors.[27]

If extra-cellular signals are not sufficient to achieve differentiation to a specific cell type, it is possible to directly activate the intra-cellular molecular mechanisms controlling the cell fate. For example, in mouse ES cells over-expression of HNF3β, a crucial transcription factor that determines liver cell fate, caused endoderm formation and hepatocyte

[21] A. J. Durston, J. P. Timmermans, W. J. Hage, H. F. Hendriks, N. J. de Vries, M. Heideveld, and P. D. Nieuwkoop, *Nature* **340**, 140 (1989).

[22] R. Dosch, V. Gawantka, H. Delius, C. Blumenstock, and C. Niehrs, *Development* **124**, 2325 (1997).

[23] N. McDowell, A. M. Zorn, D. J. Crease, and J. B. Gurdon, *Curr. Biol.* **7**, 671 (1997).

[24] M. Schuldiner, R. Eiges, A. Eden, O. Yanuka, J. Itskovitz-Eldor, R. S. Goldstein, and N. Benvenisty, *Brain Res.* **913**, 201 (2001).

[25] B. E. Reubinoff, P. Itsykson, T. Turetsky, M. F. Pera, E. Reinhartz, A. Itzik, and T. Ben-Hur, *Nat. Biotechnol.* **19**, 1134 (2001).

[26] M. K. Carpenter, M. S. Inokuma, J. Denham, T. Mujtaba, C. P. Chiu, and M. S. Rao, *Exp. Neurol.* **172**, 383 (2001).

[27] M. Li, L. Pevny, R. Lovell-Badge, and A. Smith, *Curr. Biol.* **8**, 971 (1998).

differentiation.[28] Methods for genetically manipulating human ES cells have recently been developed.[29] This will allow similar procedures to be used in these cells.

Protocol for Administration of Growth Factors

Growth factors are large proteins that may not diffuse freely in densely packed EBs. If a growth factor does not reach every cell, analysis of its effect can be problematic. As aggregation into EBs is an important step in the formation of the three germ layers, it is advisable that growth factors be administered by means of a two-step protocol.[4] During the first stage, EBs are formed over a period of 4 days. This period is sufficient for primary cellular interactions to occur within the EB. The second stage consists of dissociation of the EBs into single cells and adding the growth factors to the monolayer. Dissociation can be performed by suspending the aggregates in trypsin for 10 min in 37°C whilst shaking the tube every 2 min to assure dispersal of the clumps. Single cells and remaining clumps should be plated on appropriate matrices such as: collagen, gelatin, fibronectin, or poly-lysine. The following day, after cells have adhered, growth factors are added in the desired concentrations and refreshed every 2 days. Such a protocol facilitates initial formation of the three embryonic germ layers as a result of cellular interactions during the aggregation step, and homogeneous administration of growth factors to the cell population.

Assaying for Cellular Differentiation

During differentiation of human ES cells, multiple cell types appear in parallel. It is therefore important to develop methods for monitoring a desired differentiation process. The ability to recognize a specific cell type may also enable sorting into clean cultures, that is extremely important for transplantation therapy or developmental research. By facilitating the monitoring of the differentiation process we can dramatically influence the ability to optimize a differentiation protocol. Some cell types have been studied more than others or have easier features to analyze, but essentially any cell type can be identified provided there is enough information on its characteristics.

[28] M. Levinson-Dushnik and N. Benvenisty, *Mol. Cell. Biol.* **17**, 3817 (1997).

[29] R. Eiges, M. Schuldiner, M. Drukker, O. Yanuka, J. Itskovitz-Eldor, and N. Benvenisty, *Curr. Biol.* **11**, 514 (2001).

Morphology

The most straightforward aspect of differentiation to document is morphology by phase microscope. However, as most cells do not have distinct morphological features this method is restricted to a few cell types. Examples of cells with easily identifiable morphologies are neurons with their distinct axonal outgrowths or muscle cells appearing with sincitia of nuclei. In some cases, morphology is supplemented by function as in the case of pulsating cardiomyocytes. However, even in the case of these cells, morphology is not sufficient to discriminate between cell subtypes, such as dopaminergic or serotonergic neurons. Where this type of analysis is required, a molecular procedure must be sought. Several cell types form unique tissue patterns when allowed to differentiate in three dimensions such as in EBs or differentiated teratomas. Sectioning of these masses allows us to see such structures as bone or gut. Human ES cell derived endothelial cells have also been shown to form functional blood vessels when grown on a matrigel matrix or when forming teratomas.[12]

Molecular Markers

Molecular markers are commonly used to identify tissues that are difficult to distinguish morphologically. They are also used for analysis of cell subtypes. Each cell type, due to its unique function, has a specific gene expression pattern that allows us to analyze for cell specific genes or cell specific gene-combinations.

When assaying at the RNA level, it is possible to look at the entire population of cells by using RT-PCR for specific markers, or DNA microarrays if a more complete expression profiling is required. These methods document expression in large cell populations and thus do not allow us to estimate the number of cells that actually express a specific cellular marker. *In situ* hybridization can be used if quantification of specific cell types is required.

During differentiation most genes are regulated by transcription. Nevertheless, a more accurate analysis of gene expression is found at the protein level. Assaying for protein presence can be performed on whole cell populations by Western blotting of protein extracts. Immunostaining of single cells or tissue sections is a more commonly used method. In cases where the protein markers are extra-cellular they may easily be assayed for while retaining the cells viable through cell sorting methods. Such methods include magnetic sorting using metal beads tied to secondary antibodies or fluorescence-activated cell sorting (FACS) utilizing fluorescent antibodies. These methods have the additional advantage of producing clean cell cultures while also facilitating accurate estimation of cell proportions within the differentiating population. These methods are specifically advantageous for assaying hematopoietic cells, as the surface molecules of these cells have been well characterized. In human ES cells, such methods have yielded pure populations of $CD34^+$, hematopoietic progenitor cells.[19] Protein modification or localization are sometimes required for proper function so that the mere presence of proteins does not necessarily attest to their functionality. Ultimately, functional assays should be sought in order to assess correct differentiation.

An additional way of using tissue specific gene expression in order to follow up on differentiated cell types can be performed by genetically labeling the cells. For example, transfection of marker genes such as β-galactosidase or green fluorescent protein (GFP) under a tissue specific promoter, allows monitoring of cellular differentiation by following the expression of the marker gene.[27] Furthermore, GFP is an extremely efficient tool for sorting the cells utilizing a FACS. Similar methods using selection markers, such as the neomycin resistance gene, allow growth of purified cell populations but make quantification more difficult.[30,31] Extensive research on transfection of mouse ES cells has showed that various chemical reagents do not introduce DNA into the cells efficiently, while physical pressure such as electroporation or injection result in a much higher success rate. As injection requires special skills, the method of choice for genetically modifying mouse ES cells is electroporation.[8] In human ES cells, it seems that electroporation is less efficient, whereas poly-cationic reagents have been shown to facilitate DNA introduction at a rate that is sufficiently high for both transient and stable transfection procedures.[29] In addition, it has been shown that infection by lentiviral vectors is highly efficient in human ES cell lines.[32] When stable integration of DNA in the genome is required, viral infection yields much higher success rates. However, it would seem that transfection that does not introduce viral sequences has several advantages including gaining easier approval of genetically manipulated cell lines for use in clinical trials.

Protocol for Genetic Labeling of Human ES Cells

Human ES cells were shown to be efficiently transfected using the poly-cationic reagent ExGen (MBI-Fermentas).[29] One day prior to transfection

[30] M. G. Klug, M. H. Soonpaa, G. Y. Koh, and L. J. Field, *J. Clin. Invest.* **98**, 216 (1996).
[31] S. Marchetti, C. Gimond, K. Iljin, C. Bourcier, K. Alitalo, J. Pouyssegur, and G. Pages, *J. Cell. Sci.* **115**, 2075 (2002).
[32] A. Pfeifer, M. Ikawa, Y. Dayn, and I. M. Verma, *Proc. Natl. Acad. Sci. USA* **99**, 2140 (2002).

the cells are plated in 6-well tissue culture dishes at a density of $1.5–2 \times 10^5$ cells per single well. On the day of transfection, 1 ml of fresh media is placed in each well. A transfection master mix may be prepared. The procedure for a single well is to add 2 μg of DNA to 100 μl of 150 m*M* NaCl solution and vortexing. 6.6 μl of ExGen are then added to the DNA master mix (not the reverse order). This should be followed immediately by an additional vortex for 10 sec to ensure homogeneous distribution of the reagents. The mix is left at room temperature for 10 min to allow formation of transfection complexes of the desired size. The transfection solution is now added to each well while swirling to disperse evenly. The plates are centrifuged for 5 min at 280*g* and then left for 30–45 min in an incubator. As the feeder layer cells are extremely sensitive to ExGen, it is recommended to rinse the cells twice in PBS following the short incubation, and then add 2 ml of fresh growth media to each well. For transient transfection the cells are monitored between 24 and 72 hr after transfection. For stable transfection it is extremely important that selection only be initiated following passaging and disaggregation into single cells. Selection on the original colonies results in massive death of the colony cells, and does not allow propagation of single transfected cells. Following passaging into media containing the selection drug, resistant colonies arise from a single stable transfected cell approximately 10 days later. When colonies consisting of hundreds of cells become visible, each colony is transferred to a separate well of a 24-well tissue culture plates for expansion, freezing and analysis.

For stable transfection it is important to consider the growth of the cells during the selection process. Human ES cells grow on a feeder layer of primary MEF cells and not on immortalized cell lines such as STO that can be more easily manipulated to express drug resistant markers. Yet, as with the ES cells, MEF cells must also survive the selection procedure. It is possible to bypass the need for drug resistant MEFs during the selection process by growing the cells on a chemical matrix, such as matrigel, together with conditioned media from the MEFs. Conditioned media is supplemented with the necessary drug before addition to the ES cells, thus bypassing contact with the MEFs. This technique has been shown to sustain self-renewal for many passages.[33] Another possibility is to produce the primary fibroblasts from mice carrying resistance genes. Most genetically manipulated mice harbor the neomycin resistance gene. In addition, specialized mice strains carrying many drug resistances have been

[33] C. Xu, M. S. Inokuma, J. Denham, K. Golds, P. Kundu, J. D. Gold, and M. K. Carpenter, *Nat. Biotechnol.* **19**, 971 (2001).

developed. Examples include the DR4 mouse strain that can be utilized for neomycin, hygromycin, puromycin and 6TG selection.[34]

Functional Assays

In order to ascertain that proper differentiation has occurred it is necessary to look at the functionality of the cell type that has been produced. This is, of course, even more important if the purpose of the research is to produce cells for transplantation medicine. Some tissues needed for transplantation such as cartilage may not necessitate any complex function other than the ability to form an intact tissue, whereas others, such as neurons require very intricate functions that should be coordinated by proper connectivity. Methods used to assay *in vitro* functionality are tissue-specific. For example, functionality of neurons can be determined by their ability to produce action potentials in response to various signals. Methods for checking these parameters such as electrophysiology and calcium imaging have also been used on neurons derived from human ES cells. These analyses have shown that the differentiation protocol used facilitated formation of several types of functional neurons such as GABAergic, glutamatergic, dopaminergic and cholinergic.[26] The presence of functional cells does not necessarily indicate that the response is physiological and as such will allow a proper, coordinated reaction if transplanted *in vivo*. With some tissues assessment is even more straightforward. For example, the sole function of pancreatic β cells is to secrete insulin in a glucose dependent manner. When human ES cells differentiated spontaneously into pancreatic β cells, it was suggested that these cells could produce elevated levels of insulin in media containing higher glucose concentrations.[13] Functionality of human ES cell derived cardiomyocytes was assayed by observing pulsation[3] or more accurately by patch clamp electrophysiology that demonstrated rhythmic action potentials in the contracting cardiomyocytes.[20] Other options for monitoring human ES derived cardiomyocytes responsiveness to agonists and antagonist are: (1) External electrophysiology using a multi-electrode array that reads action potentials from whole population or (2) intracellular calcium imaging for single cells[14] (for summary see Table I).

Tissue Integration *In Vivo*

It must be realized that not all methods for differentiation assessment are interdependent: The fact that a cell type may have the right morphology

[34] K. L. Tucker, Y. Wang, J. Dausman, and R. Jaenisch, *Nucleic Acids Res.* **25**, 3745 (1997).

does not necessarily mean it expresses transcripts essential for its function. Moreover, correct gene expression does not always indicate protein presence or function. The same may be said for tissue integration. Even in cases where cells differentiate and function properly, they may not have the ability to integrate into the desired tissue due to lack of requisite cell surface signals. For this reason, after assessing functionality, it is important to also ascertain whether integration occurs easily upon transplantation. The ultimate standard for analyzing differentiation would use transplanted cells in a diseased animal. This requires an animal model suffering from organ impairment or cellular dysfunction, and monitoring phenotypic changes following transplantation of new cells. Improvement would indicate that the cell is expressing all the necessary cellular markers needed for function and integration. Only after such analysis will it be possible to determine the exact effect of various factors on human ES cells' differentiation. It is also important to perform this step before transplantation into human patients. So far studies of human ES cell differentiation have only shown integration in neuronal systems. Cells originating from human ES cells were indeed demonstrated to migrate and integrate in animal brains.[18,25,35] It is yet to be demonstrated that the *in vivo* transplanted cells function properly in the recipient animal. In order for such experiments to be performed and for medical transplantations to become feasible, methods for overcoming possible tissue rejection must be found.[36,37]

Conclusions

Over the past few years protocols have emerged for growing human ES cells and differentiating them *in vitro* via the production of EBs. The use of growth factors is important in order to achieve purified populations of differentiated cells and to study the role of secreted molecules in early embryogenesis. The combination of different matrices, addition of growth factors and transfection of transcription factors should yield more specific and mature cell types that must then be purified for further characterization or medical use.

[35] S. C. Zhang, M. Wernig, I. D. Duncan, O. Brustle, and J. A. Thomson, *Nat. Biotechnol.* **19**, 1129 (2001).

[36] M. Drukker, G. Katz, A. Urbach, M. Schuldiner, G. Markel, J. Itskovitz-Eldor, B. Reubinoff, O. Mandelboim, and N. Benvenisty, *Proc. Natl. Acad. Sci. USA* **99**, 9864 (2002).

[37] M. Schuldiner and N. Benvenisty, *in* "Recent Research and Developments in Molecular and Cellular Biology" (S. Pandalai, ed.), Vol. 2, p. 223. Research Signpost, Trivandrum, 2001.

Acknowledgment

This research was partially supported by funds to N. B. from the Herbert Cohn Chair (Hebrew University), by a grant from the Juvenile Diabetes Fund (USA), by a grant from the Israel Science Foundation (grant no. 672/02-1) and by funds from the United States—Israel Binational Science Foundation (grant no. 2001021). M. S. is a Clore fellow.

[32] Development of Cardiomyocytes from Human ES Cells

By IZHAK KEHAT, MICHAL AMIT, AMIRA GEPSTEIN, IRIT HUBER, JOSEPH ITSKOVITZ-ELDOR, and LIOR GEPSTEIN

Introduction

It is generally accepted that adult human cardiomyocytes are mostly terminally differentiated and possess limited regenerative capacity following significant cell losses such as those that occur during myocardial infarction.[1] Cell transplantation is emerging as a novel strategy for myocardial regeneration, but has been hampered by the lack of a source for human cardiomyocytes.[2] Similarly, the lack of a human cardiomyocyte cell line has significantly limited a variety of experimental procedures. The recent advances in human embryonic stem cells (ES) research suggests a possible solution for this cell sourcing problem, since these unique cells can be propagated in mass *in vitro* and coaxed to differentiate into the desired lineage.[3,4]

Human ES cells are continuously growing cell lines of embryonic origin that were isolated from the inner cell mass of human blastocysts.[3,4] These cells display the characterizing properties of ES cells, namely: derivation from the pre- or peri-implantation embryo, prolonged undifferentiated proliferation under special conditions, and the capacity to form derivatives of all three germ layers.

One of the most fascinating and important aspects of human ES cell lines is their ability to differentiate *in vitro* to advanced derivatives of all

[1] M. H. Soonpaa and L. J. Field, *Circ. Res.* **83**, 15 (1998).
[2] L. Reinlib and L. Field, *Circulation* **101**, E182 (2000).
[3] J. A. Thomson, J. Itskovitz-Eldor, S. S. Shapiro, M. A. Waknitz, J. J. Swiergiel, V. S. Marshall, and J. M. Jones, *Science* **282**, 1145 (1998).
[4] B. E. Reubinoff, M. F. Pera, C. Y. Fong, A. Trounson, and A. Bongso, *Nat. Biotechnol.* **18**, 399 (2000).

0076-6879/2003 $35.00

three germ layers. A prerequisite for this differentiation is the cultivation of the ES cells into three-dimensional cell aggregates.[5] These aggregates, termed embryoid bodies (EBs) contain differentiating tissue from all three germ layers. Using the EB differentiation system cardiomyocytes can be reproducibly generated from the human ES cells.[6] The formation of cardiomyocytes within the EB provides a powerful *in vitro* tool for investigation of early cardiomyogenesis, differentiation of early human cardiac precursor cells, the development of excitability, excitation-contraction coupling, and the molecular signals involved in these processes.[6,7]

In this chapter we describe the methodologies associated with human ES cell cultivation and differentiation to cardiomyocytes as well as the techniques used for studying the structural and functional properties of these unique cells.

Feeder Layer Preparation

Continuous propagation of the human ES cells without differentiation requires cultivation of the cells on mitoticaly inactivated murine embryonic fibroblast feeder layers (MEF). Further studies showed that human ES cells could be successfully propagated and cultured also on Matrigel or laminin, but still required a medium conditioned by MEF.[8] The critical soluble factors produced by the MEF that prevent differentiation of the cells are still undetermined.

Pregnant ICR mice on day 13 of pregnancy are routinely used for the generation of MEFs using the following protocol:

1. Remove the uterus and place in bacteriological dish. Wash with sterile PBS.
2. Open the uterus and release the embryos. Dissect the embryos individually and discard the soft internal organs.
3. Remove the embryos to a clean dish, and wash again with sterile PBS.

[5] J. Itskovitz-Eldor, M. Schuldiner, D. Karsenti, A. Eden, O. Yanuka, M. Amit, H. Soreq, and N. Benvenisty, *Mol. Med.* **6**, 88 (2000).

[6] I. Kehat, D. Kenyagin-Karsenti, M. Snir, H. Segev, M. Amit, A. Gepstein, E. Livne, O. Binah, J. Itskovitz-Eldor, and L. Gepstein, *J. Clin. Invest.* **108**, 407 (2001).

[7] J. Hescheler, B. K. Fleischmann, M. Wartenberg, W. Bloch, E. Kolossov, G. Ji, K. Addicks, and H. Sauer, *Cells Tissues Organs* **165**, 153 (1999).

[8] C. Xu, M. S. Inokuma, J. Denham, K. Golds, P. Kundu, J. D. Gold, and M. K. Carpenter, Feeder-free growth of undifferentiated human embryonic stem cells. *Nat. Biotechnol.* **19**, 971 (2001).

4. Mince the embryos and add 2 ml of Trypsin–EDTA (0.5% trypsin, 0.53 m*M* EDTA; Life Technologies, Inc.). Continue mincing and add additional 5 ml of Trypsin–EDTA. Incubate 10 min at 37°C.
5. Neutralize the Trypsin by adding 10 ml of culture medium (90% DMEM 4.5 mg/ml glucose, 10% heat inactivated Newborn Calf Serum, 100 unit of Penicillin and 0.1 mg/ml streptomycin). Transfer to conical tubes.
6. Allow the debris to settle and divide the supernatant to tissue culture flasks (80 cm^2 area, approximately three embryos per flask). Add 20 ml of medium.
7. Grow the cells for three days and freeze them in 70% DMEM, 20% Serum and 10% DMSO (Sigma Chemical Co., St. Louis, Missouri, USA).
8. Thaw the cell and passage routinely for up to five passages.

Before plating the human ES cells on the feeders, the fibroblasts should be mitotically inactivated to prevent overgrowth. This inactivation could be achieved by irradiation or using the following protocol with Mitomycin C:

1. Prepare 8 μg/ml Mitomycin C (Sigma) in DMEM. Add to the culture flask and incubate for 2 hr.
2. Wash twice with PBS.
3. Trypsinize the cells, resuspend in medium and centrifuge for 5 min at 500*g*.
4. Aspirate the medium and resuspend cells in ES cell medium (see below).
5. Plate the cells on gelatin coated (0.1%) culture dishes at a density of 300,000 cells/well in 6-well plates.

ES Cell Propagation

Human ES cells are grown and propagated in colonies on top of the MEF feeder layers. The cells are cultured in a medium containing 20% serum as described in the following protocol, but knockout serum replacement (Invitrogen, Carlsbad, CA) with the addition of 4 ng/ml basic fibroblasts growth factor (bFGF) have also been successfully used.[9]

1. ES cells are grown in medium containing: 79% KnockOut DMEM, 20% defined Fetal Bovine Serum (HyClone, Logan, Utah, USA), 1% nonessential amino acids, 1 m*M* L-glutamine, and 0.1 m*M*

[9] M. Amit, M. K. Carpenter, M. S. Inokuma, C. P. Chiu, C. P. Harris, M. A. Waknitz, J. Itskovitz-Eldor, and J. A. Thomson, *Dev. Biol.* **227**, 271 (2000).

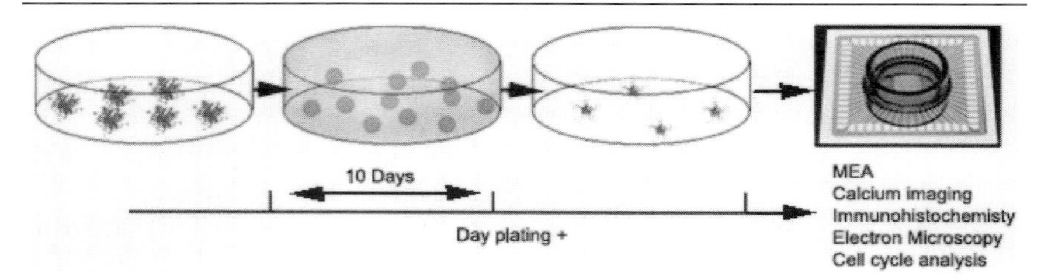

FIG. 1. Establishment of a cardiomyocyte differentiation system from the human ES cells. Human ES cells are grown on top of MEF feeder layers. The cells are transferred and grown in suspension for 10 days where they form EBs. The EBs are then plated on gelatin coated culture dishes where they continue to differentiate. Spontaneously contracting areas are dissected out from the EBs and are further analyzed by a variety of methods including MEA recordings, calcium imaging, immunohistochemistry, electron microscopy, etc.

β-mercaptoethanol (all from Life Technologies, Inc., Rockville, Maryland, USA, except when indicated).

2. Cells are split in a ratio of 1 : 3 every 5–6 days, depending on the size of the colonies.
3. Aspirate the medium and add 0.5 ml of 1 mg/ml collagenase IV (Life Technologies) in DMEM for 20 min at 37°C.
4. Scrape the cells, add 1 ml of fresh medium, and centrifuge at 100*g* for 2 min.
5. Resuspend the cells in fresh medium and plate on MEF covered wells.
6. Alternatively, the cells can be frozen in 70% DMEM, 20% Fetal Bovine Serum, and 10% DMSO.

The cultivation of the ES cells into cellular aggregates appears to be a prerequisite for differentiation. Different protocols have been used in the murine ES cells for such cultivation, including the "mass culture" technique,[10] and the "hanging drops" technique.[11] We have used direct massing of the cells to form embryoid bodies (EBs) for differentiation to cardiomyocytes (Fig. 1). Using this technique spontaneously contracting areas appear in about 10% of the EBs at day 4–20 post plating.

When grown in suspension, some of the EBs forms cysts. Both cystic and noncystic EBs can differentiate into cardiomyocytes. The EBs formed by human ES cells appear to be less organized than the ones

[10] T. C. Doetschman, H. Eistetter, M. Katz, W. Schmidt, and R. Kemmler, *J. Embryol. Morphol.* **87**, 27 (1985).

[11] A. M. Wobus, G. Walluket, and J. Hescheler, *Differentiation* **48**, 173 (1991).

formed by their murine counterparts, however they do contain differentiating tissues from all three germ layers. It is interesting to note that some human ES cell lines were reported not to form EBs.[4] It is yet to be determined whether this observation represents intrinsic differences between the cell lines, or is due to the different methodologies used. The following protocol can be used for direct cell massing and creation of EBs:

1. Grow the human ES cells on the feeder layers, as described earlier.
2. Aspirate the medium and add 0.5 ml of 1 mg/ml collagenase IV in DMEM for 20 min at 37°C.
3. Scrape the cells, add 1 ml of fresh medium, and centrifuge at 100*g* for 2 min.
4. Resuspend the cells in fresh medium. The cells can be further dispersed using a 200–1000 μl tip to form small clumps of 3–15 cells.
5. The cells are cultured at a density of 50,000 cells/ml in a 58 mm petri dish for 7–10 days.
6. Medium is replaced every 3–4 days. Aspirate half of the medium in the plate, centrifuge at 100*g* for 3 min, resuspend in fresh medium and return to the plate.
7. After 7–10 days, aspirate the medium and centrifuge at low speed (500 rpm) for 1 min. Resuspend the EBs in fresh medium and plate on gelatin coated (0.1% gelatin) 24 well plates at a density of 1–4 EBs per well.
8. Change medium every day.
9. Spontaneous contractions should appear between day 5 and 20 post-plating in about 10% of the EBs.

Immunofluorescent Analysis

The presence of cardiomyocytes and their structural development and maturation can be assessed by immunofluorescence studies. These studies can be performed on dispersed cell or on whole EBs, taking advantage of the transparency of the EB (Fig. 2a–b). Imaging of immuno-stained EBs requires a confocal microscope; alternatively paraffin sections of fixed EB can be performed and imaged using either a confocal or an epifluorescence microscope. In both cases, whole EB and dispersed cell staining, it is preferable to enrich the cardiomyocytes by dissecting out and analyzing only the contracting areas within the EB. A day in advance, glass cover slips are coated with 0.1% gelatin in 6-well plates. The EBs culture dish wells are scanned for contracting EBs and their location is marked at the bottom of

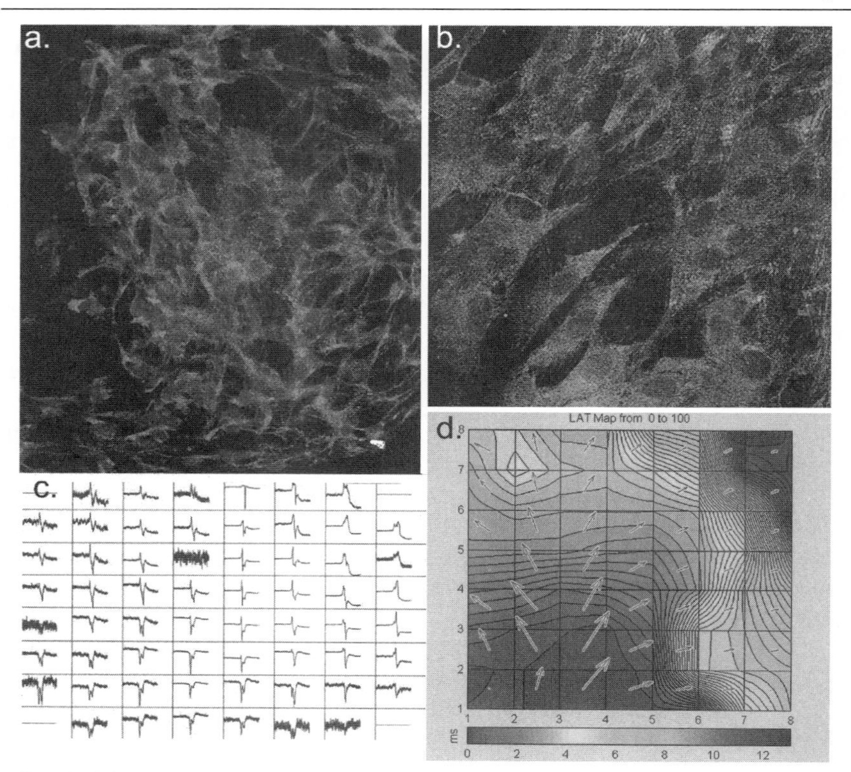

FIG. 2. (a) Immunostaining of whole EB with anti-cTnI antibodies, showing cardiomyocytes at different orientations. (b) Dual immunostaining of the EB with anti-Connexin 43 (gap junction protein) and anti-cTnI, showing gap junctions between cardiomyocytes in the EB. (c) Sixty electrograms recorded simultaneously from an EB using the MEA mapping technique. (d) The resulting activation and vectorial velocity maps. (See Color Insert.)

the plate by a marker pan. Using a binocular microscope inside a laminar flow hood, the contracting areas can be dissected out using pooled glass pipettes or small gauge syringe needles.

Several methods can be used to disperse the cells. Murine ES cell-derived cardiomyocytes were successfully dispersed in a solution containing collagenase B,[12] however adequate results can also be achieved using trypsin according to the following protocol:

1. The dissected contracting areas are aspirated with a minimal amount of medium, and transferred to a sterile tube.

[12] V. A. Maltsev, A. M. Wobus, J. Rohwedel, M. Bader, and J. Hescheler, *Circ. Res.* **75**, 233 (1994).

2. Add about 1 ml of a solution containing 0.25% Trypsin and EDTA (1 : 2000) in Puck's seline A (Biological Industries, Beit Haemek, Israel) for 1 min at 37°C.
3. Further dispersion is achieved by pipetting the EBs up and down with a filtered 200–1000 μl Gilson tip.
4. Add fresh medium to the cells, and centrifuge for 3 min at 200*g*.
5. Resuspend the cells in fresh medium at the desired cell concentration and plate on the gelatin covered glass cover slips for 24 hr.

For staining of nondispersed EBs, contracting areas are microdissected and replated on gelatin coated glass coverslips. To achieve adherence of these contracting areas, the replating should be performed with a minimal amount of medium. Small amount of medium (20 μl) should then be gradually added each hour, until the EB is firmly attached.

Immunofluorescent analysis is performed using conventional methods. EBs are fixed with 4% paraformaldehyde in PBS containing Ca^{2+} and Mg^{2+}. The cells are treated with 1% Triton X-100 (Sigma) in PBS for 10 min, and blocked in 5% nonimmune serum in PBS. The best results were achieved with overnight incubation of the primary antibody at 4°C in PBS containing 3% serum and 0.1% Triton X-100. Incubation with the secondary antibody is performed for 1 hr in the same solution at room temperature. Nuclear counterstaining can be performed using To-Pro3 (Molecular Probes, Eugene, Oregon, USA) at a concentration of 1 μg/ml with the secondary antibody. The fluorescence of this dye can be detected in the far-red channel.

Assessment of Cardiomyocyte Proliferation and Cell-Cycle Regulation

The pathways that regulate cardiomyocyte proliferation during embryonic and adult life, and cell-cycle withdrawal are largely undetermined. Human ES cell-derived cardiomyocytes represent an ideal model system for such studies. In this system, undifferentiated cells commit to the cardiomyocytic lineage, proliferate, mature, and gradually withdraw from the cell-cycle (Snir *et al.*, unpublished data). The three-dimensional structure of the EB may provide an additional advantage to such studies, since a recent report suggested that cardiomyocytes responded better to growth factors in three-dimensional tissue.[13]

[13] M. T. Armstrong, D. Y. Lee, and P. B. Armstrong, *Dev. Dyn.* **219**, 226 (2000).

Assays aiming at determining the presence of cycling cardiomyocytes can be easily adapted to the human ES culturing system. These include BRDU assay, KI-67 staining, and [^{3}H]thymidine uptake.[1] As in other systems, care must be taken to differentiate between cardiomyocytes and noncardiomyocytes. The protocol for immunofluorescence staining, described above, can be used for Ki-67 staining of both dispersed cells and of whole EBs. We have successfully used staining with a polyclonal anti-human Ki-67 antibody (Santa Cruz Biotechnology Inc., Santa Cruz, California, USA), and monoclonal anti-cardiac troponin I (cTnI; Chemicon International, Inc., Temecula, California, USA) combined with nuclear staining using To-Pro3 (Molecular Probes). The labeling index is then calculated as the percentage of Ki-67 positive cardiomyocytes nuclei. For [^{3}H]thymidine uptake assay the following protocol can be used:

1. Add [^{3}H]thymidine to the plated EB culture medium at a final concentration of 10 μCi/ml. Incubate for 18 hr at 37°C in a CO_2 incubator.
2. Wash the EBs three times with PBS containing Ca^{2+} and Mg^{2+}, mechanically dissect the beating areas and fix in 10% neutral buffered formalin.
3. Preparations are dehydrated in graded alcohols (70–100%), cleared in chloroform, and embedded in paraplast using standard techniques.
4. Deparaffinize 5 μm sections using xylene and graded alcohol, and treat with 3% H_2O_2 in methanol for 20 min in order to quench endogenous peroxidase activity.
5. Block sections in 5% nonimmune serum in PBS for 1 hr at room temperature.
6. Incubate with a cardiac specific primary antibody, such as anti-sarcomeric α actinin (Sigma) or anti-cTnI (Chemicon) overnight at 4°C.
7. Detection is accomplished with a steptavidine–peroxidase conjugate and a chromagen solution.
8. Autoradiography is performed after the immunohistochemical staining. Immerse the sections in an autoradiographic emulsion solution (LM-1, Amersham, UK) at 42°C for about 4 min.
9. Air dry and store in light tight box for 21 days.
10. Develop emulsions in Kodak D-170 developers for 7 min at 18°C.
11. Wash sections with DDW and fix in sodium thiosulfate 25% for 10 min.

Cardiomyocytes, identified by the immunohistochemical staining, are scored for the presence of silver grains in the nucleus.

Ultrastructural Assessment

The myofibril is one of the most fascinating examples of cytoskeletal assembly. When cardiomyocyte differentiate thousands of structural and regulatory proteins assemble into functional sarcomeric units. Ultrastructural analysis of differentiating human ES cell-derived cardiomyocytes is thus a powerful tool in examining the processes governing the sarcomere assembly.

Spontaneously contracting areas at different developmental stages or following different treatment protocols can be mechanically dissected from the EBs and processed for transmission electron microscopy. The following protocol has been successfully applied to human ES derived cardiomyocytes:

1. Mechanically dissect spontaneously contracting areas from the EBs.
2. Fix in 3% glutaraldehyde in 0.1 *M* sodium cacodylate buffer (pH 7.4) at 4°C for 24 hr.
3. Rinse samples in 7.5% sucrose in 0.1 *M* sodium cacodylate buffer.
4. Post fix samples in 1% OsO_4 in the same buffer for 1 hr.
5. Dehydrate samples in graded ethanol and embed in Epon 812.
6. Prepare thin (60 nm) sections using a ultramicrotome and mount on copper grids.
7. Stain sections with saturated uranyl acetate (Baker and Adamson Quality, New York, USA) and 1% lead citrate (Merck).
8. Examine using a transmission electron microscope.

Calcium Imaging

Changes in the intracellular Ca^{2+} concentration can be measured using the fluorescent probe Fura 2-AM. This technique can be used to record Ca^{2+} transients from dispersed cardiomyocytes or from cardiomyocytes in whole EBs. The technique can be used to study the development of calcium uptake and release processes and be combined with pharmacological and electrophysiological assays. The EBs are loaded with Fura 2-AM (Molecular Probes) for 25 min at room temperature (24–25°C) at a final concentration of 5 μM, in a 1:1 mixture of calcium-free Tyrode's solution. Excess Fura 2-AM is removed by rinsing twice with calcium-free Tyrode's solution. The cells are then transferred to a nonfluorescent chamber mounted on the stage of an inverted microscope (Diaphot 300, Nikon, Tokyo, Japan), and visualized with a X40 oil immersion Neoflour objective. The chamber is perfused with Tyrode's solution (1 m*M* calcium) at a rate of 1 ml/min. Experiments are performed at 37°C. The Fura 2-AM fluorescence is measured using a dual wavelength system (DeltaScan, Photon Technology International (PTI), South Brunswick, NJ, USA): light emitted from a Xenon arc lamp is fed in parallel into two independent monochromators to obtain quasi-monochromatic light beams of two different wavelengths exciting the cell at 340 and 380 nm. Either 340 or 380 nm wavelength is switched by a rotating chopper disk at a frequency enabling ratio measurements at a rate of 150 counts/sec. The two separate monochromator outputs are collected by the ends of a bifurcated quartz fiber optic bundle. The emitted fluorescence (510 nm) is collected by the microscope optics, passed through an interference filter and detected by a photomultiplier tube (710 PMT Photon Counting Detection System, PTI). Raw data can be stored for off-line analysis by the Felix software (PTI) as 340 and 380 nm counts, and as the ratio, $R = F_{340}/F_{380}$. For scaling the fluorescence ratio, cell-derived autofluorescence and noncell fluorescence are subtracted from the measured fluorescence. In these experiments both spontaneously contracting EBs and EBs stimulated at 0.5–2.5 Hz using platinum wires embedded in the walls of the perfusion chamber can be sampled. Several parameters can be derived from the measured transients including: time to peak systolic $[Ca^{2+}]_i$, the time to half-peak relaxation, and total transient duration.

Multi-electrode Array Recording

Human ES cells are not limited to the differentiation to isolated cardiomyocyte cells but rather an excitable syncytium with synchronized action potential propagation can be generated.[14] In this respect, this tissue can serve as a novel long-term *in vitro* model for human cardiac tissue. The use of a human tissue may be particularly important since it displays different electrophysiological properties from other existing *in vitro* animal models. The multi-electrode array (MEA) allows the long-term, high resolution, functional assessment of this unique tissue and can be combined with structural assessment of the tissue. Extracellular electrophysiological recordings from the EBs are performed on a PC-based MEA data acquisition system (Multi Channel Systems, Reutlingen, Germany). The MEA consists of a 50×50 mm glass substrate, in the center of which is an embedded 1.4×1.4 mm matrix of 60 titanium nitride-gold contact (30 μm) electrodes with an interelectrode distance of 100 or 200 μm. This system allows simultaneous recording of extracellular potentials from all electrodes

[14] I. Kehat, A. Gepstein, A. Spira, J. Itskovitz-Eldor, and L. Gepstein, *Circ. Res.* **91**, 659 (2002).

for prolonged periods. Data can be recorded at up to 25 kHz and band-pass filtered from 1 to 3000 Hz. Recording sessions are performed in culture medium or in a physiological solution and the MEA plates are constantly perfused with a gas mixture consisting of 5% CO_2 and 95% air while temperature is kept at $37.0 \pm 0.1°C$ by a heated plate.

Electrical activation of the EBs are recorded using extracellular unipolar electrodes that measure the potential at each site with respect to a distant reference (Fig. 2c). Local activation time (LAT) at each electrode is determined by the timing of the maximal negative intrinsic deflection (dV/dt_{min}) of the recorded electrogram.[15] The measured LATs at all electrodes are then used for the generation of color-coded activation maps (Fig. 2d). Interpolating between measurement sites, using MATLAB standard 2D plotting function (MATLAB 5.3.0, The MathWorks, Natick, MA) is used to generate the isochronal contour lines, where each contour represents the instantaneous position of the wavefront.[16] One pitfall of many interpolating algorithms is in the assumption of continuity: the data of two points is used to interpolate for all points in between. This assumption may be invalid in the presence of noncardiac tissue or in areas of block, such as in the EBs, where activation of neighboring sites may be due to different wavefronts. In order to avoid this pitfall and describe the conduction velocity in this complex tissue we have adapted an algorithm described by Bayly *et al.*[17] This is an automated method for the estimation of velocities of multiple wavefronts that is robust to errors in activation detection.

The velocity vector of a point is $v = [dx/dt, dy/dt]^T$. To estimate the vector, each active recording site z_i and its neighbors, within a Δx, Δy and Δt window, are fitted to a second degree polynomial surface using a least square algorithm. We routinely use a window that is five times the sampling interval in each dimension. This polynomial fitting is robust to missing or bad data and reduces noise. The velocity is estimated using the expression: $V_e = [dx/dT\, dy/dT] = [T_x/(T_x^2 + T_y^2) T_y/(T_x^2 + T_y^2)]$; where $T_x = \partial T/\partial x$ and $T_y = \partial T/\partial y$.

Immunofluorescent analysis of the tissue can be performed after completion of the measurements on the MEA, taking advantage of the transparent glass substrate of the array, and viewed with an inverted fluorescent microscope. For example, the EB can be stained for

[15] M. S. Spach and P. C. Dolber, *Circ. Res.* **58**, 356 (1986).

[16] Y. Feld, M. Frank-Melamed, I. Kehat, D. Tal, S. Maron, and L. Gepstein, *Circulation* **105**, 522 (2002).

[17] P. V. Bayly, B. H. KenKnight, J. M. Rogers, R. E. Hillsley, R. E. Ideker, and W. M. Smith, *IEEE Trans. Biomed. Eng.* **45**, 563 (1998).

cTnI, using the protocol described above, to identify cardiomyocytes. The resulting local velocity vectors and isochronal contour lines are then superimposed on the fluorescent image and correlated with tissue morphology.

Mechanical Activity

We have also expanded the MEA system to include an assessment of the mechanical activity of the cardiomyocytes. This method measures changes in the intensity of light passing through the tissue during the mechanical contraction. These changes can be detected using a photodiode (UV100BG, EG&G, Vaudreuil, Canada) mounted on an inverted microscope (Axiovert 135, Zeiss). The electrical signal from the photodiode is recorded with one of the MEA channels. This expanded system can be used for assessment of the excitation-contraction coupling in the tissue.

Pharmacological Studies

The combination of the human ES derived cardiomyocytic tissue with the MEA recording plates may serve as a unique platform for the *in vitro* assessment of the chronotropic, dromotropic and lucitropic effects of drugs on human cardiac tissue. Following baseline recording 20 μl of stock solution of the test drug was added to 2 ml of culture medium in the MEA plate and stirred gently. The pharmacological agents tested on this tissue, so far included tetrodotoxin (TTX; Alomone Labs, Jerusalem, Israel) at a final concentration of 10 μmol/l, isoproterenol at a final concentration of 10^{-6} M, carbamylcholine at a final concentration of 10^{-6} M, 3-isobutyl-1-methylxanthine (IBMX) at a final concentration of 10^{-5} M, forskolin at a final concentration of 10^{-6} M, and diltiazem hydrochloride at a final concentration of 10^{-6} M (all from Sigma Chemical Co., St. Louis, MO). Extracellular recordings are performed for 30 sec, at baseline and 5 min following drug application. 1-Heptanol (at 0.3, 0.6, and 1 mmol/l; Sigma) was dissolved in the medium by vigorous shaking and then added to the plate.

In order to study the effects of elevated extracellular potassium concentration on conduction properties, a baseline recording was performed in a modified Tyrode's solution (NaCl 140 mmol/l, KCl 5.4 mmol/l, glucose 10 mmol/l, $MgCl_2$ 1 mmol/l, pyruvate 2 mmol/l, $CaCl_2$ 2 mmol/l, and HEPES 10 mmol/l, titrated to pH 7.4 with NaOH). Extracellular recordings were then performed for 30 sec, at baseline and 5 min after the solution was changed to three balanced high K^+ solutions (KCl: 10, 15, and 20 mmol/l).

Concluding Remarks

The development of human ES cell lines and their ability to differentiate to cardiomyocytes may be an important tool for several research and clinical fields including developmental biology, gene and drug discovery, pharmacological testing, cell therapy and tissue engineering. We have attempted to briefly describe the basic techniques used for human ES cells propagation and cardiomyocytes differentiation, as well as methods for structural and functional assessment of these cells. These and other methodologies may be used to fully explore the enormous potential of this novel cardiomyocyte differentiating system.

Acknowledgment

This research was suported in part by the Israel Science Foundation (grant no. 520/01-2) and by the Johnson & Johnson Focused Giving Grant.

Author Index

Numbers in parentheses are footnote reference numbers and indicate that an author's work is referred to although the name is not cited in the text.

A

B

C

D

E

F

I

J

K

L

M

N

O

P

Q

R

S

T

U

V

W

X

Y

Z

Subject Index

F

G

I

L

M

N

O

V

ISBN: 0-12-182268-0

90000

9 780121 822682

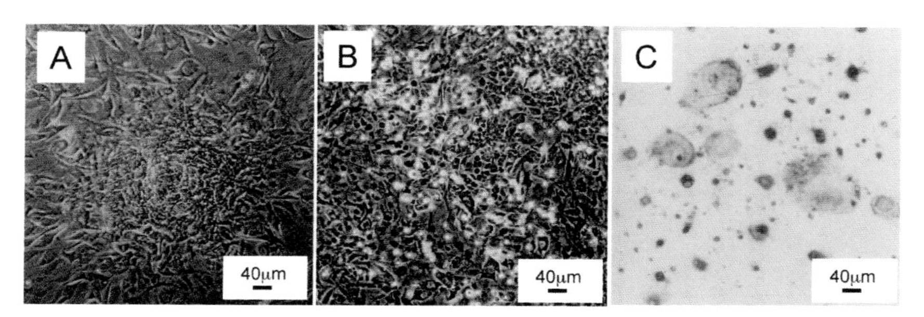

MOTOKAZU TSUNETO *ET AL.*, CHAPTER 7, FIG. 2. Appearance of step cultures at day 5 (A) and day 10 (B), and osteoclasts at day 16 (C) with TRAP staining.

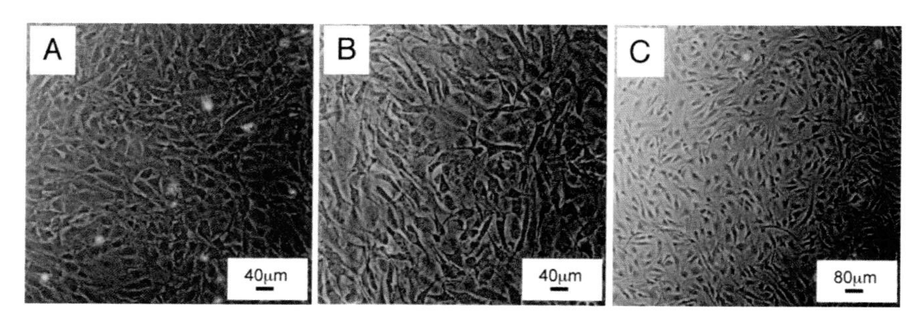

MOTOKAZU TSUNETO *ET AL.*, CHAPTER 7, FIG. 4. Appearance of ST2 (A) and OP9 (B, C). (C) is 70% confluency.

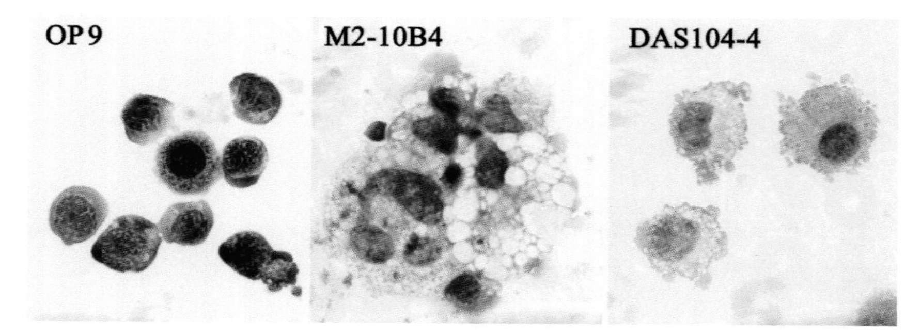

MICHAEL KYBA *ET AL.*, CHAPTER 8, FIG. 1. Morphology of Bcr/Abl-transduced cells growing on different stromal cell lines.

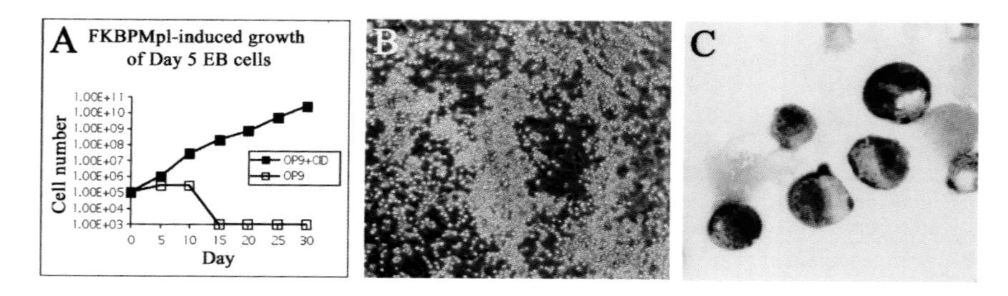

MICHAEL KYBA *ET AL.*, CHAPTER 8, FIG. 2. FKPBMpl cells. (A) Cell growth, expressed as cumulative cell number, over time, in the presence and absence of CID. (B) Colony morphology on OP9 in the presence of CID. (C) Cellular morphology of CID-induced cells grown on OP9.

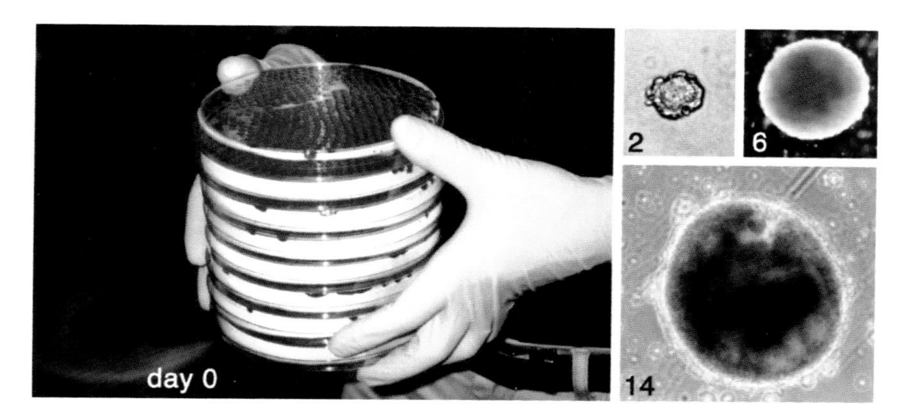

MICHAEL KYBA *ET AL.*, CHAPTER 8, FIG. 4. Hanging drop EBs. Hanging drops are plated on day 0. The EBs that form are shown at days 2, 6, and 14, postplating. The first visual sign of hematopoiesis in the form of hemoglobinizing areas can be seen under dark-field at day 6. By day 14, well-hemoglobinized blood islands are apparent.

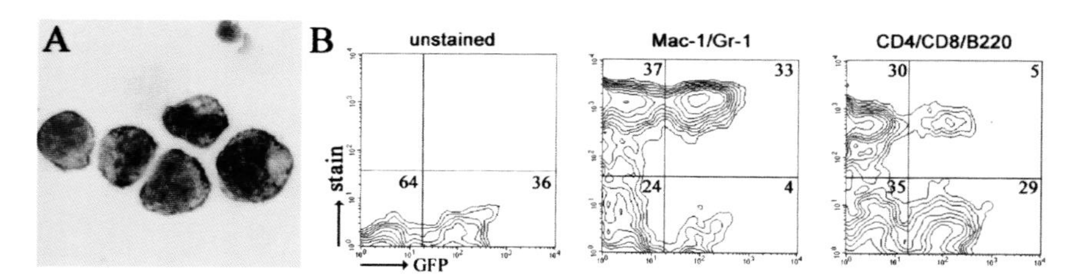

MICHAEL KYBA *ET AL.*, CHAPTER 8, FIG. 5. HoxB4-induced cells. (A) Morphology of HoxB4-induced cells grown on OP9 stromal cells. (B) FACS analysis of peripheral blood of a recipient mouse three weeks posttransplant. GFP expression, which marks donor cells, is measured on the *X*-axis. Antibody staining is measured on the *Y*-axis. The percentage of cells falling into each quadrant is shown in the upper right-hand corner. In the first FACS, cells were stained with a nonspecific antibody, which does not recognize any blood cell type. In the second FACS, a cocktail of myeloid-specific antibodies was used (Gr-1 and Mac-1) to label circulating granulocytes and monocytes. In the third FACS, a cocktail of lymphoid-specific antibodies was used (B220, CD4, and CD8) to label circulating lymphocytes.

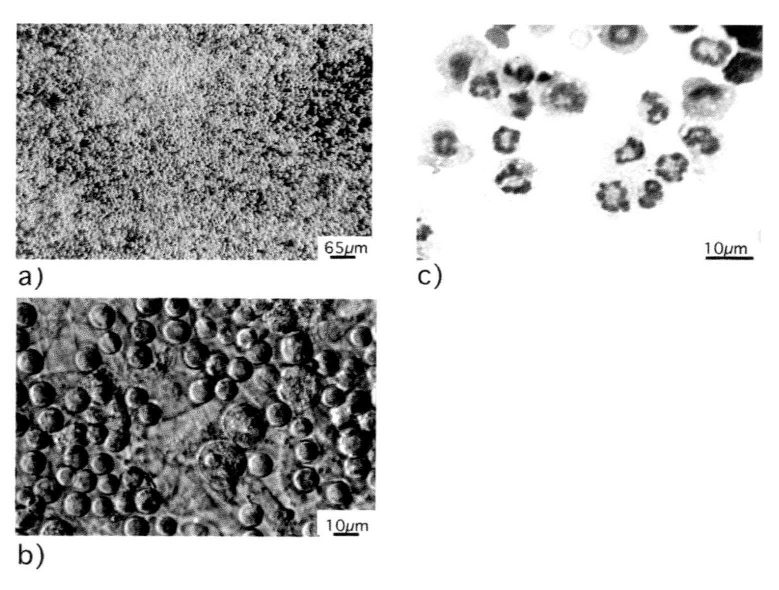

JONATHAN G. LIEBER *ET AL.*, CHAPTER 9, FIG. 4. Neutrophil generating regions contain differentiating neutrophils at all stages of maturation intimately associated with the OP9 cells. The composition of day 5 and 9 NGRs was evaluated by using fine needle aspiration to harvest loosely associated cells on the surface of these colonies. The aspirated cells were cytospun and stained with the Hema 3 staining kit (Fisher Scientific, Pittsburgh, PA). (a) Low power phase contrast photograph of portions of two NGRs from differentiating cells grown for 5 days in the Neutrophil Differentiation Mix. (b) At a higher magnification, differentiating neutrophils at multiple stages of maturation can be seen. (c) Photograph of hematoxylin and eosin stained neutrophils day 9 in the tertiary Neutrophil Differentiation Mix.

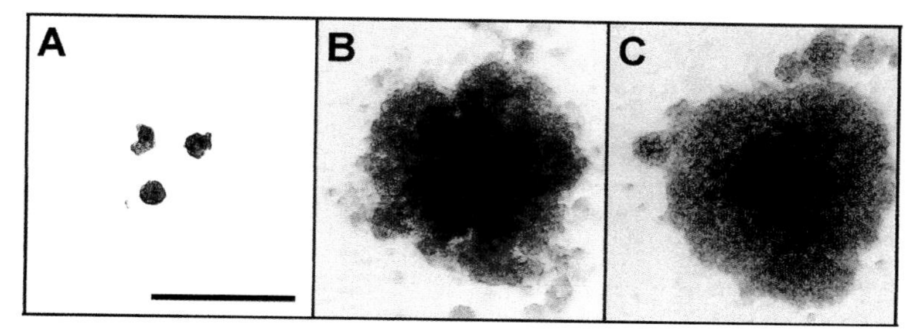

LEIF CARLSSON *ET AL.*, CHAPTER 14, FIG. 2. Morphology of different colony types appearing in the clonal assays of day 6 EB cells. Colonies shown are those containing primitive erythroid cells (A), definitive (adult) erythroid cells (B) and an HPC colony (C). Scale bar indicates 0.25 mm.

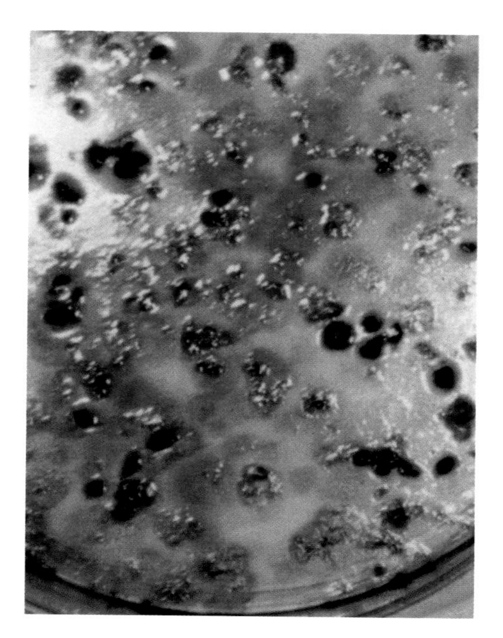

SARAH K. BRONSON, CHAPTER 17, FIG. 2. Densely plated culture to demonstrate extent of mineralized nodule formation visible after 21 days. Stained using von Kossa method with toluidine blue counterstain.

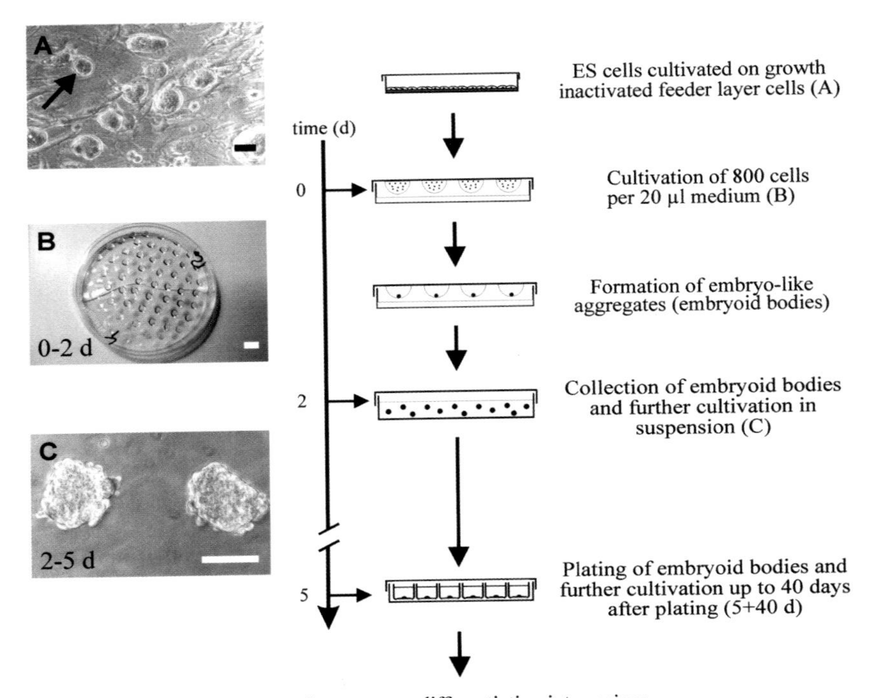

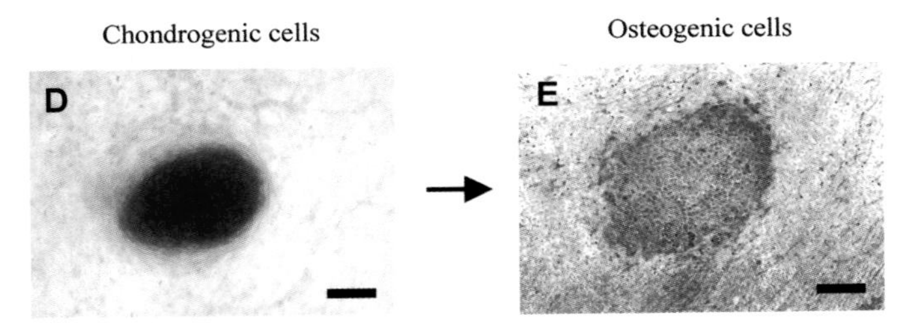

JAN KRAMER *ET AL.*, CHAPTER 18, FIG. 1. Cultivation and differentiation of ES cells *in vitro*. To keep ES cells undifferentiated and pluripotent they were cultivated in the presence of leukemia inhibitory factor (LIF) and cocultivated with growth-inactivated embryonic fibroblasts used as feeder layer (A; arrow = ES cell-colony). After trypsinization ES cells were cultivated as hanging drops (B) in differentiation medium without LIF and formed cell aggregates. These aggregates, called embryoid bodies (EBs), were collected after 2 days of hanging drop-cultivation (0–2 d) and transferred into suspension culture (C). Cultivation in suspension was performed for 3 days (2–5 d). On the fifth day of differentiation (5 d) EBs were plated onto gelatine-coated dishes. During further differentiation up to 40 days (5 + 40 d) various cell types of all the three germlayers could be detected in the EB outgrowths. For example, cartilage nodules were detected by Alcian blue-staining around 10 days after plating of EBs derived from the ES cell line BLC6 (D). At a very late stage of ES cell differentiation around 30 days after plating these nodules disintegrate and the cells express alkaline phosphatase (E) and other marker for osteogenic cells (see also Fig. 2). Bars = 50 μm (A), 1000 μm (B), 100 μm (C, D, E).

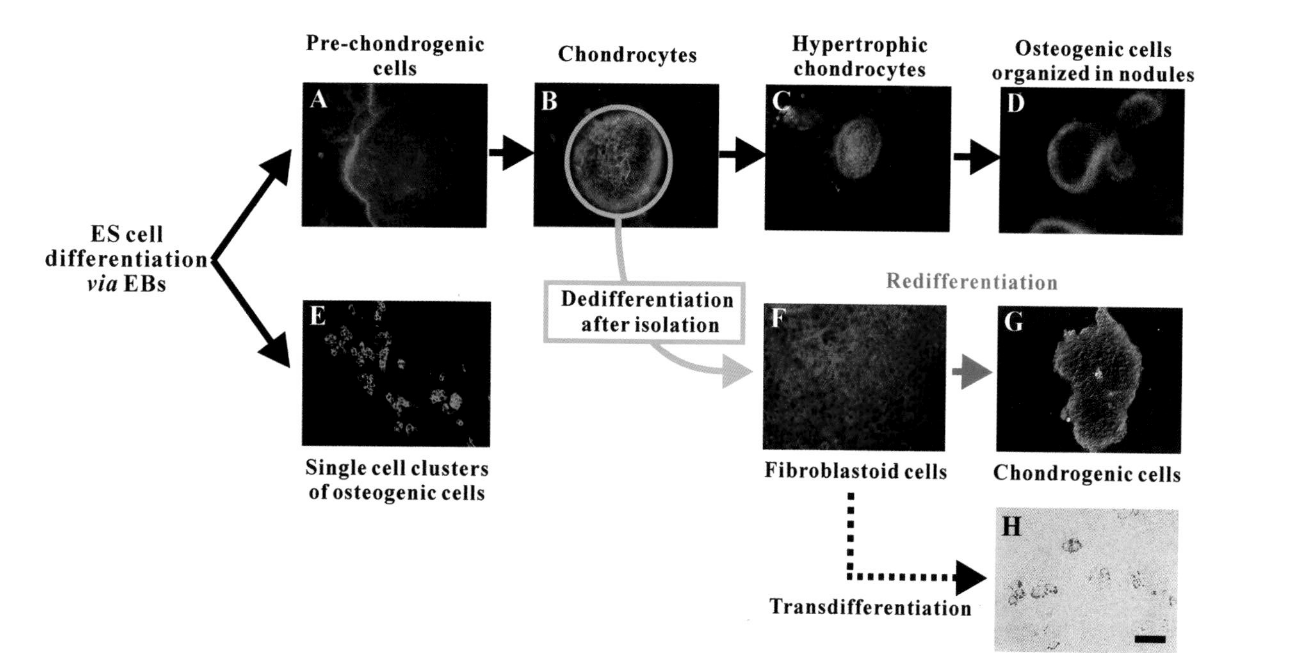

JAN KRAMER *ET AL.*, CHAPTER 18, FIG. 2. ES cell-derived chondrogenic and osteogenic differentiation *in vitro*. Differentiation of ES cells *in vitro* via EBs into chondrogenic and osteogenic cells recapitulates chondrogenic and osteogenic differentiation processes. Prechondrocytic cells expressing the cartilage-associated transcription factor scleraxis were detected in EB outgrowths (A) as demonstrated by mRNA *in situ* hybridization. These cells later formed areas of high cell density, called nodules, and expressed the cartilage matrix protein collagen II (B) as demonstrated by mRNA *in situ* hybridization for scleraxis (green) coupled with immunostaining for collagen II (red). These chondrocytes progressively developed into hypertrophic, collagen X-expressing cells as revealed by mRNA *in situ* hybridization (C). Finally, immunostaining showed that chondrocytes localized in nodules expressed osteogenic proteins, such as osteopontin (D). Furthermore, direct differentiation of ES cells into single cell clusters of osteopontin-positive cells, bypassing the chondrocytic stage, was observed (E). Chondrocytes isolated from EBs showed a distinct differentiation plasticity. The isolated cells initially dedifferentiated into collagen I-positive fibroblastoid cells (F). These cells redifferentiated in culture into chondrocytes reexpressing collagen II (green) and a reduced amount of collagen I (red) (G). However, occasionally other mesenchymal cell types such as adipocytes appeared in these cultures as demonstrated by Sudan III-staining for adipogenic cells (H). Bar = 100 μm.

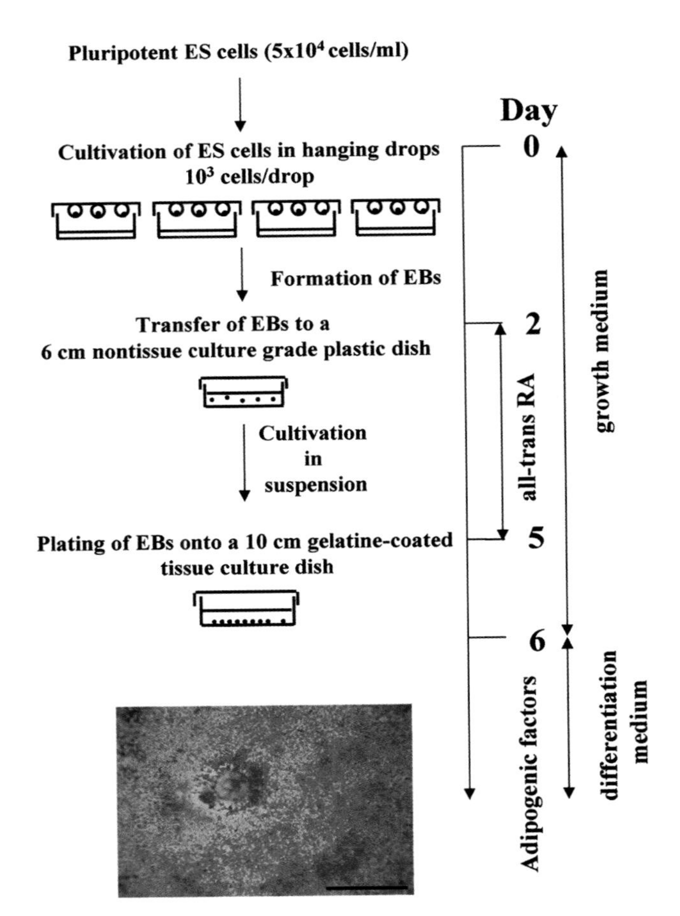

Brigitte Wdziekonski *et al.*, Chapter 19, Fig. 1. Overview of the protocol used for *in vitro* differentiation of ES cells into adipocytes and photomicrographic record of a 20 day-old outgrowth derived from RA-treated EBs. Adipocytes stained with Oil-Red O for fat droplets are shown. Bar: 1000 μM.

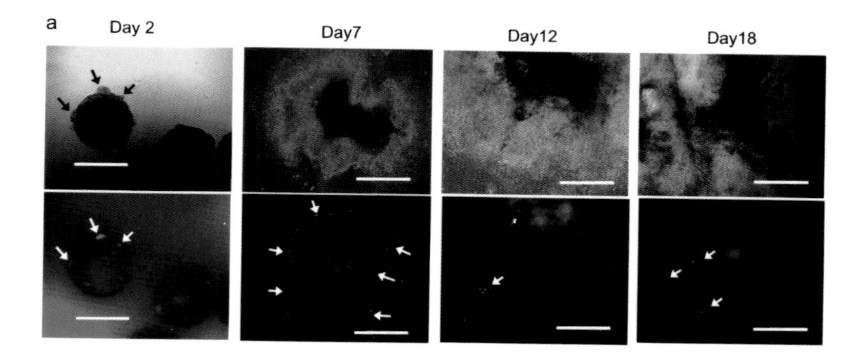

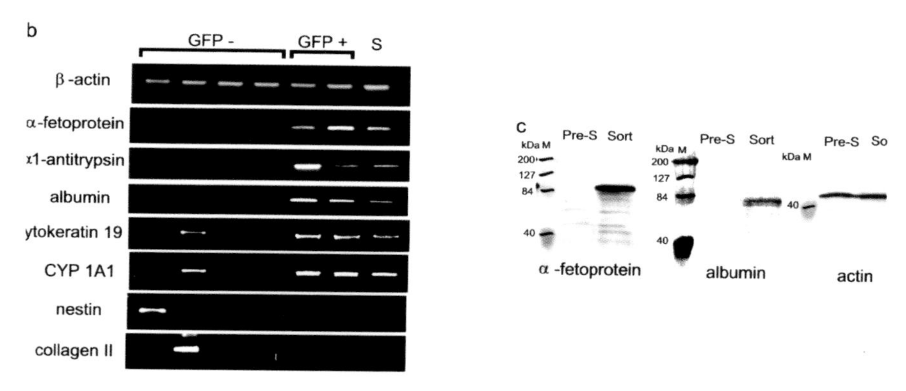

TAKASHI HAMAZAKI AND NAOHIRO TERADA, CHAPTER 20, FIG. 2. (a) Visualization of visceral endoderm and hepatic precursors in embryoid bodies. GFP expression was driven by the α-fetoprotein gene promoter (−259 to +24). GFP positive cells first appeared at the outer layer of embryoid bodies around Day 2. After embryoid bodies were attached to the culture plate at Day 5, GFP positive cells lined up as mono- to oligo-layer cells at the outer circumference of embryoid bodies (Day 7). Clusters of GFP positive cells then appeared inside the embryoid bodies around Day 12 and gradually grew until Day 18. Arrows in the lower panels indicate GFP positive cells under a fluorescence microscope, and upper panels illustrate the corresponding pictures under a light microscope. Bars: 1.0 mm. (b) Visceral endoderm and hepatic marker expression in GFP positive and negative populations. Clusters of GFP positive cells in embryoid bodies (Day 18) were directly picked under a fluorescence microscope (GFP+). As controls, GFP negative cells were similarly harvested under a microscope from a GFP negative area (GFP−). GFP positive cells were also sorted using FACS (S). The mRNAs were extracted from these cells, and expression of hepatic lineage markers was examined by RT-PCR. Nestin and collagen II genes were also examined as markers for neural and mesodermal differentiation, respectively. (c) Western blotting analysis for sorted GFP positive cells. The same amount of protein (5 μg) from the sorted GFP positive cells (Sort) and presorted Day 18 embryoid bodies (Pre-S) were loaded on 10% polyacrylamide SDS gel. Protein expression of α-fetoprotein, albumin, and actin was examined using specific antibodies.

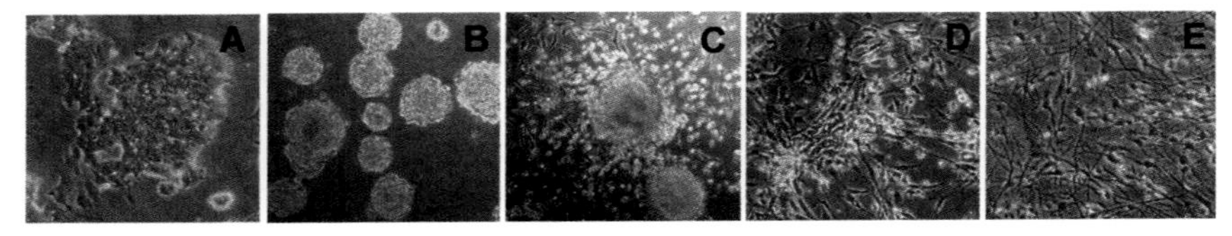

JONG-HOON KIM *ET AL.*, CHAPTER 22, FIG. 1. Procedure for induction of midbrain dopaminergic neurons from ES cells. The ES cells are taken through 5 steps or stages. In stage 1 (A), undifferentiated ES cells are cultured for 5 days in the presence of 15% fetal calf serum (FCS) on gelatin-coated tissue culture dishes in the presence of LIF (1400 U/ml.). In stage 2 (B), embryoid bodies (EBs) are generated in the presence of FCS for 4 days in the presence or absence of LIF (1000 U/ml.). In stage 3 (C), the EBs are plated into ITS medium (fibronectin, 5 μg/ml) where, over 10 days, Nestin$^+$ cells migrate from the cell aggregates. In stage 4 (D), these Nestin$^+$ cells are resuspended and expanded for 4 days in N2 medium containing bFGF, Shh, and FGF8. In stage 5 (E), the medium is changed into N2 medium without bFGF, Shh, or FGF8. These cells differentiate efficiently into neurons and astrocytes over a 2 week period.

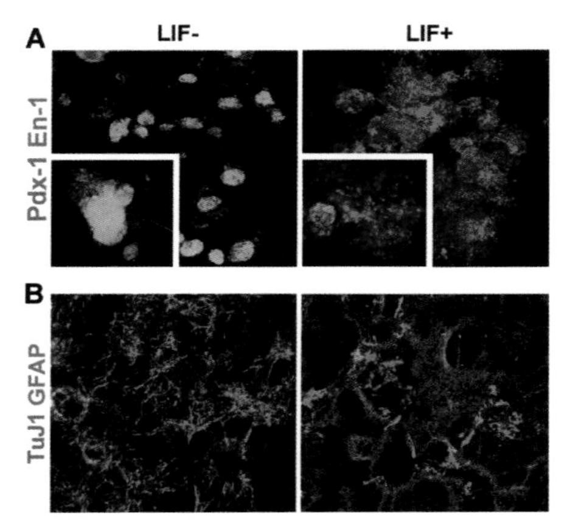

JONG-HOON KIM *ET AL.*, CHAPTER 22, FIG. 2. Neural and pancreatic differentiation of ES cells. Embryoid bodies were generated in the presence (LIF$^+$) or absence (LIF$^-$) of LIF (1000 U/ml) and differentiated. Double-immunostaining for PDX-1/En-1 (A, day 3 in stage 4) and TuJ1/GFAP (B, day 8 in stage 5). Note that LIF treatment in stage 2 (EB formation) increases the neuronal (TuJ1$^+$ cells, green) and decreases the astrocytic (GFAP$^+$, red) population. LIF treatment efficiently enhances midbrain precursor cells (En-1$^+$ cells, red) and negatively regulates pancreatic precursor cells (PDX-1$^+$ cells, green).

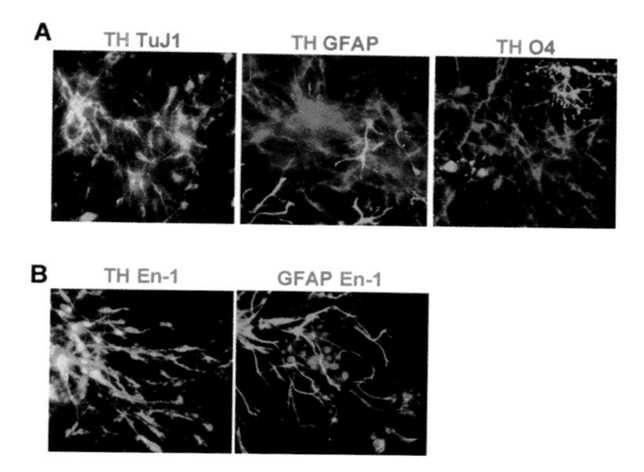

JONG-HOON KIM *ET AL.*, CHAPTER 22, FIG. 3. Differentiation into midbrain TH neurons. (A) Co-staining with TH and markers for neurons (TuJ1), astrocytes (GFAP), and oligodendrocytes (O4) shows that only neurons express TH. (B) Most TH^+ neurons express a midbrain-specific marker En-1 (TH, green; En-1, red). Some $GFAP^+$ astrocytes also express En-1 (GFAP, green; En-1, red).

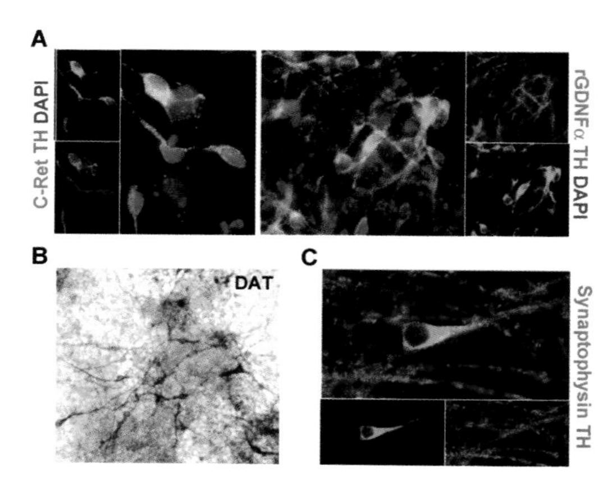

JONG-HOON KIM *ET AL.*, CHAPTER 22, FIG. 5. Characteristics of Nurr1 ES-derived TH neurons. (A) TH^+ neurons derived from Nurr1-transfected ES cells express GDNF receptor c-Ret (c-Ret, red; TH, green; DAPI, blue) and rGDNFα (rGDNFα, red; TH, green; DAPI, blue), dopamine transporter, DAT (B: brown), and synaptophysin (C: synaptophysin, red; TH, green).

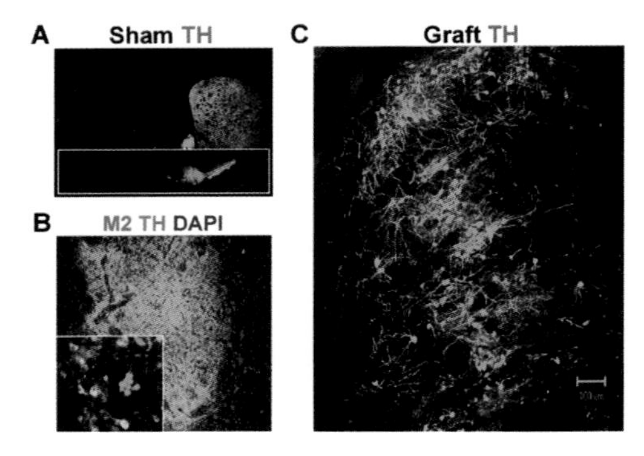

JONG-HOON KIM *ET AL.*, CHAPTER 22, FIG. 6. Integration of Nurr1 ES cell-derived TH neuron into the striatum of hemisparkinsonian rat. (A) Low power photomicrographs showing TH^+ neurons in the striatum, substantia nigra, and ventral tegmental area. Note that the lesioned side of the brain contains no positive cells. (B) Photomicrographs of the graft in the striatum. All grafts were easily detected by staining mouse-specific surface antigen, M2 (green). Many of the $M2^+$ grafted cells also expressed TH (inset: M2, green; TH, red). (C) The dorsal striatum of grafted animals showed discrete grafts containing TH^+ neurons (scale bar, 100 μm).

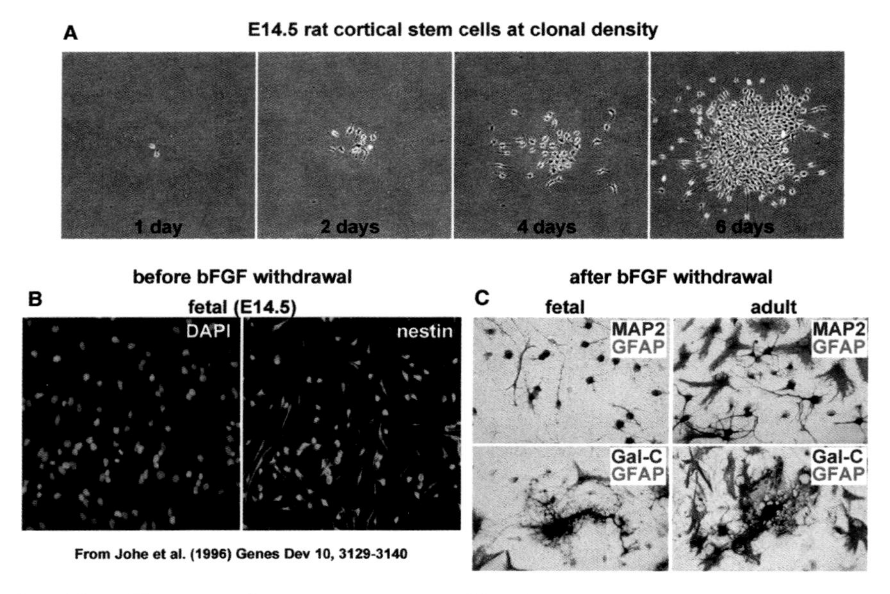

JONG-HOON KIM *ET AL.*, CHAPTER 22, FIG. 7. Proliferation and differentiation of primary stem cells from rat E14.5 cortex. (A) Cells plated at 30 cells/cm^2 generate distinct clones after expansion in bFGF. (B) Cells in proliferating clone uniformly express the intermediate filament Nestin. (C) After 7 days in the absence of bFGF, cells have differentiated into $MAP2^+$ neurons, $GFAP^+$ astrocytes, and Gal-C^+ oligodendrocytes. Both fetal and adult stem cell clones give rise to reproducible proportions of all three derivatives.

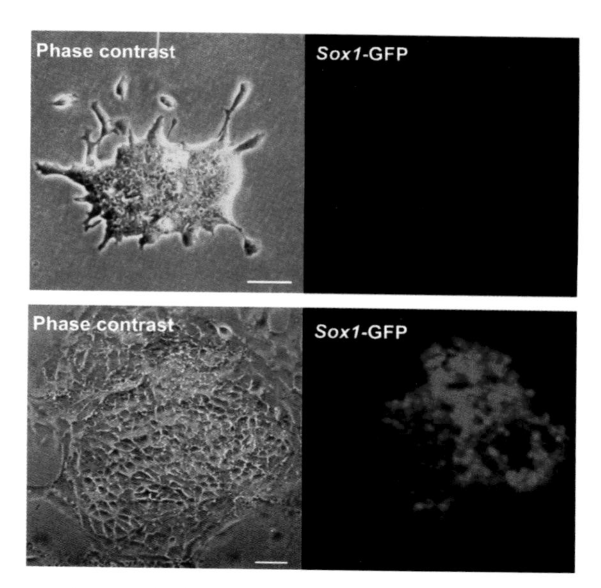

QI-LONG YING AND AUSTIN G. SMITH, CHAPTER 23, FIG. 1. Phase contrast and fluorescent images of monolayer differentiation of *Sox1*-GFP ES cells. Upper panels show a colony of undifferentiated 46C ES cells cultured in medium containing serum and LIF. Lower panels show a colony cultured in GMEM/10% FCS medium plus LIF for 2 days then transferred to N2B27 for 5 days to induce neural differentiation. Bar: 50 μm.

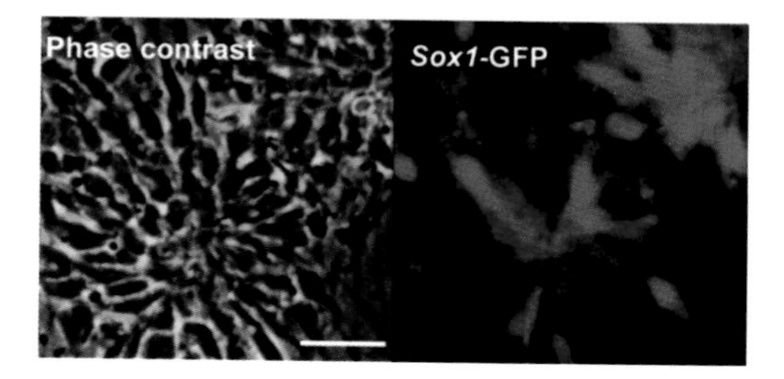

QI-LONG YING AND AUSTIN G. SMITH, CHAPTER 23, FIG. 2. Expression of *Sox1*-GFP in rosette of neural precursors derived from 46C ES cells after 7 days in monolayer differentiation. Bar: 50 μm.

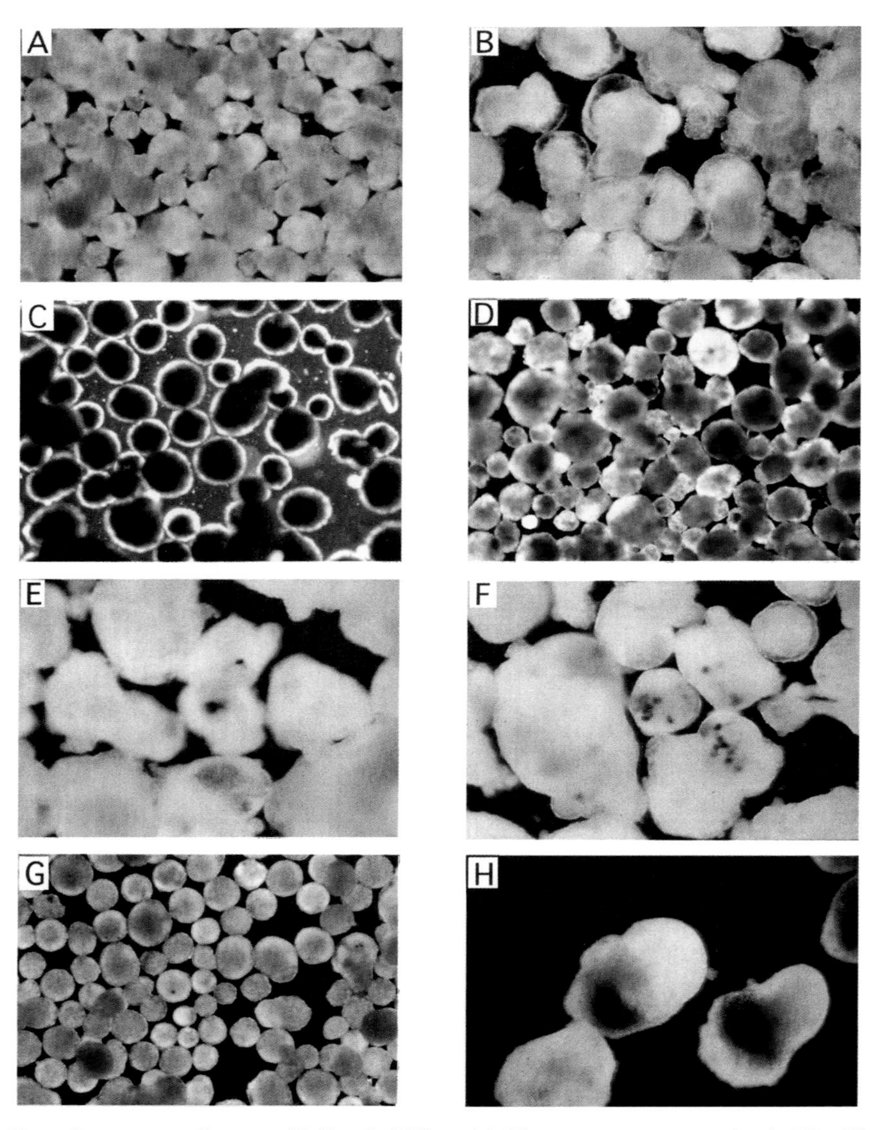

HEIDI STUHLMANN, CHAPTER 27, FIG. 3. Differential AP reporter gene expression in EBs. ES cells were infected with entrapment vectors and differentiated *in vitro* to spontaneously form EBs in suspension culture. Aliquots were harvested and stained for AP activity. (A, B) Day-4 and -7 EBs from uninfected ES cells. (C, D) Day-4 and -7 EBs from a clone infected with PT-AP. (E, F) Day-7 and -10 EBs from a clone infected with GT-AP. (G, H) Day-4 and -7 EBs from a clone infected with PT-IRES-AP.

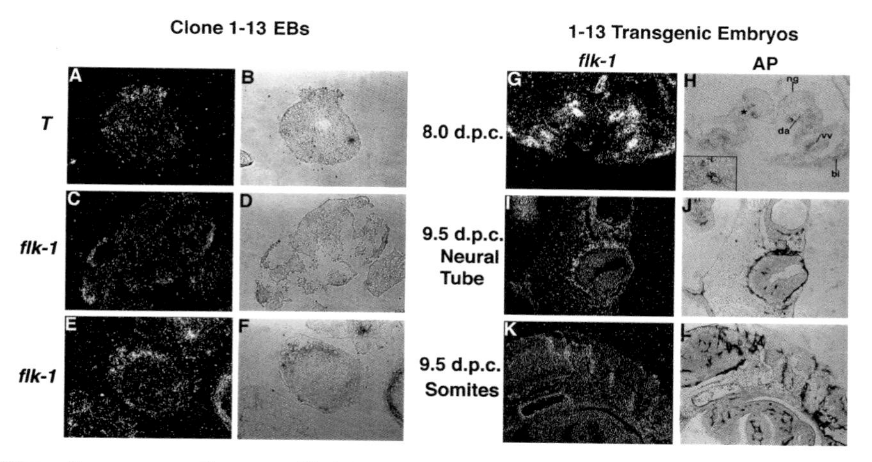

HEIDI STUHLMANN, CHAPTER 27, FIG. 4. Colocalization of *flk-1* and AP reporter gene expression in EBs and transgenic mice derived from ES cell clone 1–13. *In situ* hybridization on sectioned, AP stained clone 1–13 EBs (A–F) and clone 1–13-derived transgenic embryos (G–L). (A, B) *Brachyury* T riboprobe did not colocalize with AP reporter gene expression. Riboprobes corresponding to *flk-1* colocalized with AP reporter gene expression in (C, D) day-4 EBs; (E, F) day-9 EBs; (G, H) 8.0 dpc embryo (transverse section), insert showing higher magnification of the dorsal aorta (dorsal aorta, da; neural groove, ng; vitteline vein, vv; blood island, bi); (I, J) neural tube of 9.5 dpc embryo; (K, L) somites of 9.5 dpc embryo. Dark field images of A, C, E, G, I, and K correspond to bright field images of B, D, F, H, J, and L, respectively.

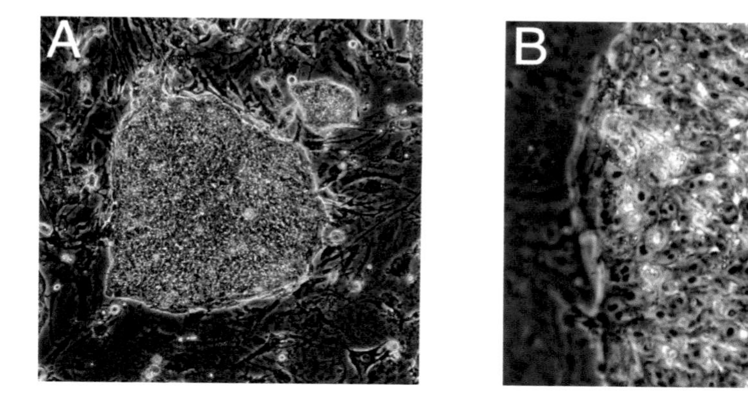

HIROFUMI SUEMORI AND NORIO NAKATSUJI, CHAPTER 29, FIG. 1. A phase contrast micrograph of an undifferentiated cynomolgus monkey ES cell colony (A), and a higher magnification view (B). Note the high nucleus/cytoplasm ratio and prominent nucleoli.

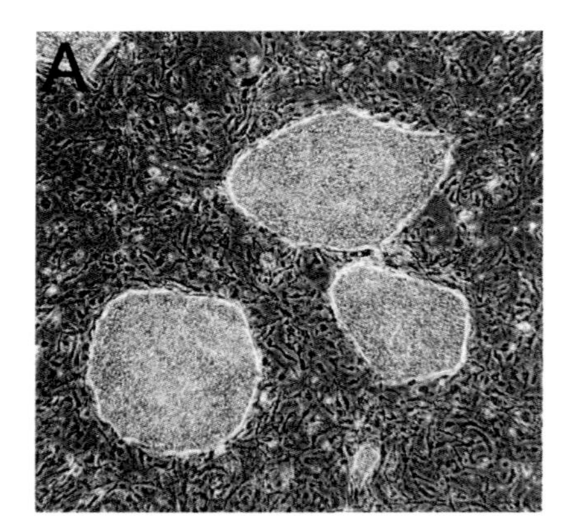

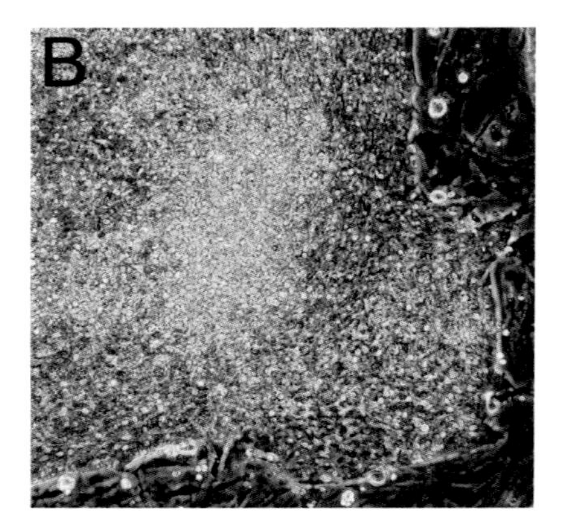

HIROFUMI SUEMORI AND NORIO NAKATSUJI, CHAPTER 29, FIG. 2. A low magnification view of a confluent culture of cynomolgus monkey ES cells (A), and a piled-up colony (B).

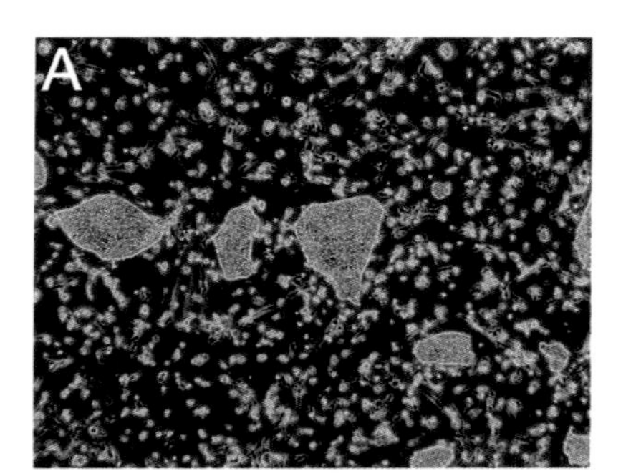

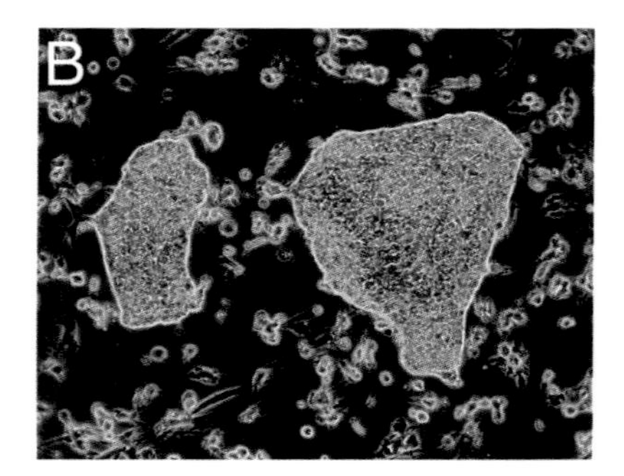

HIROFUMI SUEMORI AND NORIO NAKATSUJI, CHAPTER 29, FIG. 3. Appearance of ES cell colonies after trypsin treatment (A), and colonies detaching from the culture dish (B).

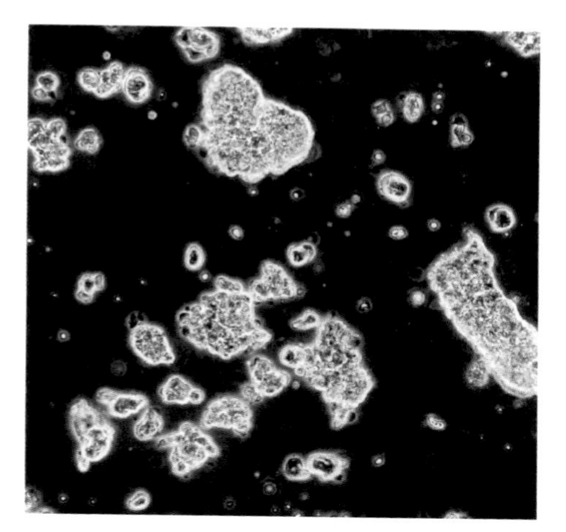

HIROFUMI SUEMORI AND NORIO NAKATSUJI, CHAPTER 29, FIG. 4. Appearance of ES cells after dissociation by pipetting. Cells must be kept as clusters of 50–100 cells, or minimum of 15–20 cells, for efficient subculture.

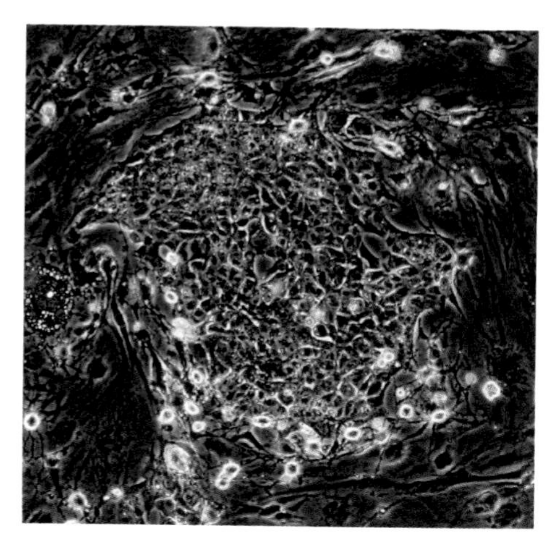

HIROFUMI SUEMORI AND NORIO NAKATSUJI, CHAPTER 29, FIG. 5. An ES cell colony one day after subculturing.

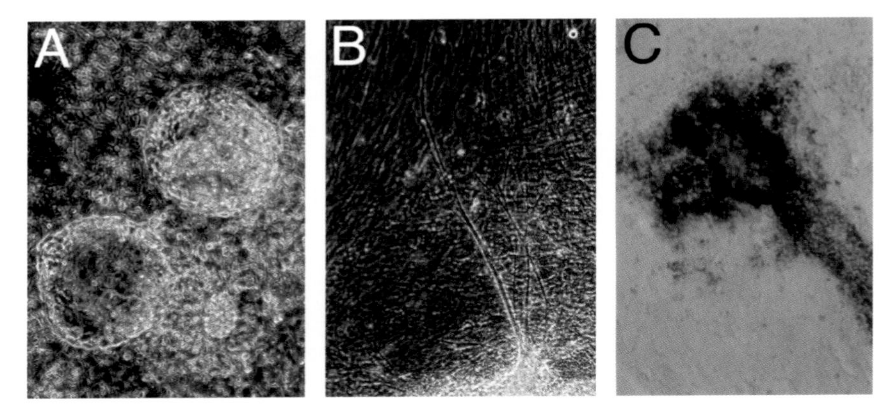

HIROFUMI SUEMORI AND NORIO NAKATSUJI, CHAPTER 29, FIG. 6. Cell differentiation *in vitro*. Yolk sac-like vesicular epithelia (A), neuronal cells extending axons (B) and pigmented epithelia (C).

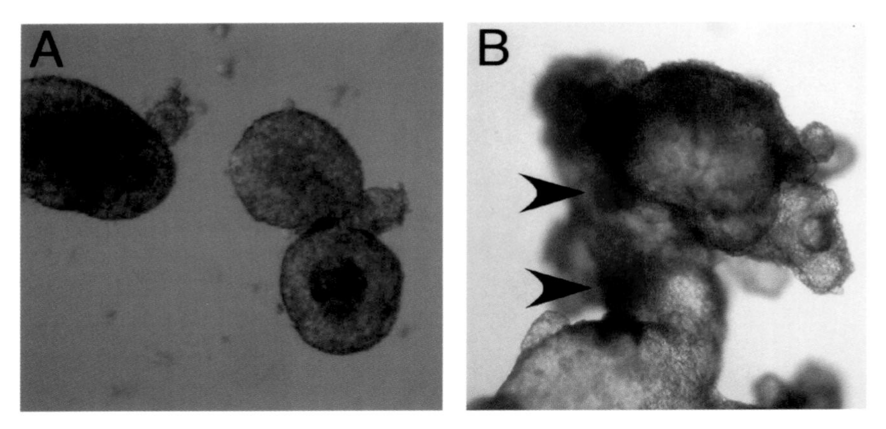

HIROFUMI SUEMORI AND NORIO NAKATSUJI, CHAPTER 29, FIG. 7. Simple embryoid bodies (A) and a cystic embryoid body (B) of cynomolgus monkey ES cells. Arrowheads indicate blood island-like structure.

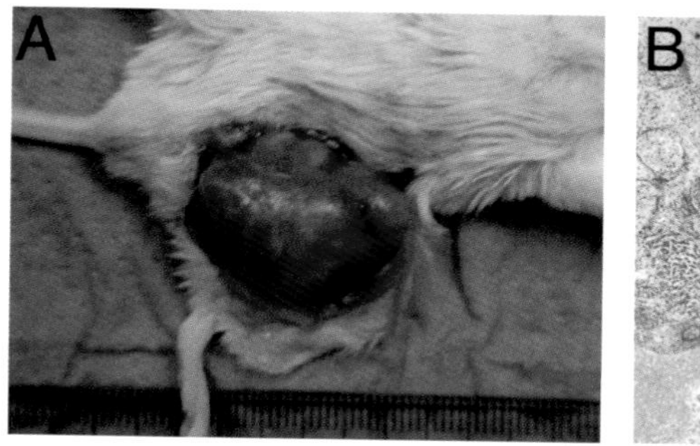

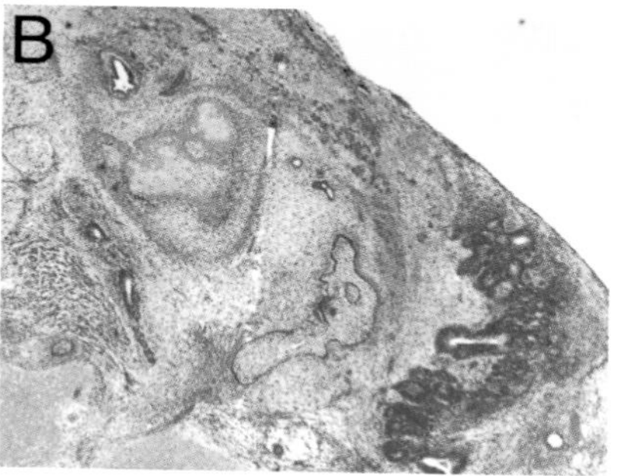

HIROFUMI SUEMORI AND NORIO NAKATSUJI, CHAPTER 29, FIG. 8. A teratoma produced by subcutaneous transplantation of cynomolgus ES cells (A), and low magnification view of a histological section of a teratoma (B).

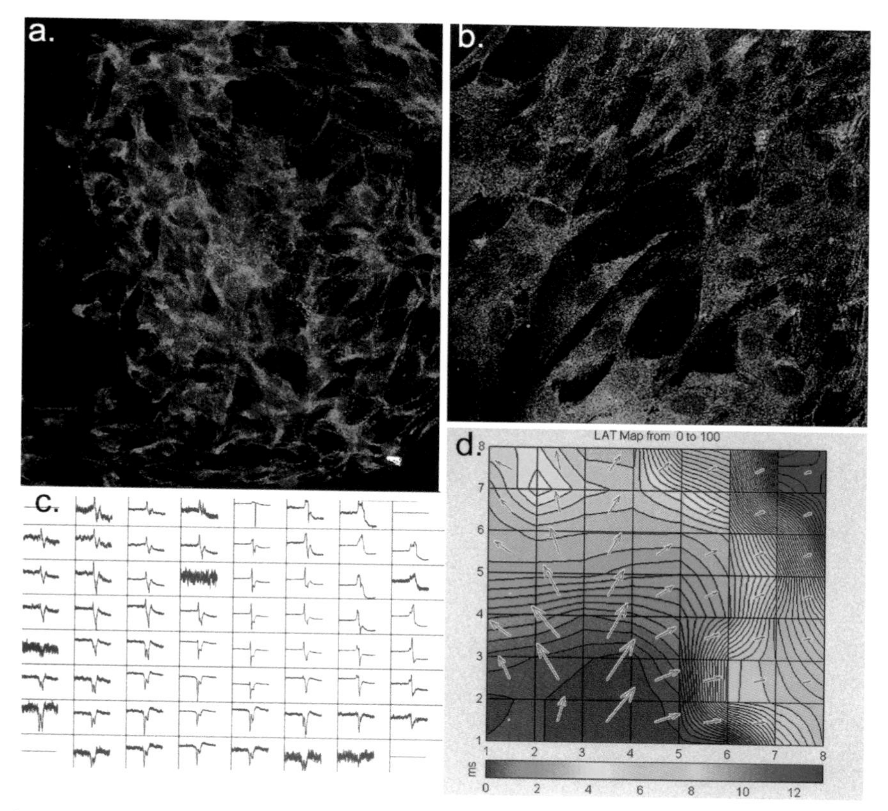

IZHAK KEHAT *ET AL.*, CHAPTER 32, FIG. 2. (a) Immunostaining of whole EB with anti-cTnI antibodies, showing cardiomyocytes at different orientations. (b) Dual immunostaining of the EB with anti-Connexin 43 (gap junction protein) and anti-cTnI, showing gap junctions between cardiomyocytes in the EB. (c) Sixty electrograms recorded simultaneously from an EB using the MEA mapping technique. (d) The resulting activation and vectorial velocity maps.